P9-EJT-680

Darkness in El Dorado

ALSO BY PATRICK TIERNEY

The Highest Altar

Darkness in El Dorado

How Scientists and Journalists Devastated the Amazon

Patrick Tierney

W. W. Norton & Company New York London

Printed in the United States of America
First Edition

For information about permission to reproduce selections from this book, write to Permissions, W. W. Norton & Company, Inc., 500 Fifth Avenue, New York, NY 10110

The text of this book is composed in Adobe Garamond with the display set in Mrs. Eaves
Composition by Allentown Digital Services Division of RR Donnelley & Sons Company
Manufacturing by the Haddon Craftsmen, Inc.
Book design by Chris Welch

Library of Congress Cataloging-in-Publication Data
Tierney, Patrick.
Darkness in El Dorado : how scientists and journalists devastated the Amazon.
p. cm.
Includes bibliographical references and index.
ISBN 0-393-04922-1
1. Yanomamo Indians—Crimes against. 2. Yanomamo Indians—Social conditions. 3. Indians, Treatment of—Amazon River Region. 4. Genocide—Amazon River Region. 5. Gold mines and mining—Amazon River Region. 6. Anthropological ethics—Amazon River Region. 7. Chagnon, Napoleon A., 1938—Influence. 8. Chagnon, Napoleon A., 1938—Public opinion. I. Title.
F2520.1.Y3 T54 2000
981'.1—dc21
00-038682

W. W. Norton & Company, Inc., 500 Fifth Avenue, New York, N.Y. 10110
www.wwnorton.com

W. W. Norton & Company Ltd., 10 Coptic Street, London WC1A 1PU

1 2 3 4 5 6 7 8 9 0

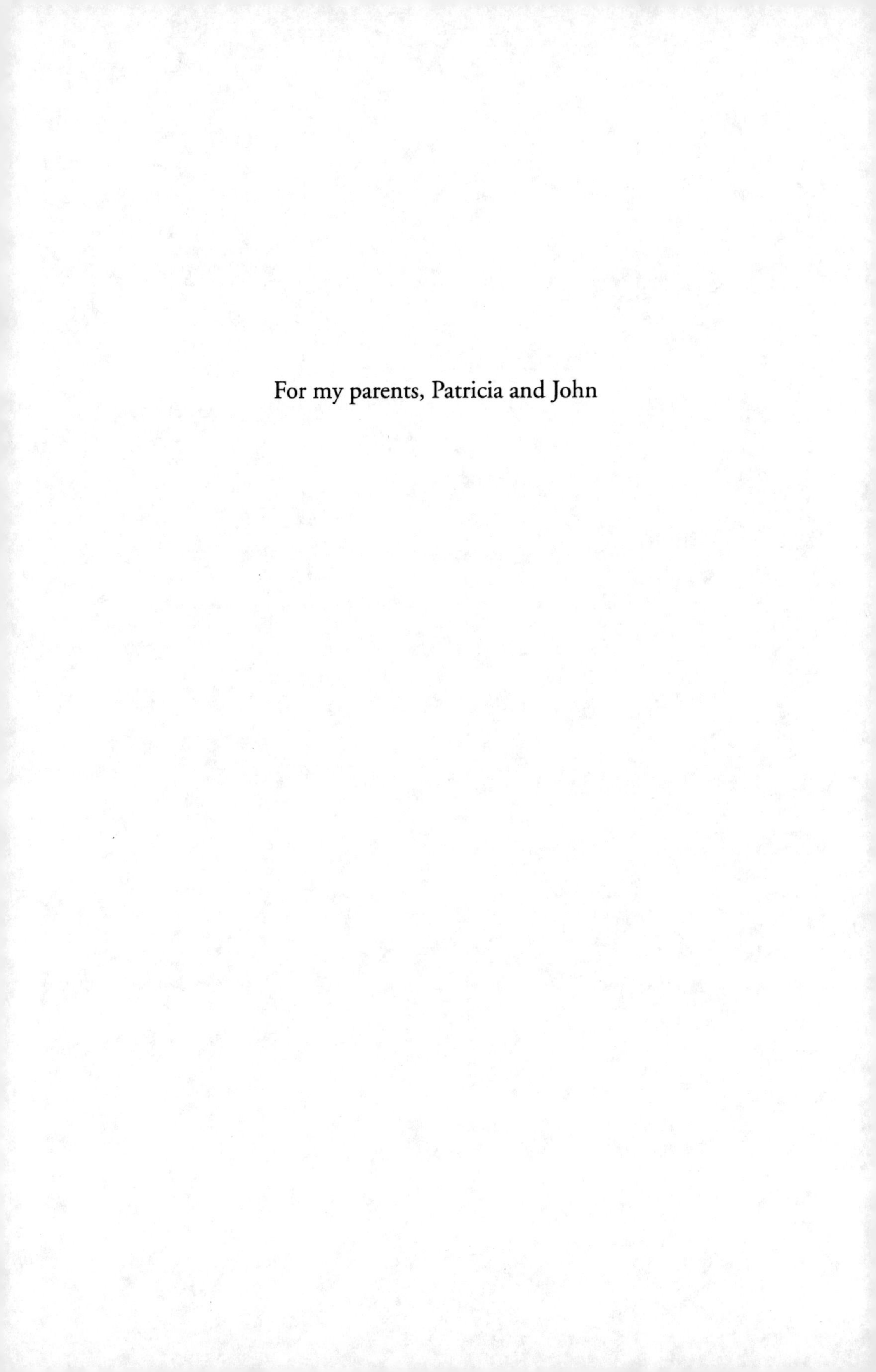

For my parents, Patricia and John

It is important to recognize that Darwinism has always had an unfortunate power to attract the most unwelcome enthusiasts—demagogues and psychopaths and misanthropes and other abusers of Darwin's dangerous idea.

—*Daniel C. Dennett,* Darwin's Dangerous Idea

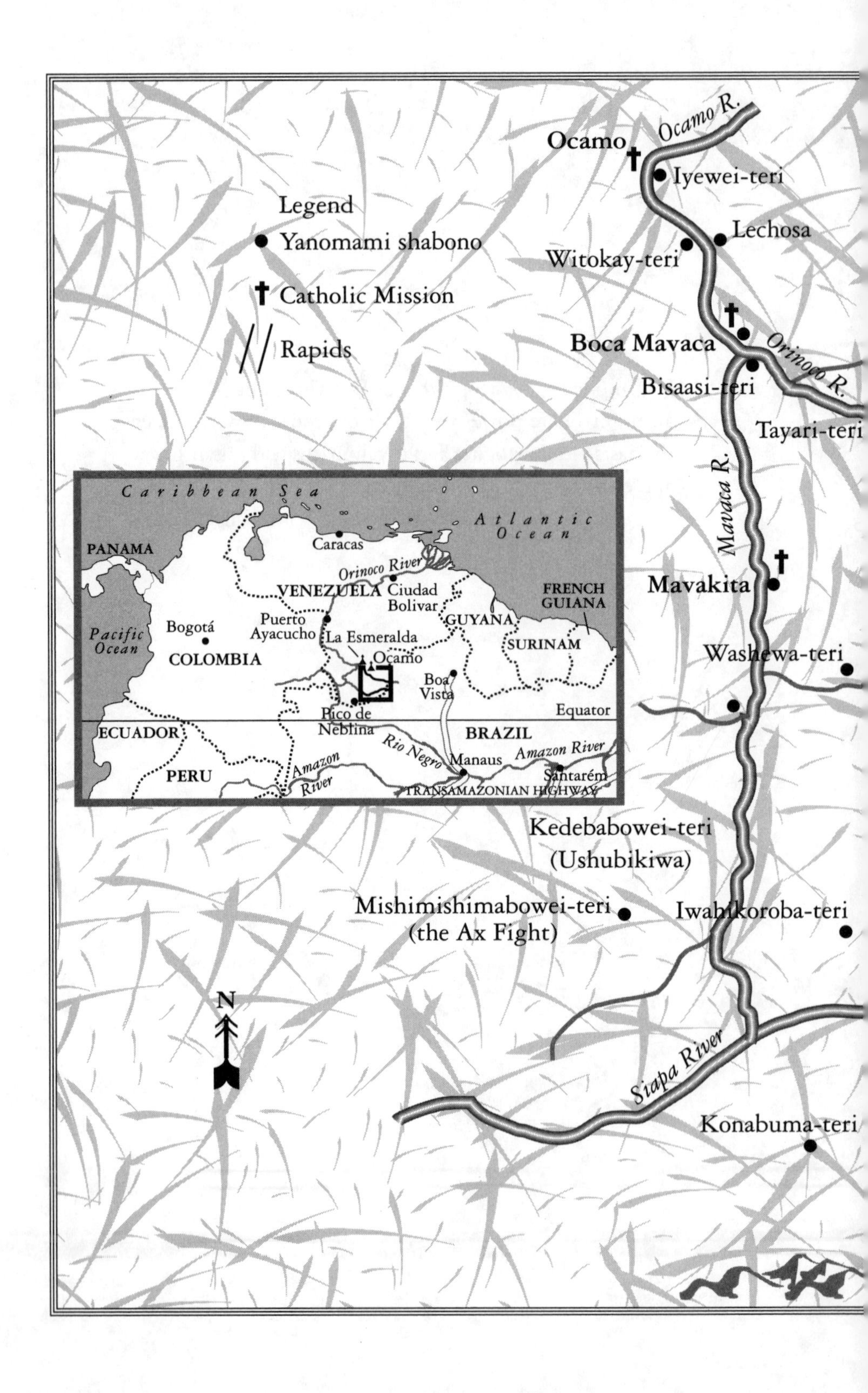
Legend
Yanomami shabono
Catholic Mission
Rapids
Ocamo R.
Ocamo
Iyewei-teri
Lechosa
Witokay-teri
Boca Mavaca
Orinoco R.
Bisaasi-teri
Tayari-teri
Mavaca R.
Mavakita
Washewa-teri
Kedebabowei-teri
(Ushubikiwa)
Mishimishimabowei-teri
(the Ax Fight)
Iwahikoroba-teri
Siapa River
Konabuma-teri
N
Caribbean Sea
Atlantic Ocean
PANAMA
Caracas
Orinoco River
VENEZUELA
Ciudad Bolivar
FRENCH GUIANA
Puerto Ayacucho
GUYANA
Pacific Ocean
Bogotá
La Esmeralda
Ocamo
SURINAM
COLOMBIA
Boa Vista
Pico de Neblina
Equator
ECUADOR
BRAZIL
Rio Negro
Amazon River
Manaus
Santarém
PERU
Amazon River
TRANSAMAZONIAN HIGHWAY

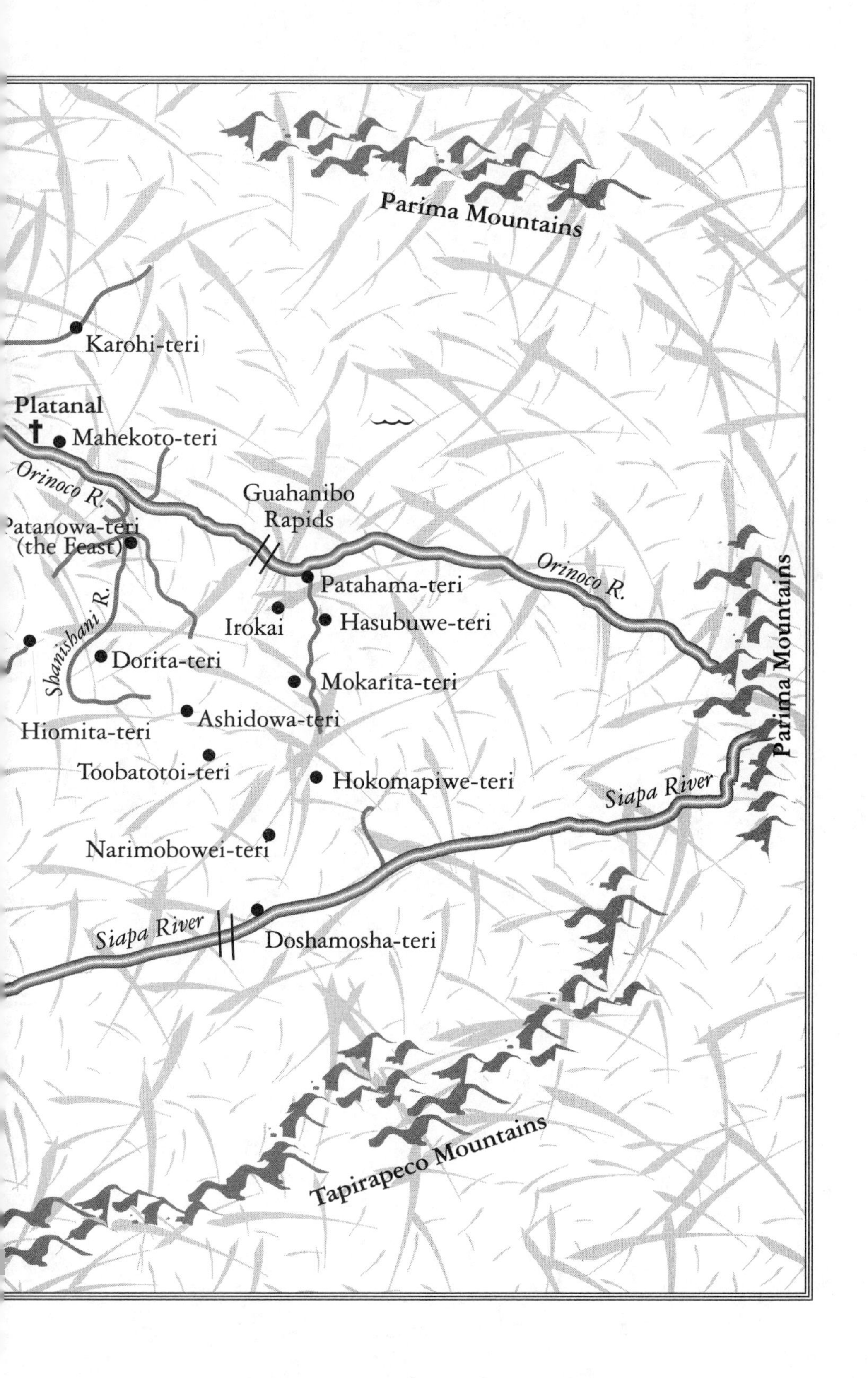

Parima Mountains
Karohi-teri
Platanal
Mahekoto-teri
Orinoco R.
Guahanibo
Rapids
Patanowa-teri
(the Feast)
Orinoco R.
Patahama-teri
Shanishani R.
Irokai
Hasubuwe-teri
Dorita-teri
Mokarita-teri
Ashidowa-teri
Hiomita-teri
Toobatotoi-teri
Hokomapiwe-teri
Parima Mountains
Siapa River
Narimobowei-teri
Siapa River
Doshamosha-teri
Tapirapeco Mountains

Contents

List of Graphs xv
Acknowledgments xvii
Introduction xxi

Part I Guns, Germs, and Anthropologists, 1964–1972

Chapter 1 Savage Encounters 3
Chapter 2 At Play in the Field 7
Chapter 3 The Napoleonic Wars 18
Chapter 4 Atomic Indians 36

Chapter 5 Outbreak 53
Chapter 6 Filming the Feast 83
Chapter 7 A Mythical Village 107

Part II In Their Own Image, 1972–1994
Chapter 8 Erotic Indians 125
Chapter 9 That Charlie 149
Chapter 10 To Murder and to Multiply 158
Chapter 11 A Kingdom of Their Own 181
Chapter 12 The Massacre at Haximu 195
Chapter 13 Warriors of the Amazon 215

Part III Ravages of El Dorado, 1996–1999
Chapter 14 Into the Vortex 227
Chapter 15 In Helena's Footsteps 243
Chapter 16 Gardens of Hunger, Dogs of War 257
Chapter 17 Machines That Make Black Magic 280
Chapter 18 Human Products and the Isotope Men 296

Appendix: Mortality at Yanomami Villages 317
Notes 327
Bibliography 385
Index 397

Photographs appear between pages 164 and 165

List of Graphs

Namowei War Deaths 34
Bisaasi-teri Mortality 51
Measles Antibody 55
Febrile Response to Edmonston B 67
Filming Deaths 121
Mortality and Mission Contact, 1987–1991 206
Corrected Mortality, 1987–1991 207
Victims in Worst Yanomami Wars 228
Stature of Amazonian Indians and Westerners 264
Yanomami Population Growth Projected at Historical Rate 269
Deaths at Kedebabowei-teri: The Impact of FUNDAFACI 324

Acknowledgments

First, I would like to offer heartfelt thanks to the guides and translators who were indispensable to both my research and my survival in the Yanomami rain forest. These included Severino Brazil, Pablo Mejía, Marco Jimenez, Alfredo Aherowe, and Jodie Dawson. Marinho De Souza, a microscopist and malaria diagnostician, was not only a great guide but also a healer for hundreds of desperately ill Yanomami Indians.

I would also like to thank the many anthropologists, doctors, and other scientists who read this manuscript. I am especially indebted to Leda Martins, who is finishing her Ph.D. at Cornell University, for her support throughout this long project and for her and her family's hospitality in Boa Vista, Brazil. Leda's dossier on Napoleon Chagnon was an important resource for my research.

I have obviously relied on Brian Ferguson's analysis of Yanomami warfare as a framework for several chapters of this book. I am also grateful to Terrence Collins of Carnegie Mellon University, Leslie Sponsel of the University of Hawaii, Terence Turner of Cornell University, Kenneth Good of Jersey State College, John Peters of Wilfrid Laurier University, Jesús Cardozo of FUNVENA, Giovanni Saffirio of the Consolata Missionaries, and John Frechione of the University of Pittsburgh for their comments and encouragement.

Mark White of the Smithsonian's National Anthropological Archives assisted my search through his warehouse retreat, where we found a box containing the tapes from the Atomic Energy Commission's 1968 expedition. Mark Ritchie sent me the videotaped interviews with Yanomami men that are transcribed in chapter 8.

The final chapter of this book, "Human Products and the Isotope Men," would not have been possible without Eileen Welsome, whose book *The Plutonium Files* opened up an entirely new perspective on the Atomic Energy Commission. She helped me contact several key individuals, including Cory Ireland, of the *Rochester Democrat and Chronicle,* who shared his research into the human radiation experiments at Rochester's Strong Memorial Hospital.

The photographer Valdir Cruz has been outstandingly kind in allowing me to use his superb black-and-white photographs without charge. Valdir worked for over eight months in Yanomami territory on a Guggenheim Fellowship. I enjoyed his company for a week around the main missions of the Orinoco, and he walked with Marinho and me to the village of Irokai after going through quarantine with us.

Kristine Dahl, my literary agent, guided the manuscript through many storms and finally to safe harbor at W. W. Norton & Company. Without Kristine's skill and determination this book would never have been published.

Norton, as everybody knows, is a brave house. Even so, Robert Weil took a courageous leap with this manuscript. He has been a discerning critic and a wise editor throughout the long preparation and legal review of this book. I know I am also speaking for Bob in thanking Rene Schwartz for her invaluable legal advice, Nancy Palmquist for her heroic patience, and Otto Sonntag for his extraordinarily detailed and helpful copyediting. Otto did wonders with the labyrinth of endnotes and sources.

My brother, John Tierney, has seen this manuscript in many different stages of evolution and has helped pull me through each one of them. *Gracias hermano.*

The University of Pittsburgh's Center for Latin American Studies gave me an appointment as a visiting scholar, which facilitated my research at the university's excellent Latin American collection. I wrote much of this manuscript while living in the quiet Pittsburgh neighborhood and in the same home where I grew up. There is no way that I can repay my parents, Patricia and John, for their loyalty and generosity. I have dedicated this book to them, but they deserve better.

Introduction

Chagnon's observations and science are basically correct. He is in the front line of modern sociobiology. Because of this, perhaps, controversy follows him.—*Edward O. Wilson*[1]

The renowned anthropologist Napoleon Chagnon appeared, unheralded, in Roraima, Brazil's northernmost state, often described as its most lawless, in September 1995. It was his first visit there in many years. Although he had helped make the Yanomami Indians the best-known tribe in the world, Chagnon faced nearly insurmountable hurdles in contacting them. In 1988, a past president of the Brazilian Anthropological Association had condemned him for portraying the Yanomami as innate killers.[2] When he attempted to visit a Brazilian Yanomami village in 1989 with a BBC film crew, Chagnon was forced to cancel the trip due to both academic opposition and a planned protest march by human rights groups.[3] And that was before Chagnon began the adventures that ended in his expulsion from the Venezuelan Yanomami Reserve by a judge on September 30, 1993.[4]

This time, however, he was able to skirt normal peer review with the help of *Veja,* the conservative newsweekly. *Veja* pressured the Indian Agency to grant special permission for a "visit of a journalistic or documentary character." The permit was actually issued to a photographer, and the anthropologist was included as a member of his "work team."[5] Chagnon had come to rely on the media both for access to the Yanomami and to communicate a shocking message—that the very people who posed as defenders of the Indians were actually destroying them. In the *New York Times,* the *Times Literary Supplement* of London, and other forums, he had already attacked "left-wing anthropologists," "survival groups,"[6] and "missionaries with secrets."[7] Many right-wing groups in Brazil, particularly miners and the military, welcomed these attacks. Indeed, a high-level intelligence report named the same "environmentalists, anthropologists and missionaries as threats to national security."[8]

It turned out, however, that Chagnon was not simply accompanying journalists searching for a feature story. The head of the Indian Agency in Roraima, Suami Percíllio Dos Santos, was surprised when Chagnon's bush pilot intercepted a shipment of vacutainers that the anthropologist planned to use for collecting Yanomami blood samples.[9] Dos Santos was not accustomed to people collecting Yanomami blood without permission from a scientific oversight committee—or informed consent from Yanomami leaders, many of who were now literate. The issue was particularly sensitive because the blood would have been harvested for a Brazilian geneticist who, like Chagnon, had participated in vast blood-collecting projects among Amazonian tribes during the worst days of Brazil's military dictatorship, the late 1960s.[10] Things had changed, however, and even *Veja*'s photojournalist, Antonio Mari, knew that this kind of research—where potentially priceless genetic material was obtained in exchange for cheap machetes—was out of bounds. "When Professor Chagnon received the test tubes and disclosed his intentions to collect blood samples from the Indians I felt betrayed and angry," Mari recalled. "I confronted Professor Chagnon and he told me that he was doing it to help the Indians. His intention, he said, was to collect blood samples to do research on different strains of malaria plaguing the area."[11] The Indian Agency's research director did not accept this explanation. Both Chagnon and the Brazilian geneticist received letters of rebuke; Chagnon was threatened with expulsion.[12] "The test tube incident caused enormous anger among FUNAI [Indian Agency] and Fundaçao Nacional de Saúde [National Health Foundation] officials," according to Mari. "Facing imminent cancellation of our permits, Chagnon gave up the blood sample idea."[13]

Native leaders were also upset. Several of them contacted Leda Martins, a Brazilian government official with three years of experience inside the Yanomami reserve, who at that time was a Fulbright scholar in anthropology at the University of Pittsburgh. They asked her for a complete background check on Chagnon. She wrote a short, annotated bibliography of the many controversies that had pursued him over the years, starting with his blood collecting for the Atomic Energy Commission and ending with his relationship with Venezuela's leading gold miner. She called it "Napoleon Chagnon: O Dossier."[14] The Indigenous Council of Roraima, an association of democratically elected Indian leaders, officially submitted the dossier to the government and asked for the revocation of Chagnon's permits.[15] In the end, Chagnon was permitted to travel to a single village, without blood sampling equipment, followed by a full-time guard—a mountainous man whose job normally was to bounce gold miners from the Yanomami reserve. Nevertheless, the brief window of opportunity enabled *Veja* to publish an interview with Chagnon, consistent with the magazine's editorial policy, that renewed Chagnon's critique against "many NGO's, anthropologists and missionaries who are most recently competing among themselves to see who can obtain the title of sole representative of the Indians to the outside world."[16]

For those of us who had seen the cataclysmic impact of the Amazon gold rush, it was both disheartening and extraordinary that Chagnon was savaging the very people who stood in the way of the Amazonian tribes' extinction. These were precisely the survival groups, missionaries, and "Marxist anthropologists"[17] who had opted to help the Indians, instead of simply studying them. One of this new breed of anthropologists was, in fact, Leda Martins, whom I had met in Boa Vista several times between 1989 and 1992. At that time, I was researching the gold rush and its chaotic impact on native peoples. It was a continental phenomenon, stretching from French Guyana to Bolivia, but I had picked Yanomamiland because of Chagnon. When I first read his ethnography *Yanomamo: The Fierce People,* he seemed preternaturally resourceful to me, a veritable hero—as he was to many other undergraduate males in the late 1960s and 1970s. In his relentless investigation of murder and tribal mayhem, Chagnon exposed the falsity of cherished myths about noble savages. I admired his philosophy, and I followed his swashbuckling example, documenting ritual murder in the Andes from 1983 to 1988. I wrote a book, *The Highest Altar: The Story of Human Sacrifice,* which cited Chagnon favorably.[18] It was a distinctly Chagnonian book, but, in a way, this was no distinction. *The Fierce People* was a narrative that launched a thousand books.

So when I decided to write about the Amazon, in 1989, I naturally gravitated toward the Yanomami and Chagnon territory. When I first went in, I thought it was a project that would last a year or so. To find contacts, and get my bearings, I contacted Chagnon, instead of his many enemies. At that time—it is hard to believe today—Chagnon's closest friends were Roman Catholic priests trained in anthropology. One of them, the anthropologist Giovanni Saffirio, was a former Ph.D. student of Chagnon's and became my closest friend.

In Roraima, I was robbed twice in 1990—once by the Yanomami and once by bandits in the city of Boa Vista. On another occasion, the federal police jailed me for a night after I had watched them try but fail to dynamite a clandestine airstrip inside the Yanomami Reserve. I had accompanied miners through little-explored swamps and mountains and, in the process, had gotten to know the frontier society much better than I wanted to. I knew which miners were paying which police officials and where the police allowed the miners to operate their clandestine landing strips. But at some point—and carrying Yanomami children in malaria comas may have had something to do with it—I came to despise the gold rush. I wanted it to end. I also grew tired of aggressive young men in civilian clothing with military crew cuts who followed me around day and night, just to make life miserable for any foreigners who were trying to witness the daily atrocities against the Indians.

In this climate, I gradually changed from being an observer to being an advocate. It was a completely inverted world, where traditional, objective journalism was no longer an option for me. My field expeditions became, increasingly, antimining expeditions. What the police were supposed to be doing, but were not, I did. I counted the miners and their machines in border areas and then submitted my findings to Brazil's powerful Procuradoria Geral da República, an ombudsman institution that refereed all other branches of government. I helped arrange a speaking tour of the United States for Davi Kopenawa, the Yanomami's most visible leader. He was an eloquent, spontaneous man, whom Chagnon called a "parrot" of Survival International.[19]

In 1992, I shifted my interests to the Macuxi Indian territory, a six-thousand-square-mile wedge of superb tabletop mountains between Guyana and Venezuela. I migrated with the miners after they were expelled from the Yanomami area. So did Leda Martins. After she came to Pittsburgh, we set up a Macuxi campaign office and collaborated with Survival International and the Rain Forest Action Network in placing articles about the Macuxi struggle around the world.[20] By the fall of 1995, the Macuxi campaign was threatening to become a full-time occupation for us. Under the circumstances, we tried assiduously to steer clear of Chagnon's controversies, which had not brought happiness to anyone who had become embroiled in them.[21] He was famously good at those competitions. And, as Leda's dossier documented, he had fought many of them.[22]

I had nothing personal against Chagnon. Actually, I did not know him personally. But as Chagnon departed from Boa Vista, I headed to New York and San Francisco to arrange protest marches supporting the Macuxi and their land rights. I decided, on October 2, 1995, in San Francisco, that I had better go and meet Chagnon at the University of California at Santa Barbara. He was becoming impossible to ignore. By now, Chagnon was one of the biggest players in the politics of the northern Amazon. I knew exactly who Chagnon's main supporters in Brazil and Venezuela were—the rogues' gallery, the principal opponents of indigenous rights and the principal cheerleaders of the Amazon gold rush. Leda Martins and I had in April 1995 written an op-ed piece for the *New York Times* that was sharply critical of Chagnon's main Venezuelan associate, one of Venezuela's leading gold miners.[23] By this time I shared the view of most indigenous and human rights organizations involved with the Amazon—that Chagnon should not be allowed back in the Yanomami territory if the elected leaders of the Yanomami themselves did not want him back. And, when Chagnon approached the State Department and U.S. embassy in Caracas for support, I also felt it was my duty to inform them that several of Chagnon's allies were under judicial investigation.[24] But I also knew there was a chasm between Amazonian reality and the outside world. The gap was so great that Chagnon could recast an illegal gold miner as a "naturalist,"* and one of Venezuela's most corrupt politicians as a philanthropist,†[25] as well as present a man charged with stealing Yanomami relief funds to Secretary of State Madeleine Albright as an Amazonian leader.‡[26] It seemed to me that Chagnon had an infinite capacity for ethnographic cleansing—of taking well-known Amazonian crooks and making them appear unsullied for the American press.

*Charles Brewer Carías (see chapter 9).

†Cecilia Matos, a fugitive from justice charged with numerous counts of corruption (see chapters 11 and 12).

‡Jaime Turón, under indictment for embezzlement.

But I felt an obligation to interview him and see whether there was some explanation for what he was doing.

So I showed up at the University of California's Santa Barbara campus for the first day of the 1995 fall quarter and made my way in and out of bike lanes to the largest undergraduate lecture hall. Nine hundred students crowded inside as Chagnon walked to the podium. His thinning white hair and beer belly contrasted with the incredibly glamorous young man—painted, feathered, and wearing a loincloth—who continued to appear on the back jacket of his picture-book format best-selling text. But there was something impressive about him. Chagnon wore the only suit and tie I had seen on campus, and he started class in a military style, ordering a dozen teaching assistants to stand up and identify themselves. "Tell them what you plan to do when you grow up," he said. When he introduced *The Fierce People,* he said, "It's by my favorite author, me." Chagnon was funny and truly self-confident.

Afterward, when I approached him, Chagnon said, "I don't want to talk to you." He swept halfway across the hall, turned around, and added, "On second thought, Mr. Tierney, I do want to talk to you. I'll see you in my office at ten A.M."[27]

I felt like an errant pupil summoned by a bullying vice principal. Oddly enough, my summons was for the exact moment when O. J. Simpson was to receive his verdict: 10:00 A.M., October 3. Most students were scurrying for television sets when I entered Chagnon's office.

"You say you're a journalist," Chagnon began. "Where is your press credential?" I told him I didn't carry a press credential. "If you say you're a journalist and you don't carry a press credential, you're a liar," Chagnon told me. "All journalists must carry a press credential."

At first, Chagnon would not let me take notes. "I don't trust you, because I think you're a witch-hunter," he said.

Things gradually got a little more civil. But Chagnon would say very little for the record. He continued to defend his association with Venezuela's leading gold miner. He also defended his decision to contact remote Yanomami villages on giant helicopters, without any quarantine precautions. Apparently, Chagnon felt he was entitled to the glory of contacting exotic natives; the Venezuelans were in charge of dealing with the diseases. He continued to blame everyone else who had worked for the Yanomami—including Survival International, Christian missionaries, and the Yanomami's best-known tribal leaders—and to demand sympathy for himself. "I'm tired of being a scapegoat," he said as I left.[28]

I went out into the California sunlight, blinking and bemused. So bemused, in fact, that when a blond student on a bicycle blurted out, "I can't believe he went free," I thought, for a moment, he was talking about Chagnon.

Part I

Guns, Germs, and Anthropologists, 1964–1972

Napoleon Chagnon and assistants at Abruwa-teri, Brazil, 1995 (photo by Antonio Mari)

Chapter 1

Savage Encounters

Every time we are making a contact, we are spoiling them.—*Charles Brewer Carías*[1]

The thunderous descent of the military helicopter at the village of Dorita-teri drove Yanomami Indian women and children screaming into the surrounding plantain gardens. Out in the jungle, panic also reigned, as macaws and parrots, deer and tapirs scrambled to escape the machine. When the dust cleared, twenty Yanomami warriors were standing in a semicircle, yelling at seven white men and one white woman who had descended from the helicopter with television cameras and sound equipment. Most of the warriors held enormous bows and arrows. The headman waved an ax.[2]

The tumultuous landing in Dorita-teri, on May 17, 1991, created an impressive spectacle for the Venezuelan television crew, which was doing a special on "the purest human groups in existence."[3] The community was located in the little-explored Siapa Highlands on the Brazil-Venezuela border, the

Amazon's last frontier. These remote mountains also concealed the last intact cluster of aboriginal villages in the world—whose inhabitants were considered living relics of prehistoric culture. The seminomadic Yanomami spent their time hunting and trekking in much the same way humanity had done for countless generations. The anthropologist directing the expedition called them "our contemporary ancestors."[4]

Although it was a novelty for the television journalists to be welcomed into an Indian village with axes and arrows in 1991, the expedition leaders Napoleon Chagnon and Charles Brewer Carías had been taking risks like this for decades. Chagnon, an anthropologist at the University of California at Santa Barbara, and Brewer, a naturalist then associated with the New York Botanical Garden, claimed first contact with 3,500 Yanomami Indians in the Siapa region alone.[5] In August 1990, their "discovery of 10 Yanomami villages they say had never been visited before by anyone except other tribal members" set off a frenzy of media competition and scientific congratulation.[6] "Stone Age Villages Found" ran a typical headline.[7]

In the economics of exoticism, the more remote and more isolated a tribal group is, the greater its market value. As the last intact aboriginal group, the Yanomami were in a class by themselves, poster people whose naked, photogenic appeal was matched by their unique genetic inheritance. Their blood was as coveted by scientists as their image was by photographers.[8] Technically, the Yanomami were defined as a virgin soil population, and there was a trace of feudal privilege in the way the visitations were doled out: ABC's *Prime Time* got one village,[9] *Newsweek* another,[10] and so it went. The *New York Times* got two villages,[11] but had to share one of them with the Associated Press.

Sometimes the media's own arrival was the real scoop. Just before visiting Dorita-teri, the same Venevisión crew had gotten exciting footage at a neighboring village, Shanishani-teri, where the helicopter landed in the middle of the circular communal house, or *shabono.* The round house's roofing was whisked up and away, like Dorothy's house in a Kansas tornado, while the Yanomami's possessions—bark hammocks, gourds, woven baskets, and bamboo arrows—splintered and shattered like Tinkertoys. The on-camera journalist, Marta Rodríguez Miranda, said, "They kindly accepted our landing in the middle of the *shabono* even though their whole roof would collapse with the downblast."[12]

Similar scenes were repeated elsewhere with different media teams. At one village, the helicopter was driven off with a hail of rocks and sticks;[13] at another, five Yanomami were injured by falling roof poles.[14] During all these adventures, only ABC's John Quiñones asked the most obvious question, one that might have occurred to any grade-school student educated about the tragic history of Indian tribes since the European discovery of America in the fifteenth century. "Aren't we doing some harm, spoiling this culture, even by coming here today?" Quiñones asked Charles Brewer, who at fifty-two, looked fit, handsome, and baby-faced behind his sprawling mustache.

"Definitely," Brewer answered. "Every time we are making a contact, we are spoiling them."[15]

In spite of the "first contact" craze, almost all of these extraordinarily remote communities had been visited before and were being reharvested after a suitable interval. In fact, Chagnon and Brewer had visited the Yanomami of Dorita-teri at another location in 1968, where they made two award-winning documentaries, which went on to become staples of anthropology classes around the world.[16] One film, *Yanomama: A Multidisciplinary Study,* dramatically illustrated the scientists' altruism in rescuing the Dorita-teri's parent village from a deadly measles epidemic.[17] The second documentary—*The Feast*—showcased Yanomami ferocity and won first prize at every film festival in which it was entered.[18] Everyone praised these films except the Dorita-teri, who apparently had a different interpretation of the scientists' camera work.

Despite their previous acquaintance, the Dorita-teri were not enthusiastic about seeing Chagnon and Brewer again. The village headman, Harokoiwa, greeted them with an ax. Swaying from side to side, Harokoiwa upbraided the scientists for driving away game with their helicopter. He also accused them of bringing *xawara*—evil vapors that, in the Yanomami conception of disease, cause epidemics. Harokoiwa angrily claimed that Chagnon had killed countless Yanomami with his cameras.[19] In reality, many of the Yanomami who starred in *The Feast* died of mysterious illnesses immediately afterward—new sicknesses the Indians had attributed to the scientists' malefic filmmaking.[20] The Yanomami abandoned the village where *The Feast* was made and never returned. Later they shot arrows into a palm effigy of the film's anthropologist—Napoleon Chagnon.[21]

Now, on Chagnon's return, the headman began swinging his ax tantalizingly close to the anthropologist's head. Harokoiwa yelled that he did not want outsiders to poison any more rivers,[22] a reference to Brewer's huge open-pit gold mines on Indian lands.[23]

Suddenly, one of the chief's sons, wielding another ax, rushed Chagnon.

As the weapon arced through the air, it appeared to be on its way to splitting Chagnon's skull when Brewer deftly intercepted the ax with one hand and, with the other, knocked the man to the ground. Adding to the confusion were screams by some of the Dorita-teri women, who begged their men not to kill Chagnon and Brewer, "because they had always brought so many presents."[24]

Under the circumstances, the scientists and television crew thought it best to leave. On returning to Caracas, Venevisión's producers shelved the footage of this confrontation,[25] though not without some pain. It was a great little scene. But it raised nagging questions that could not be answered, at least not on a show about Stone Age ancestors.

Chapter 2

At Play in the Field

For many years now anthropologists have been saying how exotic we Yanomami are. But when we finally tell our story the world will find out who is truly exotic.—Davi Kopenawa[1]

Almost every anthropology student has experienced the horror of Napoleon Chagnon's first encounter with South America's Yanomami Indians. Chagnon stumbled into a village while Yanomami shamans were blowing hallucinogenic snuff up their noses. Deeply drugged, the Indians drew their six-foot bows. "I looked up and gasped when I saw a dozen burly, naked, sweaty, hideous men staring at us down the shaft of their drawn arrows! Immense wads of green tobacco were stuck between their lower teeth and lips making them look even more hideous, and strands of dark-green slime dripped or hung from their nostrils. . . ."[2]

Yanomamo: The Fierce People, which Chagnon first published in 1968, quickly became the all-time best-seller in anthropology.[3] Four million students bought the book,[4] which is both a riveting account of warfare among Stone Age people and a sobering assessment of what life may have been like for much of prehistory. *The Fierce People* made the Yanomami the most famous tribe in the world—a model for primitive man and a synonym for aggression.[5] It made Napoleon Chagnon the best-known American anthropologist since Margaret Mead.

By the time I began studying the Yanomami, in 1989, they were caught in the middle of the Amazon gold rush, the largest gold migration in history.[6] Forty-five thousand miners were flying in and out of clandestine airstrips, bringing epidemics, alcohol, guns, and prostitution.[7] Malaria, influenza, and hepatitis were out of control: fifteen hundred Yanomami had died of infectious diseases.[8] Although I contracted malaria, my worst moments came after being robbed at gunpoint, when I found myself sleeping on the jungle floor without food and negotiating rapids in a leaky boat with gold miners who were as hungry and desperate as I was.

Before going into the jungle, I had read and admired *The Fierce People.* So it was surprising to see that the Yanomami—so terrifying and "burly" in Chagnon's text—were, in fact, among the tiniest, scrawniest people in the world.[9] Adults averaged four feet seven inches in height,[10] and children had among the lowest weight-height ratios on the planet.[11] They seemed decidedly timid compared to several other Amerindian groups with whom I had lived. The Yanomami welcomed me effusively, as they welcomed missionaries, anthropologists, gold miners, and anybody else who brought them steel, medicine, or food. But the real shock came when I visited a village on the Mucajaí River in Brazil, where Chagnon claimed to have discovered a Yanomami group that embodied the tribe's ultimate form of "treachery."[12] In reality, these Indians had lived in relative harmony for half a century.[13] I was amazed to find that Chagnon had even created his own topography—moving a mountain where one did not exist and landing cargo planes where they had never touched down[14]—while quoting people he could never have spoken to in this part of the jungle.[15] It was the Mucajaí of Chagnon's mind.[16]

When I went into the jungle, Chagnon was embroiled in an academic dispute that, like the rumblings and lightning of a distant storm, was already setting fire to anthropology journals around the world. Many of the anthropologists in Yanomami studies were denouncing Chagnon. They had accused him of inventing quotations and creating nonexistent villages—of fabricating lurid stories about Yanomami violence that were being enthusiastically broadcast by promining forces to justify the *conquista* of Yanomamiland.[17] The titles of the articles alone suggested the bitterness of the debate: "Ethnography and Ethnocide"; "The Academic Extermination of the

Yanomami"; "To Fight over Women and to Lose Your Lands: Violence in Anthropological Writing and the Yanomami of Amazonia"; "Bias in Ethnographic Reporting."[18] In the end, the journal *Science,* which had twice published articles by Chagnon, despite opposition by Yanomami field experts,[19] was forced to run a new article on the imbroglio: "Warfare over Yanomamo Indians."[20]

But no one was prepared for Chagnon's next move—which brought anthropology's war and the gold war together. In 1990, he began campaigning to turn the Yanomami's homeland into the world's largest private reserve, to be administered by himself and two controversial allies, both of whom had their own gallery of enemies inside Venezuela. One was the naturalist turned gold miner Charles Brewer, who had a police record of clandestine gold diggings on Indian lands.[21] Charlie, as everyone called him, certainly had a history as wild as that of any conquistador—Olympic swimmer, scientist, explorer, government minister—and these were just a few of his incarnations.[22] Like Chagnon, he had started off as a disciple of the great geneticist James Neel at the University of Michigan's Department of Human Genetics, where he began his romance with violent competition.[23] Like Chagnon, he loved guns and fighting.[24] Brewer ferried a surprising variety of celebrities into the forest, from Margot Hemingway to David Rockefeller,[25] and even arranged a tuxedo dinner catered by helicopters atop a magnificent, 10,000-foot *meseta.*[26] One of Brewer's jungle companions was the London *Times* editor Redmond O'Hanlon, who made Charlie the hero of his classic jungle book, *In Trouble Again.*[27] The editor of *Geo* was even more impressed. He called Brewer "the Alexander Humboldt of our time."[28]

Meanwhile, Brewer led the Amazon gold stampede.

Brewer introduced Chagnon to President Carlos Andrés Pérez's mistress, Cecilia Matos.[29] Together, they planned to control Yanomamiland—all under the auspices of Cecilia Matos's foundation, FUNDAFACI.[30] In retrospect, the bold decision to seize control of Yanomamiland—an area the size of Maine, with immense scientific and mineral resources—was the most fateful of Chagnon's career. But it made sense, from Chagnon's perspective. By 1990, he could not get research permits. South American anthropologists, Indian Agency bureaucrats, indigenous leaders, and missionaries wanted him, and the legacy of *The Fierce People,* to disappear.[31] Of course, Chagnon might have rested on his laurels, allowing his books and films to roll onward, conquering by their sheer mass and momentum. But refusing a challenge would have been contrary both to Chagnon's personality and to his theory of violence. According to him, murderers reproduced prolifically. Aggressive villages prospered. Evolution punished passivity and rewarded predation. Chagnon had no choice but to attack.

And it had to be a total war. Chagnon hoped to construct the biggest tropical research station ever in the Yanomami wilderness.[32] It would have given him unprecedented power, but it required overthrowing the legal structure already established in Yanomami territory, which in turn required a public-relations campaign. Chagnon managed this brilliantly by handing out "first contact" scoops in the hitherto-unmolested Siapa Highlands in exchange for promoting his plan and denouncing the missionaries and Yanomami leaders who opposed it. In this agile quid pro quo, reporters plugged Chagnon's plan while claiming the Yanomami were dying out at the missions at several times the rate of the remote villages—an inversion of reality.[33] (See the appendix: "Mortality at Yanomami Villages.") Thanks to Chagnon's ability at sound bites, the plan almost worked. But that was the trouble with all the plans to create a kingdom in El Dorado country. They always almost worked.

The immediate cause of Chagnon's downfall was Charles Brewer, who wanted to "administer" the same area where he had been planning one of the largest tin mines in the world.[34] The Yanomami rebelled against the proposed FUNDAFACI biosphere. And, after Pérez was impeached and jailed for corruption, Matos herself became a fugitive from justice. Among other things, the police and congress investigated her use of government helicopters to fly her friends—including Chagnon—around Yanomami territory,[35] junkets that cost millions of dollars[36] and apparently violated the law.[37]

Judge Nilda Aguilera expelled Brewer and Chagnon from Yanomami territory on September 30, 1993,[38] following public demonstrations and petitions from seventeen Indian tribes.[39] Venezuelan anthropologist Nelly Arvelo Jiménez, who has a Ph.D. from Cornell University, publicly asked about Chagnon what many privately wondered: "How could he dare become associated with . . . environment[al] predators and economic gangsters?"[40]

The aftermath convulsed the Venezuelan congress, courts, and media. Chagnon was also charged with spreading diseases through large, reckless expeditions to vulnerable, uncontacted Yanomami villages in the Siapa Highlands of Venezuela,[41] and with provoking conflict among them—to the point of setting off battles in which his own guides were killed.[42]

The scandal created concentric circles of violence, starting among the least-contacted Yanomami villages and spreading to Venezuelan national politics, where it culminated in a failed putsch, led by tanks and attack planes, against the presidential palace.[43] When I arrived in Caracas, the doors of the palace, with their bronze lions, were all shattered, and tank tracks ran down the mansion's marble stairs like the footsteps of a Hollywood star.

The fallout has also shaken American anthropology since the late 1980s. Chagnon's writings were a blessing to the gold miners invading Yanomami lands and a curse to the Yanomami political organizations trying to expel these so-called *garimpeiros* (hill bandits). Chagnon did nothing to distance himself from crude attacks against the Yanomami that utilized quotations from his books and articles—quotations so long they appeared to infringe on copyright laws.[44] He had always been a militant anti-Communist and free-market advocate.[45] Now Chagnon began lumping "leftwing anthropologists," "leftwing politicians," and "survival groups" into the same dismissive sentences,[46] while calling the Yanomami's most visible spokesman, Davi Kopenawa, "a parrot of human rights groups."[47] In the eyes of most human rights workers, Chagnon became, as a French anthropologist put it, "an intellectual accomplice of the gold miners."[48] The *Chronicle of Higher Education* now called it "Bitter Warfare in Anthropology."[49]

Terence Turner, an Amazon expert from the University of Chicago who headed a commission on the fate of the Yanomami, told colleagues in December 1994, "We have no right to castigate the gold miners, the military, the missionaries, or the governments of South America if we're afraid to look at the role of our own anthropologists in the Yanomami tragedy. Unfortunately, Napoleon Chagnon has caused a great deal of harm to the Yanomami and their chances of survival."[50]

What began as a debate about human nature has become a dispute about science at the service of ethnocide. "This is by far the ugliest controversy in the history of anthropology," commented Lesley Sponsel, a professor at the University of Hawaii who headed the American Anthropological Association's Human Rights Committee. "Nothing else even comes close."[51]

Chagnon's influence has often been compared to that of Margaret Mead,[52] whose own classic, *Coming of Age in Samoa,* was surpassed in sales and influence only by *The Fierce People.* In some ways, the current controversy started off like the one that brought discredit to Mead's writings. Mead conjured up an idyllic society in the South Pacific, whose sexual freedom coincided with the theories of her mentor, Franz Boas of Columbia

University and appealed to Mead personally. Mead managed to ignore the fact that the Samoans had one of the highest indices of violent rape on the planet.[53]

Whereas Mead continued Rousseau's tradition of pressing idealized natives into service for the left, Chagnon picked up where Social Darwinists left off. He emphasized the necessity of lethal competition in nature and the inevitable dominance of murderous men in a prehistoric society. Chagnon's ethnographic image of the ferocious Yanomami matched his own reputation for bar fighting[54] and also echoed the views of his sponsor, the great geneticist James Neel of the University of Michigan. Neel believed that modern society was going soft. From the Amazon's unspoiled inheritance, Neel hoped to find a genetic basis for male dominance—"the Index of Innate Ability"—a kind of elixir to the gene pool.[55] It was Neel who selected the Yanomami as experimental subjects and sent Chagnon to find evidence for his quixotic theory.[56]

That is how Chagnon initially found himself in remote rain forest highlands surrounded by ancient mountains of granite, called the Guiana Shield, which divide the immense Amazon-Orinoco watersheds. When Chagnon arrived in 1964, this remained one of the last unmapped areas of the Americas; the origins of rivers and the boundaries between Venezuela and Brazil were still uncertain. The highest peak, Cerro Neblina (9,889 feet), had been discovered only in 1953. White-water rapids, 3,000-foot cliffs, and swamps the size of European states had frustrated conquerors of all countries since the sixteenth century, making these redoubts a perfect blank slate for wilder hopes than the leadership gene. Here, Sir Walter Raleigh unsuccessfully searched for a second Cuzco and then wrote a popular book in 1601 about an Inca city of solid gold, Gran Manoa, built next to "a mountain of crystal."[57]

The tectonic plates of all European empires collided here, too, creating the splinter states of the Guyanas, and a cartographer's dream labeled El Dorado. Spectacular tabletop mountains added to the mystery. Sir Arthur Conan Doyle used one of them, Mount Roraima, for his fictional Lost World, an Edwardian Jurassic Park inhabited by ape-men, continuing Raleigh's image of Indians "who dwell upon the trees."[58] The German naturalist Alexander von Humboldt concluded that delusions came with the territory. "Above the great cataracts of the Orinoco a mythical land begins, the soil of fable and fairy vision."

But the reality Chagnon described was in some ways stranger than the tradition of projection and fantasy. He focused on the seemingly compulsive violence of the Yanomami, whose 25,000 members made up the world's largest intact aboriginal culture. As Chagnon went farther into uncharted territory, he had a Conradian sense of going backward in evolutionary time to an awful, almost apelike existence. The foreword to *The Fierce People* characterized the Yanomami as a "brutal, cruel, treacherous" people whose morality was the antithesis of the "the ideal postulates of the Judaic-Christian tradition."[59]

The Yanomami had no metallurgy and little social hierarchy. They slung bark hammocks around the periphery of communal round houses with open centers, called *shabonos.* Personal possessions were almost nonexistent. Although the Yanomami practiced slash-and-burn gardening, they spent much of their time on long treks hunting and gathering, the way of life that predominated for most of humanity's prehistory. The Yanomami did not use canoes and had little use for clothes, other than a cotton waistband for women and a penis string for men. They practiced ritual combats that no other Amazonian group shared—a graded series of exchanges starting with chest pounding and followed by duels with long poles.[60] Even their blood was different. The Yanomami have a private gene mutation not found in any other human population. They also lack the Diego factor, an antigen found in all other Mongoloid peoples, including other Amerindians.[61]

Presumably, the Yanomami are also of Asiatic origin, but their skin is often lighter and their eyes are hazel, characteristics that have earned them the name White Indians. Some scientists believe they are descended from the first paleo hunters who crossed the Bering Strait at least thirteen thousand years ago[62] (and whose few skeletal remains suggest Caucasoid features). In Chagnon's evocative writing, the Yanomami became both unique and normative, a one-of-a-kind tribe held up as a model for humanity's earliest type of warfare, sexual competition, and economy. It was as close as an anthropologist could come to discovering El Dorado.

The Yanomami may always remain an enigma. But James Neel certainly picked the wrong place and the wrong people to try and prove his quirky ideas about hierarchies of violence and genetic selection. The Yanomami have a low level of homicide by world standards of tribal culture and a very low level by Amazonian standards. Compared to other tribes, they are fearful of outsiders, especially when it comes to dealing with aggressors like gold miners. As Chagnon noted in his Ph.D. thesis, "the Yanomamo are not brave warriors."[63]

The attempt to portray the Yanomami as archetypes of ferocity would be pathetic were it not for its political consequences—and for the fabulous distortions this myth has perpetrated in biology, anthropology, and popular culture. The ripple consequences of the Yanomami fantasy can be seen from the film *The Emerald Rain Forest* (where an apelike group called "the Fierce People" create indiscriminate mayhem), to the Harvard primatologist Richard Wrangham's recent book *Demonic Males,* which has a whole section about "Yanomamo Indians and Gombe chimpanzees."[64] Just as Mead's beliefs about sexual freedom and child rearing worked their way into public-policy debates, Chagnon's ferocious Yanomami have become proof to some social scientists that ruthless competition and sexual selection cannot be legislated away by idealistic do-gooders. The Yanomami are the Cold Warriors who never came in from the cold.

Unraveling this academic distortion might have been as significant, say, as Derek Freeman's book *Margaret Mead and Somoa: The Making and Unmaking of an Anthropological Myth.* But, as I began investigating on the Upper Orinoco, I found that things were both stranger and more complicated than I had expected.

One of the oddest things I uncovered among the most remote Yanomami villages was a pattern of choreographed violence, dating back to the early, internationally acclaimed films of the Yanomami made by Chagnon and Timothy Asch, and continued to the present by *Nova* and the BBC. As a missionary who accompanied me said, "It's amazing how many alliances were created and villages were built just to satisfy the film crews."[65]

A case in point was the recent *Nova*/BBC documentary "Warriors of the Amazon," which has aired many times in the United States in the late 1990s. The hour-long piece was a dramatic narrative about an unnamed *shabono* that was said to be ceaselessly warring against another unnamed *shabono.*[66]

In reality, the hosts had not had any wars in years, until the film crew arrived and built a new *shabono,* negotiated a new alliance, and helped create a feud that has split the former community apart.[67]

Anthropologists have left an indelible imprint upon the Yanomami. In fact, the word *anthro* has entered the Indians' vocabulary, and it is not a term of endearment. For the Indians, *anthro* has come to signify something like the opposite of its original Greek meaning, "man." The Yanomami consider an *anthro* to be a powerful nonhuman with deeply disturbed tendencies and wild eccentricities—an Olympian in a funk.[68]

It is no exaggeration to say that the Yanomami are ethnographic experts on the madness of anthropologists. A German anthropologist from a prestigious Max Planck Institute near Munich committed suicide at the Yanomami village of Patanowa-teri after his Yanomami lover deserted him.[69] A French *anthro* in the Parima Mountains had to be disarmed, tied up, and carried off by parachutists after he tried to kill one of his colleagues with a knife.[70] Chagnon, according to videotaped testimony by his own principal informant, played the role of a shaman who took hallucinogens and incorporated the most fearsome entities of the Yanomami's spirit pantheon.[71]

And these were the normal anthropologists.

Well, yes, when compared with Jacques Lizot, a University of Paris anthropologist and disciple of Claude Lévi-Strauss. Lizot lived for thirty years with the Yanomami, far longer than any other anthropologist, and served as principal consultant for *Nova*'s recent documentary. Yet, in some ways, the Yanomami he conjures up would appear utterly alien to the Fierce People. Lizot has portrayed the Yanomami as sexual innovators of stunning sophistication, an Erotic People—a product better appreciated in the French cultural sphere.[72]

According to the author Mark Ritchie, Lizot was not altogether lacking in sexual imagination himself. Through transcribed testimonies from a variety of Yanomami sources, Ritchie recounts, in his book *Spirit of the Rainforest,* Lizot's exotic career. Lizot was identified by his Yanomami name, Bosinawarewa—which Ritchie renders as Ass Handler[73] (literally, Anus Eater). At the same time that Lizot became the acknowledged expert on Yanomami language, the French scientist indirectly expanded the Yanomami lexicon. In some villages, the Yanomami word for anal intercourse is *Lizo-mou:* "to do like Lizot."[74]

Not even the most inventive chronicler could have envisioned a mythology as florid as that which scientists and filmmakers have scripted for the Yanomami and enacted for themselves. Above the great cataracts of the Orinoco is the strangest story in the history of social science—the Bermuda Triangle of anthropology.

"We started calling the Upper Orinoco 'Macondo,' the surreal world of Gabriel García Márquez," says Jesús Cardozo, president of the Venezuelan Foundation for Anthropological Investigation (FUNVENA). "When I first started doing research among the Yanomami, I was told, 'Lizot is going to kill you.' I thought it was a joke, you know. But I found out that the most insane things were going on. I mean, anthropologists were chasing each other around with shotguns. Each had his own fiefdom. Villages were named for Lizot and Chagnon, as though they were great Yanomami chiefs. And the anthropologists' villages took on their personalities. Chagnon's Yanomami were more warlike than any other group; Lizot's village became the capital of homosexuality. Of course, there's a serious question of human rights violations. But what interests me is not bringing Lizot or Chagnon before an international tribunal. I just think it's important for anthropology, for the history of science, to understand how this happened, and what role the media played in creating this strange new world. Because the more weird Chagnon and Lizot became, the more they were worshiped as celebrities."[75]

In the discourse of the Upper Orinoco, words like "paranoid,"[76] "sociopath,"[77] "loathsome"[78] and "criminal"[79] are commonplace—especially when anthropologists talk about Chagnon. Chagnon, for his part, has written that one of his critics is "fucked,"[80] and he has dismissed the rest as skunks.[81]

"I think so many anthropologists went bonkers among the Yanomami because there were no limits, no rules among the Yanomami," says the missionary Michael Dawson, who has spent forty years among the Indians. "With their tools and guns, they were like Connecticut Yankees at King Arthur's court. They could become whatever they wanted to become. They became gods."[82]

For the Yanomami, as for the Greeks, a proof of a god's power was the ability to bring epidemics. This is unquestionably the most impressive legacy of scientists and journalists on the Upper Orinoco. On the basis of the scientists' own detailed records, hundreds of Yanomami died in the immediate wake of exploration and filming.[83]

I was surprised to learn that the Atomic Energy Commission (AEC) had lavishly funded the earliest and deadliest expeditions. As I requested AEC documents through the Freedom of Information Act, I found that the AEC had used the Yanomami as a control group, comparing their rate of genetic mutation with that of the survivors of the atomic bombs in Hiroshima and Nagasaki. The Department of Energy wrote to me, "The results of this research have contributed to our understanding of the natural development of gene mutations in man and have helped to bridge the gap between mutagenesis studies in experimental animals and observations in people."[84]

To complete these unique studies, which helped the AEC set radiation standards in the United States, the AEC needed great amounts of Yanomami blood, all purchased with steel goods. The researchers were particularly interested in Yanomami responses to disease pressure: "How disease as well as warfare decimates the population."[85]

The Venezuelan Yanomami experienced the greatest disease pressure in their history during a 1968 measles epidemic. The epidemic started from the same village where the geneticist James Neel had scientists inoculate the Yanomami with a live virus that had proven safe for healthy American children but was known to be dangerous for immune-compromised people.[86] The epidemic seemed to track the movements of the vaccinators. An estimated 15 to 20 percent of the Venezuelan Yanomami died of measles in the months following vaccination.[87]

I sensed that the injustice done to the Yanomami was matched by the distortion done to science and the history of human evolution. Yet the incredible faith the sociobiologists had in their theories was admirable. Like the old Marxist missionaries, these zealots of biological determinism sacrificed everything—including the lives of their subjects—to spread their gospel. A fascination with this fanaticism led me to places I did not intend to go—including the National Film Archives in Washington. After a week of searching through a collection no one else had ever examined, I found a dusty box labeled "Very valuable. Original sound tracks of 1968 expedition." I also found myself trekking, *shabono* by *shabono,* mountain range by mountain range, into the Siapa Highlands, and into an Amazonian heart of darkness where scientists and journalists were the chief protagonists.

The risks of such a crusade should have been obvious. After all, plenty of cautionary tales dotted the Yanomami landscape. "You really have to really feel sorry for the Yanomami," said the anthropologist Kenneth Good, a former Ph.D. student of Chagnon's who has lived longer among the Yanomami than any other American anthropologist. "The United States sent Chagnon. France sent Lizot. In Caracas they added on Charlie Brewer. *Jesus.* You better be careful down there yourself. Yanomami studies make people crazy."[88]

Was it heredity or the environment? Did the Yanomami's culture, or something in the Orinoco's water, drive researchers mad? Or was it a self-selecting group of born misfits? Luckily, I did not have time to worry about such trifles as I conducted censuses and advanced farther into the jungle, always making good progress. After a particularly exhausting trudge through the Siapa wilderness, I was only slightly disconcerted when the Yanomami at the village of Mokarita-teri all dispersed in terror at my arrival. The first man who returned came back warily, as if approaching one of the dangerous ghosts, called *bore,* who show themselves to the Yanomami at night.

"Are you Chagnon?" he asked.[89]

Chapter 3

The Napoleonic Wars

The village I'm living in really thinks I am the be-all and end-all.—*Napoleon Chagnon, 1965*[1]

The wars that made Chagnon and the Yanomami famous—the ones he wrote about with such relish in *The Fierce People*—began on November 14, 1964, the same day the anthropologist arrived with his shotguns, outboard motor, and a canoe full of steel goods to give away.[2]

"A war started between groups which had been at peace for some time on the very first day Chagnon got there, and it continued until he left," said Brian Ferguson, a Rutgers anthropologist who is an expert on violence in primitive societies. "I don't think that was an accident." Ferguson's book *Yanomami Warfare,* published in 1995, is perhaps the most comprehensive account ever written about tribal conflict. Two of its chapters are devoted to Chagnon's own role in fomenting warfare among the Yanomami.[3] "I originally considered calling my book *The Napoleonic Wars,*" Ferguson said.[4]

Ferguson's work is part of a growing consensus that Westerners, including scientists, profoundly disrupt tribal health, life, and politics on arrival. The 1998 Pulitzer Prize for nonfiction went to the UCLA medical researcher Jared Diamond and his book *Guns, Germs, and Steel: The Fates of Human Societies,*[5] a meditation on the worldwide spread of Eurasian war, disease, and trade goods. No tribal society could withstand their onslaught. Historians who have revisited the role of European scientists in the exploration of Africa (in *Dark Safari*) and New Guinea (in *First Contact*) have documented widespread devastation, caused almost unconsciously by specialists convinced of their own objectivity. In some cases, an expedition was not needed. Diamond, who did field research in the South Pacific, recounted how a single British sailor, Charlie Savage, drastically altered Fiji society in 1808 with the help of a couple of old muskets. "The aptly named Savage proceeded single-handedly to upset Fiji's balance of power . . . His victims were so numerous that surviving villages piled up the bodies to take shelter behind them, and the stream beside the village was red with blood."[6]

Far-traveled Carib tribes that gave their name to the Caribbean once settled the Orinoco. They lived in large, fortified towns and plied the great river in hundred-foot canoes. The wars and disease that accompanied spasmodic efforts of Europe's colonial empires to locate El Dorado exterminated their civilization. An enterprising Dutch governor of Suriname, Gravesande, launched the final quest in the first half of the eighteenth century. He formed a military-slaving alliance with a Rio Negro tribe, the Manau, whose leader, Ajuricaba, styled himself king of Gran Manoa (an alias for El Dorado) while flying the Dutch flag.[7] Brazil's colonial authorities sent an army that crushed the Manau, captured Ajuricaba (who committed suicide by leaping into a river in chains), and supplanted the Dutch as the leading entrepreneurs in the slave business. The Portuguese kidnapped or purchased over five thousand Indian slaves between 1725 and 1750 on the Upper Orinoco alone.[8]

The earliest mention of the Yanomami came from a multidisciplinary expedition of engineers, surveyors, naturalists, and artists who worked for the Portuguese boundary commission. In 1786, Lobo de Almada described the Yanomami as the "remnant" of a "nation," whose survivors were "still living" in the inaccessible headwaters between Venezuela and Brazil. Almada, who brilliantly directed the collection of new plant species and the cataloging of Indian cultures, contributed to genocide by relocating the Yanomami's eastern neighbors, the Macuxi, to a reservation a thousand miles away, where most of them died in what the historian John Hemming styled "a grotesque experiment."[9]

Other experiments were also underway. In 1784, Alexandre Rodrigues Ferreira, whom Hemming called "the first great naturalist to study the Amazon," began an ambitious scientific enterprise that conscripted hundreds of native guides, porters, paddlers, and servants. "Such navigation is fatal to the Indians, most of whom generally die or are incapacitated for life," a local bishop complained. Thousands perished or fled from the main rivers to escape Ferreira's botanical enthusiasm, leaving whole stretches of the Rio Negro on the Yanomami's southern border devoid of inhabitants.[10]

The Yanomami's strategy of hiding in the hills was more successful—reportedly some three thousand managed to escape the slavers and naturalists and the plagues that accompanied both of them.[11] It was an accident of geography. Europeans failed to reach the source of the Orinoco because the river narrowed to a swift, stone channel—a granite flume only six feet wide in stretches, with wild waters broken by impassable cataracts.[12] Farther upriver, after the portaging of a dozen waterfalls but still far from the source, the going became excruciatingly slow. The Orinoco spilled out into a labyrinthine swamp, choked by rotting logs and densely matted vegetation, where navigation was out of the question. It was miserable going. Even in the so-called dry season, from January to March, downpours fell forty-eight days out of fifty.[13] There were few stands of rubber or cacao trees, nothing to excite collectors. And, along the Orinoco's final stages, the only thing that obviously glittered was mica, fool's gold. Given the surfeit of pain and the apparent absence of reward, every European expedition for two centuries turned back without reaching the Orinoco's source. It was actually a tiny catch basin, a few feet in diameter, situated on the rim of a dark, steep gorge above granite escarpments, and it was finally located in 1951 by a Franco-German expedition. Nevertheless, earlier explorers left their mark and contributed to the creation of a Yanomami myth of unbridled ferocity.

The first American to attempt the Orinoco's origin was the noted geographer Hamilton Rice, on assignment for the Royal Geographical Society. He camped above the turbulent Guaharibo Rapids, considered the border of Yanomamiland, on January 21, 1921. There, seeing his abundant supplies, a group of about sixty Yanomami came begging for food and trade goods. This was the Yanomami's typical approach to outsiders,[14] but it startled Rice, who decided to take no chances. He opened fire with his Thompson machine gun and did not bother to count the dead. The Rice expedition fled downriver. He later wrote in Royal Geographical Society's *Journal* that the Yanomami were cannibals who ate raw flesh and that, given the danger of becoming dinner, it had been "necessary to fire to kill."[15]

The next incursion of Americans on the Upper Orinoco came during World War II. A team of U.S. Army engineers and surveyors did a feasibility study on a super-canal to join the Amazon and Orinoco watersheds.[16] Although the canal, which would have dwarfed Panama's, was never built, the friendly engineers got along well with the Indians. The Yanomami eagerly ate the army people's leftovers and shared their cigarettes before returning home with priceless machetes—and deadly contagion. The new respiratory diseases decimated *shabonos* far from the Orinoco, while sparking wars over witchcraft accusations, the double whammy that outside infections have historically brought to Amazon tribes.[17]

The wars and epidemics that shadowed these expeditions profoundly altered the Yanomami landscape. According to local colonists, Rice's machine-gun "massacre against unarmed Indians"[18] provoked Yanomami raids against the only remaining settlements on the Upper Orinoco between 1921 and 1931.[19] Although the Yanomami did not kill any whites, they stole all the steel goods they could find and wreaked so much havoc that colonists abandoned the area altogether. For the first time since the Spaniards arrived in 1750, there were no garrisons or trading posts within hundreds of miles. The jungle reclaimed old towns, missions, and forts. Another American geographer, Earl Hanson, reported on the phantasmagoric victory of the rain forest. "It is probable that the present regression of the region is the most complete in its history since the first advent of the Spanish," he wrote. "An interesting spectacle is taking place . . . affording an opportunity for some ethnologist to record a brand-new primitive culture in the making."[20]

There was no one better equipped than Chagnon to record this "brand-new primitive culture in the making." *The Fierce People* was written in a fresh, unfettered voice. After giving an account of a man who beat his brother with the blunt edge of an ax, Chagnon confided that the victor was "one of few Yanomami that I feel I can trust."[21] The anthropologist admitted he would have preferred studying some other, kinder group, but cautioned, "This is not to state that primitive man everywhere is unpleasant."[22] He described Yanomami women over the age of thirty as having "a vindictive and caustic attitude toward the external world." There was no puritanical preaching, no concession to the ideal of the Noble Savage. Another reason for the book's popularity was that Chagnon combined two favorite undergraduate themes—violence and sex—into a single theory about Yanomami warfare: Yanomami men fought over women, a message that has resonated on American campuses.

Chagnon survived a nighttime murder attempt by his hosts, whom he frightened off with his flashlight, and a close encounter with a jaguar, which sniffed him in his hammock. He hollowed out his own log canoe to ride down the Mavaca River, after a Yanomami guide abandoned him, and pushed on into unknown territory in spite of repeated death threats. You had to admire his courage—though it was harder to admire the Yanomami as Chagnon depicted them. By the end of the story, many readers concluded the Yanomami were, well, pretty awful.

Perhaps Chagnon's most brilliant achievement was fitting his grimly fascinating adventures into a clear, simple Darwinian framework that seemed to shed new light on human origins. *Time* magazine summarized Chagnon's theory: "the rather horrifying Yanomami culture makes some sense in terms of animal behavior. Chagnon argues that Yanomami structures closely parallel those of many primates in breeding patterns, competition for females and recognition of relatives. Like baboon troops, Yanomami villages tend to split into two groups after they reach a certain size."[23] You had to be fierce to survive and reproduce.

Chagnon said he had to become fierce himself in order to survive among the Yanomami: "I soon learned that I had to become very much like the Yanomamo to be able to get along with them on their terms: sly, aggressive, and intimidating." Otherwise, they would have pushed him around unmercifully and stolen him blind. He learned to shout at them "as loudly and as passionately as they shouted at me." "I had to establish my position in some sort of pecking order of ferocity at each and every village."[24]

Pecking orders of violence were popular in the 1960s, in part because of Konrad Lorenz's influential book *On Aggression,* published in 1966. Lorenz, a Nobel Prize–winning biologist at the Max Planck Institute, made many crucial contributions to understanding the behavior of rats and geese—two very aggressive animals—and a few equally crucial mistakes in applying his laboratory observations to human behavior. He concluded that humans were a simian species gone awry, great apes deformed by hunting and technology to kill without inhibition unlike any other animal.[25] Thus, original sin was reinvented, and man became known as a killer ape. Chagnon's Fierce People resembled killer apes: Amazonian primates, similar to baboons, whose perfect amorality turned murder and treachery into tribal ideals.

Today even Chagnon's strongest supporter, the Harvard sociobiologist Edward O. Wilson, recognizes that humans are probably less violent than any other species, at least as measured by common homicide and infanticide: "The murder rate is far higher than for human beings, even taking into account our wars."[26] Humans are not killer apes, nor are the Yanomami "fierce people."

There are Amazonian tribes, like the Huarani and Achuar, that have levels of violence far higher than that of the Yanomami.[27] Among the Huarani, for example, over 60 percent of all men are killed,[28] compared with 30 percent among the Yanomami Chagnon studied.[29] But the Yanomami have four regional dialects and are spread out over 80,000 square miles. All other Yanomami subgroups have homicide levels much lower than those Chagnon recorded.[30] The adult male homicide rate for the entire tribe might be 12–14 percent. There are villages where no one has been killed in generations and others where a high percentage of the men have been slain.[31] Therefore, rates of adult male war deaths could be engineered in a range from zero to over 40 percent, depending on the village and the time frame. And, if the approximate Yanomami homicide rate appears high when compared with *domestic* rates for wealthy, democratic societies, it is unfair to say, as Chagnon often does, that the Yanomami have a higher homicide rate than the city of Detroit.[32] Such comparisons are dubious, not only because the data is so uneven but because tribal violence conflates war and common murder—categories that modern societies keep strictly separate. (If murder rates for the Soviet Union or Poland were computed like the Yanomami's—tallying *all killings* over several generations—they would also be high, since they would include millions of male "murders" during World Wars I and II.) In any case, the overall level of violence among the Yanomami is undoubtedly modest for a tribal society without written laws or police.[33]

The question is no longer why the Yanomami are so fierce, but why Chagnon's Yanomami have homicide rates so much higher than those of other Yanomami groups. Although Chagnon has portrayed his home village, Bisaasi-teri, as a "typical Yanomamo village,"[34] it was exceptional. By the time Chagnon met the Bisaasi-teri, they were living at the juncture of the Orinoco (400 yards wide) and the Mavaca River (100 yards wide). From the air, the area looks lovely, with its riverine forests shading the muddy Orinoco's banks and granite foothills fingering their way out of the luxuriant growth. But the aerial view is deceiving, for this is a miserable, sticky, malaria-ridden place. No traditional Yanomami village was located anywhere near such a wide stretch of river.[35] Archaeological excavations at Bisaasi-teri have uncovered pottery shards and manioc strainers commonly used in Carib cultures but unknown in the Yanomami's mountain redoubt.[36] The Yanomami penetrated this far down the Orinoco only because the Carib tribe that traditionally dwelt there was driven off or enslaved. Two thousand Carib speakers were pressed into servitude by a band of adventurers while an energetic Frenchman set up a trading village in the 1830s at the same spot where Chagnon met the Bisaasi-teri.[37]

All experts, including Napoleon Chagnon, agree that the existing Yanomami groups originated in the Parima and Siapa highlands, which they populated "during untold centuries." The first scientist to live with the highland Yanomami was a University of Pittsburgh geographer, William Smole. After experiencing the Yanomami in their ancestral habitat in 1969–70, Smole began to publicly dispute the "fierce people" appellation for the Yanomami. In the Parima Mountains, Smole settled near a large village that had been at peace for two generations. There were no headmen to speak of, and much less squabbling over marriageable females. Whereas Chagnon's villages had a dramatic shortage of women, the highland villages had a slight surplus. Sorcery was the main cause of what warfare did occur; capturing women was secondary.[38] Smole concluded that Chagnon's Yanomami differed so markedly from the villages in the tribes' more tranquil homeland that they could not be considered traditional Yanomami at all.

Even within the subgroup where Chagnon worked, there is a sharp split between highland and lowland villages. In fact, when Chagnon surveyed five mountain *shabonos* from his own linguistic group in 1990–91, he learned that only about one-fourth as many men had participated in killings as among the lowland groups (11 percent as opposed to 44 percent).[39] Chagnon has yet to reveal the actual homicide statistics for these mountain villages contiguous to the Mavaca River. Nevertheless, his more recent findings confirm what William Smole has been saying for decades—that violence is spacially variable in the Yanomami world, the villages living at low elevations along the Orinoco-Mavaca drainage being the most violent known. These dozen *shabonos,* with a population of 1,394,[40] comprise less than 6 percent of the 25,000 Yanomami alive today. As Smole put it, "Certainly the Yanoama who have moved to sites on or near navigable water are not representative. They are outside their niche in the broadest sense, caught in a squeeze between various adverse influences of 'civilization.' "[41] Smole believed that steel goods, disease, and the divisive influence of outsiders altered such émigré groups beyond recognition. Ferocity and fighting over women "might apply to a lowland zone of acculturation and acute cultural instability."[42]

Almost all subsequent researchers have echoed Smole's criticisms, including most of Chagnon's own students. Chagnon blamed these attacks on romantics trying to create a prettified version of Yanomami culture. Having met with some assaults myself for graphic description of violence among Amerindian groups, I initially sympathized with Chagnon. Once I reached Yanomamiland, however, I found it increasingly difficult to accept Chagnon's version of their culture.

Brazil's nomadic gold miners, whose cross-continental wanderings have brought them into contact with dozens of tribes, have often remarked on how friendly the Yanomami are compared with other Amazonian Indians. The Yanomami at first welcomed me with a kindness that was disconcerting—tied my hammock, brought me food and water, lugged my heavy equipment, and lit lights all around me at night. I soon realized they were desperate for medicines and for someone to take their dying children upriver to a medical clinic.

Later, I was robbed at gunpoint by several young Yanomami who were working with gold miners. Had I wanted to render a heroic, Chagnonian version of the incident, it might have gone like this:

> When we came to the big curve of the Mucajaí River, white water sprayed us as we dodged in and out of eddy currents. Just after escaping the last whirlpools, we confronted a new danger: a canoe of belligerent warriors heading straight toward us. They pulled even. Then a vicious and powerfully built man leapt into our boat and pointed a shotgun at my head. "I kill gold miners!" he shouted as he beat his chest to establish his dominance. He swayed from side to side, proclaiming his murderous intentions. I stared him down, knowing that a true warrior will never display fear. I also knew the real motive for the treacherous assault: the Yanomami's perpetual suspicion that outsiders wanted to steal their women.

Actually, I had happened on a Yanomami funeral ritual, in which the ashes of the dead are taken out and shared in a tribal communion, a time when the feared ghosts from the past are honored and when old scores can be settled. My boat was boarded in midstream by a twenty-year-old who was drunk on imported whiskey, and he was soon supported by other drunken youngsters with guns of their own. The gold miner, named Cícero Hipólito dos Santos, and I were forced out of the boat at gunpoint, and, as a crowd of warriors and women gathered around, there ensued a debate about whether they should kill us. The surprising thing was that the Yanomami did not kill the gold miner, or me, for that matter; they just stole all our stuff. The local chief, painted red and black, with macaw feathers in his ears, planted himself between the adolescents' guns and me. He yelled, "Go away! Go down the river! The Indians here are all drunk. Indians are very dangerous when they're drunk."

I realized that my own actions, as well as the Yanomami's needs and the bizarre twists of the gold rush, had created situations from which I could have fashioned either a romantic or a Darwinian image of the Yanomami. Of course, either one would have been a distortion, like the portrait in *The Fierce People.*

The Yanomami I met on the Mucajaí were certainly no proverbial saints. But in sixty years they launched only two raids; on two other occasions, a few Mucajaí men joined allied raiding parties. That was it. Yet Chagnon took one of the two raids that the Mucajaí people initiated, and turned it into both the prime example of Yanomami treachery and a case study of a war fought exclusively to capture women. In *The Fierce People,* he claimed that the Mucajaí Borabuk "had a critical shortage of women" and proceeded to describe "the treacherous means by which the group alleviated its problem":

> The headman of the group organized a raiding party to abduct women from a distant group. They went there and told these people that they had machetes and cooking pots from the foreigners, who prayed to a spirit that gave such items in answer to the prayers. They then volunteered to teach these people how to pray. When the men knelt down and bowed their heads, the raiders attacked them with their machetes and killed them. They captured their women and fled.
>
> Treachery of this kind, called nomohori (dastardly trick) is the ultimate form of violence.[43]

But the demographer John Early and the sociologist John Peters, who spent over eight years on the Mucajaí, have put this raid into a wholly different perspective. In the first place, the Mucajaí Borabuk (People of the Waterfall) were not trying to capture women. "They did not view themselves as having a sex ratio problem as such."[44] It is true that they had fewer women than men, but they were not overly concerned about it, because the

Yanomami can acquire wives through trade and bride service (such as providing game for a marriageable woman's parents). So the temporary imbalance, common in tribal populations, was taken in stride. In the meantime, they simply shared wives. Chagnon perceived "a critical shortage of women," but the Borabuk did not. In fact, they had not raided anyone in over twenty years.

What really disturbed them were the devastating illnesses that came with first contact, which had arisen from their desire for steel goods. "Previously such tools had been obtained by exchange with or raids upon other indigenous groups." But the Mucajaí group had been isolated since their last raids to obtain steel in the mid-1930s. "The tools they had were wearing out and in need of replacement. They had moved to the banks of the Mucajaí River in the hope of making contact with Brazilians from whom they could obtain the tools. At the time of contact this appeared to be their most preoccupying problem."[45]

In 1955, an amazing event changed their lives: missionaries flew a small plane over the Mucajaí Borabuk and dropped fishhooks. The Borabuk sent a party of men in search of the source of steel. They built canoes for the first time and dispatched them far downstream, where, in late 1957 and again in late 1958, they made contact with Brazilian peasants and received some trade goods. Unfortunately, on both occasions the Mucajaí people also "contracted respiratory infections from the Brazilians and many died after they returned upstream. They had no immunity due to their previous isolation." "The resulting sickness and death was a new and frightening experience for the Mucajaí community."[46]

Two months after the second wave of imported illness, the missionaries arrived, who treated the sick and contained the epidemics. Then the second act in the tragedy of contact began. On the pretext of going on a long hunt, the Mucajaí Borabuk borrowed a gun from the unsuspecting missionaries and traveled upriver, searching for the sorcerers who they believed had caused all the deaths. They investigated the Marashi-teri, on the Couto de Magalhães River, who accused another distant group, the Shiri-teri, of being the agents of witchcraft against the Borabuk. Finally, in a confused encounter characterized by mutual misunderstanding, the Borabuk and the Marashi-teri attacked the Shiri-teri, although they did not kill them with machetes as Chagnon reported. One Shiri-teri was actually shot with a gun, showing how radically the impact of first contact had changed warfare on the Mucajaí. Their tricking the Shiri-teri into "praying" for metal goods also underscored what a strange new brew of outside influences was working on the Borabuk.[47]

Shortly afterward, the Borabuk sent peace offerings to the Shiri-teri, and they have been on good terms for the last thirty-five years. Although the Borabuk live in some half-dozen different *shabonos* spread out over a wide area of the Mucajaí River, with a population of over three hundred, there have been no raids between any of these *shabonos.* Between about 1935 and 1985, a total of three Borabuk men were killed violently; two others disappeared.[48] By the standards of the Amazon, or the world, the Borabuk form a fairly peaceful tribal society.

I also visited over thirty Yanomami *shabonos,* including several in the Parima Mountains. Of all the varied landscapes of Yanomamiland, I loved these inaccessible highlands best. The *altiplano* has majestic scenery, splendid waterfalls, and a blessedly temperate climate. Mosquitoes are not as horrible a nuisance there as elsewhere. Until recently, the Parima Yanomami did not suffer from colds or malaria.[49]

Why, then, did the Bisaasi-teri end up at a malaria trap exposed to Western diseases on the main course of the Orinoco?

The Bisaasi-teri splintered from a larger block, the Namowei, which had been torn apart by the respiratory infections that coincided with the U.S. Army expedition of 1942–43. The outbreak killed off most of the tribal elders, giving power to immature and aggressive young men—like the ones who robbed me on the Mucajaí—who plunged the Bisaasi-teri into a fratricidal war. As usual, the killing started over suspicions that rival Yanomami had sent lethal new diseases through witchcraft. But the strife also involved competition to secure the trading routes to a new Protestant mission—the first permanent source of steel goods in Yanomamiland—which opened in 1948.[50]

Defeated by both disease and war, the Bisaasi-teri relocated and adapted to river life, learning canoe travel and line fishing. As upland Yanomami, they did not even know how to swim. Nor did they have any clothes to keep off the clouds of gnats and dive-bomber mosquitoes. Bisaasi-teri was a village created by the catastrophe of first contact, and it first coalesced, six years before Chagnon's arrival, in 1958, around a government malaria post, without whose medicine the Bisaasi-teri could never have survived the unhealthy lowlands.[51]

Chagnon's exciting narrative edited out these unfortunate details. Prior to the arrival of the U.S. Army and Protestant missionaries in the 1940s, the

Namowei Yanomami had lived in peace for a generation. Their only raiding parties had gone out searching for whites in order to steal machetes. But since there were no whites in the area, nothing happened. Other Yanomami journeyed three hundred miles to the Rio Negro in order to steal *madohe* (stuff): axes, machetes, knives, pots, and cloth.[52]

On one of these epic forays near the Rio Negro, the raiders captured a young white girl, Helena Valero, while she was traveling with her family on a hunting trip. "It was not to rob women but to seize the goods my family was carrying; they were not interested in women," Valero recalled. "They carried me off because they found me abandoned. But the Indians did not want to capture women, just *madohe.*"[53]

During her twenty-four years as a wife and mother among the Indians, from 1932 to 1956, Valero witnessed the epidemics that carried off the Namowei leaders and the subsequent killings over sorcery suspicions. She described how the Namowei's young men had to be trained in the art of raiding because they had never fought anyone. In the beginning, they were comically incompetent, unable even to locate enemy *shabonos.*[54] But in the terrible struggle that followed the arrival of the first missionaries, Helena Valero's husband became the Namowei war chief. He was murdered in 1949; the other war leaders were all killed by 1951. The group split up into two villages—Bisaasi-teri and Patanowa-teri. Peace ensued.

After settling on the Orinoco, the Bisaasi-teri gained fitful access to Western manufactures. They traded a trickle of metal goods to villages in the hill country and received a bounty of young brides in exchange. From 1951 to 1964, no Namowei were killed in warfare. Then Chagnon arrived.

During Chagnon's brief, thirteen-month residence, ten Yanomami were killed in a war that once again pitted the people of the Bisaasi-teri alliance against their old Namowei cousins, the Patanowa-teri. These deaths constituted a third of all the war fatalities over a fifty-year period for the Namowei villages, according to Chagnon's Ph.D. thesis. All of the remaining male war deaths in these villages occurred during another brief period, 1949–51, when Protestant missionaries first established their bases on the Upper Orinoco.

The missionaries initially made serious mistakes. They distributed machetes to win converts and unknowingly provoked bloody battles for monopoly rights to their supplies. But they eventually brought stable trading relations and good medical care to the Indians. They also actively intervened to stop fighting.

Chagnon could not provide ongoing medical attention or stable terms of trade, not because his intentions were less good but because his research, which will be examined in the next chapter, required him to collect thousands of genealogies and blood samples in a short period of time. He had to buy the Yanomami's cooperation in scores of villages across an area larger than New York State.

Chagnon arrived with a boatload of machetes and axes, which he distributed within twenty-four hours; the delighted recipients of this instant wealth immediately left the village unattended and went to trade with equally delighted allies. For the steel-poor villages of the Yanomami hill country, Chagnon was a one-man treasure fleet. The remote villages of Patanowa-teri and Mishimishimabowei-teri began sending messengers begging Chagnon to come and visit,[55] but their ambassadors were driven away by Bisaasi-teri and its closer allies, who fought to maintain their monopoly of Chagnon's steel wealth.

Within three months of Chagnon's sole arrival on the scene, three different wars had broken out, all between groups who had been at peace for some time and all of whom wanted a claim on Chagnon's steel goods. "Chagnon becomes an active political agent in the Yanomami area," said Brian Ferguson. "He's very much involved in the fighting and the wars. Chagnon becomes a central figure in determining battles over trade goods and machetes. His presence, with a shotgun and a canoe with an outboard motor, involves him in war parties and factionalism. What side he takes makes a big difference."[56]

Chagnon has dismissed this charge as "the 'bad breath' theory of tribal warfare."[57] Yet Chagnon brought more than breath with him into Yanomami territory. He introduced guns, germs, and steel across a wide stretch of Yanomamiland—and on a scale never seen before. The Yanomami's desire for steel is as intense as our longing for gold. Westerners became the Yanomami's metal mines, local El Dorados that dispensed machetes, axes, and fishhooks that instantly increased agricultural production by 1,000 percent and protein capture by huge amounts. Yanomami groups made heroic odysseys in search of a single secondhand machete. Remote groups traded their daughters for a worn machete or a blunt ax. Villages with more steel always acquired more women. The sociologist John Peters, who lived among the Brazilian Yanomami for eight years, was offered two young girls in exchange for a couple of stainless steel pots. He refused the offer.

Chagnon did not wait to be asked, according to his closest friend and main informant—Kaobawa, the Bisaasi-teri headman, who was videotaped by Mark Ritchie, author of the 1995 book *Spirit of the Rainforest.* Kaobawa's picture formerly graced the cover of *The Fierce People,* where he held a pole with his right hand and jabbed an angry right index finger at the world. Chagnon has long considered this "unobstrusive, calm, modest, and perceptive" man as "the wise leader" of the Bisaasi-teri. Kaobawa's decision to help Chagnon sort out his interviews with dozens of informants "was perhaps the most important single event in my fieldwork," Chagnon wrote, adding, "Kaobawa's familiarity with his group's history and his candidness were remarkable. His knowledge of details was almost encyclopedic."[58]

Therefore, Kaobawa's videotaped statements raised a number of questions—about both men. Kaobawa claimed that Chagnon offered him a special deal. "That's my picture there," Kaobawa said when Mark Ritchie showed him a copy of *The Fierce People.* "When he was taking my picture he said, 'If you'll really help me, I'll give you a motor. . . . He said, 'Father-in-law, I'm going to really be a Yanomami and you're going to get me a wife.' That's what he said. But although he said that, he just left. . . ."[59]

According to Ritchie, "The story of Chagnon trying to get a wife from Kaobawa is a comedy of errors. As Kaobawa explains it, Shaki—Chagnon—wanted to buy a wife from a distant village, and Kaobawa kept trying to stop him because Kaobawa didn't want Chagnon and his trade goods to move away. Apparently, Chagnon wanted a Yanomami wife, but far enough away from the missionaries so that they wouldn't find out."[60]

Chagnon suddenly went from being an impoverished Ph.D. student at the bottom of the totem pole to being a figure of preternatural power. His first letter from the field revealed this: "The village I'm living in really thinks I am the be-all and the end-all. I broke the final ice with them by participating in their dancing and singing one night. That really impressed them. They want to take me all over Waicaland to show me off. Their whole attitude toward me changed dramatically. Unfortunately, they want me to dance all the time now. You should have seen me in my feathers and loincloth! They were so anxious to show me off that they arranged to take me to the first Shamatari village so that I could dance with them."[61]

Chagnon's status was enhanced by a pair of shotguns. The geneticist James Neel described Chagnon firing off his gun preemptively to scare off young men they suspected might steal some goods. "At dusk Nap casually blasted the tips of a tree branch overhanging the *shabono* where we were sleeping, and we retired with the shotgun leaning against his hammock—to a quiet night."[62] Of course, this was an old conquistador strategy, one employed over the centuries to keep the natives cowed. In 1531, when Francisco Pizarro reached his first Inca city, Tumbes, at the Bay of Guayaquil, a soldier named Pedro de Candia "astounded the inhabitants by firing an arquebus at a target."[63] For the Spaniards, it became a standard technique of forced entrance.

The American Anthropological Association first got word of Chagnon's shotgun diplomacy when, in 1991, the anthropologist Terence Turner, head of its Yanomami survival commission, interviewed Davi Kopenawa, the Yanomami's most visible spokesperson and a winner of the UN Global 500 Award for defending the rain forest. Kopenawa told of reports that had come to his community of Chagnon's threatening behavior—walking around villages brandishing firearms and showing himself as a warrior. "Chagnon is fierce," Kopenawa said. "Chagnon is very dangerous. He did crazy things. *Ele tem a própria briga dele."* This literally means "He has his own personal war."[64]

That is what Brian Ferguson concluded.

"Chagnon's role is a strange thing for me," admitted Ferguson, whose *Yanomami Warfare* breaks a professional taboo by scrutinizing a field-worker as though he were a native. "One of the things I'm saying is that anthropologists need to be looked at. Anthropologists have been trained to screen out their own effects on their subjects. Their behavior is also a fit subject for investigation. The influence of Chagnon in the Yanomami area is a fit subject for investigation."[65]

Chagnon found himself in a difficult predicament, having to collect genealogical trees going back several generations. This was frustrating for him because the Yanomami do not speak personal names out loud. And the names of the dead are the most taboo subject in their culture.

"To name the dead, among the Yanomami, is a grave insult, a motive of division, fights, and wars," wrote the Salesian Juan Finkers, who has lived among the Yanomami villages on the Mavaca River for twenty-five years.[66]

Chagnon found out that the Yanomami "were unable to understand why a complete stranger should want to possess such knowledge [of personal names] unless it were for harmful magical purposes."[67] So Chagnon had to parcel out "gifts" in exchange for these names. One Yanomami man threatened to kill Chagnon when he mentioned a relative who had recently died. Others lied to him and set him back five months with phony genealogies. But he kept doggedly pursuing his goal.

Finally, he invented a system, as ingenious as it was divisive, to get around the name taboo. Within groups, he sought out "informants who might be considered 'aberrant' or 'abnormal,' outcasts in their own society," people he could bribe and isolate more easily. These pariahs resented other members of society, so they more willingly betrayed sacred secrets at others' expense and for their own profit. He resorted to "tactics such as 'bribing' children when their elders were not around, or capitalizing on animosities between individuals."[68]

Chagnon was most successful at gathering data, however, when he started playing one village off against another. "I began traveling to other villages to check the genealogies, picking villages that were on strained terms with the people about whom I wanted information. I would then return to my base camp and check with local informants the accuracy of the new information. If the informants became angry when I mentioned the new names I acquired from the unfriendly group, I was almost certain that the information was accurate."[69]

When one group became angry on hearing that Chagnon had gotten their names, he covered for his real informants but gave the name of another village nearby as the source of betrayal. It showed the kind of dilemmas Chagnon's work posed. In spite of the ugly scenes he both witnessed and created, Chagnon concluded, "There is, in fact, no better way to get an accurate, reliable start on genealogy than to collect it from the enemies."[70]

His divide-and-conquer information gathering exacerbated individual animosities, sparking mutual accusations of betrayal. Nevertheless, Chagnon had become a prized political asset of the group with whom he was living, the Bisaasi-teri. He took a Bisaasi-teri raiding party partway to their enemies' *shabono* with his outboard motor; later he helped Bisaasi-teri's allies leapfrog their enemies and avoid an ambush. By making one man, Kaobawa, the principal funnel of his largesse, Chagnon effectively created him "headman," a pattern he would repeat at other villages. With Chagnon established at Bisaasi-teri, minding the store with his shotguns, the Bisaasi-teri could raid other groups at a much greater distance because Chagnon made them immune to attack. Chagnon gave one of his shotguns to a Bisaasi-teri guide who was afraid of traditional foes nearby. "I had two shotguns. . . . I gave one of them to Bakotawa, along with a dozen or so cartridges and a quick lesson in how to load and shoot a gun."[71]

Another time Chagnon helped his Bisaasi-teri allies recapture a woman, Dimorama, whose abusive husband, Shiborowa, had shot her in the stomach with a barbed arrow. "They were going to Momaribowei-tedi to take Dimorama away from her protectors by force, if necessary, and asked me to come along knowing that I always traveled with a gun, presuming that my presence, with a gun, would aid in their objective." Their presumption was correct. They recaptured the girl and gave her back to Shiborowa.[72]

Although any Westerner bringing piles of steel goods would have disrupted Yanomami culture, Chagnon's role was arguably unique. Not only did the Bisaasi-teri have first choice of Chagnon's seemingly endless supply of steel goods; they also had a Western chief of sorts. "Dancing in another village is a part of politics—one way of displaying strength," Ferguson noted. "The participation of a white man in feathers and loincloth, virtually declaring his identification with Bisaasi-teri in intervillage relations, would represent a major coup." He added, "And it was during these first months of Chagnon's fieldwork that the Bisaasi-teri's conflicts with the Shamatari and Mahekoto-teri transpired. . . . But while he was behaving more like a Yanomami big man in his interpersonal relations, his other actions—his quest for the taboo names of the dead ancestors, his moving back and forth between antagonistic villages, and, above all, his being the source of Western goods that every village wanted to monopolize—created a very different and 'un-Yanomami' context for his behavior. Chagnon thus became something of a wild card on the local political scene."[73]

It is precisely the "un-Yanomami" context of the Napoleonic wars that makes them so problematic. Chagnon now recognizes that Yanomami violence is "actually quite low" by world standards of tribal culture.[74] And it is undeniably connected to the fluctuating impact of Western technology and disease. Whatever else can be said about Yanomami warfare, it is not "chronic," as hundreds of articles, documentaries, and books still insist. All of the violence among Chagnon's subjects can be spelled out in two stark spikes, both corresponding to outside intrusion. This is the picture of

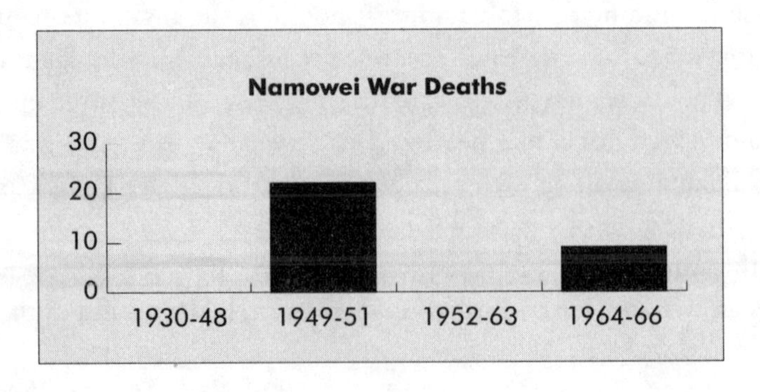

Yanomami ferocity that actually emerges from Chagnon's own Ph.D. thesis, the only complete accounting of Yanomami war deaths he has published for any group.[75]

An "uncertainty principle" pertains to these wars. Would they have occurred at all without the germs, steel, and guns brought by strangers?

Chapter 4

Atomic Indians

I announced I would take blood samples that afternoon and evening. . . . As soon as I moved my sampling equipment over to an unoccupied section of the village, I was surrounded by about 200 pushy, impatient, angry, shouting people, each determined to get a particular item. . . .—*Napoleon Chagnon*[1]

A full moon was in a clear sky as we traveled the Padamo River easily under a natural spotlight. It was September 1996, and the rainy season was just easing up. I lay in the prow, looking for logs, while Agustín, a Yanomami man in his midtwenties, handled the motor. At full throttle, a collision with a submerged tree could have given us a severe jolt or broken the propeller. Fortunately, the Padamo, even this high on its course, was placid, wide, and deep; and Agustín was a master of the outboard who knew every rock in the main channel. As we came around a final bend, we could make out Mount Marahuaca, a tabletop over nine thousand feet high, overlooking the dark forest like a ghostly altar. Closer by, we saw a striking stiletto peak, with a curious rock resembling a bird in flight. It was Toki, Eagle Mountain.[2]

Toki is a Maquiritare village that has an indoor theater, where a mixed crowd of six hundred Yanomami and Maquiritare were gathered to watch *Yanomama: A Multidisciplinary Study,* a film by James Neel, Timothy Asch, and Napoleon Chagnon.[3] When the Atomic Energy Commission's name came up in the titles, several viewers rose to their feet.

"Why did the Atomic Energy Commission want to study us?" a man named Antonio asked.

Students of *The Fierce People* have gotten only the vaguest inkling about why the agency that manufactured atomic bombs spent large sums studying the Yanomami. Chagnon reduced the AEC to a single footnote. And, for anyone who bothered to read the fine print, the AEC's Yanomami research acquired both a humanitarian glow and an otherwordly abstraction.

Chagnon reported collaborating with "a medical-genetics group whose responsibility it was to treat the survivors of the nuclear bombings at Hiroshima and Nagasaki."[4] This was true, in part. The geneticist James Neel was on the Atomic Bomb Casualty Commission; his team studied mutations. Neel never mentioned treating any survivors. Neel wrote, "One of the most frequent Japanese complaints has been that we (the ABCC) only examined them (like guinea pigs), but did not offer treatment in the event of findings of medical significance. The fact is that the terms under which the ABCC operated did not permit treatment. . . ."[5]

Why include the Yanomami in such a study? According to Chagnon, it was "justified in the sense that both the Japanese and the Yanomamo reproduce according to cultural rules."[6] Yes, but so does everyone else in the world. In 1966, in letters to missionaries responsible for Yanomami health care, Chagnon explained that the geneticists' primary concern was "in the well being of the Yanomamo, and a number of other tribes, particularly from a medical point of view." They wanted "to study the epidemiology, genetics, and disease problems of the American Indians."[7]

The suggestion that Neel's geneticists were trying to solve the Indians' critical "disease problems" persuaded both Venezuelan health officials and the Yanomami themselves to collaborate. "Whenever we asked Shaki [Chagnon] why he wanted so much blood from us, he would said it was to help us, to find cures to our diseases," said Pablo Mejía, a Yanomami leader on the Padamo River.[8]

Chagnon was actually the advance man for a new order of scientific adventure, the most comprehensive study of a tribal society ever undertaken. This project was conceived by James Neel, a doctor who helped found the modern science of human genetics. While still a medical student, in 1938,

Neel explained how red hair is inherited. He first won international recognition for discovering the gene for thalassemia, a frequently fatal form of anemia that occurs among those of Greek or Italian descent. Later, he helped demonstrate that sickle-cell anemia, which affects tens of millions of sub-Saharan Africa, as well as thousands in the United States, is a positive adaptation against malaria, a deep insight into the dynamics of disease and natural selection. After leading the atomic bomb survivor studies, he founded the first human genetics department in the United States, at the University of Michigan, now considered one of the world's finest.[9] Finally, Neel received several prestigious awards for his lifelong work. He conceivably could have won the Nobel Prize himself, like several of his colleagues and students whose research he furthered, but his openly eugenic views made him something of a pariah outside his specialty.[10]

Neel's autobiography, *Physician to the Gene Pool,* published in 1994, frankly explained his concern about the gene entropy of modern society. He was convinced that democracy, with its free breeding for the masses and its sentimental supports for the weak, violated natural selection. Even Neel's fellow conservatives on President Reagan's National Council on Aging were startled by Neel's refusal to support genetic research to extend life expectancy. Neel objected that it could "only result in an increase in the number of senior citizens and exacerbate the problems already manifest in the emerging gerontocracy." He resigned rather than compromise his principles. Neel is probably the only geneticist of his reputation in the post-Nuremberg world to praise the early eugenicists for their "concern for the future" of the gene pool. He has also criticized other scientists for "fearing the opprobrium of an eugenic label" and refusing to take strong political stands designed to improve the gene pool.[11]

Neel was obviously not afraid of being called a eugenicist; the title of his autobiography, *Physician to the Gene Pool,* is a good definition of one. Sir Francis Galton, Charles Darwin's cousin, founded eugenics in the 1880s. It became a political-scientific movement to weed out undesirable traits from the gene pool, and encourage desirable ones. Eugenicists led campaigns for mass sterilization of the unfit. Neel had a career-changing moment when he visited the Eugenics Records Office in New York in 1942 and realized how much work it would take him to make eugenics a true science.

In Michigan, Neel campaigned for statewide screening of defective fetuses and did a cost-benefit analysis showing how much money each abortion would save the state ($75,000). Where Neel differed from most people, and most scientists, was his belief that fetuses with some easily curable defects should be aborted—because they would ultimately run down the gene pool by passing on the undesirable trait.[12] Like Galton, Neel stood in self-confessed "awe" of the process of evolution, and horror of modern society's attenuation of competition. Galton preached a crusade to promote Social Darwinism and went as far as suggesting that "a missionary society" be founded "with an enthusiasm to improve the race."[13] In some respects Neel's Department of Human Genetics became this missionary society.

But while European eugenicists saw northern blonds as the pinnacle of creation, Neel felt a romantic attraction to tribal societies. By 1957, he had begun speculating that primitive tribes optimized selective breeding. In 1962, he visited Brazil's Xavante tribe, where he had an almost conversion experience while hearing their shamans chant around night fires. "Suddenly the thought came to me that I was witness to a scene which, in one variation or another, had characterized our ancestors for the past several million years. The sudden realization of this contact with the thread of evolution resulted in another of those very emotional professional moments; this time I could feel the hair on the nape of my neck stirring. . . ."[14]

In fact, the Xavante had been caught up in contact and conflict with Brazilian armies, missionaries, traders, and raiders for over two centuries. Most historians would agree with Claude Lévi-Strauss, the world's preeminent cultural anthropologist, that tribes like the Xavante are "not examples of archaic ways of life that have been miraculously preserved for millennia, but the last escapees from the cataclysm that discovery and subsequent invasions had been for their ancestors." Lévi-Strauss compared the Amazon's so-called Stone Age societies to "scattered groups of survivors after an atomic holocaust"[15]—the very groups Neel had studied in Japan.

Terence Turner, an Amazon specialist now at Cornell, recalled an encounter with Neel: "In 1963, James Neel brought me out from Harvard to give a talk to Chagnon and a few other students working for him about doing field research in the Amazon. I'd just returned from being with the Kayapo [Indians of Brazil]. So I went, and after the talk there was a little reception where I heard James Neel say to his group of researchers, 'Good. Now we'll have a chance to find the leadership gene.' I was amazed. I turned to him and said, 'You must be joking. You don't seriously believe there's a gene for leadership?' Things became hushed, and I realized this was an important belief of theirs. Neel answered, 'I don't think it's so silly. I think there might be a genetic basis for dominance.' And he went on to explain his theory that in Amazonian tribal societies headmen battle for control of the greatest number of women, and their fights select out the genes most fit for survival. So in these small breeding populations, leadership has a chance to reproduce itself genetically. It's like the baboon model of alpha males who keep other males from their harem. In Chagnon, this works out to killers who differentially reproduce themselves. And he's been loyally pursuing this ever since."[16]

Neel's autobiography confirmed that he had originally planned to send Chagnon to live with the Kayapo. And, although he never used the phrase "leadership gene" in his writings, he proposed a genetic "Index of Innate Ability." Neel believed that this Index of Innate Ability (IIA), located at paired alleles along the DNA chain, became concentrated in the offspring of dominant, polygamous chiefs, just as Turner recalled.[17]

Scientific discovery is driven by strange dreams. These, too, can break hearts. Neel never did locate the elusive alleles for his Index of Innate Ability. "That was the greatest disappointment of my life," he told me.[18]

In Neel's great quest, Napoleon Chagnon became "the indispensable cultural anthropologist."[19] For the twenty-six-year-old undergraduate, it was an opportunity to escape not only academic obscurity but also the bleak poverty he had experienced all his life. Chagnon was born in the tiny frontier town of Port Austin, Michigan, the second of twelve children. His childhood home did not have plumbing.[20] He was a boy of small stature from a low-status family who had inherited a big name from his French-Canadian grandfather. It is not hard to guess how this affected him in a tough little town in rural Michigan, where differences were not welcomed, where xenophobia, linked to anti-Communist feeling, ran high, and where Senator Joseph McCarthy enjoyed strong support.

Chagnon had to fight to keep afloat in more ways than one. In high school, he earned five dollars to date his sweetheart—whom he later married—by embalming an extra body at his father's undertaking business. He started higher education as a scholarship student at the Michigan College of Mining and Technology, spending his summers as a surveyor.[21] The detached efficiency he learned while draining body fluids as an undertaker's assistant later proved invaluable in the AEC's vast blood-collecting project. So did his surveyor skills, once he ventured beyond Venezuela's official maps.

From the mining school, and from his austere childhood, Chagnon retained a sympathy for industrial progress. He also retained a deep dislike of anything remotely related to Communism, so much so that he wanted to become a physicist to help battle the Soviet Sputnik threat.[22] Throughout his studies and his career, Chagnon demonstrated a commitment to hard work that was wholly admirable. As the filmmaker Timothy Asch wrote, "Chagnon was the perfect Horatio Alger, American hero: the boy from the sticks who, through his own diligence and perseverance, competed and made good."[23]

In *The Fierce People,* James Neel has also been reduced to the barest of footnotes ("Neel et al., 1971"). Neel, who needed no further testimonials, may well have approved. For thirty years, Chagnon has insisted that he arrived in Yanomamiland as a convinced cultural anthropologist who expected wars to be related to the environment, not to reproduction. He told *U.S. News & World Report,* "I went down there looking for shortages of resources, but it turns out they are fighting like hell over women."[24]

If Neel had sent Chagnon to Brazil's Kayapo Indians, a tribe whose muscular warriors have a homicide rate about three times as high as the Yanomami's, the history of anthropology might have been quite different. (Several of Chagnon's most outspoken critics have admitted that it would have been much harder to refute his theories if he had draped them upon other Amazonian tribes.) But a funny thing happened in 1964 while Neel was en route to Brazil. During a layover in Caracas, the Brazilian military staged a coup. Neel had to wait three days. And during this time he met Charles Brewer Carías. "Charlie occupies a special place in my heart," Neel wrote. "He wished to join forces with us, so I brought him to Ann Arbor for a year of training." With Brewer as part of his "forces," Neel decided to make the Venezuelan Yanomami the object of his genetic studies. Chagnon agreed.[25]

Neel had given himself a ten-year deadline to document Darwinian sexual selection among tribal people. He and Chagnon believed that their informants were destined to die out or suffer such alteration through contact with missionaries and other outsiders that they would in a few years become useless to science.[26]

There was also a political program underlying this scientific endeavor. Though Neel's politics were too extreme for Reagan's council on aging, he was liberal compared with Brewer and Chagnon. As I will explain later, Brewer organized his own gang of thugs to attack the Marxist government of Guyana. Chagnon's early patriotism blossomed into personal hatred against hippies. In 1984, according to the anthropologist Jesús Cardozo, then a graduate student of Chagnon's at the University of California at Santa Barbara, a ponytailed student raised his hand and asked Chagnon, "Aren't there any pacifists among the Yanomami?"

"You mean cowards?" Chagnon shot back. "I don't go to the Amazon to study cowards."[27]

It is easy to forget that *The Fierce People* had its genesis during the Vietnam War and its cultural equivalent on the University of Michigan's Ann Arbor campus, where hippies in tepees chanted slogans like "Make love, not war." The whole point of Neel's genetic perspective was that you had to make war in order to make love—that violence was part of the natural order.

Yet, despite their philosophical embrace of violence, neither Neel nor Chagnon took the opportunity to fight in the wars of their times. During World War II, Neel was assigned to a medical school hospital in Rochester, where other scientists began studying the effects of radiation three years before the bomb fell on Hiroshima. Neel saw the Yanomami expeditions as a way of testing himself after a "safe life."[28] Chagnon, who was a married graduate student as the Vietnam conflict began, also continued his studies during the war.

As a Cold War metaphor, the Yanomami's "ceaseless warfare" over women proved that, even in a society without property, hierarchies prevailed. Thus, Communism was unnatural because even the most classless of societies had pecking orders in reproductive matters. The underlying ferocity and deceit that fueled the Yanomami's successful military strategy also offered a kind of parable: the ruthless Communists were going to win if long-haired hippies did not rejoin the march of Darwin.

The agency behind the atomic bomb was not on the side of the hippies. The AEC's decision to promote Yanomami films, including at-cost distribution in the United States, was rational and fruitful, given the AEC's political bias. The AEC-financed expeditions produced over fifty scientific articles, a host of popular articles, and a paradigm shift that is still with us. Chagnon recently admitted, in a newspaper interview, that he had hoped, in writing *The Fierce People*, to correct "all the garbage about the Noble Savage," and he was glad he had ushered in a "revolution in writing anthropology."[29]

Most anthropology students are not out to start revolutions. So, from the outset, Chagnon's scientific mission differed from that of a typical anthropology student—and in several important ways. Most anthropologists develop friendships with a small circle of trusted informants, but this was not an option for Chagnon. He was an employee of Neel's, and his job was to prepare thousands of Yanomami at dozens of villages to receive teams of scientists.

Many Yanomami who attended the screening of *A Multidisciplinary Study* at Toki wondered what had happened to all the blood Chagnon collected.

In Yanomami mythology, blood is a dangerous, taboo substance. The first men were born of blood dropped by Moon Spirit, who was a cannibal. The Yanomami cosmos teems with voracious beings hungry for blood. It was interesting that Chagnon, when he took hallucinogenic snuff, identified himself with Vulture Spirit, Rahakanariwa, the most fearsome cannibal entity of them all.[30] But none of these mythical creatures matched the appetite of the Atomic Energy Commission.

Data I have obtained through the Freedom of Information Act shows that, between 1965 and 1972, the AEC funded James Neel's genetic study, which compared "the survivors of the atomic bombings in Japan" with the Yanomami and other, less intensively studied tribes. The AEC spent $2,289,279 to "determine the mechanisms by which radiation induces changes in the genetic material of cells," using the Yanomami as the principal virgin population control group. The AEC wanted thousands of Yanomami blood samples, together with their corresponding genealogies, to determine mutation rates in a completely "uncontaminated" population. "The general approach was to search for and characterize mutations among 35 different proteins found in blood." Yanomami blood thus provided the baseline for "our understanding and characterization of the health risks associated with exposures of people to energy-related radiation and chemicals."[31] It was not an abstract, academic exercise. And, though it helped set radiation safety standards in the United States, the Department of Energy did not pretend that the research benefited the Yanomami in any way.

Today, Venezuelans charge that the AEC took what it wanted without explaining what it was doing. "Using the Yanomami as a control group to compare them with the survivors of the atomic bomb is absolutely unjustifiable from an ethical point of view," said Alejandro Arenas, the physician who heads Venezuela's Yanomami Health District. "If an experiment is done with a closely controlled protocol for the benefit of the same community where the study is performed, then I think it might be valid, provided it didn't take human lives or negatively affect the quality of life of the participants. But if the experiment offered no solutions to the community itself, then it was basically criminal. It would be like using a human being as a guinea pig or a laboratory rat."[32]

In fact, the Department of Energy's admission that its Yanomami research "helped to bridge the gap between mutagenesis studies in experimental animals and observations in people"[33] raised questions of medical violation. "In 1957 the U.S. Supreme Court ruled that scientists had to receive 'informed consent' from experimental subjects," said John Earle, a medical historian at the University of Pittsburgh. "It would be interesting to know how they obtained informed consent from the Yanomami."[34]

"I think this Atomic Energy Commission study was wrong from the start," said Ysbran Poortman, a biologist at the University of Utrecht and president of the European Alliance of Genetic Support Groups, a two-million-member organization that is developing ethical guidelines for teaching genetics at European Community schools. Poortman said the AEC Yanomami project violated the 1947 Nuremberg Code, whose first statute prohibited experimentation on people without their full knowledge and agreement. "Taking blood from people without informed consent is theft," he said.[35]

The AEC Yanomami research took place at the height of the Cold War and the Vietnam conflict, a period when the agency building nuclear weapons was not especially scrupulous about its experiments. An AEC study at Vanderbilt University subjected 850 pregnant women to daily doses of radioactive iron; in other cases, hospital patients were injected with plutonium and uranium and retarded children were fed radioactive oatmeal. None of these people knew what was going on.[36] In 1996, the federal government agreed to pay twelve people who had been injected with radioactive plutonium or uranium an average of $400,000 each. Secretary of Energy Hazel O'Leary said, "This settlement in no way fully compensates families for what they have suffered and for what they haven't known about their suffering."[37]

Admittedly, Chagnon and Neel faced uncommon difficulties in obtaining "informed consent" from Yanomami Indians, who had no idea of what the atomic bomb was—or radiation, or Japan, or the deadly risks associated with even brief encounters with outsiders. Yet Chagnon's closest friends and field collaborators—including New Tribes missionaries, the Catholic Salesian order, scientists at the Venezuelan Institute of Scientific Investigation, and members of the government Malaria Department—were all distressed to learn that the AEC had conducted a huge experiment comparing Yanomami blood samples with those of atomic bomb survivors.[38] "Chagnon always sounded so interested in helping the Yanomami," said Gary Dawson, a Protestant missionary who has lived on the Padamo River near Toki for forty years.[39]

It was true that Neel and Chagnon had their own research agendas. Chagnon wanted a Ph.D.; Neel sought the Index of Innate Ability. But the AEC administrators were not whistling Dixie. They paid for a blood study to determine radiation pathology, and they got it.

It was also true that, as a graduate student, Chagnon could not control the inherent flaw of the AEC protocol: the decision to use the most remote tribe in the world as a laboratory to benefit American military and industrial pursuits.

As Chagnon recounted in 1997, he was a subordinate to senior geneticists in an expanding project that turned into "a bad nightmare." Conflict was built into a "study that by nature required that maximum data and samples be collected from each village in the shortest amount of time possible."[40] This type of maximization followed its own inevitable logic, one closely patterned on James Neel's experience as the military officer who directed the A-bomb studies in Japan. Initially, Chagnon prepared the way for a team of eight researchers—a biologist, a linguist, a dentist, three geneticists, a filmmaker, and himself.[41] But each researcher needed many native assistants, to collect everything from beetles to hallucinogenic plants. As time went on, Neel enthusiastically added new shifts: when the first multidisciplinary team was finished, a second team landed on the same cargo plane that evacuated Team A (from the airstrips an engineer enlarged). Finally, Neel had three teams coming and going in a single season.[42]

Bloodletting was their most crucial ritual, and it was expensive. "Thus, to assure the complete cooperation of entire villages for some of our studies, I had to give goods to men, women and children," Chagnon recalled. He frantically traveled ahead of the geneticists, preparing dozens of *shabonos* where he "had gotten agreements from the Yanomamo to provide endless outstretched brown arms into which many needles would be stuck for the next weeks."[43]

According to Brian Ferguson's book *Yanomami Warfare,* Western influences were disruptive at their inception and termination. Long-term Western influences, like missions, could give rise to peace and stability, depending on circumstances. But short-term, fluctuating infusions of trade goods always created conflict. Gold miners were a fluctuating Western presence that caused huge disruptions in terms of trade and war.

The geneticists' fluctuations were as uncertain as any miner's. A single visit from the geneticists or filmmakers could multiply a village's steel goods a hundredfold overnight. Sudden downpours of steel presents created a brief euphoria, followed by the inevitable letdown. "The distribution of trade goods would always anger people who did not receive something they wanted, and it was useless to try to work any longer in the village."[44] Chagnon always came with promises of "big" presents; he always left with angry, disappointed people. Then he went back to the village's nearest enemies, to read the names of the dead and check his data against their enraged reactions. "More than the depressing constancy of giving goods and withholding them as each situation developed, the threats are what ultimately wore me down. I particularly disliked threats to my life."[45]

It was not much fun for the Yanomami either, judging from the reactions of those who watched *A Multidisciplinary Study.* For some of members of the audience, the film brought nightmares to life. It took nearly two hours to get through the fifty-minute documentary because speaker after speaker stood up to interrupt and object.

Pablo Mejía, who at age forty-five is now literate and fluent in Spanish, first met Chagnon when Mejía was about twelve years old. "I was in Momaribowei-teri. That's the first village where Chagnon arrived after he established himself at Bisaasi-teri. He thought he would become a sorcerer [*brujo*]. In order to be a sorcerer, he asked the other *brujos* to teach him. When he arrived at the village, he had his bird feathers adorning his arms. He had red *onoto* dye paint all over his body. He used a loincloth like the Yanomami. He sang with the chant of his shamanism and took *yopo* [a powerful hallucinogen used by Yanomami shamans that alters vision and self-awareness]. He took a *lot* of *yopo.* I was terrified of him. He always fired off his pistol when he entered the village, to prove that he was fiercer than the Yanomami. Everybody was afraid of him because no one had ever seen a *nabah* [white man, outsider] acting as a shaman. He would, say, ask, 'Who was your dead father?' He said to my brother Samuel, who was the headman, 'What is your mother's name?' My brother answered, 'I don't want to say her name. We Yanomami do not speak our names.' Shaki [Chagnon] answered, 'It doesn't matter. If you tell me, I'll pay you.' So, although they didn't want to, people sold their names. Everyone cried, but they spoke them. It was very sad. I remember well because I was about ten or twelve years old. That's how things were with Shaki. He said, 'I want to be a shaman who works only for your village. Go ahead and teach me.' He would say this to the old ones, the shamans. But they were afraid. Later he went to Mishimishi, where they taught him. Shaki had his own shaman circuit. He would say, 'I am the *cacique* of all the Yanomami.' He played everything, risked everything. I'm not the only one who heard—everyone heard him. He can't deny it. When he would come to our village, all the children would run into the forest screaming with fear. I've never seen anything like it."[46]

The Yanomami were in awe of the *nabah* (outsiders) and their technology. "If I had just suddenly flown straight up into the sky, they would have been surprised, but it wouldn't have come as an utter shock to them," wrote Kenneth Good. "They don't know about *nabuh,* about what they can do. They know *nabuh* have incredible powers. . . . Even after a year and a half of living with the Hasupuweteri, with all the friendship and understanding we had developed, I was still not clear to what extent they even considered me a human being. They would still ask questions that you wouldn't ask a fellow human being. 'Are you going to die sometime?' the would say."[47]

In *The Fierce People,* Chagnon described taking hallucinogens with the Bisaasi-teri. Chagnon took off all his clothes except a swimsuit, decorated himself with feathers, and took a hit of *ebene,* a generic name, used interchangeably these days with *yopo,* for any one of the mind-altering snuffs—derived from two different trees—that the Yanomami use on the Upper Orinoco.[48] As the green powder penetrated, Chagnon began dancing and chanting. "Wild things passed through my mind," he wrote. His song, however, disturbed some of the Yanomami warriors. They "hid the machetes and bows, for I announced that *Rahakanariwa* [Vulture Spirit, a cannibal entity usually invoked to kill enemy children] dwelled within my chest and directed my actions, and all know that he caused men to be violent."[49]

Chagnon's shamanism was an extension of warfare by other means. He struck out against other spirits on the horizon, threatening them with arrows, which he broke over his head. "As my high reached ecstatic proportions, I remember Kaobawa and the others groaning as I broke the arrows over my head and pranced wildly with the shambles and splinters clutched tightly in my fists. . . ."[50]

Why was Kaobawa groaning as Chagnon's ecstasy reached its climax? Kaobawa explained his consternation in a 1995 interview with the author Mark Ritchie. "Shaki [Chagnon] started eating Yanomami souls," Kaobawa said. "Shaki said, 'Come here Shoriwe [Brother-in-Law] and help me eat this child.' Before he killed the child I said, 'You know what you're going to do? You're going to bring retaliation to me.' So I said, 'Go ahead. But somebody from my village will have to pay the price.' So he killed the child."[51]

Gary Dawson, who was also present when Chagnon took *ebene* on this occasion, translated for Mark Ritchie during this interview with Kaobawa.

Ritchie wanted to know whether the child Kaobawa referred to had physically died, or whether this was a spiritual event. But, as Dawson explained, the distinction between physical and spiritual did not make much sense to Kaobawa. "In Kaobawa's mind," Dawson said, "and in anybody else's in that village, Shaki killed that kid with the spirit, probably without the foggiest idea of what he was doing. He was on drugs, dabbling with something in someone else's culture when you don't have the slightest idea of what you're doing."[52]

Although it might appear that these were simply the antics of an ego out of control, there was a logic to Chagnon's anthropological methods. He had two nearly insurmountable problems. The first was how to get the Yanomami to divulge their tribal secrets and give up their blood; the second was how to protect his treasure of trade goods long enough to buy off the Yanomami's objections to his research. He was alone with all this metallic wealth. Just as firing off his guns frightened the Yanomami into leaving his trade goods untouched, so his shamanic pretensions also strengthened his hand. The Yanomami believed that white men were supernatural beings who had the power to send terrible epidemics. Chagnon's guns and claims of magical power were necessary correlates of the AEC's high-pressure research agenda. And, just as Pizarro had started the tradition of firing guns off to impress South American Indians, Cortés had begun the equally venerable conquistador strategy of posing as a native deity. Cortés was Quetzacoatl, the Plumed Serpent. Chagnon was the Vulture Spirit.

At Toki, I heard the word *anthro* for the first time. It was an attack word, both epithet and noun—like "communist."

"These *anthros* come, they take pictures, collect blood, carry them off to their countries, sell them, and make money," Mejía declaimed before the gathering at Toki. "And we get nothing. We have to stop this study of the Yanomami. They are like miners, and we are like their gold. Why do they want to study us so much? *Nabah* [foreigners] have a brain; Yanomami have a brain. *Nabah* have two eyes; Yanomami have two eyes. *Nabah* have five fingers; Yanomami have five fingers. Why are they so interested in studying us?"[53]

As James Neel explained it in the film, other tribes had been studied, but the result had been an unsatisfactory "mosaic of unrelated findings." Neel wanted the definitive "population structure" of the world's least acculturated tribe. "Population structure includes the totality of factors that determine how genes are transmitted from one generation to the next. . . . Since transmission of genes of course depends on survival well into adulthood, we feel we must understand many aspects of the disease pressures to which these Indians are subjected. . . . Since the Yanomama are extremely warlike, we must understand how warfare, as well as disease, decimates the population."[54]

This was a coded message. Neel hoped to prove that the Yanomami "population structure" was the one dictated by natural selection: a society dominated by aggressive, polygamous chiefs, where very few people reached the age of fifty. His core belief was that modern society's gene pool problems arose "primarily from abandoning the population structure and the selective pressures under which humankind evolved."[55]

While most people saw draft deferments for the handicapped as a humane necessity, Neel detected an "agent of negative selection."[56] While almost everyone applauded the democratic freedoms that allowed women to choose their own mates, Neel glumly concluded that the "loss of headmanship as a feature of our culture, as well as the weakening of other vehicles of natural selection, is clearly a minus." This loss could "contribute to lowering IQ, and by a fairly direct extension, to undermining human abilities." Discovering "test procedures to determine whether and to what extent the headman really is characterized by a high IIA [Index of Innate Ability]" became, for Neel, "the number one objective."[57]

In his autobiography, Neel regretted that his Amazonian insights had been "the most difficult to project to those who are not professional geneticists." That is why, in the film *A Multidisciplinary Study,* Neel did not come out and say that students at American universities were all products of a mediocre gene pool—"a large, increasingly homogeneous, quivering blob of jelly"—incapable of reproducing the superior qualities of Yanomami warriors.[58]

Like the early Jesuit missionaries who sought to create perfect Christian communities among unspoiled natives, Neel saw salvation coming from uncontaminated Yanomami genes. From Neel's standpoint, the Yanomami's "extremely violent" society was not pathological. But American society *was*—with its costly medical interventions, draft deferments for the handicapped, and other well-intentioned kindnesses that defiled the gene pool. Neel defined his own culture as "dysgenic." There was quite a bit of self-hatred in this picture, since Neel thought doctors were among the weakest specimens of our pampered society.

While Neel saw salvation coming from Yanomami chromosomes, the

Yanomami had other goals. They were searching for metal goods, not realizing that their reliance on Westerners for steel also made them susceptible to new diseases that only outsiders could cure.

If germs, steel, and guns were the Three Horsemen of the Amazonian Apocalypse, germs had the greatest impact, and they were intimately associated with steel. Some 95 percent of all New World Indians died of Old World diseases.[59] These were usually disseminated far beyond the immediate circle of contact by the natives' desire to obtain and barter precious metal. The Yanomami shared this fatal attraction for steel gifts.

Although the AEC protocols admirably maximized data collection, they also maximized exposure to a host of new germs. In some ways, the assembly-line blood-collecting routine was a formula for disease propagation, starting with the arrival of scientists from major cities around the world, who were ferried by speedboat to isolated Yanomami *shabonos* in the company of Yanomami guides from the missions. At no time in their films, or in any of the voluminous writings of the scientists who participated, is the question of how their own presence among barely contacted villages added to "the disease pressure" that "decimates the population."[60]

It was not as if Chagnon was unaware of the problem. He frequently warned about missionaries or tourists spreading deadly illnesses among the vulnerable Yanomami. In *The Fierce People,* he wrote, "I once put the hypothetical question to a Protestant missionary: 'Would you risk exposing 200 Yanomamo to some infectious disease if you thought you could save one of them from Hell—and the other 199 died from the disease?' His answer was unequivocal and firm: 'Yes.' "[61]

In spite of their early mistakes, during the 1940s and early 1950s, the missionaries soon lost their taste for the adventure of first contact. By the time the AEC-financed expeditions started, the missionaries had settled into established posts from where they spent at least as much time ministering to the Yanomami's health needs as proselytizing. Consequently, the Yanomami population boomed around the missions. Even discounting immigration, the demographic explosion at all the Yanomami missions, Protestant and Catholic—in Brazil as well as in Venezuela—amply testifies to what Ferguson has called "the sphere of mission beneficence."[62]

The story was completely different for ephemeral visits by gold miners, tourists, or scientists. Just as brief outpourings of steel goods sparked wars, brief exposure to outside pathogens without medical follow-up sparked uncontrolled epidemics. "In my paradigm there is a close correlation between large expeditions and epidemics," observed Ferguson. "In the 1950s, you had big, military-style expeditions on the Upper Orinoco, and it was a time when sickness ran very high. The more people who go in without being screened, the more likelihood of disease. Chagnon is the one who's saying that people are spreading diseases among the Yanomami through trade goods, and here he's got all these trade goods and people are vying for them, carrying the goods and diseases all over the place."[63] In fact, Chagnon criticized missionaries for giving steel machetes to the Yanomami, because the trade in steel could innocently disperse colds and other illnesses from mission stations to the Yanomami hinterland.[64]

Yet the AEC purchased twelve thousand Yanomami blood samples, dispensing a steel gift for each vial of blood—along with thousands of other presents for other services. Today, those vials are located in an old refrigerator at Penn State University, where Chagnon once taught, and are the property of the Human Genome Diversity Project of the U.S. government.[65]

Of course, steel was not the AEC's only gift to the Yanomami. Chagnon's data showed that infectious disease, including colds and malaria, unknown in the Yanomami highlands, caused 70 percent of all the deaths at his home village, Bisaasi-teri.[66]

War accounted for only 15 percent of the adult mortality. Less than 2 percent of all adults died of old age, a catastrophic profile common at first contact. Yet the true picture was actually much worse, since Chagnon did not

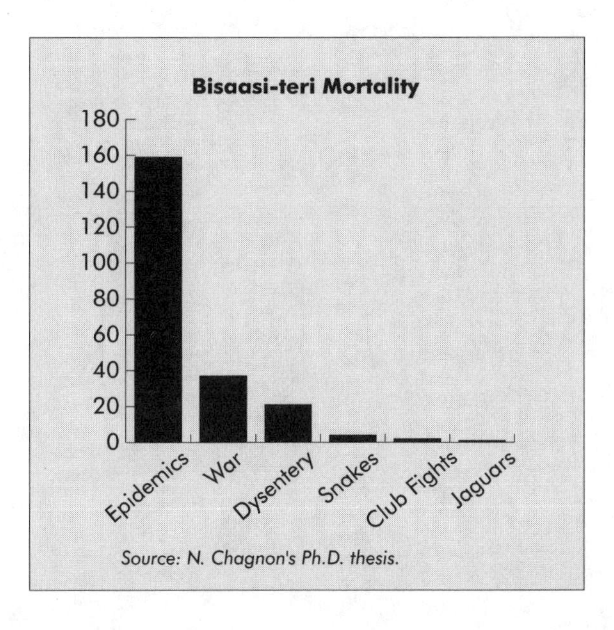

Source: N. Chagnon's Ph.D. thesis.

take infant mortality into account (and deaths of young children almost always outnumbered those of adults). A less ideological researcher might have called the Yanomami the Sick People.

The wars Chagnon documented, and the sorcery accusations that inspired many of them, took place against a backdrop as terrifying as Europe's fourteenth-century Black Plague. By 1966, Chagnon's base camp had become a deadly repository of imported illnesses; it soon emerged as a hub for distributing those diseases to remote villages. Pathogens followed the AEC's steel paths.

Chapter 5

Outbreak

The reaction to measles vaccine without gamma globulin had been, in some cases, as severe as the disease itself among Caucasian children.—*James Neel*[1]

Near the juncture of the Orinoco and Ocamo rivers, by a dirt airstrip at a Catholic mission, there lies an unmarked grave. Thirty years ago, a small cross, befitting a child's burial, was erected at this spot, but the tropical weather made a quick casualty of the wooden memorial. With clouds of gnats by day and mosquitoes by night, it is not a pleasant place to live, or to die, or even to be buried. Today nobody except Roberto Balthasar's parents remembers where he was interred or what killed him.[2]

Yet, according to mission records, Roberto Balthasar died of measles, on February 15, 1968. Hundreds, perhaps thousands, of others also died of measles that year on the Upper Orinoco. Two things made Roberto Balthasar's death notable: his was the first clearly diagnosed case of measles among the Venezuelan Yanomami. And, according to the boy's father, Napoleon Chagnon vaccinated him.[3]

As it turned out, Chagnon experienced two of the most unfortunate coincidences in Yanomami history. By his own account, the wars that made the Yanomami famous began on the day he arrived in the field, November 14, 1964;[4] also by his own account, their worst epidemic started the day he returned after a long absence, on January 22, 1968.[5]

On the latter date, he carried live measles vaccine virus into Yanomami territory and within forty-eight hours had helped vaccinate Indians at the Ocamo mission—from where the epidemic began and spread "as a wave away from the original point."[6] This second coincidence, corresponding to the exact time and place of the application of a live measles virus to a susceptible population, is striking.

Scientists had been competing worldwide to observe measles in a "virgin soil" population. By the late 1960s, it was impossible to find societies outside the Amazon whose adults lacked measles antibodies. Even the most isolated enclaves of the world, like Micronesia and Iceland, experienced periodic outbreaks of measles.[7] Because measles attacked everywhere with such predictable ferocity, geneticists expected that a measles contagion in an Amerindian tribe could allow them to measure the difference in inherited immunity between New and Old World people—a key factor in natural selection. So it was fortuitous that James Neel and his team of geneticists were the only scientists who observed a measles epidemic among Amerindians without previous exposure. Neel was justifiably proud of this medical scoop. He boasted that its replication "under identical conditions will probably never be possible."[8]

Although I agree with Neel that the Yanomami measles epidemic was unique in medical history, my reasons for thinking so are different from his. To understand the outbreak's origins, one must consider three questions: What kind of vaccine was used? When and where did measles first appear on the Upper Orinoco? How did this outbreak spread across dozens of villages and thousands of square miles?

The history of the 1968 epidemic, as told by Chagnon and Neel, has varied. The number of vaccines they brought to the Upper Orinoco was either 1,000,[9] 2,000,[10] or 3,000,[11] depending on the article or on the edition of *The Fierce People.* Neel's team spent either two weeks or a month or six weeks fighting the epidemic. It battled the outbreak either with the help of local missionaries[12] or, in Chagnon's most recent revisions, alone and in spite of the "callous" indifference of the missionaries.[13] In general, Chagnon and Neel have grown more heroic in their combat of the 1968 epidemic as time has passed.

Yet, throughout these various accounts, the AEC researchers have never explained their choice of vaccine: the Edmonston B live virus. It was one of the most primitive measles vaccine, first developed in the late 1950s. From the beginning, it was described as "a new disease" with serious symptoms.[14] In 1959, researchers in Panama hospitalized nine children after vaccinating them with the Edmonston B; they advised against using it anywhere without emergency facilities.[15] Among Canadian children, 60 percent of the Edmonston vaccinees contracted fevers over 103 degrees Fahrenheit.[16] These results looked surprisingly like natural measles. No rigorously controlled study of the Edmonston B and wild measles was ever conducted, because it would have meant denying children aspirin and antibiotics. In general, the Edmonston virus raised temperatures about four degrees; wild viruses, about five degrees. However, one way of measuring the virulence of measles viruses was to compare antibody titers, which reflected the intensity of fevers and the

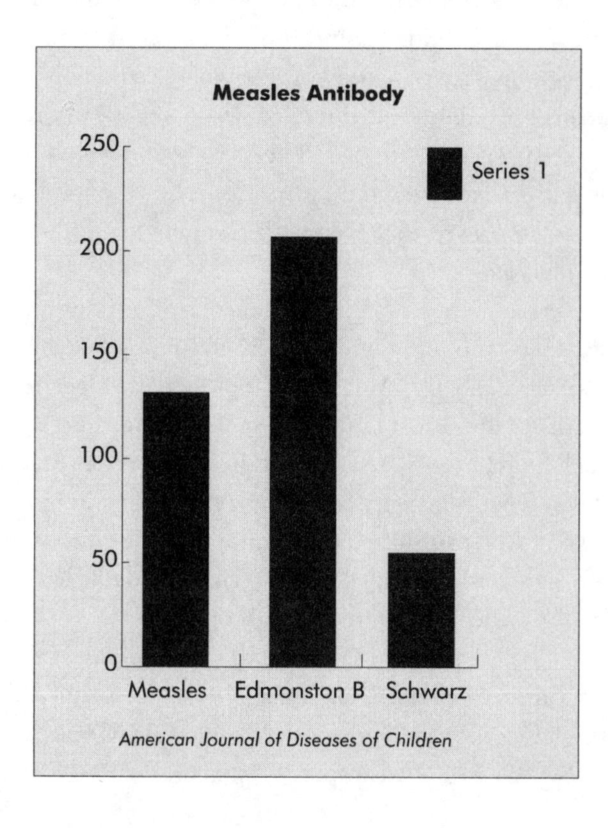

American Journal of Diseases of Children

body's immune response. Surprisingly, in a baseline study at NYU, antibody levels proved *higher* among Edmonston B vaccinees than among those with wild measles. Later studies found antibody levels to be identical to those of wild measles.[17] All subsequent measles vaccines, including those currently in use, generated far lower antibody responses—about one-quarter as high as that of the original Edmonston virus.[18]

In 1961, the National Institutes of Health sponsored a conference on the Edmonston vaccine. The keynote speaker was G. S. Wilson, head of England's Public Health Laboratory Service, who warned of possible fatalities. And, in unusually blunt language, he said the test of a vaccine was whether "the disturbance caused by the vaccination" was "greater than that caused by the disease itself." With most vaccines, the difference was obvious; in the case of the Edmonston strain, however, Wilson thought the difference between the disease and the vaccine virus was "not so clear."[19] The consensus of participants was that "the occurrence of fever of this magnitude and incidence . . . will be deleterious to public acceptance and widespread use of the vaccine in its present unmodified form."[20]

In summarizing these early measles studies, a standard text on immunization states, "The Original Edmonston A and B vaccine strains caused quite severe reactions akin to natural measles. These could only be overcome by simultaneous use of immunoglobulin, which was clearly not a practical procedure for routine vaccination."[21] The use of gamma globulin cut the reaction rate in half, but the procedure of double vaccination was cumbersome, and several studies showed that even with gamma globulin the Edmonston B virus was dangerous.

The greatest risks were among Amerindians and immune-compromised people. Researchers who tested measles vaccine among Native Americans—in Alaska, Panama, and the Amazon[22]—reported stronger reactions than among Caucasians; all warned against using the Edmonston B vaccine, even with gamma globulin, among Amerindians in remote areas. Among Eskimos, controlling high fevers for two weeks following vaccination posed a "major problem."[23] For anyone suspected of suffering from immune deficiency, the Edmonston B was contraindicated. The vaccine killed one leukemic child in spite of a dose of protective gamma globulin.[24]

There was a close tie between the acute vaccine responses of American Indians and immune-depressed patients. Most scientists believed that Amerindians had lower genetic defenses against Old World contagions. At issue here was whether the catastrophic impact of so-called European crowd diseases, which contributed to the collapse of both the Inca and the Aztec empires, was due to genetic or to cultural factors. These new viruses—including those of measles, mumps, smallpox, diphtheria, and influenza—were originally incubated in domesticated animals. They mutated into forms that could infect human hosts and then required large, urban populations to permanently sustain them. All had arisen in the Old World, quite recently in evolutionary terms—during the last five thousand years or so.[25] Would such a short period of time be sufficient for inherited immunity to arise in the Old World gene pool? It certainly looked that way. These imported diseases decimated Native Americans, while the conquistadors who brought them soldiered on and on. Although Spanish steel, guns, and military discipline were decisive forces in defeating native armies numbering in the hundreds of thousands, superior technology would never have sufficed without smallpox, measles, and influenza—the three principal scourges of the Americas. Of these three, smallpox was eradicated by the 1960s and influenza had evolved so many mutant strains that it was a tricky disease to study.

Measles, however, was tailor-made for experiments. It was both the world's most universal disease and the easiest one to diagnose, given its uniform symptoms.[26] Of all extant crowd diseases, it had the highest attack rate—100 percent in areas where the disease was not endemic[27]—and conferred the most certain, lifelong immunity.[28]

There was also a mystery about measles propagation. Prior to general vaccination, an estimated four million cases of wild measles in the United States brought five hundred deaths per year—one in eight thousand.[29] In stark contrast, South American Indians averaged about 25 percent loss of their total population at the first onset of measles, a rate two thousand times higher than that in the United States.[30] Even if one allowed for the impact of modern medicines, there appeared to be evidence of differential adaptation, possibly genetic selection. In 1961, G. S. Wilson challenged NIH doctors to solve this puzzle.[31]

By 1965, the intense measles-vaccine reactions seen among Amerindians had gone a long way toward confirming the theory that Native Americans were more susceptible to Eurasian epidemics. Francis Black, a medical researcher at Yale, was keenly involved in these studies. Pursuing measles reactions from Iceland to the Amazon, he determined that the Tiriyo Indians of the Brazil-Suriname highlands had the world's lowest antibody responses to a wide variety of diseases, including measles. By vaccinating the Tiriyo with a vaccine "model of natural measles" Black hoped to resolve the debate about genetic selection because "enhanced susceptibility to measles would be reflected in more severe reactions to vaccine." However, noting a previous study of Alaskan Eskimos, he warned of "the risk of severe febrile response to Edmonston B level vaccine with gamma globulin."[32] Black opted for the more attenuated Schwarz vaccine, which had been available from 1965 onward. The Schwarz virus produced half the febrile response of the Edmonston and one-tenth as much rash but conferred the same lasting immunity. It was also less dangerous than the Edmonston B plus gamma globulin—producing fewer fevers over 104 degrees Fahrenheit and a fraction of the painful rash.[33]

"Choosing the Schwarz over the Edmonston B with gamma globulin wasn't just a personal decision," Black recalled. "It has been universal. The Schwarz was cheaper, easier to administer, and had fewer reactions."[34]

When Black vaccinated the Tiriyo in 1966, he discovered that they had by far the world's highest febrile response to the Schwarz vaccine—on the average almost three times as high (171 percent) as that of other groups tested by the World Health Organization (WHO).[35] Black's results were remarkable and suggested that utmost caution should be used in vaccinating Amazonian tribal groups with no histories of measles.

When I told Francis Black that James Neel had administered the Edmonston B vaccine to the Yanomami in 1968, he did not believe me. "That happened around 1964," he corrected me. "It would have been contraindicated any time after about 1967."[36] Samuel Katz, an internationally respected authority on measles who helped develop the Edmonston B vaccine at Harvard, added, "By the late sixties, the Edmonston B was practically extinct."[37] One million children in the United States received the Edmonston B vaccine with gamma globulin in 1968—about one-fifth of the vaccinated cohort. It was thus still legal in the United States, even if considered obsolete by experts. The manufacturers of the Edmonston B's in North America were practically the only companies in the world still producing it, and they were rapidly phasing it out. (In 1969, production fell precipitously and then disappeared.)[38] By 1968, two additional antimeasles vaccines had become available—the combined measles-smallpox vaccine and the Moraten measles vaccine—both of which, like the Schwarz, had much lower reaction rates than the Edmonston vaccine.[39]

Actually, the WHO had advised doctors to switch from the Edmonston B vaccine strains in 1965 because "severe reactions are too frequent to permit their wide general use." The WHO found that "the further attenuated vaccines are preferable to Enders Edmonston B vaccine plus gamma-globulin."[40] Neel was the only known scientist who had ever exposed a "virgin" Amazon tribe to the Edmonston B vaccine.

It was strange that Neel, with millions of dollars from the Atomic Energy Commission, would bring a dinosaur vaccine with him to the Yanomami. His teams set new standards in genetic research, ethnographic filmmaking, and the computer modeling of tribal demographics. They arrived on C-130 cargo planes, with tons of equipment.[41]

The choice of vaccine was particularly odd because administering the Edmonston virus required twice as much work as administering any of the safer strains (because of the extra shot of gamma globulin). Yet, in spite of the risks to the Yanomami and the inconvenience to his own medical team, Neel obtained the Edmonston vaccine from Parke Davis Laboratories, Philips Roxane, and Lederle, rather than seeking the more attenuated measles vaccine viruses.[42]

Why did Neel do it?

Although I can only speculate about Neel's personal motives, opting for the Edmonston vaccine was a bold decision from a research perspective. Obviously, the Edmonston B, precisely because it was primitive, provided a model much closer to real measles than other, safer vaccines in the attempt to resolve the great genetic question of selective adaptation.

And, despite all the evidence to the contrary, Neel simply did not believe "the medical dogma that the isolated tribal populations . . . have a special inborn susceptibility" to diseases like measles.[43] The consensus of scientists is that tens of millions of American Indians, from the Mississippi valley to Tierra del Fuego, died of "Old World germs to which Indians had never been exposed, and against which they therefore had neither immune nor genetic resistance."[44] This conclusion, from UCLA's professor of physiology Jared Diamond, has been echoed by thousands of observers.

But James Neel disagreed. He believed that the Yanomami were models of good health. The adult men appeared to be in excellent shape. "The males, in general, present a picture of exuberant vitality, an impression confirmed by their dancing and chanting frequently extending through most of the night."[45] But watching adult men dance in their ceremonial feathers and paint says little, if anything, about the actual health of a tribal community. The WHO considers anthropometry, the study of weight and height, a much surer guide to assessing overall well-being.[46] The only systematic anthropometric study on the Yanomami, done in 1980, found widespread malnutrition in children.[47] Significantly, children are often the ones who are at greatest risk during epidemics; Yanomami children make up 80 percent of the victims in some outbreaks.[48] Also, Yanomami adults' tiny stature (under five feet) constitutes strong, if indirect, evidence of "long-term nutritional inadequacy or generally poor environmental conditions, especially ones in which chronic or repeated infections are prevalent."[49] The Yanomami along the Upper Orinoco suffered terribly from chronic malaria, which contributes to immune depression.[50]

It is difficult to imagine a group at higher risk to a live measles virus than the Yanomami. Yet Neel's inoculation protocol was even more puzzling than his choice of vaccine. Yanomami at the Ocamo mission received the Edmonston B without the recommended gamma globulin coverage, which doubled the risk of reaction.[51] The vaccinators were Napoleon Chagnon and a respected Venezuelan doctor named Marcel Roche.[52] "Marcel and Nap were at Ocamo but somehow came without the gamma globulin that should have been administered with the vaccine," Neel recalled.[53] It was a mistake, but measles had already broken out and they had to act quickly, Neel explained.

Nevertheless, Neel also sent Edmonston B vaccines without gamma globulin to two Indian villages in Brazil, where missionaries administered them. If vaccinating without gamma globulin was an unfortunate accident, it was repeated three times, producing valuable data in each case.[54]

At the Ocamo mission, Chagnon and Roche vaccinated forty people. Thirty-six Yanomami at this same village did not receive the vaccine. If they were inoculating in an emergency, as Neel claimed, why only half the village?

Neel and Chagnon have dramatically described how they raced against the contagion, trying to inoculate as many villages as quickly as possible around Ocamo. In reality, no more vaccinations were given for two full weeks.[55] Even more curious was the fact that Neel never vaccinated the other half of the Ocamo village,[56] even though he arrived on February 4 with both vaccine and gamma globulin, which he and Roche administered to some of the surrounding villages.[57]

There were only two possibilities. Either Chagnon entered the field with only forty doses of virus; or he had more than forty doses. If he had more than forty, he deliberately withheld them while measles spread for fifteen days. If he came to the field with only forty doses, it was to collect data on a small sample of Indians who were meant to receive the vaccine without gamma globulin. Ocamo was a good choice because the nuns could look after the sick while Chagnon went on with his demanding work. Dividing villages into two groups, one serving as a control, was common in experiments and also a normal safety precaution in the absence of an outbreak.[58]

Although Neel gathered data from three Indian villages where the Edmonston B vaccine was given without gamma globulin, he still ran out of gamma globulin before he ran out of vaccine.[59] Marcel Roche told me that the expedition had traveled to the mission of Platanal, where they studied the Yanomami but did not vaccinate anyone—which surprised him since, as he put it, "supposedly this was a preventive thing."[60] The extra injection of gamma globulin complicated everything. This was one reason the Edmonston B was not considered a practical vaccine, especially for remote regions. Gamma globulin could spoil or get lost or might be used to provide temporary protection against measles. It might also be given in larger doses to pregnant women and infants under the age of one year in lieu of vaccination. On February 17 Neel paid an emergency visit to the unvaccinated villagers of Ocamo who had come down with measles, where he administered gamma globulin as protection.[61]

Whether the crisis depleted his supplies or whether he made a mistake he could not correct, Neel found himself short of gamma globulin. He was careful never to vaccinate without gamma globulin himself, but he was less scrupulous about his subordinates, like Chagnon, or the Brazilian missionaries, or his Venezuelan colleagues. The Venezuelans never suspected that gamma globulin was needed, because no up-to-date measles vaccine in Venezuela, or anyplace else in the world, demanded gamma globulin coverage.

According to Neel's account in the *American Journal of Epidemiology*, a fourteen-year-old Brazilian worker at the Ocamo mission brought measles to the Yanomami. "By chance, the arrival of our expedition in Venezuela coincided with the introduction of measles to the Yanomamo of the Upper Orinoco by a young Brazilian." (The boy, and seven other Brazilians, had come from a distant jungle outpost to extend the Ocamo airstrip for the researchers' cargo plane.) However, in this original version of the epidemic, Neel acknowledged that the Brazilian teenager never showed a measles rash. ("The characteristic morbilliform rash never developed. . . .")[62] That was peculiar. One hundred percent of measles victims develop a rash, according to most medical texts.[63] That is why measles is the most readily diagnosable disease. Samuel Katz recently observed that the only exceptions he knew of were people with depressed immune systems, but that would be unlikely in a healthy fourteen-year-old brought in to work on an airfield.[64]

Neel admitted that the diagnosis of measles was "uncertain" because the boy's symptoms—high fever and bronchopneumonia—could not be distinguished "from any of a variety of jungle fevers." He shifted the responsibility for making the crucial, but "tentative," diagnosis of measles onto his Venezuelan colleague, Marcel Roche, who was not a coauthor of the article.[65]

I was so intrigued by this that I sought Roche out in Caracas. He was smoking a cigar at the office of Venezuela's leading science magazine, *Interciencia,* of which he was then the editor. When I read him the Neel-Chagnon version of the measles epidemic, his cigar almost fell out of his mouth. *"Me?* I only gave that vaccine because Neel said to. The plan was to vaccinate them whether they had measles or not." About the fourteen-year-old's symptoms—fever, pneumonia, absence of a rash—Roche shrugged. "A nonspecific case." Asked again whether he had diagnosed measles before beginning his vaccinations at Ocamo, Roche replied, "No. I don't remember having done that." He had never seen the article before.[66]

Roche said Neel had given him the vaccine without discussing its side effects or the history of its use in other parts of the world. He assumed, like the missionaries and health workers who helped vaccinate the Yanomami with the Edmonston B, that Neel had brought the best and safest vaccine available. *"A veces uno comete errores."* Sometimes one makes mistakes.

I was able to obtain access to the Salesian mission records. There existed three separate sets of diaries for the period of the measles epidemic, kept by the nursing nuns at the Ocamo and Mavaca missions, respectively, and by the priests at the Mavaca mission. The Ocamo mission chronicle reported that Roche had not yet arrived on January 22, the day Neel and Chagnon claimed that Roche diagnosed measles.[67] Although Chagnon, Asch, and Roche had planned to reach Ocamo a day earlier, Neel had given them so much equipment to unload that they were stuck at a jungle airstrip twenty-five miles downriver, according to Asch's narrative.[68]

And once Roche reached the mission, he would have needed special abilities, beyond logic or common sense, to diagnose measles in a boy without any of the disease's characteristic symptoms. Not a single instance of measles had ever been seen on the Upper Orinoco. But cases of high fevers accompanied by pneumonia were the principal cause of death among the Yanomami. There had been hundreds of cases like the young Brazilian's over the years. Five Yanomami died of bronchopneumonia in the months prior to the AEC's expedition.[69]

And there were dozens of individuals with precisely these symptoms when

Roche and Chagnon reached the field. "The doctors recognize it as a strong epidemic of bronchopneumonia," wrote one of the nuns.[70] The "doctors" she referred to were Marcel Roche, Inga Steinvorth Goetz (who, like Roche, was from the Venezuelan Institute of Scientific Investigation, IVIC), and a Frenchman, Jean Pier Poirier. Dr. Poirier added his own annotation: "The doctors of IVIC [Roche and Goetz] gave everything they could, but nothing worked against the bacterias."[71] As far as the missionaries and the French doctor knew, Roche diagnosed bronchopneumonia. Poirier, new to the Yanomami, was surprised at how ineffective antibiotics proved against respiratory infection.

It was a critical situation—the height of the malaria season complicated by severe colds—but one that was almost normal along the main course of the Orinoco. The odd fact was that the Yanomami population at the missions was booming in the midst of chronic malaria and respiratory infections, a paradox sustained by continuous medical assistance. Similar circumstances obtained in parts of tropical Africa, where measles vaccination required the additional precaution of malaria screening. Since Chagnon and Roche did not screen for malaria, some of the people they vaccinated almost certainly had it. A few probably had both malaria and bronchopneumonia.

But measles was a different matter. A measles outbreak was the single worst disaster that could befall an Amerindian community. It would have required an instant mobilization of missionaries and health workers, with an automatic quarantine. Yet neither the Ministry of Health, nor the Malaria Department, nor the Salesian bishop of Puerto Ayacucho (who was the federal government's legal guardian of the Yanomami) had any record of measles breaking out during the week of January 22. Chagnon and Asch also went about their business, traveling far inland to the most isolated *shabonos,* hiring many porters from mission posts, seemingly insensible to the danger of decimating the highland villages.[72]

In the mission archives, the word *morbillo* (measles) first appeared on February 4. And it concerned the Yanomami's response to a new wave of vaccinations at the Ocamo mission: "Dr. Roche applies the measles vaccine, but the reactions are a bit strong."[73]

In several cases, the reactions may have been fatal. Eight days after the first vaccinations, on February 1, the Ocamo mission chronicle noted, *Sono morte due bambini.* "Two babies died." Forty people had been vaccinated; by February 2, thirty-six people were in the Ocamo infirmary. One sick child was sent to the Mavaca mission, whose diary for February 15 read, "At 13 hours the little one-year-old boy, the son of the worker Vitalino of the Ocamo residence, breathed his last. He was brought here by his parents in critical condition—measles, bronchopneumonia—he had every medical attention possible."[74]

I spoke to Vitalino, the baby's father, at his small house in the city of Puerto Ayacucho. Vitalino, a small, sturdy man with light brown skin, was the administrator of the Ocamo mission. He was a Brazilian of mixed Indian, African, and Caucasian background, who married an Indian woman on the Orinoco. He possessed a quiet dignity and a memory that matched the mission records in almost every respect. "I was vaccinated with my children and my whole family," he said. "First, I got it [measles]. I was in bed about a week. I started recovering, little by little. The boy was tranquil. Since he was always with me in the hammock, I infected him, and since this is contagious right away, he got fever. He got fever, and in the end you could see he had measles. . . . His skin turned reddish." When I asked him who vaccinated his one-year-old, he said, "Napoleon." Like most of the people on the Upper Orinoco, Vitalino has no idea what a live virus is; he still presumes that the injection he and Roberto received was medicine, not a form of measles. He blames himself for the boy's death and feels that taking the boy by boat to the Mavaca mission, where the French doctors were, may have contributed to it. Of course, Vitalino's testimony raised the question of whether or not vaccinated individuals became active transmitters of measles. However, if his son died "about a week" after vaccination, as Vitalino recalled, it was another case of death occurring surprisingly close to the peak reaction time for the Edmonston virus, but several days before wild measles would normally manifest.[75]

Regarding the death of Vitalino's child, the priest at the Mavaca mission, José Berno, wrote, "What a coincidence! . . . As soon as the Atomic Energy Commission arrives, IVIC with its best doctor and professors from the University of Michigan, we immediately have *three deaths.*" The emphasis was Berno's. For a mission as small as Ocamo, with fewer than a hundred people, three deaths were equal to the total mortality expected in a year.[76] Berno wrote that there had never been anything like it "in the whole history of the mission." Another comment of the French doctor was also in the ledger: *Mal papel, decía Poirier con los tres muertos.* "A bad job, said Poirier because of the three deaths."[77]

Father Berno sounded angry in his mission chronicle, and he was still angry when I spoke with him in an enclosed rose garden adjacent to the cathedral of Puerto Ayacucho. "What I want to know is: With whose permission did they give that vaccine? And what kind of a vaccine was it? Neither *Sanidad* [the Ministry of Health] nor *Malarialogía* [the Malaria Department] knew anything about it. The mission sure didn't know anything about it. They just arrived, did what they wanted, and later made up this story about the Brazilians bringing measles. Who knows who brought it?"[78]

That was the question. Measles killed in two ways—with the first onset of high fever or with later complications, most frequently pneumonia.[79] The two babies who died at Ocamo were apparently in the eighth day of vaccination, which is the peak day of response to the Edmonston B virus. That caused Samuel Katz of Duke University to observe, "It's very suspicious when you say that they gave the vaccine and eight days later they were ill."[80]

Pneumonia, if it was brought on by measles, occurred after the telltale rash. But, according to Neel and Chagnon, the Brazilian teenager had bronchopneumonia without any measles symptoms. "If the Brazilian boy developed pneumonia without first showing rash, then he almost certainly did not have measles," said Carlos Botto of CAICET.[81]

Neel never identified the origin of the measles in another epidemic area. "If they didn't establish a chain, it would be hard to prove that a boy without rash was the source," said Francis Black. "I would have expected them to establish the origin of the epidemic."[82]

The Brazilians had been summoned to the Ocamo airstrip from a frontier outpost, San Carlos del Río Negro, where fewer than a hundred people lived. There was no measles outbreak at San Carlos while the Brazilians were there; none had been there for many years. It was the most isolated spot on the Venezuelan map, connected to the Orinoco and rest of the country only through the Casiquiare Canal, *la monstruosité en géographie,* which had given Humboldt the most painful passage of his career. In 1968, not a single person lived along the banks of this treacherous, insect-plagued waterway.[83] The Brazilians navigated for a week through the 227-mile-long Casiquiare with a tiny outboard motor, traversing uninhabited wilderness. How could they have picked up measles en route to Ocamo?

They didn't, Neel told me. "He [the Brazilian teenager] wasn't coming from a measles epidemic."[84] Neel's theory, as he explained it to me, was that the boy contracted measles at some unknown point and unknown time previous to his trip, and the infection "simmered" subclinically until it erupted shortly after he reached the Ocamo mission. This was within the reach of possibility, but just barely. Subclinical measles is extremely rare, according to a recently written world history of the disease;[85] transmission of measles by a subclinical carrier has never been proven, according to a widely used medical text on vaccination procedures.[86]

I have found only one case of a person suffering from "subclinical" measles, where it "simmered" for months. This happened to a boy with leukemia, who was innoculated with the Edmonston B vaccine virus—not natural measles. The boy went twenty days without showing rash, then burst into a full-body eruption that lasted weeks. When the skin lesions vanished, the disease did not. He died three months after vaccination, with Edmonston virus in his throat and conjunctivae. That meant not only that the vaccine virus killed him (his leukemia was in remission and did not return) but that it had moved to a portal—the respiratory tract—where he could infect others.[87] John Enders of Harvard University, the creator of the Edmonston vaccine, conducted an autopsy. It revealed gaping inner wounds caused by the virus.[88]

Measles is a horrible way to die. The outer signs, the pimple-like inflamations, are just hints of an inner war. The virus makes its appearance inside the mouth with red sores—Koplik's spots—that spread to the throat and lungs and, eventually, can attack the eyes and intestines. In fact, people suffering from immune depression often have severe internal hemorrhaging, causing blood flows from the nose, mouth, and anus.

According to Neel's chronology, the fourteen-year-old boy passed measles on to another Brazilian adult, who also had no rash. (Botto was extremely skeptical of this, since almost all adults from Brazil's Rio Negro have had measles.) Only then, fifteen days after Roche made his "tentative" diagnosis, did measles spread to the Yanomami.

If the Brazilians were all immune depressed—the only explanation for the absence of rash—they should have died or developed lingering, debilitating illnesses. Instead, they quickly recovered and went back to heavy work without ever showing signs of measles, inner or outer.

In the absence of visible symptoms, and without a known origin for the epidemic, it was extraordinary to think that Roche could have diagnosed measles as Neel claimed. A case of "measles" like this could have been detected only by a blood test, which Roche did not perform.

And there was another unaccountable detail. Neel wrote that Roche and Chagnon vaccinated "Brazilians" at the Ocamo mission against measles.[89] Vitalino Balthasar, the mission administrator, confirmed this. "The fifteen-year-old boy was vaccinated."[90] But what doctor would vaccinate someone with late-stage measles against—measles?

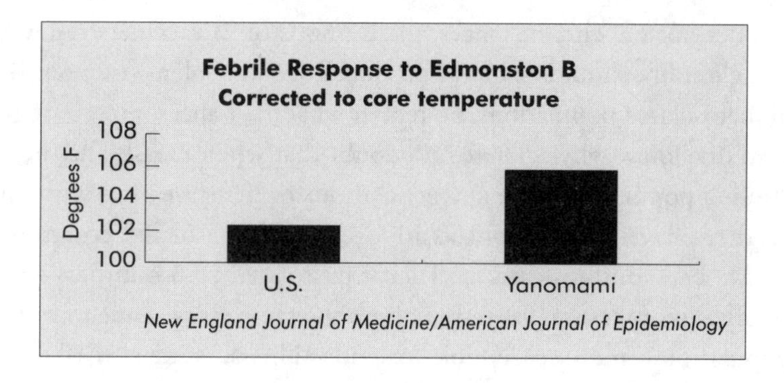

There was a much simpler explanation for the measles epidemic, however, and it was also implicit in the original account by Neel and Chagnon. According to them, the Yanomami first vaccinated at Ocamo "had definite rash" in strong reactions that began six days after vaccination and continued for more than ten days (January 29–February 8). Significantly, "a few reactions were indistinguishable from moderately severe measles."[91] There was no doubt, then, that a full measles rash and fevers first appeared among the Ocamo Yanomami within a week of the Indians' vaccination. Prior to the Yanomami's severe vaccine reactions, according to Neel's own chronology, no one had seen the disease's telltale lesions.

The Yanomami children vaccinated at Ocamo had their temperatures taken only once, but it was enough to show that they had the highest fevers ever recorded for the Edmonston B vaccine. Neel reported an average temperature of 39 degrees Celsius (102.2 degrees Fahrenheit) by the axillary method, which is obtained by placing a thermometer under the armpit. This method very greatly understates real or core temperature—by an average of 3.6 degrees Fahrenheit. He neglected to make the standard adjustment, to 105.8 degrees Fahrenheit, which other experts considered obligatory. Even by the most conservative measure, the temperatures for the Ocamo Yanomami were at least equal to those of wild measles. "I'd say that if somebody is 39 [102.2 F] axillary he's at least 40 [104] core temperature, and that is high," said Samuel Katz. "That's way off where it ordinarily occurred." In South American history, widespread death occurred whenever Indians were stricken with measles of this intensity.

Katz said he was perplexed by the Yanomami's response to the Edmonston vaccine. He speculated that either a concomitant exposure to another disease or malnutrition might be a factor in the high temperatures, though he had not observed anything like it among malnourished children in Africa. He felt that, under normal circumstances, the Edmonston B vaccine, even without gamma globulin, should not have produced such a violent reaction. But he noted that isolated populations, no matter what their ancestry, were at greater risk. No one knew why. "There's no doubt that when measles has been absent from a population for many generations or many years as with Greenland—there have been outbreaks in Greenland in isolated communities, there have been outbreaks in the Pacific in isolated communities—the disease is often more severe. Even though it's the same virus, somehow or other the population is more susceptible to serious illness." Katz also asked, "Was this group for whatever reason, whether they have some genetic locus in their immune system that's not normal or what, were they responding aberrantly to the vaccine?"[92]

While Neel touted the epidemic as a conclusive refutation of the long-held beliefs about Amerindian immune depression, IVIC's Dr. Inga Steinvorth-Goetz witnessed a tragic absence of resistance during the epidemic:

> Many Indians fled deep into the forests, but for most of the people living near the Orinoco it was already too late. They carried the disease germs with them, infecting others and dying by the score. They had absolutely no resistance. Only a very few even developed the characteristic rash, which is a sign of the skin's fight to throw off the disease. Mucous membranes became horribly inflamed, with extreme toxic vomiting and diarrhea. Many had hemorrhaging of the inner walls of the larynx. Many developed pneumonia and died from it. All too often even relatively mild cases failed to respond to penicillin.[93]

The Yanomami's collective agony in some ways paralleled the suffering of the leukemic child: visceral wounds and no external rash. The Yanomami's defenses collapsed. The inner walls of the larynx hemorrhaged. Like Goetz, missionaries also mentioned, with repeated consternation, the failure of antibiotics. At the village of Mahekoto-teri, measles still raged sixty days after its onset, in spite of massive doses of penicillin and vitamin C, and repeated visits from emergency medical teams.[94] This protracted infection differed decidedly from the course of measles in Eurasians but also recalled the ordeal of the leukemic child that died ninety days after vaccination.

"It is very interesting that the Edmonston B vaccine proved fatal to a leukemic patient," said Dr. Carlos Botto. "We found that when the Yanomami are exposed to a wide variety of diseases, including tuberculosis and hepatitis their antibody responses are much weaker than [those of] a Westerner exposed to the same disease. They undoubtedly have much lower disease resistance. At the same time, if they had been vaccinated in the middle of another epidemic, their defenses would have been weaker still."[95]

In recent years, French researchers in tropical medicine have joined CAICET in measuring Yanomami immune reactions. Together they have found that the Yanomami have decidedly weaker defenses against imported illnesses like hepatitis, a finding that most doctors would have predicted.[96] But they also showed less resistance to onchocerciasis, a parasitical infection similar to African river blindness, which has been present for many generations in the Parima Highlands.[97] These clinical studies corroborate the hundreds of anecdotal reports, starting with the first medical observers in the eighteenth century, that Amazon Indians have less resistance to their own tropical illnesses than do recently arrived Europeans.[98]

Chagnon and Neel described an effort to "get ahead" of the measles epidemic by vaccinating a ring around it.[99] As I have reconstructed it, the 1968 outbreak had a single trunk, starting at the Ocamo mission and moving up the Orinoco with the vaccinators.

Hundreds of Yanomami died in 1968 on the Ocamo River alone. At that time, over three thousand Yanomami lived on the Ocamo headwaters; today there are fewer than two hundred. "We used to visit the villages on the Upper Ocamo in the early 1960s," said the missionary Gary Dawson, whose family lived on the middle reaches of the Ocamo. "They had huge *shabonos* with over three hundred people in some of them. Now those *shabonos* are all abandoned. We have the last survivor of one of them at Kosh. When we went up the Ocamo with him, we found the site of his old village—hardly anything was left of the *shabono.* It made me want to sit down and cry. When we tell the government that three thousand Yanomami have died on the Upper Ocamo, they just laugh at us. 'That's impossible,' they say. 'The Yanomami are nomads, and they've just gone somewhere else.' Yes, they have gone somewhere else—they're dead. And I think it began with the measles epidemic."[100]

This pattern of decimation and dislocation was widespread, and I recorded many eyewitness accounts of mass funerals, where the Yanomami cremated their bones and drank the ashes of the dead. Vitalino Balthasar, the former Ocamo mission administrator, recalled, "They left after vaccination, but among them they already carried the disease. They left with fever, and they returned two months later. Few of them returned, because the majority died. And then, the month afterwards, they went in search of the cadavers and the bones, and they brought them in baskets to Ocamo. They brought them to burn them there in the village. I saw the burning. . . . *No basto nada la vacuna.*"[101]

"That vaccine wasn't worth anything."

The nurse Juan Gonzalez, who was in charge of the Ministry of Health's Mavaca malaria post, simply called it "the bad vaccine." He said, "Up to the present, they [the Yanomami] say they died from that vaccine. That's why even now some Yanomami don't want to be vaccinated. They're afraid because so many died." Gonzalez, who is a big bear of a man, seemed to shrink when he remembered the measles epidemic. He shook his head. "I don't know how to explain it myself, because we initially believed that that first vaccine had come to help us. Instead, it came to destroy us. At that time that was what we already suspected."[102]

Maria Wachtler, a nun who is a nurse, also witnessed the vaccinations at Mavaca. "I remember it very well," she said. "They vaccinated on a Thursday, and by Sunday everybody had measles."[103]

I was surprised to learn that she was right. Neel vaccinated at Mavaca on Thursday, February 15. By Sunday, February 18, there was a full-blown measles epidemic around the mission,[104] one documented in synchronous sound for the AEC.[105]

I spent more than a year searching for the outtakes—the unedited film and sound—from Neel's 1968 expedition. I knew, like anyone who has ever studied anthropology in college, that Timothy Asch and Napoleon Chagnon had made several films in 1968. One of them, *Yanomama: A Multidisciplinary Study,* actually featured radio conversations about the spreading course of the measles outbreak. As I learned more about the Yanomami's version of the epidemic, from my fieldwork in 1996, I became convinced that the AEC's version of events was upside down. The Yanomami said their villages were destroyed by the measles vaccine; the AEC film presented Neel and his fellow scientists as courageous heroes, sacrificing their research priorities to fight the epidemic—and rescuing everyone they vaccinated.

I devoted months to measles, reading several books and several hundred articles on early vaccination experiments. And the deeper I immersed myself in the subject, the stranger Neel's choices appeared.

But even after I had gained access to the Salesian mission records, with their terse witness of vaccine deaths at Ocamo, I still longed for the unedited reels that had rolled while measles spread. I contacted the distributors of the Asch-Chagnon films, but they could not help me. I bought old catalogs of the films—also useless. Then, in May 1997, I called Patricia Asch, Timothy Asch's widow, who told me that her husband had bequeathed all his papers and tapes to the National Archives in Washington. Finally, in September 1997, I traveled to Washington and spent two weeks going through a world of boxes and papers that no one else had ever examined: the Timothy Asch Collection. It was frustrating. Asch had documented every detail of his life, including how much he spent on film stock and airline tickets, but I found nothing new on the 1968 expedition.

Until Wednesday, September 10, 1997, when archivist Mark White and I discovered the sound recordings of the 1968 expedition. No one had ever heard them at the Smithsonian before. It was possible even Asch had forgotten what they contained. But his handwritten note on the dusty box, "Very Valuable," showed that he had an inkling of their historical importance. Sound Roll 3 was labeled "Mavaca Measles Epidemic."

The tapes are not a consistent or complete record, but I consider them priceless all the same. Neel complained, at one point, that Asch was determined to record every tiny detail of the expedition, including Neel's excretory functions.[106] Yet February 18, 1968, marked the first mention of measles on the AEC tapes—twenty-seven days after Marcel Roche reportedly diagnosed it at Ocamo.

The action began with Neel excitedly giving Asch directions on how to film the measles outbreak. "Let me tell you what we want to get—extremely severe morbilliform rash. Can you get this? Can you come in on this?"

"Yeah."

"Both eyes. He has the typical morbilliform rash on both cheeks. There is a little rash on his chest, but the rest of his body is essentially without rash. Now, this would be considered a moderately severe case of measles, such as we might expect to see in an adult in the United States or in Europe. Okay. Now, anything further you want to do to get a close-up and jump to the next major group. And the thing I want to make sure we record—you'll see, I hope not, but I'm afraid you're going to, some severe cases of measles. And you will also see some cases—get it back maybe—because you'll also see some cases that are, do not have as much rash as this man. So we're going to be able to document the whole gamut of measles in this group."

Chagnon observed, "His nose doesn't seem to be running as much as the ones at Ocamo."

"No," Neel agreed, "but the coryza [runny nose] comes early—that's the first thing; the runny eyes and the runny nose and the rash comes later."

"Is he an effective transmitter right now?" Chagnon asked.

"I'm afraid he's a very effective transmitter right now. The most, as far as we know, they're most contagious I believe when the eyes and nose are running. But he's still a very effective transmitter. Now you may have to clean me up. Remember, I'm no pediatrician. I'm no measles specialist."

"Don't worry," said Asch. "Don't worry."

"A lot of this we may have to put in later," said Neel.

"You don't have to worry about a thing, all right?" Asch repeated. "I know what I'm doing on my end."

Neel's excitement was understandable. Witnessing measles as it infected an aboriginal group was a once-in-a-lifetime event. It seems to have been the only time in recent history when scientists were present at such an outbreak. And Neel was on hand with a documentary filmmaker to capture the scenes, the exact progression of the disease, and the panic response, as they unfolded. It was as if the sound and video had been suddenly added to the sixteenth-century Spanish chronicles, at once opening a window on the past and resolving a twentieth-century genetic puzzle.

But who were these people with measles rash? There were two Yanomami groups at Mavaca on February 18—the Bisaasi-teri, who were vaccinated on February 15, and the Witokay-teri, a *shabono* on the Orinoco just upriver from the Ocamo mission. Neel vaccinated the Witokay-teri during the second round of inoculations around Ocamo, between February 4 and 8. Chagnon's guide, Rerebawa, from the village of Karohi-teri, on the Manaviche River, was also at Mavaca, with his hammock slung inside Chagnon's house.

The government nurse Juan Gonzalez recalled, "Rerebawa was going around with Chagnon and was vaccinated by him. He'd been recently vaccinated but not properly immunized. He caught the measles in Ocamo and came back infected—fever, fever, fever—to Mavaca, where it spread to all the Yanomami who came to visit, and they all got it."[107] Rerebawa's role illustrated how the AEC's guides from distant *shabonos* accelerated the disease's spread as the scientists shuttled between infected villages.

The tapes confirmed Gonzalez's recollection that Rerebawa returned to Mavaca with measles. Neel was keenly interested in filming Rerebawa's unfolding symptoms, particularly the first signs of measles, mouth sores called Koplik's spots. He directed Chagnon and Asch.

"The man in your front hall [apparently Rerebawa]. Now is the time, before the rash, when he might show the Koplik's spots of measles. Go back and check that."

But Chagnon, who was normally a dutiful subordinate, balked.

That was because while Neel had been giving Asch camera directions, Chagnon had a parallel conversation, in Spanish, with a man named Rousseau,* a technician at the Venezuelan Institute of Scientific Investigation, who was in charge of radio communications with Caracas and Puerto Ayacucho.

"Let me explain to you again, Rousseau, that we need two doctors."

"Yes. They already know that. At one o'clock, I'll tell them again about Bicillina [a form of penicillin] because of the effects of the vaccine," said Rousseau. "Now, if because of the effects of the vaccine we get outbreaks of measles . . ."

"Well, the vaccine gives an effect almost equal to it," Chagnon said.

"It's the same, isn't it?" Rousseau rejoined. "If there are outbreaks, right, we'll see. If the doctor is here, we can either take the doctor to them or bring the sick."

So when Neel told Chagnon to go and film Rerebawa's Koplik's spots, Chagnon had another, more immediate concern. The enormity of what he and Rousseau had just said was percolating through, along with a sudden recognition of danger: Rousseau was about to tell doctors in Caracas and the Catholic bishop of Puerto Ayachucho that the measles epidemic at Mavaca and Ocamo was a vaccine reaction.

"Oh, could you explain a minute," Chagnon asked Neel. "If anybody breaks out with these runny eyes and rash. . . . Does . . . does the vaccine give runny eyes?

"The vaccination occasionally. . . ."

"I've just explained to him that a few people out the vaccinated group will get a clinical case."

"Right," Neel said.

"But he's trying to interpret all of them to mean that it's a reaction to the vaccination, which I don't think is a wise thing to do. And I think that even . . ."

"Right," another expedition member interjected.

"I hope that's right," Neel said. "But, uh, we . . ."

It is impossible not to sympathize with Neel at this point. He had downshifted, in a matter of a minute, from a film director on an adrenaline high, to someone who sounds as if he was at the end of his rope. "I hope that's

*Pronounced Rusó on the tapes. Spelling uncertain.

right." He stressed the word "hope," not the reassurance Chagnon was looking for. And Neel's admission that the vaccine was "occasionally" giving the symptoms typical of highly infectious carriers—running noses and watery eyes—was equally worrisome. With measles, a few clinical cases were more than enough to infect a whole population. But Neel's vulnerability lasted only an instant.

"The vaccination with gamma globulin gives sometimes a little fever, a little runny eyes, but if he sees somebody with a real rash . . ."

"Get him out," Chagnon suggested.

"Well, get him out where?"

"Out here."

"From where?"

"From, away from the group," Chagnon said with obvious exasperation. "He's going to go every day to visit Patariwe's group [a Bisaasi-teri *shabono* just across the river where Rerebawa had in-laws]."

"Okay. Okay. By that time, by the time he sees somebody like that, that person will have contaminated the entire group. That is how contagious measles is."

"Well, the point is. . . ."[108]

Neel just didn't get the point. Chagnon was trying to nudge his mentor toward taking quarantine precautions. The proposed quarantine included Chagnon's own guide and other vaccinees. In fact, everybody with measles at this point was at a vaccination center, either Ocamo or Mavaca. Neel seemed to take it as a matter of course that they were all infected and contagious. Rousseau was "trying to interpret all of them to mean that it's a reaction to the vaccination"—an interpretation that would make no sense unless they had all been vaccinated. Chagnon wanted to send a doctor back to the Ocamo mission "in the event that those three visitors from Iyewei-teri [the mission group] that went to Shubariwa-teri have in fact infected the Shubariwa-teri with measles."[109]

Unfortunately, a doctor was not sent in time, and the Mavaca mission chronicle reported that all of the Shubariwa-teri died, except for a few rescued by Protestant nurses.[110]

By pinpointing the carriers who spread measles on the Ocamo River, Chagnon's recorded statements substantiated the eyewitness testimony of Vitalino Balthasar—who told how recently vaccinated Yanomami from the Iyewei-teri took the disease upriver. But the sound record also underscored how the AEC's distribution of steel goods contributed to the vectors of infection. At Ocamo, the Iyewe-teri sold their blood to Neel's researchers. They obtained knives, machetes, and other goods and then journeyed to barter their acquisitions with the Shubariwa-teri, who were wiped out by the additional gift of measles.

Other communities shadowed the expedition like pilot fish. After being vaccinated, the Witokay-teri, who lived near Ocamo, trekked to the AEC's Mavaca base camp, hoping to pick up steel scraps. Chagnon warned that the Witokay-teri were now contagious. He was also telling all the Yanomami in the area, including the Bisaasi-teri, that they could not accompany the expedition farther inland, because they might infect remote villages; that did not stop them, as I will show in the next chapter. Meanwhile, the impasse at the radio continued as Rousseau prepared to inform Venezuelan authorities about the expanding course of the vaccine reactions.

Charles Brewer broke in decisively. "I am going to tell him [Rousseau] that the vaccination doesn't give pimples. If [you see] anyone who has pimples, bring him here." Brewer then switched to Spanish: *La vacuna esta supuesta a no dar . . .*[111]

Brewer's solution was drastic and simple—the vaccine did not produce any rash. (Brewer's translation of "pimples" was technically correct: "measles rash" is really an eruption of "raised lesions,"[112] not just discoloration.) This message was obviously untrue, but it was the only story available to them. And it was only one of many miscommunications occurring around the radio transmitter. Neel misled Brewer when he claimed the vaccine with gamma globulin gave only "a little fever, a little runny nose." In any case, they had started vaccinating *without* gamma globulin (which Brewer probably did not know either). Chagnon had admitted, in Spanish, that the vaccine was "almost the same" as measles. But when Chagnon translated his conversation to Neel, he sanitized it, saying, "I told him that a few members of the vaccinated group will get a clinical case."[113] Rousseau insisted that the vaccine reaction was identical to measles, but he was uncertain that the reactions could spread beyond the vaccinees to become a true epidemic.[114]

Chagnon showed genuine concern for the Yanomami. At the same time, he moved quickly toward a cover-up. *But he's trying to interpret all of them to mean that it's a reaction to the vaccination, which I don't think is a wise thing to do. And I think that even . . .* Chagnon was quickly supported by another expedition member, who added: *Right.* Everyone looked to Neel to deny Rousseau's interpretation, but this was as foreign to Neel as aborting his research because of the epidemic. Neel finally conceded that measles carriers could be brought back to the Mavaca mission for medical care. But he quickly overruled any delayed departure.

Neel: We may be moving out one o'clock to get to Platanal.
Brewer: The river is very bad right now.
Neel: We don't have much choice. Maybe under these circumstances we can draw on the boats of Malariologia. Instead of going up in two heavy bongos, we can go up in four or five lighter boats with plenty of people to get us across the sandbars. Right?[115]

Neel's solution to the Yanomami measles crisis was to call Caracas and move on within half an hour, taking with him the only doctors in the region and one of the two government nurses—Danny Shaylor—whom he needed as a translator. Neel also wanted to borrow the boats that were the remaining nurse's only means of transportation, and recruit a group of Yanomami workers to help the expedition get upriver. Fortunately, Chagnon dissuaded Neel from borrowing the boats and hiring additional guides. But the American medical team deserted the desperately sick Yanomami, who waited for over a week until the next plane from Caracas arrived with additional doctors.[116]

When Neel explained why the Yanomami at Ocamo were vaccinated without gamma globulin, he blamed Chagnon (and Roche).[117] Yet it was Neel's decision to shove off that sealed the fate of many villages. A single doctor or nurse could save dozens of lives—as was demonstrated by the death of all the Shubariwa-teri except those who were treated by evangelical nurses.[118]

Why was Neel so intent on going upriver when the sick were all around him?

As Neel explained it to Brewer, who had taken the radio phone from Rousseau, he wanted to get upriver to vaccinate the Yanomami at Platanal, the Mahekoto-teri, "and maybe catch it on the Upper Orinoco, and then the Patanowa-teri, who are the principal inland village we might get to." In reality, only 250 vaccines were left, according to what Chagnon told Brewer,[119] and the medical team could have administered only a portion of these because there was much less gamma globulin. So Neel could in no way have vaccinated the Mahekoto-teri (numbering 125), the Hasupuwe-teri on the Upper Orinoco (over 300), and then the Patanowa-teri (220). If Neel had followed vaccination norms at this point, he would have stayed with the infected groups at Mavaca and Ocamo, at least until help arrived from Caracas. And he would have spent his limited gamma globulin stores to reduce the severity of the attacks in his immediate vicinity. He should not have continued vaccinating, especially after having seen the violent responses to the Edmonston virus.

However it was viewed, Neel's explanation that he was going upriver to vaccinate the Mahekoto-teri was misleading. The scientists traveled from Mavaca to Platanal on February 18, and spent the night.[120] The following morning, Asch recorded a brief conversation between Chagnon, Brewer, and the local missionary, Father Sanchez, in which the expedition members said nothing about the measles outbreak.[121] Why? I do not know for sure. But Neel could hardly tell Father Sanchez the same story he had just told the Venezuelan bishop over the radio at Mavaca—that he was going upriver to vaccinate the groups around Platanal—because Neel was *at Platanal.* And that is not what he did. Neel and Chagnon have always maintained that they did not vaccinate any of the Yanomami at Platanal.[122] And they could not tell the priest they were saving their limited store of gamma globulin for the Patanowa-teri—a group that was extremely remote from the danger of contagion—because that made little sense (except, as I will show, from the perspective of a planned film). Sanchez would not have understood why Neel's expedition, with its two doctors and a nurse and 250 doses of vaccine, passed through Platanal, invited all of the Mahekoto-teri to a filming event, but failed to vaccinate them, as it had promised Venezuelan authorities. It is difficult to understand that decision today, knowing that 25 percent of the Mahekoto-teri, about thirty individuals, died of measles.[123]

Chagnon and Neel later claimed that either one or two Brazilian workers brought measles to the Mahekoto-teri.[124] There were two Brazilian men who assisted Father Sanchez. But both Brazilians can be seen, in a single freeze-frame in *A Multidisciplinary Study,* standing right next to the AEC boats at the Platanal mission on February 19, two weeks before, Chagnon later claimed, one of them came down with measles. Although Neel had 250 vaccines on board, he made no attempt to inoculate the two men. Obviously, if Neel or Chagnon believed that someone from the Brazilians' party had brought measles to the Yanomami at Ocamo, their top priority should have been vaccinating everyone else in the group, especially these men, who were surrounded by Yanomami. But the notion that the Brazilians were the source of the outbreak had not yet occurred to anyone.

Of course, it is now easy to condemn Neel and Chagnon with perfect hindsight and peripheral vision. The AEC expedition had been in the field for almost a month, and, as I will show in the next chapter, the scientists were exhausted, sick, and increasingly disgruntled. Most of them were in the jungle for the first time, and each had a demanding research agenda. Their scientific hopes were all pinned on reaching the remote village of Patanowa-teri. It was hard to turn back to care for sick Indians, especially when the scientists, like the missionaries, were still not sure what was going on.

On February 18, Brewer, Chagnon, and Neel were just recognizing that they had a measles epidemic on their hands. Up until then, they and the missionaries alike had believed that the extremely high fevers seen at Ocamo were reactions to the vaccine. Brewer told the Salesian bishop by radio phone, "The situation is getting worse all the time." But Brewer was still hesitant about how to describe "this epidemic, or this attack, of measles." The bishop was shocked to hear the news. He said, in summation, "Our subject this morning has been problems that had not occurred there before. . . ."[125]

Chagnon whispered to Brewer, "Now the situation is more complicated, more critical. We need as soon as possible . . ."

"More vaccine?"

"No. Not more vaccine. We have enough vaccine. But it's certain that now we have measles at Mavaca and Ocamo and I don't know where else it is—and I don't know when it arrived."[126]

Chagnon's admission—that he didn't know where measles had come from or when—contradicted his later accounts. Moreover, if a Brazilian boy had really been diagnosed with measles four weeks earlier, then Rousseau's conclusion—that the vaccine was responsible for the outbreak—would have been absurd. And Chagnon, instead of turning to Neel to find out whether the vaccine virus could become contagious, would have answered, "Look, Rousseau, your own colleague from IVIC, Marcel Roche, diagnosed measles when we first entered the field at Ocamo."

In the Mavaca mission chronicle, Father José Berno offered three rumors of how the epidemic started. One was that a scientific expedition—Inga Steinvorth-Goetz's—had brought measles. Another possibility was that a man from Karohi-teri who had been to the Padamo River returned with measles. (He was referring to Rerebawa, from Karohi-teri, who traveled with the AEC to the Padamo, where measles also followed vaccination.)[127] The most original interpretation, voiced by the Yanomami, was that new metal objects were responsible—black magic transferred by the recent arrival of zinc roofing material.[128]

Neel's history of the epidemic was equally fantastic. Although he claimed the Brazilians brought measles to the Yanomami, he included all kinds of details—from the absence of clinical symptoms in the Brazilians to the astonishingly high vaccine response of the Yanomami—that undercut his argument. Looked at in the cold light of day, Neel gave the world the Immaculate Epidemic—a contagion brought by a carrier without symptoms from a place without measles and spread by adults who were almost certainly immune. One of the many errors in Neel's account was his claim that he could distinguish the vaccine virus from "wild measles" by a comparison of antibody titers.[129] Mark Papania, a measles specialist at the Centers for Disease Control in Atlanta, said, "The wild measles virus and the Edmonston B strain are so similar the [antibody] responses are the same. There's no way of looking at antibody and knowing to which one they were exposed."[130]

Strange, too, that the *American Journal of Epidemiology* published such an improbable story. After all, the same journal had run articles about the dangers of inoculating Amerindians with the Edmonston B strain. But collegiality was a powerful bond, and Amazon adventures have never been subjected to the same scrutiny as events closer to home. In this way, the Yanomami measles epidemic went from being the worst tragedy in the tribe's history to the most convincing proof of Neel's theory about native disease resistance. It became another of Humboldt's "fairy visions" haunting the Orinoco.

What made Neel eventually hit on the Brazilian adolescent as the origin of the outbreak? His autobiography, *Physician to the Gene Pool,* presented another version of the epidemic's origin. There Neel finally admitted that measles was not recognized "immediately, in its preeruptive stage," in the Brazilian boy.[131] Originally, Neel had written that "the characteristic morbilliform rash never developed."[132] The new rash, added twenty-four years after the original, might be dismissed as another contradictory element of the saga—the world's longest delayed outbreak of measles rash in an infected individual. What Neel omitted from his latest rendering, however, was that the boy had already been vaccinated when Marcel Roche first saw him.[133] So the Brazilian adolescent developed a rash after being vaccinated—the normal response to the Edmonston B virus. This told Neel that the boy was a susceptible, and a convenient, scapegoat.

Only Neel was famous then, but the achievements of the 1968 expedition would make several of its members leaders in their fields. For a few frightening moments, however, they all saw their bright prospects threatened. If Rousseau had delivered his message—*Por los efectos de la vacuna ahora si vienen brotes de sarampión*[134]—they would have been blacklisted for life as the men who had caused the vaccine epidemic. They would not have become icons of Amazon romanticism—Asch the world's most highly regarded ethnographic filmmaker, Chagnon the heroic anthropologist, and Brewer the daring explorer. In fact, they would never have gotten scientific permits to go anywhere in South America again.

Their problem was that Neel had chosen a relatively obsolete vaccine and his scientists had administered it in a way contrary to safety norms at the spot where the epidemic ignited. The body of Roberto Balthasar would have been hard to explain, too. And the team would also have had to contend with the ghost of G. S. Wilson, the measles researcher who had warned, back in 1961, that the Edmonston B vaccine might prove worse than wild measles.[135]

Could the Edmonston B virus have become an epidemic? That was the question Chagnon and Rousseau discussed, and I put it to Mark Papania of the Centers for Disease Control. "Sure it's possible and, especially talking historically, the thing that would support your hypothesis is that this was obviously not a very attenuated vaccine." He said, "It had a much higher side effect than currently used vaccine. It is not used anywhere in the world. All the vaccines now are further attenuated."[136]

I also asked Papania how he would interpret an outbreak where no signs of measles rash appeared until after vaccination. He said, "It's fascinating. It's really unfortunate. It's hard to say. I think it's possible that it relates to the actual vaccination. It's also possible that some of the vaccinators were infected. It's also possible there was a concomitant exposure. It may be that the greatest possibility is in the vaccine."[137]

Papania was only speculating, of course, and not referring specifically to the Yanomami case, of which he had no particulars. Samuel Katz has emphatically denied the possibility of transmission: "Vaccine virus has never been transmitted to susceptible contacts and cannot cause measles even in intimate contacts."[138] When I first spoke to Papania he looked over the literature on transmissibility of the Edmonston B and found that infection had never been documented, but that sensitive studies had not been performed. "I guess we didn't look very hard," he said.[139]

In any case, it is not clear that tests done on populations in the U.S. or any other part of the world are equally valid for the Yanomami. In 1971, James Neel sent an experimental protocol to Dow Chemical for testing a triple vaccine—measles, German measles, and mumps—on the Yanomami, but the director of biological research at Dow Chemical rejected the study, noting that "safety data accumulated in the U.S. population may not be fully applicable to the Yanomami. The cultural and epidemiologic differences are obviously very great. . . . It is obviously best that the proposed triple vaccine study in the Yanomami be abandoned."[140]

Many experts are understandably worried about the implications for world vaccination programs of admitting any possibility of a vaccine epidemic[141]—the very possibility that Rousseau, Chagnon, and Neel discussed at their radio conference. I am also concerned about this, and it is unclear whether the Edmonston B became transmissible or not. That was the question that perplexed the expedition. The chaos and deaths that followed vaccination at Ocamo, Mavaca, and elsewhere can be explained in terms of the extraordinarily high vaccine reactions, coupled with simultaneous exposure to malaria and bronchopneumonia. All of these apparently worked together to create a fatal response. The Edmonston B had proved safe for many millions of children in the United States and elsewhere, but it was a particularly poor choice for the Yanomami, not only because of its high rate of fever, but because of its high incidence of rash. The Yanomami believed that some new form of black magic must have brought on the disease and, at the first sign of measles rash, they panicked and fled from their homes into the forests, away from food, family support, and further medical attention. It may be simplistic to speak of only of a "measles epidemic." At first, this was a measles vaccine–bronchopneumonia-malaria epidemic that triggered the Yanomami's well-known panic response, adding a strong cultural component to the incredibly complex epidemiological one.[142] Beyond the missions, little is known about the cause of death.[143]

However, the fact that Neel was able to locate measles antibodies in individuals far beyond the vaccination circle shows that someone somehow started to transmit measles.[144] I do not know who that person was. The simplest explanation may be Rousseau's—that one of the vaccinees, in spite of all previous expectations, became infectious. Chagnon's guide Rerebawa was one candidate. Neel described him as "a highly effective transmitter right now."[145] Father Berno and Juan González both thought Rerebawa spread measles, but Berno believed Rerebawa had picked it up on the Padamo River, where he had traveled with Chagnon—a real possibility, since that is the frontier of the Yanomami world, with many more outside contacts and chances for contagion. I am happy to leave this epidemiological puzzle to the medical historians.

None of these issues can mitigate the AEC's behavior during the epidemic. In January 1968 the Venezuelan government had begun a national vaccination campaign with the Schwarz strain, given in three diluted doses, on the advice from the Centers for Disease Control in Atlanta. It was one of the safest vaccination programs in the world. According to the director of Venezuela's vaccination department, Dr. Adelfa Betancourt, Neel vaccinated without her department's permission. She said, "Now, if it occurred to somone to vaccinate the Yanomami with another strain without permission or accompaniment from the Ministry of Health, we have no knowledge of that."[146]

Neel had no reason to think the Edmonston B could become transmissible. The outbreak took him by surprise. Still, he wanted to collect data even in the midst of disaster. I believe Neel's desires were divided—like the Yanomami community at Ocamo, half of which was vaccinated and half of which was not. At times, Neel genuinely wanted to help the Yanomami and sincerely thought he was doing so. An outbreak of measles at two Brazilian missions on the extreme eastern and southern periphery of Yanomami territory had occurred in September 1967, and, though the outbreaks were contained, the Yanomami's demonstrated vulnerability was probably one of Neel's motives in wanting to vaccinate the Venezuelan Yanomami.[147] Clearly, he and his doctors distributed medicine and cared for some of the sick they encountered. But his choice of vaccine suggested that he wanted new data, and his impatience with Venezuelan authorities meant that he had no backup from government doctors when crisis occurred. Moreover, Neel barely slowed his pace of blood-collecting or filming, both of which required massive payments of trade goods, a reckless policy during an epidemic.

Over the decades, measles had broken out in Yanomami villages far from the Orinoco on several occasions, but the cultural barriers between villages stopped the spread of disease. (See Appendix: Mortality At Yanomami Villages.) One reason the 1968 epidemic was unique was because Neel's expedition attracted whole villages to blood-collecting centers and arranged new alliances, which facilitated the spread of disease. The scientists kept moving on and the epidemic moved on with them.

Chapter 6

Filming the Feast

And I have to do so many work about the feast and about the blood.—*Charles Brewer Carías*[1]

The first person in Yanomamiland who was killed in the line of photographic fire was reportedly a Protestant missionary. He navigated up the Ocamo River in 1951, accompanied by his ten-year-old son. It was part picnic, part tour of duty. When they reached a remote Yanomami village, the Indians set to stealing the missionaries' goods. Unconcerned, the missionary pulled out his camera, attached a flash, and took his last picture. The explosion of white light was interpreted as an attack, and one of the Indians put a six-foot bamboo arrow through the man. The missionary's son ran off into the forest, where he lost his way. A rescue team of rubber tappers found the blond boy days later, by which time he had also temporarily lost his mind.[2]

The Yanomami believe cameras kill. This is not an aesthetic judgment, like Susan Sontag's in her book *On Photography.* It is a technical evaluation, having to do with emanations that the Yanomami perceive shooting from photographic devices. For them, cameras are like sci-fi ray guns, whose energy envelops and steals its targets' spiritual essence, *noreshi.*[3] What remains after this photographic kidnapping has occurred are people devoid of vital force—empty shells, body husks. Cameras are soul cannibals in the Yanomami's version of *Body Snatchers.*

Yet the Yanomami, like many other impoverished artists, have reluctantly embraced Hollywood. They are "the most filmed non-Western, non-industrialized society in the world."[4] As early as 1978, there was a Paris film festival devoted exclusively to the Yanomami[5]—who were either masters of the spirit world, according to French filmmakers, or making war, according to American artists. As documentary crews, news teams, and ethnographic filmmakers competed for authentic footage, cinema verité became the principal source of income for many Yanomami along the Orinoco.[6] These acculturated Indians became professionals at creating sets and enacting violent scenes for the camera.[7] The cost of a new *shabono* or a new war rose over time, driven upward relentlessly by demand. Eventually, a few Yanomami acquired enough skills to begin filming the filmmakers, which has created a new genre—Yanomami noir.[8]

Napoleon Chagnon was a pioneer in this frontier of film, and he was perhaps inevitably drawn into its controversies. He started taking 35mm photos and found that the Yanomami "are not fond of being photographed." Women and infants were particularly fearful. Fortunately, he was able to enlist young helpers who began "dragging various people out of their hammocks to have me take their photographs. . . . They assumed the obligation of assuring that nobody escaped, which resulted in a 100 percent photographic sample of the village in a matter of hours." As the anthropologists' photographic interests expanded, so did the opposition: "I have been chased around the village on a number of occasions by irate people wielding clubs and firebrands, people who were very upset because I was attempting to photograph specific events—particularly cremations."[9]

Chagnon saw himself as recording "specific events." The Yanomami recall his staging them. For example, a man named Waloiwa was Chagnon's first guide in the field, taking the anthropologist inland from Bisaasi-teri in early 1965 on his initial foray into the inland communities. "When Shaki arrived, I was visiting here at Bisaasi-teri at the time," he told me at the village of Guarapana, located on the Mavaca River not far from the Orinoco confluence. "He took me as his guide on this trail here to my village, and I walked with him. Momanipue was the village. Then Chagnon told me to dress up in my ritual paint, the way my ancestors did, and he asked me to get the whole village to dress up like that. Since he promised us axes, machetes, and knives, we did it. Chagnon really took a lot of pictures; he even had me shoot a branch way up on a tree."[10]

Waloiwa was not blaming Chagnon. But his perspective was fairly common among Yanomami at villages where Chagnon had worked. Eventually, these films seriously complicated Chagnon's relations with the Yanomami. Some Yanomami threatened to kill him over his filming.[11] "You have made films . . . with many bleeding Yanomami," wrote Cesar Dimanawa, a leader of the Yanomami trade cooperative, United Shabonos of the Upper Orinoco.[12] Diminawa told me that he was conveying a message from the men of Mishimishimabowei-teri who had been paid by Chagnon to fight for a film.[13]

"I think it is highly unlikely Chagnon paid them steel goods to fight," said Patricia Asch, an ethnographic filmmaker at the University of Canberra, Australia, who is Timothy Asch's widow. "But they may have thought they'd get paid more steel goods if they acted more violently."[14]

Similar allegations surrounded *Dead Birds,* a celebrated documentary about New Guinea's Dani tribesmen, filmed in 1962 by Robert Gardner and Karl Heider of Harvard's Peabody Museum. "We were accused of starting wars so that we could film the Dani in battle," said the anthropologist Heider, who has steadfastly denied the Dani's account. But Heider now admits, "Despite what we said, our very presence at the battles gave a kind of implicit approval to these events."[15]

Dead Birds was Chagnon's model, and he took his first footage to Harvard's Gardner for advice.[16] But whereas Gardner had filmed magnificent battles on open fields, with gory killings carried out by hundreds of warriors in full ceremonial regalia, Yanomami warfare proved far more elusive. No scientist has ever witnessed a Yanomami war death despite two hundred years of cumulative observations. This was frustrating for Chagnon. Doctors at the University of Michigan who did not consider his anthropological studies to be real science constantly taunted him.[17] He had found the Fierce People but no proof they actually fought.

Working on his own, with a 16mm Bolex camera, Chagnon filmed a raid in June 1965. (This sequence, Chagnon explained, was spliced into *Yanomamo: A Multidisciplinary Study.*)[18] Although Chagnon never identified the raiders, the Yanomami I interviewed described them as coming from a "mixed-up village," with members of three groups who had had formerly lived in separate *shabonos.*[19] Shortly after Chagnon took up residence at Bisaasi-teri, they all came together to build a giant new village, protected by a palisade.[20]

Although Chagnon mentioned the new *shabono,* and its protective wall, he did not mention his role in building the round house. In the film, however, six big pots, bright new, without smoke stains, graced the center of the structure. These were among Chagnon's contributions to an inaugural feast celebrating the triple alliance.[21] He also supplied the warriors with red cloth.[22] One of the evangelical missionaries in the area, Joe Dawson, recalled, "Chagnon invited me and my sons to come and visit the new *shabono* he'd built for the Bisaasi-teri. He said he built it because he wanted them to hold their dances with other groups, so he could film them."[23] This new *shabono* was the largest near the main course of the Orinoco, with 202 members[24]—three times the average village size.[25] It was also the only village Chagnon studied that had a surplus of females (106 to 96),[26] a surplus achieved by exporting steel to the interior villages. According to Chagnon's theory of war, villages without enough females must fight to obtain them.[27] Yet Bisaasi-teri, in spite of its surplus of women, launched, after Chagnon's arrival, one of the most intense wars in Yanomami history against Patanowa-teri, a village with a sharp deficit of women.[28] This war ultimately produced more than twenty-five raids against Patanowa-teri, as well many counterraids, without the capture of a single woman by any war party on either side.[29]

Chagnon's first raid footage was from an attack in which he played an active role.[30] Through a frame-by-frame analysis, I have been able to identify the raiders and to match the action with one of the most debated episodes of *The Fierce People,* a war whose ostensible purpose was to avenge the recently killed headman of Monou-teri.

> . . . the [Monou-teri] raiders lined up, shouted in the direction of Patanowa-teri, heard the echo come back, and left the village to collect their provisions and hammocks. I allowed them to talk me into taking the entire raiding party up the Mavaca River in my canoe. There, they could find high ground and reach the Patanowa-teri without having to cross the numerous swamps that lay between the two villages. There were only ten men in the raiding party, the smallest the war party can get and still have maximum effectiveness. As we traveled up the river, the younger men

began complaining. One had sore feet, and two or three others claimed to have malaria. They wanted to turn back because I had forgotten to bring my malaria pills with me as I had promised. Hukoshikuwa, a brother of the slain headman, silenced their complaints with an angry lecture on cowardice. . . . They unloaded their seemingly enormous supply of plantains. . . .[31]

For years, anthropologists have speculated about why Chagnon interjected himself into this war. "This assistance gave the Monou-teri a significant advantage. Because of the impassable swamps, the intended victims would not be taking the precautions of a village expecting raiders."[32] Even more puzzling, Why would the raiders need incentives, like transportation and malaria pills? Until now, no one has realized that this raid was also Chagnon's film initiation. The raid sequence spliced into *A Multidisciplinary Study* consisted of a ten-man war party from Monou-teri, led by Hukoshikuwa, whose black body paint exactly matched a 35mm photo included in *The Fierce People.*[33] Most of the warriors wore new loincloths fashioned from Chagnon's red cloth.

Despite Chagnon's encouragement, the Monou-teri came home without ever firing an arrow in anger against Patanowa-teri. They were not so terribly overwrought about the death of their leader, after all. It appeared they were just doing Chagnon a favor: "It was then that I discovered they were dallying, trying to be polite to me," he wrote.[34] In the film clip, as the young men lined up for their "attack," they kept looking right at the camera, embarrassed and curious and seemingly waiting for the cue that they were free to go. Alfredo Aherowe, who comes from one of Bisaasi-teri's allied villages, Mahekoto-teri, said, "This film is not true. They are just making a show. . . . Shaki [Chagnon] tells them, 'Do this. Do that.' And they do it."[35] Like César Dimanawa, Aherowe is an elected leader who is one of Chagnon's political opponents; his remarks have to be understood in the context of that polemic.

By the same token, however, Chagnon's political problems in Yanomamiland are inextricably wound up with his filmmaking. Aherowe's opposition to Chagnon is based in large part on his childhood experience that the anthropologist's filming created havoc at his village, Mahekoto-teri, and other communities.[36] "There is no way I will allow this kind of filming to go on again," Aherowe said.[37]

The "mixed-up" alliance that Chagnon filmed and help form in 1965 briefly made his friend Kaobawa the head of the largest and most militarily potent village on the Orinoco. Within weeks of Chagnon's departure, the triple alliance splintered; half the Bisaasi-teri moved to another *shabono.*[38]

What endured was a formula for Yanomami filmmaking. The way to make a successful Yanomami movie was to build a new *shabono,* sponsor a feast, create a new military alliance, and record a raid by the newly created power. A frequent sequel to this stock sequence was an epidemic, which might kill a quarter or more of the Yanomami actors.

In January 1968, Timothy Asch and Napoleon Chagnon arrived in Yanomamiland, intending to film a feast between two Yanomami villages. They wanted to illustrate feasting as a dangerous political-military event, where participants risked ambush for the sake of new alliances to launch wars. Prior to this, aboriginal feasting had been portrayed as a celebration of crop fertility.[39]

By shifting from the magical to the Machiavellian, Asch and Chagnon broke new ground in anthropology.[40] But the strategic analysis Chagnon and Asch applied to Yanomami feasting also applied to their own drama. By the late 1960s, cameras had changed the way football was played, presidential campaigns were run, and the war in Vietnam was fought. It also changed fieldwork on the Orinoco.

Asch and Chagnon flew in on a Venezuelan C-34 transport, a flying boxcar. They needed space for all the expeditionary equipment and heavy trade goods they brought with them. Asch wrote, "In the face of a coming rain storm, Napoleon hurriedly separated the mountain of belongings the airplane had disgorged, and I loaded them on a huge hand truck. . . . Though I had worked in tropical climates before, this was really the worst fatigue I had ever experienced. Napoleon is such an eager athlete, I worked hard to show that I could carry my share of the load."[41] Asch nearly collapsed from heat prostration.

The next day, they traveled upriver to Mavaca, where Chagnon was reunited with his old friend Kaobawa. Though they were happy to see each other again, Asch was frightened at the change he saw come over Chagnon. It was not just his loincloth and body paint. Chagnon's personality seemed to morph, making Asch feel he was alone in the jungle with aliens.[42]

The Yanomami were also capable of instant changes. On seeing Chagnon's boatload of trade goods, Kaobawa immediately announced that a feast was scheduled at an allied village, Reyabobowei-teri. The coincidence took Asch by surprise—"never dreaming the opportunity to film a feast would come so soon. . . . Starting off the next day for the feast, the Yanomamo were put off by all the camera equipment they had to carry, with the promise of some trade goods they were willing."[43] In fact, fifteen Yanomami accompanied the filmmakers,[44] and they left Mavaca, without quarantine precautions, at the height of an epidemic of bronchopneumonia that had already taken five lives,[45] to visit a remote interior village that was extremely vulnerable to outside disease. The very same day Asch and Chagnon left Mavaca,[46] there were more than thirty Yanomami with temperatures above 104 degrees Fahrenheit, "seven of them very grave and in danger of death."[47] This critical picture does not take into account the malaria problem, or Chagnon's later claim that Marcel Roche had diagnosed measles forty-eight hours previously at a mission twenty miles downstream.[48]

According to Chagnon, he and his fellow scientists "ended up spending most of our research time trying to vaccinate a 'ring' around the [measles] epidemic."[49] In fact, Chagnon and Asch were wholly engrossed in their filmmaking itinerary. They did not vaccinate anyone at Mavaca, or at the interior villages of Momaribowei-teri or Reyabobowei-teri. Their trip, which lasted "over six days," pushed Asch to the limit of his endurance. Several times the Yanomami stole away, leaving the filmmaker completely lost in the forest. He found that "the Yanomamo had some devilish streaks,"[50] but he did not care for Chagnon's fieldwork methods either. Their final approach to the Reyabowei-teri *shabono* resembled a stealth attack. "We snuck up on the village and only at the last moment announced, by shouting, that we were nearby. Napoleon and the headman [Kaobawa] who had guided us, knelt at the entrance to the village, with his shotgun over his knees. . . . Napoleon looked up at me and said, 'It's alright, I only have to shoot one of them.' . . . That really terrified me."[51]

There was no feast at Reyabobowei-teri.

Asch made a short film, *Kaobawa Trades with the Reyabobowei-teri*. The actual trading, however, occurred "without much enthusiasm because both hosts and guests are annoyed that the feast could not be held for lack of meat."[52] Kaobawa traded the AEC's steel presents for arrows. (The film showed only the traditional Yanomami gifts Kaobawa received in return.) By then, however, Asch was exhausted. On his first night at the village, he just fell down on the ground inside his limp hammock, too tired to tie it up. Nearby, Chagnon adopted the posture of a visiting chief, hands over his face.

Chagnon did not move to help Asch up. Instead, he shout-whispered, "For God's sake, get that hammock strung up! You're embarrassing the whole expedition."[53]

There were other embarrassments. In a fit of enthusiasm, Asch decided to go naked, like the Yanomami. Or so he thought. But the Yanomami women hid themselves in the bushes and the men began screaming at him.

"You're naked!"

"What's wrong with that? You people run around like that. Why can't I?"

"Yes," Chagnon interpreted. "But they say your penis is dangling down."[54]

Once Asch tied up his penis to a Yanomami waist string, he was socially acceptable. On returning to the Mavaca mission, Chagnon stayed at their base camp, while Asch, a young missionary, Danny Shaylor, and three Yanomami porters went on an even longer trek, to the highland village of Patanowa-teri.

> The next time we filmed a Yanomamo event was in a village Napoleon told me to find. He had known Patanowa-teri before, but it had now moved to a mountain hide-out to escape its myriad enemies. . . . After finding the Patanowa-teri we set up an aircraft radio and reached Napoleon at Mavaca. He convinced the headman to move back to an old garden near the Orinoco River where the genetics expedition could work with them and we could take film. The radio connection was full of static and Napoleon's voice cracked and sputtered. The headman behaved as if he were talking to a ghost and said later, throwing down the microphone, "Eech, that awful machine—did you hear it coughing? It's going to give everyone in the village a cold."[55]

Asch's guide and translator, Danny Shaylor, recalled, "I went with Asch to find Patanowa-teri—that was the craziest trip of my life. We got lost for I don't know how many days. We had some carriers from Mavaca loaded down with pots and other gifts for the Patanowa-teri. They got tired and threatened to go back home. So I said, 'All right—you can go back, but I'll tell everyone you couldn't make it when I return to Mavaca.' They continued."[56]

Nobody knew exactly where the Patanowa-teri were hiding out. All told, Asch, Shaylor, and their three porters wandered for eleven days. Food was scarce and wild game scarcest of all. Asch fainted from hunger on the trail; if the Yanomami hadn't revived him, he would have died. When the Indians finally killed a small bird, they gave all the meat to Asch.[57]

Except for Helena Valero, the white girl kidnapped by the Yanomami in 1932, Danny Shaylor and Timothy Asch were the first twentieth-century explorers who ventured into the mountain redoubt known as the Siapa Highlands. It is an ecological island, similar to the Parima Highlands, but even more inaccessible, separated from the surrounding highlands by the Orinoco, Siapa, and Mavaca rivers. The Yanomami were occupying the Siapa Highlands in the eighteenth century when the Portuguese explorer Lobo de Almada first heard of them, and the earliest maps place the "Yanomami Nation" here—inside a triangle formed by the Orinoco, Mavaca, and Siapa rivers.[58] All the groups Chagnon studied lived there prior to 1950,[59] though all modern groups had originated even farther away, in the Parima Mountains.[60] When Chagnon finally flew into the Siapa Highlands, in 1990, he invoked its "mythical" quality.[61] The highlanders were left unmolested for so long because all previous explorers stuck to the navigable rivers. Maybe the fact that Asch and Shaylor nearly died persuaded Chagnon to wait for helicopters.

At this time, Patanowa-teri was perhaps the most marginalized Yanomami village, with enemies to the east, west, and north, completely cut off from the lowland *shabonos*.[62] They had gotten the worst of the Napoleonic wars against Bisaasi-teri (three Patanowa-teri and one Bisaasi-teri died). When Chagnon left the field, in January 1966, he made a testable prediction: Patanowa-teri warriors would not cease raiding the Bisaasi-teri alliance until Patanowa-teri had avenged its war dead.[63] But, contrary to this confident expectation, there had been no more killing. By retreating into the more peaceful highlands, the Patanowa-teri became inaccessible.[64]

There was another advantage to this strategy. While the Yanomami along the wide Orinoco had steel and medicine, they also had malaria and influenza, both of which were running rampant in 1968.[65] The Siapa Highlands had neither—as yet. Nevertheless, the Patanowa-teri had experienced the onslaught of respiratory infections during the 1940s, when the U.S. Army Corps of Engineers was active nearby. That is why the old chief was fearful that the "ghost" in Chagnon's radio would give them all colds. In accepting Chagnon's invitation to relocate for the sake of filming the event, the Patanowa-teri also shared responsibility. They knew they should beware of gringos bearing gifts—that their joint feast might prove fatal.

Asch was guarded about his actual payments of metal pots, axes, and machetes to the Patanowa-teri. Their gleaming presence, however, can be seen all over the Patanowa-teri *shabono* in *A Multidisciplinary Study* and *The Feast.* Guarded though it was, Asch's memoir prompted scholars in recent years to politely question the authenticity of *The Feast,* as the film scholar Jay Ruby did in an issue the *Visual Anthropology Review* dedicated to Timothy Asch.[66]

Chagnon denied that he had choreographed events. "I intended, with Asch, to film a feast that year and knew the Patanowa-teri would be having one with one of its allies, not knowing which one. They eventually held a feast for the Mahekodo-teri. I did not 'stage' this—it happened naturally. They could not have cared less about our interests in filming and are the kind of people who would not do something this costly and time consuming for two whole communities simply to accomodate the filming interests of outsiders. . . ."[67]

For anyone who knew the history of what trade goods can buy in Yanomamiland, this was a bitterly comic overstatement of Yanomami cultural purity. But anyone who read Chagnon's Ph.D. thesis would have had reason to doubt his explanation of *The Feast.* When he left the field, Chagnon wrote that the Patanowa-teri had a single ally—Ashidowa-teri.[68] Therefore, he could not have guessed that Patanowa-teri would be having a feast "with one of its allies, not knowing which one." There was only one. And Ashidowa-teri was located so far into the Siapa Highlands that it remained beyond the reach of Chagnon's research for another generation.

To attract the Patanowa-teri toward the Orinoco, Chagnon altered the political landscape. He brokered a new alliance with one of Patanowa-teri's "myriad enemies,"[69] the Mahekoto-teri. They were the closest village to an old Patanowa-teri *shabono* site on the Shanishani River, an Orinoco tributary located an hour and a half by motorboat from the mission post of Platanal.

The Mahekoto-teri, like the Patanowa-teri, were fearful when Chagnon and Asch proposed the new alliance. They finally accepted on two conditions: a machete or an ax for every man and boy, and Chagnon's guarantee of their safe conduct—with the expedition's shotguns as surety.[70]

Even so, there were snags. Yanomami feasts were usually prepared with months of anticipation and required vast amounts of food to meet native standards of generosity. Whatever the spiritual significance of bananas and fresh meat in these seasonal celebrations, one needed a lot more of them than could be gathered on short notice, as the aborted festivities at Reyabobowei-teri showed.[71]

Moreover, Yanomami feasts were preceded by a formal invitation—*texe-momou*[72]—extended by a messenger from the host village. But the Patanowa-teri refused to send a messenger to Mahekoto-teri, which forced Chagnon to assume this role. "Shaki [Chagnon] told us not to be afraid, that he would talk to the Mahekoto-teri," said a Patanowa-teri elder, Kayopewe. "He spoke to us by radio and promised a huge amount of trade goods—a machete for each man. Each leader and elder got a cooking pot as well as an ax. He also promised to hunt with his shotgun and provide the meat himself. So we went to our old *shabono* at Patanowa, which we had thrown away, and fixed it up like new for the film."[73] Kayopewe said Chagnon initiated the rebuilt *shabono* with a small mound of machetes that he dramatically gave away as a sign of his sponsorship.

But why did Neel and his team have to induce Patanowa-teri, such a distant village, to come out of the hills? Why did they not just pay one of the forty-odd villages where they collected blood to put on a feast? I believe they tried to arrange another feasting event near the Mavaca mission, at Karohi-teri, on the Manaviche River. The Karohi-teri took me to an abandoned *shabono* that they had built at the request of the scientists. "Neel and Asch paid us to build the *shabono* and to do our dances here," said Jepewe, a middle-aged man with a wispy beard who had been an adolescent during the 1968 expedition.[74] We stood in the undergrowth that enveloped the ruined round house, one much smaller than normal and apparently never occupied for anything but theatrical purposes. Jepewe was angry at Chagnon—whom he blamed for having brought the measles vaccine that he said killed many people—but still fond of Neel, who paid the Karohi-teri all the trade goods promised for their performance.

It must have been a poor affair, like the failed feast at Reyabobowei-teri. In this way, through trial and error, starts and stops, Patanowa-teri became, in Neel's words, "the principal objective of this year's work." Patanowa-teri, located "in the very heartland of the tribe,"[75] was certainly a more dramatic site for filming.

The AEC had agreed to finance two different films—one on Neel's biological program and one on a Yanomami feast.[76] To simplify matters, Asch decided to make both films at Patanowa-teri. The documentary on Neel's work, *A Multidisciplinary Study,* began with the expedition motoring up the Shanishani River, Brewer and Chagnon each piloting a boat.[77] The next cut showed Chagnon painting himself and entering the village as a visiting headman—with his bow and arrow and shaman's monkey headdress. It was a rousing and apparently spontaneous arrival. But it had taken more than two days to orchestrate.[78]

The Patanowa-teri had traveled slowly from Sheroana, where Asch found them, to their old *shabono* near the Shanishani. I have walked there. You can make it in one day if you are a young Yanomami male who has no burdens and can go at a jog trot for about ten hours. You can make it in two days at a brisk walk, which is what I did. You can make it in three or four days if you have children to feed on the way. Since Asch had asked the whole village to relocate for filming and blood taking, the Patanowa-teri were obliged to make frequent detours for food gathering. And when they finally arrived, they found their old *shabono* in a collapsed state.[79] It required extensive repair in order to pass muster for the film. When the two AEC boats made it to their take-out point on the Shanishani River, the morning of February 19, the Patanowa-teri were still not ready to produce the cheering, theatrical entrance Asch finally obtained for Chagnon. Even so, he had to micromanage the event, which occurred at noon, on February 21.

"Come on, come on. Come all at once! Don't look in that camera, okay?" Asch yelled.

A scientist remarked, "They aren't ready. Give it half an hour."

Neel rebelled at the choreographing: "He has me sitting in the sun, and I can't take it."

"You can't do that!" Asch exclaimed, as someone apparently broke ranks.[80]

While the Indians repaired their round house, Danny Shaylor, the AEC's missionary translator and guide, came down with a severe case of malaria. Shaylor was so sick that Chagnon had to take him back downriver to the Mavaca mission. "That was the worst case of malaria I ever had," Shaylor told me.[81] He thought it was falciparum malaria, a potentially fatal strain that attacks the brain. But he was unsure because nobody was doing malaria microscopy on the expedition. In a radio conference, Chagnon remarked that everybody in the expedition had come down with "the same thing."[82] Neither Neel nor Chagnon even knew what the correct malaria medication was when Shaylor nearly died—they can be heard scrambling around, calling to another doctor, trying to figure it out. And this was after being in the field for a month at the height of the malaria season, moving from infected village to infected village along the main course of the Orinoco. If all the AEC researchers came down with malaria like Shaylor, the doctors became carriers to the villages in the healthier hill country.

Malaria proliferation has been one of the two most common causes of death to Amazonian Indians at first contact. As medical researcher Daniel Reff has noted, "There have been numerous instances in which malarial epidemics, some of which have claimed thousands of lives, began after one or more individuals harboring plasmodium entered an area where anopheles mosquitoes and susceptible hosts coexisted in large numbers."[83] Similar tragedies were common outside the Amazon. Hudson Bay Company trappers brought malaria to California in 1833, killing twenty thousand Indians. The historian Albert Hurtado has written that "a decade later there still remained macabre reminders of the malaria epidemic: collapsed houses filled with skulls and bones, the ground littered with skeletal remains."[84]

Common colds closely followed malaria as a cause of death among the Yanomami.[85] A Venzuelan doctor who spent many years helping the Yanomami said, "The Indians simply have no resistance while among us these diseases are relatively benign."[86] Though Patanowa-teri's headman feared that Chagnon's radio would give them all colds, it was probably the expedition's cook who did so, because he shared the scientists' leftover food and cigarettes with the Indians. As he cooked, the Yanomami gathered around, waiting for handouts. The gregarious cook puffed on his cigarette and then shared it with the Indian next to him. This scene made *A Multidisciplinary Study*'s final cut. As Asch put it, "While the scientists work, Juan, the cook, teaches the Yanomamo to smoke."[87]

Four days after Neel's team arrived at Patanowa-teri, a loud coughing could be heard from a Yanomami man. Chagnon called to a doctor, Willard Centerwall, "Hey, Bill, there's a sick human being down here."

As Centerwall responded, Asch moved closer, picking up severe coughing and retching. He tried to film the scene, but Neel rushed over, enraged.

"Not the picture of the physician ministering to his flock. This is very d——al [static] to the expedition."

"You said," Asch was caught off guard. "What percentage did you say?"

"I said none of this, from the beginning."

"Well, what percentage of film did you want. You said eighty-twenty, or seventy-thirty?"

"I don't want any of this," Neel repeated. "You're here to document the kind of a study we're trying to make. Anybody can walk into a village and treat people. This is not what we're here to do. Now, I don't know how I can be more definite about it."[88]

Asch was understandably confused. Just five days earlier, at Mavaca, Neel had been equally adamant about capturing "the whole gamut of measles."[89] Now he didn't want any of this. Neel had decided that showing sick Yanomami was very detrimental to the expedition. He also did not want to waste any time treating the mundane health problems of the Yanomami. The contempt in Neel's voice was thick.

The "minister and his flock" was, for a eugenicist like James Neel, the emblem of dysgenic behavior—the cosmologically confused priest who foolishly "helped" the weak by violating the laws of Darwinian selection. At Patanowa-teri, Neel was far from the missionaries and out of range of the Venezuelans, except for his disciple Charles Brewer. After a stressful month in the rain forest, time was running out for the 1968 expedition, and Neel's civilized façade was cracking.

Asch did not yet know James Neel, so he had some trouble figuring out what the geneticist objected to so vehemently. Neel was by no means opposed to featuring routine medical interventions. In the multidisciplinary film, Asch showed doctors vaccinating the Indians immediately following Chagnon's entrance. Neel then appeared on camera, radioing to Protestant missionaries. "But what is also needed is a good supply of antibiotics in order to treat the secondary infections that so often go with measles. And I hope that you will let us know as soon as possible . . . so we can begin to think how we can help with the program of medication. Over to you."[90] The film concluded with Neel's passionate plea to meet "the humanitarian challenge—to protect these people from against the medical and cultural deterioration which has so often been the lot of primitive man in the past. . . ."[91]

From the film, as well as from books and articles the AEC scientists have produced, one would readily expect that, once they had completed vaccinating the Patanowa-teri, they moved on to the next village—caring for the sick, administering antibiotics, continuing their humanitarian efforts.[92] Certainly, their assistance was desperately needed. On the day following Chagnon's stirring entry into Patanowa-teri, on Thursday, February 22, Robert Shaylor, father of the AEC's sick translator, called with more bad news. Protestant missionaries had applied the Edmonston B vaccine on the Padamo River—sometime between February 4 and February 15[93]—and another measles outbreak had followed, just as it had at Ocamo and Mavaca. "They used up all that was sent there and were able to get quite a few of the folks done. But of course the measles have broken out there. And we're *pendiente* to see how they've gone and how many villages have possibly been infected."[94]

In the film, this statement from the missionary Shaylor was edited out, and a voice-over simply asserted that the scientists were vaccinating a ring around the epidemic, saving all the groups they managed to inoculate. When the voice-over ended, Neel was heard and seen speaking over the radio, offering to help with more vaccinations. "I'm sorry to hear this. Now, when we come out to get the blood to the plane, we will after that work on the Padamo if possible. And if Danny is still badly I would try to get down to Tama Tama to see him."[95] But Danny Shaylor did not see Neel again that year.[96] Nor did Neel's doctors join the missionaries and government doctors in controlling the epidemic on the Padamo River or anywhere else.

According to the sound tapes, the scientists left Patanowa-teri on Saturday, February 24. They traveled two days by boat to the Esmeralda airfield, where they met a plane on Monday, February 26.[97] They were rushing to get their blood out of the tropical heat. In all, the scientists purchased a staggering amount of blood, urine, and saliva at nearly forty Yanomami villages during their three weeks on the Upper Orinoco in 1968.[98] Thousands of Yanomami were placed on what Asch called "a production line: numbers are assigned to them: specimens of their blood, saliva and stools are collected; impressions of their teeth are made; and they are weighed and measured by the physical anthropologists."[99]

Even as Neel and Chagnon at least feared their vaccine reactions might turn into an uncontrolled epidemic, they tried to attract hundreds more Yanomami to their blood-collecting station at Patanowa-teri.[100] Chagnon promised the Bisaasi-teri at Mavaca that he would return downriver, pick them up, bring them to the feast at Patanowa-teri—and have them go several days off into the jungle to extend an invitation to the distant village of Ashidowa-teri. Chagnon also hoped to draw blood from another group, the Hasupuwe-teri,[101] who had two *shabonos* with a total of over three hundred people above the Guaharibo Rapids.[102] On February 18, Neel had told the Venezuelans that he was going to vaccinate the groups on the upper reaches of the Orinoco, but he never vaccinated any of them. Under the circumstances, it was just as well that he did not do so. But his misinformation distorted the rescue plans of the Venezuelan emergency medics, who left the Hasupuwe-teri to their own devices. About one hundred Hasupuwe-teri died of measles.[103]

Although I believe Neel was sincere when he told the missionaries he would join them in battling measles on the Padamo River, the logic of his immense scientific enterprise followed a momentum all its own. Asch explained that "the blood and other samples must be quickly taken out to laboratories to prevent spoiling due to heat. Patanowa-teri was the last village they were able to inoculate."[104]

The split between the on-camera and the off-camera James Neel was only one of the film's anomalies. Why were the AEC doctors vaccinating at all?

They were at an extremely isolated village, with no medical backup. Vaccination with Edmonston B required at least fifteen days of continual care after inoculation.[105]

Panic and dispersal had followed their only other vaccinations, around the Ocamo and Mavaca missions. The same day the expedition left Mavaca, about seventy Yanomami who had just been vaccinated ran off into the jungle, where the missionaries retrieved them ten days later all "very sick."[106]

Chagnon had admitted that the vaccine was almost as bad as measles.[107] Neel's data showed that the vaccine reactions were indistinguishable from severe measles.[108] But, once they had told the Venezuelan authorities that their vaccine produced no rash, they stuck to their story so tenaciously that they apparently believed it. The expedition physician Will Centerwall looked the camera in the eye and said, "This kind of measles, especially the vaccine, is very unlikely to cause any trouble. Okay?"[109]

This was a confusing testimonial. What "kind of measles" was he talking about? *Especially the vaccine?* Asch cut this, along with the fact that Centerwall decided not to vaccinate pregnant women.

"Aaah. Let's put it this way. I think this is the lesser of the evils."

"Then give her the vaccination?"

"I think so. I'll just give her two, three, I'll give her three cc's of gamma globulin, which means that if measles does hit her it'll be moderate."[110]

Some doctors felt it was better to suspend vaccinating in the middle of an epidemic and to provide gamma globulin coverage only. But Neel's decisions were increasingly driven by the logic of the film rather than by safe medical practice. He was trapped by promises to vaccinate "a ring around the epidemic" and by the pretense that the vaccine was harmless. The film became a defense against the unraveling of Neel's story, a way of canonizing the geneticist's version of reality. By filming the Patanowa-teri being inoculated, Neel justified his earlier vaccinations as well as his decision to leave the sick at Mavaca behind. "This village was fortunate," Neel narrated. "It was vaccinated in time."[111]

It was not. Neel did not have nearly enough gamma globulin to vaccinate the whole village of Patanowa-teri. In the outtakes, Chagnon is heard admitting that they had been unable to finish the job. Worse, Yanomami exposed to measles from Mavaca trekked through the forest trying to rejoin the expedition. Only one of them made it to Patanowa-teri. The others were too sick—they dropped off in the forest. By radio, Chagnon told the missionaries that he was trying to quarantine the Patanowa-teri from the sick man who had arrived from Mavaca. "We'll have to try and isolate them as best we can. Meanwhile, we've gone ahead and vaccinated all of the rest of the Patanowa-teri that we had vaccine for."[112]

In fact, they had enough vaccine, but not enough gamma globulin. They never admitted this in repeated radio conferences, however, perhaps to keep the Venezuelans and missionaries from realizing that the expedition was giving out an antiquated vaccine. Even a poor country like Venezuela had by then switched to the Schwarz virus, which did not require gamma globulin with vaccination.

The expedition simultaneously exposed the Patanowa-teri to malaria, bronchopneumonia and, depending upon which group they were in, either the Edmonston live virus or the germs of a carrier from Mavaca. As time went by, more and more people started coughing. None of this was shown, per Neel's instructions.

Instead, the Yanomami were presented as pictures of exuberant health. "This is the chief here," Centerwall gushed. "He certainly is a fine specimen of a man."[113] (Actually, he weighed 108 pounds, but he was big by Yanomami standards.) Centerwall was equally enthusiastic about Yanomami urine. "You know these urine specimens, ah, Tim, are a beautiful assortment of yellows and ambers. . . ."[114] Neel found the fecal samples "remarkable." Charles Brewer, a dentist, praised the natives' teeth. "They are perfect. No decay or accumulation of debris." Brewer attributed their good health to a high-fiber, sugar-free diet. Yet Neel warned that the Yanomami's idyll was ending. "Each year extends further the tentacles of civilization. . . . The health of primitive man usually quickly deteriorates in the course of acculturation."[115]

At Patanowa-teri, acculturation and deterioration were well underway.

The expedition had trouble simply feeding the Indians from one day to the next. By moving toward the Orinoco, the Patanowa-teri left their producing gardens far behind. Their nearest one was a three-hour walk away.[116] They could not feed themselves, much less supply a feast for 125 guests.

Although Chagnon had promised to provide all the meat for the feast, it was not an easy task, even with the AEC's two shotguns. "Look at this," Charles Brewer complained on returning from one hunt. "I have been out since six o'clock this morning, or five-thirty, I don't know what time I got up. And I have to do so many work about the feast and about the blood. And this guy took me for a five-minute ride to do some hunting, and there I am getting up at one o'clock now. Well, you know I went to do some hunting for them because they were really hungry."

"And I got several candies," Brewer added in disgust, referring to the expedition's sweets, which were now littered all over the place. "They asked me to pick them up after I point them, and here they are."[117]

Thousands of candies were being paid out to Yanomami women and children. One of the women's tasks was to collect beetles for the expedition's biologist, who soon found himself inundated with the same species and unable to pay them the lollipops they were demanding.

"You told them if they brought in beetles you'd give them candy," Chagnon said, sticking up for the women.

"Tell them there must be some adjustment," the biologist insisted. "If the beetle is not the kind I want, then I cannot pay them for it. . . . Come to my rescue. Tell them at this point I would settle for something different and bigger. But I can't take any more of these beetles. My bottles are all full."

"Unless we change this into a bonafide beetle hunt, there's no way of stopping the flow once you've asked for it," Chagnon replied.[118]

It was hard to stop. Everything was falling apart. The scientists were also hungry. "But there is it: we have to eat also in spite of all the scientific work we have to do," Brewer said.[119] Their shotguns quickly drove off game; they killed a few birds and finally a pregnant monkey, which Brewer would never have shot had game not been in short supply. Not surprisingly, the Patanowa-teri's feast suffered from a lack of meat—just like the improvised feast at Reyabobowei-teri. In the film, Chagnon blamed the village headman. "His hunters have done so poorly that he must make the meat go further than it should."[120] But the chief's hunters were Chagnon and Brewer.

"Before apologizing for this, I am not to blame, you know," Brewer began.

"You're not?" Neel asked, who sounded amused. He had a soft spot for Brewer, but he was tougher on Chagnon. He complained that Chagnon hadn't hit anything with his shotgun the whole expedition.

"You're a sad crew, you guys," Asch said.[121]

In all fairness, both Chagnon and Brewer were overburdened with other tasks. Chagnon supervised the making of the ceremonial plantain soup that was the main beverage at the feast. This meant hauling a ton of plantains from a distant garden, a thankless task that would never have been necessary at a normal feast, held at the harvest season near a producing garden.

According to the film, the Patanowa-teri headman was Kumaiewa. "He's *the* big man."[122] However, the Patanowa-teri elders told me the real village headman was an older shaman, Shamawe, who was less pliable to Chagnon's desires than the younger Kumaiewa. The competition over Chagnon's favor was evidently a source of internal conflict at Patanowa-teri, as it had been at Bisaasi-teri.

This could be seen from a brief conversation between Chagnon and a Yanomami woman. As translated in *The Feast,* the woman asks, "Shaki, are you my older brother? Tell me you are my friend."[123] Now, this would be a surprisingly forward thing for a Yanomami woman to do, publicly asking a *nabah* to be her "friend" and "older brother." Quite a few eyebrows would have shot up around the campfire.

The Yanomami text is quite different, however. *Shaki wa wohimai ya irawe* really means, "Shaki, do you love your brother?"[124] What she wanted to know was whether Chagnon would favor her husband, Hotihewe, when the anthropologist distributed trade goods—as he did on a regular basis, marking the Yanomami with different colors once they had been paid for their various tasks and bodily fluids.

Trade goods played a key role in the film's assessment of Yanomami politics. "The trade goods help bind the alliance by creating obligations which the visitors must discharge at a return feast."[125] This was certainly true of the AEC's goods, the true basis of the alliance. Chagnon skillfully plied the Patanowa-teri with presents and the promises of presents to keep them working. While hauling plantains some people seem to question making the film, and Chagnon apparently tempts them with the vision of *madohe totohiwe,* "beautiful trade goods."[126] But here, too, supplies proved inadequate to the size of the project. When his trade goods ran out, Chagnon radioed for another planeload.[127]

Asch was surprised when, en route to get food at another garden, the Yanomami burst into a frenzied dance, "screaming at the top of their lungs, waving branches of leaves in the air." Asch filmed them, believing it was "a garden ritual." When the exhausted, sweating Yanomami finally stopped, Chagnon asked them, "What was that all about?"

They were mystified. "Isn't that what you just asked us to do?"[128]

There was a question of how much Chagnon was really able to communicate with the Yanomami. He had spent a total of fifteen months in the field, and no one has become proficient in Yanomami in such a short time. At one point, he said to Asch, "Shoot that scene over with me in it. My Yanomamo is a little rusty."

"That was kind of nasty and not really called for, you know," Neel interrupted. "If your Yanomamo is rusty now, you ought to be should be ashamed of yourself."[129]

The Yanomami understood that Chagnon wanted scenes of violence. Asch also got that message. Chagnon's preference was the subject of an article Asch wrote, "Bias in Ethnographic Reporting," excerpted in the April 1995 memorial issue of the *American Film Quarterly.* Asch said Chagnon became "bitter" if Asch trained his camera on anything but aggressive behavior. Chagnon thought nonviolent episodes were a waste of valuable film. When Asch urged Chagnon to film women's activities, Chagnon "whipped around" and asked, "What makes you think there are any women's activities?"[130]

Chagnon, who narrated *The Feast,* explained, "Women are rather inconspicuous at political events such as these."[131] This was true at the AEC's feast, but it was not true for Yanomami feasting in general. Normally, women begin the festivities with marathon, call-and-response chanting called *amoamou.*[132] This often becomes a long, friendly competition between the women hosts and visitors.[133] Later, the women from the visitors often danced with the men from the hosts, in a spectacular performance, *hakimou,* that sometimes ended in sexual dalliances.[134] But the women of Patanowa-teri were terrified that enemies might attack them at any moment (a fear that had caused them to move away from this site). In fact, Dr. Centerwall noted that the lovely colors of the women's urine were caused by dehydration—they were afraid to venture down to the nearby creek to drink.[135]

It was violence and the expectation of violence that appealed to film juries and students and that gave *The Feast* its edge. "The Patanowa-teri have been raided twenty-five times in the previous sixteen months, with a total of ten deaths," Chagnon narrated.[136] Actually, the Patanowa-teri had been raided twenty-five times during Chagnon's fieldwork, from November 1964 until January 1965, with a total of ten deaths. But there had been no deaths since Chagnon left the field.

Nevertheless, Chagnon was right when he said, "Many of the Patanowa-teri still regard the Mahekoto-teri as enemies. They are fearful, as are their guests, because they know that any feast can end in violence. . . ."[137]

Any feast can turn violent. Very few actually do. In this case, the atmosphere was strained because neither group wanted to be there in the first place. With the AEC's sponsorship and Chagnon's shotguns, however, there was no danger of violence occurring during the feast. The film achieved the illusion of immediate conflict by mistranslation. In the film, a lead dancer for the Mahekoto-teri entered the Patanowa-teri plaza dancing ecstatically and shouting, "Fight! Fight! Fight!"[138]

That was the film's translation. Actually, his chant was, *Mita mitahe*—"Look! Look!"[139] He was not threatening anyone.

The real danger lay elsewhere. Neel explained why: "Feasts are also the occasion of a joint raid on enemy villages. . . . Sometimes villages will unite at a feast, drink the cremated bones of a common friend or relative killed in war, and leave in a group to raid a mutual enemy." This is precisely what happened after the filming of *The Feast.* The Patanowa-teri and Mahekoto-teri united to attack the village of Yabitawa-teri, where they killed an old woman, an unusual event in Yanomami warfare. But since this whole feast was inspired, paid for, provisioned, and arranged by the diplomacy of the AEC scientists, this put the resulting raid and death in a dubious light.

"It might seem like a great idea to bring two groups together and make peace," said the missionary Mike Dawson, who has lived for over forty years among the Yanomami. "But the Yanomami don't wage war that frequently, and their way of avoiding war is to move apart, so they don't have any more contact with their enemies. When you bring them together to film an alliance, you're naturally going to make these two groups remember all their previous hostilities, and just about the only way they can channel that is to launch an attack against a third group. And this has happened more than once after filming."[140]

A handful of warriors, Asiawe told me, went on a joint raid. My sense is that the young men who helped broker the feast and act as as intermediaries for the outsiders were the only participants. It was a sign of the social disruption that outsiders always brought, promoting youths before they had matured in tribal traditions. Given the size of the feast—attended by 340 individuals—a normal raiding party would have been far larger—60 or 80 men.[141] And, since this was a new alliance, the older male leaders would have been virtually obligated to participate along with everyone else.[142] But the headmen did not join the attack.

Nor did the newly allied villages perform any of the ritual preparations for a raid, which center on the sharing of funeral ashes in a sacred meal. The women keep these precious ashes; significantly, only they are allowed to imbibe the ashes, in a plantain soup, before the warriors depart. It is also at this mortuary meal that shamans take hallucinogens and divine the spiritual enemies whom the warriors should attack. Although the raid which followed *The Feast* does appear to to have had purely material motives, based on trade advantages, it was a feast alliance unlike any other described in the extensive

Yanomami literature.[143] The raid was, like the feast, an event without a sacramental center, and it happened just after the filmmakers left.

Chagnon appeared disappointed with the film's weak finale. "Come on, I've pleaded with you to put the fucking recorder on," Asch snapped. "There was a lovely little kid just standing there lounging."

"Look, Tim, I'm in a bad mood," Chagnon answered darkly.

"Then you're in a bad mood on the most important day of the filming."

"There's nothing here that's that important."[144]

After shooting this last scene, Chagnon and Asch left the village. As they headed downriver, the raiders jogged off into the jungle, led by Asiawe, son of the Mahekoto-teri headman.

"We went with the Patanowa-teri to raid Yabitawa-teri, where we killed an old woman with arrows," Asiawe told me when I interviewed him at the village of Mahekoto-teri. "The next day we returned to Patanowa. I was beginning to feel sick by then, and so were some of the others from my village. Then we left the Patanowa-teri and returned to Platanal, but by that time many of us were very sick. Four of my people died, and González [the government nurse] helped me to hang them up in the jungle." Asiawe was referring to the Yanomami's custom of leaving their dead inside hammocks, or baskets, or on top of platforms, high above the forest floor, when many people die at once during an epidemic and no one has the strength to perform the cremation ceremonies. "Then I moved across the river," Asiawe continued. "And when I was across the river, more of my people died. We hung them out in the jungles there. Then I moved downriver a little bit, and more of my people died again, and they kept dying. Then we moved back upriver, and we started getting better. During this time, González helped us hang our people out in the jungle and gave us medicine. There weren't very many Shashanawa-teri [another group the AEC took blood from but did not vaccinate]. They had only four leaders—and they all died off. While I was tying my dead out in the jungle, they were doing the same thing: they just tied their dead up in hammocks inside the *shabono*. . . . After I drank my bones then, we went up there and drank their bones. Shashanawa-teri first came to our cremations at Mahekoto; then, when they did their bones, we helped them."[145]

According to Chagnon, 25 percent of the "Platanal Yanomamo" died of the measles.[146] What he never admitted is that the Platanal Yanomami are no other than the Mahekoto-teri, and that they died immediately after *The Feast* was filmed. Timothy Asch was the only member of the expedition who eventually acknowledged the sad truth.[147]

The death of 20–30 percent of Indian tribes at first contact was normal over the centuries. The first English colonists at Roanoke, Virginia, noticed that every time they entered an Indian community "within a few days after our departure . . . the people began to die very fast." This mysterious phenomenon was initially blamed on "the Eclipse of the Sun," but was later attributed to divine providence.[148] Nobody took responsibility for these acts of nature or of God. Thousands died, but nobody was to blame. Finally, it was decided, more or less universally, that the Indians were destined to die off.

Big expeditions always left the dead behind as the explorers went on to win knighthoods or estates or Ph.D.'s. Even when historians or Indians complained, the real evidence was gone: the dead could not speak.

That's why Asch's unedited footage was invaluable. Indians about to die are complaining on tape about the visitors who have come with germs, guns, and steel. At one point, on February 27, just prior to the feast, shamans can be heard conjuring away sickness and people are heard coughing.

A woman weeps and shouts, *Hariri*—disease. Another woman apparently does not want to join the feast with the Mahekoto-teri, because they are fierce: *Mahekoto-teri waiteri.*[149] Meanwhile, people are yelling at the cameraman Timothy Asch and hurling rocks at him. Asch apparently responded—I have not seen the film's outtakes—by throwing a rock back and hitting a dog, setting off yelps from the dog and a chorus of cries from the Yanomami.[150]

Asch: "Actually, it isn't good."

Chagnon: "I think that's enough, Tim."

[Women are coughing loudly and spitting]

Asch: "It's mean. . . ."[151]

It *was* mean, but they had to keep choreographing everything. Asch wanted a shaman to repeat something: "I wonder if he would do that again without that kid in there."

A Yanomami man began to intone Asch's name. *Ashe, Ashe.*

Asch, who took no notice, said, "Those are wild sounds to go with the cotton scene, but they may be too . . . not quite what . . ."

At one point a man muttered a sentence including the word *horemu*, meaning "lying" or "faking."[152] Though the tapes still await competent translation, this was the same word the Patanowa-teri elders repeated, over and over again, when they saw a screening of *The Feast* in September 1996. They felt the film was undoubtedly a *horemu*, a fake.

It would be an equal deception, however, to think that any of us would have done things differently. I know I would not have done things very differently. Not at the age of twenty-nine—as both Chagnon and Asch were—with new film equipment and orders to record a military aliance. Maybe I would have done a few things otherwise. I would have organized a much bigger, better finale and made sure that all the raiders went off with an enormous cheering section. If the Yanomami men wanted their axes, they would have had to put on a finer performance.

But if the edited film was a *horemu,* the unedited *Feast* truly broke new ground. It brought all the unconscious horrors of contact into the open. The cook from Caracas passed his cigarette and shared his food with the Yanomami, possibly sharing respiratory illness. The missionary translator Danny Shaylor contracted malaria on the Orinoco's main course and brought it with him to Patanowa-teri. The doctors applied a dangerous vaccine and then abandoned the Indians. An infected man from Mavaca searching for steel presents stumbled out of the jungle with measles. James Neel became infuriated at filming wasteful acts of altruism. Meanwhile, industrial quantities of blood, beetles, urine, and plants were collected, miles of film rolled, and food ran out. The scientists had "so many work to do,"[153] as Brewer lamented, and so little time.

All prior studies of first contact had been, to use Neel's apt phrase, "a mosaic of unrelated findings."[154] In bequeathing the National Archives his takeouts, Asch left the definitive documentary on how disease and acculturation were introduced to a vulnerable tribe. At Patanowa-teri, all the skeletons from the past came out and danced for *The Feast.*

Shortly after *The Feast,* the surviving Patanowa-teri joined another village, Iwahikoroba-teri, in making an effigy of Chagnon. They set it up and shot it full of their long arrows. Both groups blamed Chagnon for having worked black magic against them; both relocated far away from Mavaca, to escape the anthropologist's deadly powers. Chagnon noted merely that he was "annoyed"[155] that his former friends had participated in such a ritual—the only time a non-Yanomami has been targeted in this way. But when Chagnon tried to revisit Patanowa-teri, in 1969, his guides forced him to turn back. "His informant warned him as they went upriver, 'We can't go on, doctor. They're going to put an arrow through you,' " said Sister Felicita of the Ocamo mission. "They had made a doll [of Chagnon] out of banana and palm leaves. When Chagnon came back to the mission, he was almost in tears."[156]

Chapter 7

A Mythical Village

I was their village. Their village was me.—*Napoleon Chagnon*[1]

While the Yanomami who had been filmed in *The Feast* were fleeing in panic and abandoning their dead to improvised funeral platforms in the jungle, and as measles spread to villages all over Yanomamiland, Napoleon Chagnon began the most challenging adventure of his career. During the second week of March 1968, he traveled up the Mavaca River to explore villages that "had never seen a foreigner other than me in their entire history."[2] These villages belonged to a Yanomami subgroup—the Shamatari. "My subsequent work among the Shamatari would lead me to describe them as the 'Fiercer' people."[3]

Fiercer, farther, Chagnon was always pushing himself to new limits, going where no other anthropologist had gone before. For the real addict of El Dorado—conquistador or explorer, scientist or journalist—the quest never ends, though it always disappoints. In a sense, El Dorado was history from the beginning—a history of civilizations that had ceased to exist. The Spaniards kept looking for the same pristine places they had already erased. El Dorado, a high, cool city ruled by a runaway Inca, sounded a lot like Cuzco, where the Spaniards had an unforgettable and unrepeatable looting party.

American anthropology was born of a similar nostalgia. Just as "wild Indians" had been wiped out or reduced to reservations, scientists conceived a desire to recover them. That is why there was a stampede of publicity and scientific hoopla when a solitary survivor of the Yahi Indians emerged in 1911. His name was Ishi and he had been hiding for forty years in the Sierra Madres of northern California. Cartoonists drew Ishi as a Stone Age man with a club, capturing white women and dragging them off by the hair. (In fact, Ishi had been celibate all his life; he had no culturally acceptable partners because vigilantes, ranchers, and government agents had hunted four hundred members of his tribe to extinction.) Ishi became a living display at the University of California at Berkeley's museum and, in a real sense, its foundation sacrifice. Thousands lined up to see him every Sunday. Ishi was photographed and filmed so often that he became an expert in posing and lighting, able to suggest the right props and angles to prospective picture takers. Within weeks, he contracted pneumonia. The scientists were aware of the risks; in 1897, amid extraordinary fanfare, Admiral Perry had brought six Eskimos to New York, where four of them died of tuberculosis. In the end, Ishi also died a lingering death from tuberculosis, hastened by deathbed interrogations from America's leading linguist.[4]

If a single Yahi Indian after the turn of the century could launch a major museum and catapult his discoverers to national prominence, the scientific potential of totally uncontacted villages in the late 1960s was incalculable. For an enterprising man like Chagnon, it was also irresistible. He honestly admitted that his motive was "scientific curiosity."[5] Like his predecessors at the University of California, he saw this as a final opportunity for science. "The Yanomamo, like all tribesmen, are doomed, and soon they will be swept aside and decimated by introduced diseases as Western civilization penetrates deeper and deeper into the remaining corners of the world where it has not extended itself."[6]

Chagnon had been trying to contact the Shamatari Yanomami since his first months in the field. "These were the people against whom Kaobawa and his people had waged ceaseless war for half a century. . . ."[7] When the Shamatari heard that Chagnon had arrived with his bounty of steel goods, they sent messengers asking him to come and visit them. In fact, they began migrating from the Siapa River to the Mavaca headwaters shortly after Chagnon set up camp at Bisaasi-teri. Although they were receiving handed-down axes and machetes, they wanted an unmediated relationship with the anthropologist.[8] Chagnon eagerly accepted.

The Bisaasi-teri opposed Chagnon's plan with arguments, delays, threats, and war. When Chagnon first attempted to travel up the Mavaca, the Bisaasi-teri lined the banks and screamed at him not to take his steel presents to the worthless, treacherous Shamatari, who were going to kill him anyway. All of Chagnon's Bisaasi-teri guides abandoned him two days upriver, forcing the angry anthropologist to return.[9] Bisaasi-teri allies launched preemptive attacks at the largest Shamatari village, killing one man, the first death in the war between the two villages in five years, but keeping them at bay. When, against all odds, Chagnon tried to contact them on foot, he became violently and mysteriously sick after eating food his Bisaasi-teri friends gave him.[10]

This clash of wills naturally soured Chagnon's relations with his host village and with its headman, Kaobawa. Chagnon resented the fact that the Yanomami saw him only as a dispenser of metal goods, not as a friend. But the Bisaasi-teri could not understand that their growing acculturation, which Chagnon had done so much to accelerate, made them less valuable to him as informants and film subjects.[11]

By 1968, Chagnon had found a way to move on. He hired a boy, Karina, who had been raised at one of the villages Chagnon wanted to visit. Although Karina was now living with one of Bisaasi-teri's allies, he was treated as "an outcast" in the village. "The boys of his age also teased him mercilessly, and the adults ordered him around as if he were a recently captured enemy child."[12]

Chagnon repeatedly risked his life in this journey to the edge of the world. Or so it seemed. No one had traveled the Mavaca in seventy-five years. Its headwaters were off the map. As it progressed upriver, Chagnon's party found a profusion of wild peccaries and turkeys—an almost Edenic scene of abundance. Against this idyllic backdrop, however, lurked the ever-present threat of death from the terrible Shamatari and from other, mysterious forces. Karina had to be reassured against *Raharas,* mythological serpents that inhabited unexplored rivers. Chagnon combated the Yanomami myth with one of his own: he told Karina he had killed many Raharas in his youth and had a special weapon for them. Chagnon demonstrated how he would shoot them. "Right here! In the neck!" At last, he reached the "almost legendary village of Mishimishimabowei-teri." When his guide stole his trade goods and boat,

Chagnon hollowed out a canoe and radioed to missionaries for help. The missionaries left Yanomami villages afflicted by measles on the Padamo River in order to rescue the anthropologist.[13]

Chagnon shrewdly understood the appeal of the virgin frontier to American audiences. And he skillfully turned what is normally a long day's run up a deep river with no rapids into a harrowing, three-day trip. (When I checked one of Chagnon's handwritten maps, published in *Studying the Yanomamo,* I saw that it took him exactly eight hours to reach a point a few miles below Mishimishimabowei-teri, but he had also stopped for two hours to talk with an informant.)[14] The Mavaca is such an easy waterway that it was a major route for rubber traders in the nineteenth century. They had a post on the Upper Mavaca, and hauled rubber overland to the Siapa River, on its transcontinental journey to Manaus, Brazil. Though Chagnon claimed to be the first to travel the Mavaca River in a century, the explorer Carlos Puig had reached the Mavaca headwaters in 1941,[15] as had the government malaria service in 1962.[16]

Chagnon's most suspenseful drama, making "first contact with Mishimishimabowei-teri," was also questionable. Helena Valero, the white girl previously mentioned, lived with the group for most of 1933. She ran away to one of their allies, but continued to see them at feasts for about a decade.[17] Actually, while Chagnon and Asch filmed *The Feast,* the Venezuelan government nurse Juan González took two Bisaasi-teri guides up the Mavaca River, where he claimed to have vaccinated some Shamatari with the government's more benign Schwarz vaccine.[18] That might explain why the village that Chagnon contacted was apparently not hit by measles. Whereas Chagnon constantly emphasized his own anxiety and the risk of death at the hands of the Shamatari, González said he felt no fear, though as a nurse his humanitarian mission was very different from Chagnon's. Later the Shamatari took González on foot all the way to the Siapa River.

There was a limited sense in which Chagnon made "first contact" on the Upper Mavaca. Until his trip, no one had used the name Mishimishimabowei-teri; Juan González said he visited the Mowaraoba-teri in 1968. That was the name this group had used for about three decades. I have not found any references to Mishimishimabowei-teri prior to Chagnon's 1968 visit. In 1967, they had been living in two separate communities in the Siapa River valley. By 1968, only eighty of them were at a place called Mishimishimabowei-teri. On hearing that Chagnon had come with fifteen machetes, six axes, and twelve pots,[19] and promised to return with gifts for everyone, other villages from the Siapa River valley immediately pulled up stakes and joined their cousins. Salesian mission records initially describe it as a hodgepodge of five different villages.[20] As late as 1972, the priest at the Mavaca mission, José Berno, was unsure whether this was "a great tribe or five tribes who live together."[21] The unusual village of four hundred—the largest ever reported for the Yanomami—apparently coalesced around Chagnon. And it remained intact, like the triple alliance of the three previously separate villages around Boca Mavaca described in the last chapter, only as long as Chagnon's extended visits lasted (1968–72).[22]

The Mishimishimabowei-teri acknowledged this with a remarkable gesture. Chagnon had christened the new, five-tribe village "Mishimishimabowei-teri." The villagers returned Chagnon's compliment by bestowing their new name on him. They called *him* Mishimishimabowei-teri. "I was their village," Chagnon wrote. "Their village was me. That is about as high an honor a Yanomamo can achieve."[23]

The "ceaseless warfare" between Kaobawa's village and the Mishimishimabowei-teri was another exaggeration. Helena Valero, who remained in the region until 1956, witnessed a decade of peace between the groups in the 1930s and early 1940s. That tranquil period ended when the leader of Mishimishimabowei-teri's parent village was accused of causing the epidemic that followed the U.S. Army Corps of Engineers' foray into the Upper Orinoco. Some Bisaasi-teri and some members of its close ally, Wanitama-teri, massacred six of the Mishimishimabowei-teri. But many members of the two villages vehemently denounced the massacres.[24] Then another seven or eight years passed without violence until, within months of the permanent establishment of a Protestant mission, a complex alliance of villages killed somewhere between eleven and fifteen Bisaasi-teri. According to Chagnon, all of the actual killings in 1950 were accomplished by one of Mishimishimabowei-teri's allies, while the Mishimishimabowei-teri themselves played an ineffectual role.[25]

Whatever the reason for their old wars, the Bisaasi-teri had achieved clear military dominance since joining missionaries, malaria workers, and anthropologists on the banks of the Orinoco. They killed three Mishimishimabowei-teri in 1960, and the Mishimishimabowei-teri made no response.[26] In 1965, when another Mishimishimabowei-teri man was killed by Bisaasi-teri allies, they also failed to retaliate.[27] Instead, they retreated into the Siapa Highlands, where they had spent most of the preceding decades. This region, according to Chagnon, has poorer food resources[28] and less access to metal goods than lowland villages.[29] Typically, the Siapa villages were militarily weak and lost women to resource- and metal-rich villages near the Orinoco.[30] Mishimishimabowei-teri was no exception. It suffered from "a severe shortage of females"[31] and had a pathetic dearth of metal tools, both characteristic of the vulnerable highland villages.[32]

Their known warfare from the mid-1930s until 1968 suggested that the Mishimishimabowei-teri were one of the least efficient groups of warriors in Yanomamiland. Against the Namowei—the group including Wanitama-teri, Bisaasi-teri, and Monou-teri—their record stood at ten war deaths to zero.[33] And it was about to get worse.

In 1970, Chagnon decided to help foster a new alliance between Bisaasi-teri and Mishimishimabowei-teri. He had some initial misgivings: "This was risk taking in spades. . . . I was also worried I might be a contributor to an enormous disaster." Nevertheless, he felt he could help end "twenty years of war." In June 1970, Chagnon ferried Kaobawa to Mishimishimabowei-teri, where he witnessed "the social and heroic ingredients of Neolithic Peace." Once again, Chagnon's 16mm camera and shotgun played key roles. As Kaobawa prepared to meet the Mishimishimabowei-teri, Chagnon was prepared to fire in his defense. "I recall," he wrote, "how difficult it is to be ready to shoot, but yet try to look friendly and nonchalant, pretending that your weapons were not really ready to shoot THEM. . . . Kaobawa shouted that I was with him and we were friendly. He was extraordinarily alert, like an animal who had detected either prey or a predator, his eyes dodging rapidly back and forth scanning the dim, gray jungle ahead. . . ."[34]

Although peace would have appeared out of the question, and death almost certain, what actually happened was very similar to the alliance building captured in *The Feast.* Not only did the Mishimishimabowei-teri have no problem with welcoming the Bisaasi-teri; they had no problem with being filmed together in a remarkable ritual that Chagnon made into another award-winning film, *Magical Death.*[35] This took place in the late spring of 1970.[36] The Mishimishimabowei-teri "began an elaborate two-day shamanistic attack on the souls of children in Mahekoto-teri." Their purpose was "to make friends" by killing enemy babies "and stealing and eating their souls."[37] The ritual involved taking hallucinogens, chanting, and enacting a pantomime of devouring the children of Mahekoto-teri (the guests at *The Feast* who lost a quarter of their people to measles).[38]

Timothy Asch did not participate in this film. Indeed, he hated it. He begged Chagnon to remove it from circulation because he had found that his students at USC were horrified by the Yanomami's symbolic cannibalism. Eating enemy children, even in the spirit, appeared psychotic to southern California undergraduates, according to Asch. Chagnon attributed this to jealousy on Asch's part; after all, Chagnon had made the film all by himself, and it won a blue ribbon at the American Film Festival.[39] In spite of the film's initial accolades, the anthropologist Linda Rabben, of Amnesty International, has recently echoed Asch's complaints about *Magical Death.* "They [students] watch green mucus pouring from the nostrils of Yanomami warriors dancing and chanting under the influence of a hallucinogenic powder. All the scholarly explanations (and the sight of Chagnon himself, befeathered and painted, prancing about in a drug-induced trance) cannot eclipse that image."[40]

That image, however, was less immediately relevant than the new power arrangements that Chagnon helped consecrate. In 1968, Chagnon and Asch brokered a new alliance between the Mahekoto-teri and the Patanowa-teri, creating a formidable military force—one that immediately attacked a nearby village and killed an old woman.[41] Now Chagnon was participating in a new peace that also meant a new war—one that would pit *The Feast* allies, 350 strong, against the *Magical Death* allies, some 500 strong, in an innovative regional war fought with shotguns and outboard motors.

Chagnon merely wrote, "A peace had been forged and a new era of visiting and potential alliances had opened up. The Mishimishimabowei-teri were invited to his [Kaobawa's] village to feast and dance, and they agreed to come."[42]

But they received a little help from their friends. On June 28, 1970, Father Berno wrote in the Mavaca mission chronicle that he "was invited by Dr. Chagnon to accompany him" four hours up the Mavaca to the Mishimishimabowei-teri, who were then ferried to the Orinoco for visits to three different villages, with dancing, drug taking, and ritual fighting.[43] The rituals turned ugly. According to Chagnon, one of the men from Mishimishimabowei-teri died after he was beaten with an ax.[44]

So "a new era of visiting and potential alliances had opened up"[45] with Mishimishimabowei-teri, but the opening brought the first violent death between the two villages since 1960. And it led to another war, this time against Patanowa-teri, which the new alliance promptly attacked. A picture of the Bisaasi-teri/Mishimishimabowei-teri raiders preparing for their first attack—the largest Chagnon had ever witnessed or filmed—was featured in the 1997 edition of Chagnon's textbook.[46] It was part of "a new chapter that discusses how a dramatic alliance between the Mishimishimabowei-teri emerged, ending a war between them that lasted over 20 years."[47] In the new war, however, Bisaasi-teri raiders would blow off the head of Patanowa-teri's headman, Kumaiewa, and kill one other member of his village with a shotgun.[48]

Chagnon blamed these shotgun attacks on missionaries who unwittingly lent guns to the Yanomami, ostensibly for hunting purposes. But the real problem was that villages on the Orinoco could barter goods to buy shotguns. In fact, the first shotgun killing by a Yanomami was committed by Heawe, son of the Mahekoto-teri headman, after *The Feast* in 1968.[49] A new shotgun was worth somewhere between six and ten new pots;[50] an old shotgun, much less. The tremendous windfall in steel wealth the AEC expedition dispensed for its filming event, including a large number of new pots, could easily have allowed the Mahekoto-teri's headman to buy his family a gun. By this time, the Yanomami were able to buy guns from many sources.[51] Whatever the immediate source of their shotguns, it is a fact that the two first-recorded shotgun killings were carried out by villages where Chagnon brokered large film productions. It is also interesting that the killings targeted leaders of the rival film teams. The Mahekoto-teri blew off the head of a close relative of Rerebawa, who starred in *Magical Death.* The Bisaasi-teri blew away the Patanowa-teri headman, who starred in *The Feast.*

In 1971, Timothy Asch joined Chagnon at Mishimishimabowei-teri. (*Magical Death* had not yet been released, and they were still on good terms.) The year 1971 became the annus mirabilis in ethnographic filmmaking. Asch and Chagnon took twenty-two miles of footage and made twenty-six films. It was astonishing how productive they were. It was even more astonishing how accommodating the ferocious Mishimishimabowei-teri were.

Within twenty-four hours of Asch's arrival at Mishimishimabowei-teri, on February 26, 1971, a fight broke out. Chagnon had advance warning of who was going to fight, and where. "Bring your camera over here," he ordered Asch. "It's going to start."[52] A flurry of blows, shouts, and duels followed, involving about fifty people in a madcap sequence. This gave rise to the most popular and enduring ethnographic film ever made, *The Ax Fight.* It was their third film to win first prize at the American Film Festival.

One of the novel features of *The Ax Fight* was its inclusion of dialogue between the filmmakers as the events unfolded. First, a viewer saw a frantic scramble, people threatening each other with poles, machetes, axes. Yanomami of all ages and both sexes flailed about, screaming and shouting.

But the camera picked up only a piece of the action, and a very inconclusive piece at that. What happened?

When the *shabono* plaza finally cleared, Chagnon appeared, a pipe in his mouth and a 35mm camera around his neck, looking very pleased. He explained, "Well, two women were in the garden, and one of them was seduced by her 'son.' It was an incestuous relationship and the others found out about it, and that's what started the fight."

"No kidding!" Asch said, equally pleased. "So this is just the beginning of lots more."[53]

But, as Asch edited the film, he deconstructed this simple, sexual explanation. Incest had nothing to do with it, after all. Chagnon's first informant had been incorrect, and, as the film developed, Chagnon realized the fight really started because a young man hit his aunt, who had refused to give him some plantains. "Ethnographic filmmakers had never before been so honest about the difficulties of fieldwork," according to Peter Biella of USC's Center for Visual Anthropology.[54]

But the film was not totally honest, even about this initial misunderstanding. In reality, no one told Chagnon the fight started over incest. Chagnon mistranslated the Yanomami word *yawaremou* as "sexual incest."[55] The word really meant "improper behavior toward a blood relative." By glossing over Chagnon's difficulties at understanding the subtleties of Yanomami language, *The Ax Fight,* like *The Feast,* fostered a comforting illusion that the anthropologist in the wild knew what he was talking about.

Arguments over food were not uncommon among the Yanomami, though they rarely became full-blown village nightmares like the one witnessed in *The Ax Fight.* Chagnon could not accept the idea that a disagreement over food, followed by a blow to a female relative, could convulse the whole *shabono.* After viewing the footage over and over again, Chagnon developed a wholly different theory about the fight: it was actually a conflict between two patrilineal descent groups for dominance of the village.

A group of guests, the Ironasi-teri, had refused to leave the village at the accustomed time. This often happened to Chagnon. His policy of distributing trade goods at the end of each visit gave a strong incentive for everyone to stick around. Chagnon was eventually driven from the Mishimishimabowei-teri after the headman, Moawa, threatened to put an ax in his head—unless Chagnon distributed his machetes to the men Moawa indicated. "Distribute all of your goods and leave," he told Chagnon.[56]

Yet Chagnon treated these trade disputes as secondary. Sex and dominance were always in the forefront of his thinking. "You know the joy of *The Ax Fight* is that because Chagnon was so stuck in simple theories that, right away, the film became a real joke," Asch said in an interview. "It is funny with its simplistic, straightjacketed, one-sided explanation . . . I was feeling, you know, halfway into making the film, this great suspicion of the whole field beginning to fall apart before my eyes as I was putting *The Ax Fight* together. I had a powerful piece of material and it was suddenly looking kind of foolish. . . . I felt it was a little bit like a gargoyle at Chartres . . . one of those strange things that stick out and you say, what's this?"[57]

The Mishimishimabowei-teri offered me other interpretations of the film and the fight. Gustavo Konoko, one of the adolescents who joined the ruckus, claimed he and the other *huyas* (young men) were encouraged to start "una pelea horemu," a fake fight. "He [Chagnon] said, 'Fight with poles! We're going to film, and then I'll pay you. I'll give you whatever you want.' When he said that, many young men bloodied each other, playing. 'Hit each other! Be fierce! Argue! When the young men play, let the women begin to scream at them.' That's what he said." Konoko claimed he and other young men each received a machete, a knife, and red cloth.[58]

Personally, I think the dispute that triggered the ax fight was real and that Chagnon was not, in spite of Konoko's account, coaching matters so directly. But I also accept Konoko's statement as a real reflection of his state of mind and that of other young men at Mishimishimabowei-teri. Without their desire to earn trade goods, the family squabble over plantains would probably not have boiled over into a public free-for-all. I think Chagnon's informants realized that this private fracas was a valuable offering for the film crew. By now, they were all veterans of *Magical Death.* So they expertly rescheduled the fight and relocated inside the *shabono,* several hundred yards away. "It's very strange that Chagnon knew when and where the fight was going to take place," said the anthropologist Leda Martins, who spent three years directing Yanomami health programs for the Brazilian government. "The Yanomami are spontaneous, and, when they fight, they don't send a messenger to the nearest white person to have him come and film it."[59]

Almost certainly, different people in the film had different motives. Some were really angry; others were acting out, hoping for trade goods. At first, the combatants, an uncle, Uuwa, and his young nephew, Mohesiwa, deliberately missed each other half a dozen times. Then, after the older man landed a minor, glancing blow, his nephew got angry and chased him. Things began to take on a different color. "Some people started to get mad," Konoko recalled. "The fight almost became real."[60]

Most of the people in the fight maintained a distance that suggested they did not take it seriously. Moawa, the headman whom Chagnon called the most violent man he had ever met, took no interest in the fight, even though his own blood relatives from Ironasi-teri were beaten. The only thing that concerned Moawa was the camera. The great headman turned his back on his embattled relatives, posed for Asch, and then turned around and went back to his hammock.

A little later, the plaza cleared, and all the others returned unhurt to their hammocks. But a group of seven men surrounded the cameraman Timothy Asch and the soundman Craig Johnson. These Yanomami men were all laughing. One of them, wearing a bright red loincloth, took a new machete, brandished it at the film crew, and pretended to rush them. At the last minute he pulled back. The Yanomami all laughed even harder, though Johnson was terrified. Another man took a pole and deliberately drew a line in the ground in front of the filmmakers, seemingly excluding them from the *shabono.*

"Notice how completely out of their social relationships [we are] that they can kid us about it," Asch observed.

Johnson was still in shock. "Some guy came up with a machete and . . ."

"Yeah, but he was joking!"

"I know, but I didn't know that."

"But they were all—they were all joking! We're really, we're really out of it!"[61]

One of the best jokes about *The Ax Fight* was its solemn title. According to Chagnon, Mishimishimabowei-teri did not have any real machetes and only two old axes when he first met them. In the film, new machetes and axes were everywhere.

Although the images of *The Ax Fight* were confusing and ambiguous, Chagnon's narrative was spellbinding. A 1997 CD version of the film, put together by two USC film professors, included a wealth of unedited material that showed how Chagnon kept rhetorically ratcheting up *The Ax Fight.* He gave one account at Mishimishimabowei-teri, another at a Harvard sound lab, and yet another in an article.[62] Chagnon's first thesis was incest. His second thesis was that closely related men were vying for control of Mishimishimabowei-teri. This latter interpretation matched that of his Ph.D.

dissertation, which sketched Yanomami war in terms of fratricidal conflict over reproductive resources. Chagnon had observed that brothers competed for the same pool of available women,[63] and, as the men got older, they drew closer to their in-laws.[64]

But in 1975 E. O. Wilson published *Sociobiology.* A renowned international authority, Wilson explained how biological relatedness was the key to evolution. In this new conception of biological competition, it was less orthodox for brothers and cousins to fight each other over reproductive resources. In 1977, Chagnon then reworked the entire ax fight into a battle between two groups whose hostility was an exact function of kin distance. The members of Team A were more closely related to one another than to those of Team B. The members of these teams gravitated to one another—like atoms—by the very weight of their biological proximity. Chagnon brought a mathematician on board, performed Olympian feats of genetic looping, and claimed that his revised version was based on conclusive evidence from still photographs only he possessed.[65]

The 1997 *Ax Fight* CD untied all the genetic loops. Of the seventeen individuals Chagnon identified as one "team," only eight actually behaved in the way Chagnon aligned them.[66] A number of closely related males, including three uncles of Mohesiwa, acted directly opposite to Chagnon's detailed descriptions. Asch was right about Chagnon's mental straitjacket.

But the wealth of new material raised new questions. For instance, it was not possible to determine from the freeze-frames and accompanying still shots whether anybody was ever struck with an ax, much less whether, in the climactic moment of the film, the youngster Torawa was "knocked unconscious" or "almost killed," as Chagnon asserted. This was the traumatic blow, never seen, never filmed, that has made students cringe all over the world—as they heard the horrific thud of the Atomic Energy Commission ax descending on Torawa. But Asch admitted in 1992 that he had created the sickening sound of impact by striking a watermelon.[67]

It was a case of the incredible, shrinking *Ax Fight.* Asch had suspected as much, and he designed *The Ax Fight* to undermine any easy interpretation, by including evidence that, as he put it, made the film "unintentionally postmodernist."[68] Unfortunately, his dialogue with Johnson flashed by so quickly that only a few viewers ever shared his moment of epiphany—when he realized the Yanomami were "all joking" and the filmmakers were "really, really out of it."[69]

Student surveys found that a large majority saw *The Ax Fight* as a traditional chronicle of savagery. A sophomore at USC reacted typically: "The only thing I know about the Yanomami is that they act on their raw passions. They are very primitive people. It seems that they don't even think before they act. They are very violent people that just go raiding other villages. They take drugs and they freak out on drugs, and on drugs they've been known to attack people."[70]

Chagnon, in his textbook narrative, also looked upon the Mishimishimabowei-teri primarily as threats. "My study of the Shamatari groups began with threats to my life and ended that way."[71] There was no question of how Chagnon's expeditions and their germs might have threatened the lives of the Mishimishimabowei-teri and the other Shamatari.

During the filming of *The Ax Fight,* the village of Bisaasi-teri was experiencing its worst epidemic since the measles outbreak of 1968. Again, there was a double outbreak of malaria and respiratory disease. Falciparum malaria claimed six lives out of about three hundred at the Mavaca mission; four persons died while trekking to another village, so the missionaries could not immediately medicate them. It was terrible, and other villages along the Orinoco experienced similar outbreaks.[72] But it was a only fraction of the loss the Mishimishimabowei-teri experienced.

In the middle of the double epidemic, Chagnon took Bisaasi-teri guides up the Mavaca River.[73]

With sickness raging at the mission stations, there should have been an absolute ban on travel to the inland villages except for express emergency relief. But the widespread sickness was also related to the frenetic pace of scientific research in 1971. During this year, Chagnon gathered blood at more than a dozen villages on the Ocamo River alone,[74] made first contact with another huge village on the Upper Mavaca (where he also collected blood),[75] and shot sixteen miles of film at Mishimishimabowei-teri. He kept traveling through the malaria and cold epidemics sweeping the mission bases, picking up guides, paying everyone in steel, and never stopping for quarantine controls. Sometimes he had to travel at full throttle at night. He couldn't stop. This was the year Neel sent more geneticists into the field than ever before—three complete expeditions, one after the other. And this was the year Asch received a grant from the National Institute of Mental Health to film the Yanomami as never before. Asch needed half the village of Mishimishimabowei-teri to carry all their gear.[76]

In this way, the worst epidemics to hit the Upper Orinoco coincided with the AEC's two most productive years, 1968 and 1971.[77] Sickness soon spread to Mishimishimabowei-teri. According to Chagnon, one man from Ironasi-teri, Mohesiwa's group, died of respiratory disease right after *The Ax Fight.*[78]

Many other deaths followed. "A month after Chagnon left the Mishimishimabowei-teri in 1971, I was fishing on the Mavaca River," recalled Juan Finkers, a Salesian brother who has lived on the Mavaca River since 1971, where he collected plants and myths and wrote a book, *The Yanomami and Their Food System.* "Two Yanomami, a man and a boy, came downriver in a boat made of bark, like a shell, that they use. 'Where are you going?' I asked. 'We're all dying, and we don't want to die,' they answered. 'We're at the Moshata River [a tributary of the Mavaca], and we're all very sick.' That's the first time I met the Mishimishimabowei-teri because the mission had never gone that far. I went back to the mission to get the nuns who were nurses, and then we went up the Mavaca, where we found a group of twenty-eight Mishimishimabowei-teri. We cooked and took care of them while the government malaria team, which landed by plane in Mavaca, came upriver by boat. They found that twenty-four had falciparum malaria. Others had hepatitis. Many others had either died or fled into the mountains because they go to the mountains in small groups to get the spirits off their trail, so that they can't make them sick any more."[79]

Students who see *The Ax Fight, Magical Death,* or any of the twenty other films about the Mishimishimabowei-teri have not been burdened by the knowledge that the community was decimated shortly after the filmmaking. Chagnon employed the same distancing device he had used to soften the death of so many Yanomami filmed in *The Feast.* He changed the name of the village, again.

The Yanomami who died after the filming of *The Feast* became "Platanal Yanomami," instead of Mahekoto-teri.[80] The dead Mishimishimabowei-teri became "Village 16." In an obscure journal, Chagnon wrote about an epidemic that devastated Village 16—a *shabono* of nearly 400 individuals on the Upper Mavaca that he had first contacted in 1968.[81] Disease wiped out 27.4 percent of its members, 106 people. Because of its location, its size, and the time frame, the village could only have been Mishimishimabowei-teri. To clinch the matter, Chagnon identified Mishimishimabowei-teri as "Village 16" in an appendix to his book *Studying the Yanomamo.*[82]

Chagnon has maintained that respiratory epidemics decimated Village 16. He has admitted he has only a vague notion of when this might have happened—sometime in 1973 or 1974—because he was gone for two years.[83] He cites Salesian nuns as his sources. The missionaries have not surprisingly

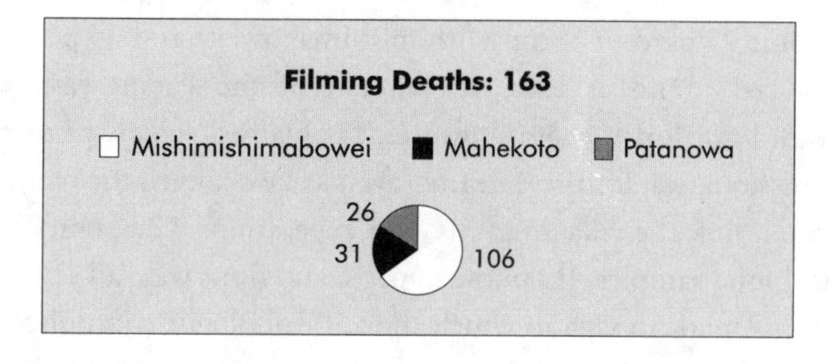

pointed the finger back at Chagnon, saying his expedition was probably responsible.[84]

The mission records support Finkers's account to a significant degree, but not perfectly. After *The Ax Fight* was filmed, Finkers did go up the Mavaca River, where he found, as he claimed, twenty-four of twenty-eight Mishimishimabowei-teri extremely ill from falciparum malaria and hepatitis.[85] However, this happened three months after Chagnon left the field, not one month. And it appears that a later epidemic, in the fall of 1973, was responsible for about 40 percent of the total deaths at Mishimishimabowei-teri.[86]

But whether the Mishimishimabowei-teri died in 1973 or 1971—or sometime in between—the fact is that Chagnon's procedures of "first contact" and alliance making opened up a new era of epidemics. (See the appendix: "Mortality at Yanomami Villages.") Chagnon attributed the deaths at Village 16 to intervisitation with the Mavaca mission.[87] Elsewhere, he took credit for brokering that intervisitation.[88]

When he decided to arrange an alliance between Bisaasi-teri and Mishimishimabowei-teri, Chagnon knew "this was risk taking in spades." He correctly feared that he "might be a contributor to an enormous disaster."[89]

Today, anyone who brings a remote group into permanent contact with the outside world and outside disease is held accountable, at least to the anthropological community, for providing ongoing medical care. "All newly contacted native groups should be provided with immediate, long-term access to modern medical care," according to a 1989 *National Geographic Research* article by the anthropologists Kim Hill and Hillard Kaplan. "Once new diseases are introduced, intervisitation among groups leads to massive epidemics. If untreated, a third or more of the population can die within a very few years. . . . [O]ften, the groups are neglected after the initial excitement associated with contact wanes."[90]

The protagonists of Chagnon and Asch's most famous films all met with disaster. Some 27 percent of the Mishimishimabowei-teri,[91] 25 percent of the Mahekoto-teri,[92] and at least 12 percent of the Patanowa-teri died.[93] Chagnon did not forget them, however. He blamed others, principally the Salesian missionaries, for these deaths, even as he changed the names so that no one could link the villages to his own expeditions. Chagnon's computer printouts, blood samples, ID photos, maps, and films were all scientific supports for an American saga in which the anthropologist triumphed over intransigent Indians and the Indians politely died off-camera. Watching *The Feast* and *The Ax Fight,* knowing that many of the dancers and fighters will soon be dead from imported disease, gives these documentaries the feel of unintended snuff films.

Part II

In Their Own Image, 1972–1994

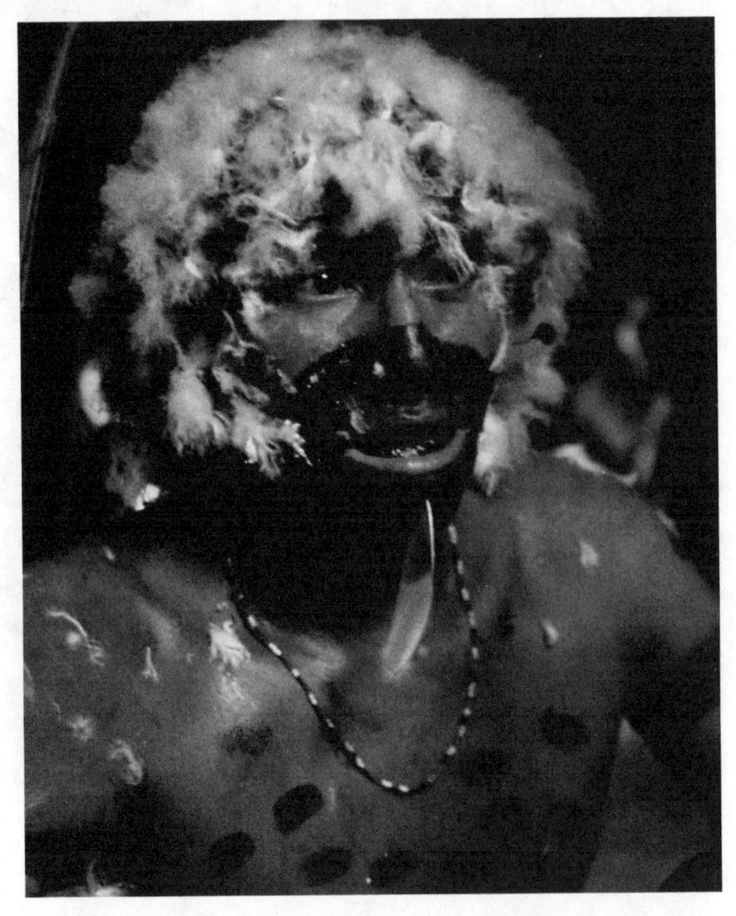

Mario, headman, and shaman near the Catrimani River, 1975 (photo by Giovanni Saffirio, courtesy of the Carnegie Museum of Art)

Chapter 8

Erotic Indians

I was a go-between in most of the love stories I set down, and sometimes I was a witness.—*Jacques Lizot*[1]

The Yanomami at Adulimawa-teri in the Parima Mountains saw a strange portent in the sky in June of 1969. It looked like one of the balloons the missionaries' children played with, but it was launched from an airplane, like a seedpod released from the high forest canopy. As the white sphere floated nearer, they made out a creature tied up under it, a new kind of man. When he emerged, he had a fire weapon, a big knife, and a rope, and the parachutist needed all three to subdue the French anthropologist Claude Bourquelot, who had tried to kill his colleague Jacques Lizot with a machete.[2]

Bourquelot's sudden madness was a mystery. He was a newcomer who did not know anything about the jungle or about Jacques Lizot. By one account, Lizot took Bourquelot into the bush and then left the tenderfoot to his own devices.[3] Bourquelot panicked, got lost for a week, and, when he managed to find the village again, conceived a single desire: to split Lizot's head open. What sounded like insanity to the authorities in Caracas, who received Lizot's rescue call, was to Bourquelot an obvious solution. The problem, the poison, was inside Lizot's head. "Lizot does very cruel things sometimes," Napoleon Chagnon observed in regard to Lizot's treatment of Bourquelot. "I don't know why he does such cruel things."[4]

Jacques Lizot was the ultimate outsider among the Yanomami—Gypsy, homosexual, and Parisian. He was an orientalist who came to South America with a mastery of linguistics and French culinary arts and soon established a reputation for both ferocity and erotic energy that surpassed any Yanomami's. Although he was repeatedly denounced for child molesting, he served only a short stint in jail, and was quickly released at the insistence of a Venezuelan congressman.[5] Lizot's connections to the University of Paris, where he studied with no less than Claude Lévi-Strauss, guaranteed a kind of immunity in Venezuela, which remains a Francophilic country. French anthropologists are held in particular esteem because, outside the United States, Lévi-Strauss's structuralism rules supreme.

Structuralism is a glass bead game of elegant symbols and aching ambivalence. As a college freshman, I was introduced to structuralism through Lévi-Strauss's *Tristes Tropiques,* an immensely erudite, often witty history of the anthropologist's long odyssey from a Paris high school to the Brazilian jungle and back again. Lévi-Strauss's ambition was to study a whole continent. To this end, he enlisted three other researchers who explored the Brazilian interior, spending several weeks per tribe, to gather as many artifacts and as much linguistic information from as many groups as possible. With ample funds from the French government, they traveled the wilds of Brazil like the pioneers of America's West, in a wagon train, accompanied by fifteen mules, fifteen muleteers, and thirty oxen.[6]

Near the Brazil-Paraguay border, Lévi-Strauss discovered an entire tribe, the Caduveo, that had "a strong dislike for procreation. Abortion and infanticide were almost the normal practice, so much so that perpetuation of the group was ensured by adoption rather than by breeding, and one of the chief aims of the warriors' expeditions was the obtaining of children."[7] The small percentage of children who were not murdered at birth were promptly painted black and given away to some other family. The Caduveo rejected parental responsibilities to devote themselves to sculpture and painting, not unlike a colony of displaced Parisian artists. Lévi-Strauss compared the Caduveo's artistic genius to the Spanish Baroque masters.[8] He saw them as murderer-aesthetes in "some romance of chivalry, absorbed in their cruel game of prestige and domination in a society which . . . created a graphic art which is quite unlike almost everything that has come down to us from pre-Columbian America. . . ."[9]

Lévi-Strauss did not offer a single footnote—no data—to buttress his phenomenal find. A tribe that murdered almost all of its children and gave the rest away appeared to contradict the basics of both natural selection and natural affection. Lévi-Strauss's Caduveo seemed, in their fanatical cultural determinism, as inhuman as Chagnon's Yanomami in their biological determinism.

Lévi-Strauss feared that the Amazonian tribes he studied were "miserable creatures doomed to extinction."[10] Nevertheless, he firmly eschewed activism on their behalf because it would have shattered his scientific mirror of contemplation and objectivity. "Never can he [the anthropologist] act in their name . . . such a position could not but prejudice his judgement."[11] This detached ambition did not endear Lévi-Strauss to contemporary anthropologists struggling for Amazonian native rights, like Linda Rabben of Amnesty International. She recently asked of Lévi-Strauss, "Why bother to learn about Indians, after all, if they are doomed to extinction and we to passivity?"[12]

But Lévi-Strauss was not entirely passive in the Yanomami's political history. He took what turned out to be the momentous initiative of steering Jacques Lizot away from Eastern studies—in which Lizot had obtained a Ph.D.—and reorienting him toward the Yanomami. Lévi-Strauss encouraged Lizot to join a large French expedition to Yanomamiland in 1968.[13] Lizot arrived on a flying boxcar with Timothy Asch and Napoleon Chagnon, accompanied by several French doctors, mounds of fresh vegetables and trade goods that included plastic dolls.[14]

Lizot initially established himself at Bisaasi-teri, the AEC's base camp and the site of two missions. Lizot traveled in the company of boys, who began accumulating trade surpluses. This was not altogether surprising, however, since other anthropologists also preferred hiring young boys as guides. In fact, Lizot appeared to be an enthusiastic heterosexual, and initial concern was focused on Lizot's pursuit of young girls. The nurse Marie Dawson recalled how the women of the village dealt with Lizot's sexual advances. "One day they all came running to my house, with Lizot running after them," she said. "When I let them inside my gate, Lizot looked at me and walked away."[15]

A Yanomami shaman has described the same incident from the Indians' perspective:

> [A]ll the women of the village ran yelling and screaming into the naba's house, almost breaking the door. They jumped over tables and chairs in a race to hide behind Keleewa's [Dawson's] wife. Some were even trying to get their heads under her skirt. Because Keleewa's wife couldn't understand our talk yet, she thought it was a raid. Then the naba, A.H. [Ass Handler, Lizot], appeared and stuck his head in the door. He saw the women hiding under things and behind the white girl. The women hoped that A.H. wouldn't bother them when there was another naba around who wouldn't be afraid of him.[16]

Mark Ritchie, a businessman turned author, recorded this testimony for his book *Spirit of the Rainforest.* Ritchie is a Christian evangelical who is not sympathetic to homosexuality—or to anthropology. His book attempts to let the Yanomami speak for themselves and, though most of his sources are also evangelicals, does break new ground. What the Yanomami say about outsiders in general—including a few missionaries, both Catholic and Protestant—is not flattering. But what they say about Jacques Lizot sounds as unlikely as Lévi-Strauss's description of the Caduveo—a play of ambition and domination so cruel it could be true only in a medieval legend.

Ritchie has designated Lizot as A.H.—Ass Handler. In fact, "Ass Handler" is a polite rendering of Lizot's Yanomami name, Bosinawarewa—literally, "Anus/Vagina Devourer." Ritchie's translator was the missionary Gary Dawson, who also served as *National Geographic*'s translator,[17] and possibly speaks Yanomami better than any other white person. At the Padamo mission, Dawson showed me the transcripts of the interviews about Lizot. "I have no problem identifying 'Ass Handler' " Dawson said. " 'Ass Handler' is Jacques Lizot."[18]

Gary Dawson clearly does not like Lizot, whom he knew at Bisaasi-teri. Lizot was expelled from Bisaasi-teri because he beat a thirteen-year-old boy, whom Ritchie calls Youngbird (the same "outcast"[19] who had been Chagnon's guide to Mishimishimabowei-teri). Lizot reportedly punished Youngbird because he was angry that someone—he didn't know who—had stolen his food, a common source of conflict between Yanomami and their visitors. Youngbird, an orphan with no male relatives, had no one to protect him when Lizot snuck up by night and savagely beat him in his hammock. Gary Dawson, who was seventeen years old at the time, had to be restrained (by his father, the missionary Joe Dawson) from going after Lizot with a shotgun.[20]

After the beating, the government medic at Bisaasi-teri, Juan González, treated Youngbird, whose eyes were swollen shut. Then González, accompanied by the elder Dawson, went to Lizot's hut, where the anthropologist had locked himself in. González banged on the door and ordered Lizot out of the village. Lizot threatened to throw González into the river. But González weighed about 250 pounds; Lizot, though wiry and tough, was a small man. González repeatedly jabbed his finger into Lizot's chest. "*You* are going to throw *me?*"[21]

Lizot left. He traveled farther up the Orinoco to the Catholic Platanal mission, where the Yanomami boys of Mahekoto-teri began to acquire a windfall of *madohe,* foreign stuff. A Salesian missionary, Father José González, confronted Lizot in the *shabono*'s central plaza.

"What are these boys doing for you that is worth all you are paying them?" González asked. "I hear you're paying the boys to use them for sex."

"That's a lie! Where could you have ever heard a thing like that?"

"From this boy right here."[22]

González asked Lizot to leave Platanal; Lizot refused, and threw a punch at the priest. A fistfight ensued, in which Lizot was knocked out. He left, expelled again, the wandering anthropologist.[23]

Lizot then moved downriver to the small community of Tayari-teri, on the main course of the Orinoco between Platanal and Bisaasi-teri. He selected a site up a short distance from the Orinoco, on a creek, where he built a house and began an international campaign against the Salesians.

There were legitimate grievances against Padre González and his mentor, Padre Cocco. Both of them were beloved figures among the Yanomami and were tolerant of Yanomami customs like polygamy and the use of hallucinogens. (Cocco's views of the Yanomami were portrayed sympathetically by Asch and Chagnon in the film *Ocamo Is My Town.*) Essentially, Cocco and González viewed the Yanomami as undernourished, impoverished people—not unlike the ghetto children in Milan whom the Salesians' founder, John Bosco, had set out to help with vocational education.[24] Unfortunately, both priests supported the Venezuelan government's plans to introduce peasants from the Orinoco region among the Yanomami.[25]

Lizot threatened to burn down the Salesian missions. The Salesians, who took the threat seriously, petitioned the government for Lizot's expulsion. But Lizot had become the champion of Yanomami culture to many people, including the French embassy. Whatever Lizot's motives, he successfully turned back a colonization plan that amounted to ethnocide.

Napoleon Chagnon also supported Lizot in this battle. "I wrote a letter to the Venezuelan government supporting his right to do research after he had threatened to burn down the Salesian missions," he said.[26] At this time, the two were close friends, united in their opposition to González's plans.[27]

When Chagnon reentered the field in 1976, he arrived with a growing international reputation and a $260,000 grant from the National Science Foundation, primarily to study "Mortality and Divorce in Yanomamo."[28] This was about $1 million in today's currency, and it significantly raised Chagnon's status, enabling him to hire two senior consultants, one being Robert Carneiro of the American Museum of Natural History, and three graduate students.

One of the graduate students was Kenneth Good. Until then, Good had been a good friend and protégé of Chagnon's. They got to know each other at Penn State University, where Good was Chagnon's drinking buddy. "We used to go down to bars and drink together," Good recalled. "It was an embarrassment, but I did it because he was going to be my chair. He was the type of guy who had German shepherd attack dogs, and he'd have people come over to his house in the afternoon and he'd have the students dress up in padded suits and have the dogs attack them. Oh, yes. They'd have to put out an arm or a leg and the dog would attack. Students could get injured. And he used to like taking the attack dogs—whose names were Gus and Parma—into bars so he could corner big, 200-pound-plus weightlifter types."[29]

To prepare his students to deal with the Yanomami's supposed extreme violence, Chagnon obtained extra-strength chemical mace from the Pittsburgh Police Department (which Good re-labeled "Center County Dog Repellent" in order to pass customs).[30] Chagnon also armed Good with a double-barreled Winchester shotgun.

Once in Caracas, Venezuela, members of the expedition fanned out across the city for supplies. "We had to get barrels of trade goods and big sacks of rice. . . . We bought axes, machetes by the box, loincloth material by the roll, fishhooks by the thousands. All this was added to the tons of equipment we had shipped down in advance, in addition to what we had brought with us on the plane. It was an incredible operation. I had to admire Chagnon's energy and persistence. The man was driving, driving, driving, all the time."[31]

Although Good had a severe cold, they rushed into the rain forest on a DC-3, which dropped them at the Ocamo mission. "We started dumping our things off the plane, a sight that left the little crowd of missionaries and

Indians that had gathered wide-eyed with astonishment. There were fifteen army trunks bursting at the seams, all of them painted different colors as part of a coding scheme. . . . Then four outboard motors came out of the cargo door, all of them packed in wood-framed, protected crates. . . . The knot of onlookers couldn't conceive that such a stupendous accumulation could possibly belong to such a small number of people."[32] It took four hours for a priest and his tractor to haul all their stuff to the banks of the Orinoco, where it was loaded onto a covered riverboat that Good dubbed the *African Queen.*

They shoved off immediately and headed upriver, traveling until midnight, when the riverboat ran aground on a sandbar. But they pulled free in the early morning and reached Lizot's camp that same day. In previous years, Chagnon had used missions for supply depots,[33] but now he wanted to show his solidarity with Lizot. As Good explained it, "Lizot was currently at war with the missionaries, and he didn't want us—colleagues from the anthropological community—to have anything to do with them either."[34] Chagnon got a room in Lizot's house. The students slept in huts.

During his first, nervous night in the jungle, Good was terrified when two screaming men burst inside, pushed him into a table, and ripped his mosquito netting. In the ensuing tussle, all three men wound up sprawling on the ground, bruised and covered with mud, but not before Good recognized his assailants as Chagnon and another anthropologist, both drunk. Good, a tall, husky man, was so angry he threw Chagnon, who is much smaller, over an embankment.

"*Tranquilo,* Ken," Lizot said, as he helped bring peace.[35]

Fortunately, Chagnon could not remember what had happened to him when he woke up, rather bruised and muddy, the next day. Good never forgot the experience, however. It was the only time anyone ever attacked him in Yanomamiland. "In my twelve years, I witnessed only one raid."[36]

In the end, Good turned his chemical mace against bats. "In my opinion, the Fierce People is the biggest misnomer in the history of anthropology," Good said. "The Yanomami were quaking with fear when they met me. They were in awe of foreigners, the *nabah.*"[37] Chagnon's other students would also report much lower levels of violence than their mentor found among the Yanomami.[38] But Good was the only one to write a book about his fieldwork experience—*Into the Heart*—which has been well received by anthropologists and translated into eight languages, becoming the most widely read account of the Yanomami after *The Fierce People.* With Good's arrival on the scene, it is as though a granite boulder by the banks of the Orinoco had been lifted up, revealing a teeming underworld.

Strangest of all was Lizot's village of Tayari-teri. Chagnon built a storage depot at Tayari-teri that Good visited from time to time. One of the first things to strike him was that the men greatly outnumbered the women, and that the women were not allowed to approach Lizot's bungalow, where the anthropologist lived with a community of boys. The women had a separate path around Lizot's compound. And the boys had odd chores, including the tending of a marijuana patch.

Lizot's boys behaved differently in other ways, too.

"All the guys were smoking and wearing deodorant and stuff," recalled Good. "It was disgusting. Apparently it was a bunch of queens or something. Oh God. Yeah, the kids of Tayari-teri. They used to have fifteen things of beads around their necks. Oh God. I don't know what all the reasons were. They were Lizot's guys—they were eating spaghetti. Yeah, kilos of spaghetti, and they'd cook it up and the kids would go down and wash all the pots. They were like his house boys. And he felt he was paying them well. Not only spaghetti and deodorant and cigarettes but machetes and all this stuff. One time I opened a big, thirty-gallon, waterproof drum, and it was all full of packs of cigarettes—there must have been thousands of packs of cigarettes. It was really impressive. I've never seen so many cigarettes in one place."[39]

Lizot's war with the Salesians continued and then abruptly ended. Lizot won. Father González's church superiors ordered him to leave Platanal mission. Profoundly depressed, González traveled downriver to the Ocamo mission and went to a New Year's Eve party with a Salesian brother, Emilio Fuentes, which lasted long into the night. As a grand finale, Fuentes took González riding on a tractor, very fast while they were still drunk, and somehow González fell out and broke his neck. "I was with Lizot when we got the word that Father González was killed," Good remembered.[40] Brother Fuentes left the Salesians after this misadventure, and Lizot arranged a scholarship for him to study anthropology in France.[41]

Good's relationship with Lizot was also turbulent. "He used to get pissed and write letters to Chagnon when I used to stay there for more than a couple of days that I was bothering him—that I was interrupting. And he kept thinking I was going around asking about him whether he was smoking dope and I wasn't. The guys would just come and talk to me, and I would put two and two together. They said, 'The *policía,* the police are coming and smoking this grass stuff; the police they don't like that.' I never asked them whether he was ass-fucking these guys or anything. . . . This is what Eibel [-Eibesfeldt] wanted me to do."[42]

Irenaus Eibl-Eibesfeldt is Konrad Lorenz's successor as the head of human ethology at the Max Planck Institute and Germany's best-known evolutionary biologist. He began studying the Yanomami in 1971 and hired Good as a researcher after Good's relationship with Chagnon soured in 1977.[43] Eibel-Eibesfeldt was reportedly horrified by the stories about Lizot's pedophilia and wanted Good to document it. But Good was reluctant to get involved. He and Lizot had become allies of sorts. They jointly wrote a letter to *Science,* accusing Chagnon of presenting a Maquiritare Indian village as a Yanomami one in an article about protein consumption.[44] (This led to a sharp break between Lizot and Chagnon.) Shortly afterward, Eibel-Eibesfeldt began working with Lizot himself.

In the end, everybody worked with Lizot. "I didn't like going to Tayari-teri, but that's where I went for R and R," Good said. "I didn't know that Lizot had turned Tayari-teri into Sodom and Gomorrah with mutual masturbation, anal sex, and a system of paying for these favors with trade goods—and all of this among a people, who, as far as I know, never practiced homosexuality until Lizot."[45]

Meanwhile, Lizot began writing the ethnography of Yanomami sexuality. Sodomy was normal for children. "One can frequently see boys of all ages simulate it publicly in their games; often brother-in-laws are involved, for these are usually devoted to each other through mutual and lasting affection. Homosexual practices, though more frequent in this kinship category, are not exceptional between brothers or first cousins. If it is scandalous to 'eat the vagina' of a sister . . . there is no shame in 'eating the anus' of one's brother."[46]

According to Lizot, there was no shame and no blame for any kind of sexuality, even animal intercourse. Children practiced sodomy and bestiality as training for adult mating. In Lizot's world, the Yanomami were also ingenious masturbators. They used everything from holes in the ground to tree stumps and dead animals.[47]

There are Amazonian tribes where homosexuality is common, but even among them it is usually discreet.[48] According to both Helena Valero and Father Luis Cocco—two sources considered authoritative by the anthropologist Brian Ferguson in his survey of Yanomami literature[49]—homosexuality among the Venezuelan Yanomami is rare, brief, and culturally unacceptable.[50] One Yanomami legend mentions homosexuality, but treats it as dirty and nonhuman.[51]

Alcida Ramos, a Yanomami specialist at the University of Brasilia who holds a Ph.D. from the University of Wisconsin, was the first to politely argue that Lizot's erotic Yanomami were a projection of his own personality. "Discretion and naturality . . . are overriden by Lizot's voyeurism. . . . Having the ever-present narrator hovering over them has the effect of lending the Yanomamo an unreal quality, as if they were characters in a play. . . ."[52] Ramos also ventured that Lizot's unfettered love stories would illicit "a chuckle of disbelief" from the Yanomami—particularly the passages about achieving orgasm in broad daylight inside the communal *shabono.*[53]

I have yet to speak to a Yanomami specialist who ever observed the kind of openly displayed sexuality Lizot described—except at Lizot's village.[54] "I never saw homosexuality among the Yanomami," said Good, who spent seventy-one months in the field. "The lewd atmosphere at Tayari-teri—where kids were doing all this ass-grabbing and talking dirty—was Lizot's creation. Where Lizot describes this type of sexual behavior, he's describing what he created. We're learning about Lizot, not the Yanomami. It was a terrible place, Tayari-teri. I don't know, it was like the Yanomami there had been corrupted or something. I was always relieved to get back to Hasupuwe-teri, where they didn't have any of this."[55]

But Hasupuwe-teri, located above the formidable Guaharibo Rapids, did not remain a haven for long. After Good left the remote village, to take a furlough in Caracas, he returned to find that Lizot had been visiting in his absence and pursuing Hasupuwe-teri's teenage boys with gifts and threats. "Most of it was just told to me by the guys—how they'd run off in the gardens when he came, how he'd grab them and he'd threaten them. First he'd offer them a loincloth, and then he'd get his shotgun and he'd threaten them and they got away. And they were scared to death of him."[56]

New words were apparently invented for this phenomenon. In some villages, sodomy became *Lizot-mou,* "to do like Lizot."[57] At most *shabonos* throughout the region, however, an excruciating new compound verb appeared: *Bosinaware.* If the Yanomami conceive of death as spiritual cannibalism, their verb for intercourse, *naware,* literally means "eating the vagina." Broken down, Lizot's nickname translates as ass *(bosi)* vagina eater *(nawarewa).* The suffix *wa* turned the verb into a male name. According to Good, "Ass Fucker" is the most accurate translation.[58]

Good recorded one of the Hasupuwe-teri boys' accounts of Lizot's attempted molestation. "When Lizot heard about it, he got defensive," Good said. "So, to try and smooth things over, he invited me to a really posh restaurant in Caracas, and, as we ate, he explained to me that he didn't really have anal intercourse with the Yanomami. They just practiced mutual masturbation. I guess he thought that was okay."[59]

This conversation took place at Porto Vino, still an excellent Italian restaurant. "He really spent his wad on me—there was all this expensive food and wine. And while I'm putting pasta in my mouth, he tells me he's only having mutual masturbation. 'Oh.' What am I supposed to say?"[60]

Gary Dawson and Mark Ritchie videotaped three Yanomami men who gave hearsay accounts of mutual masturbation at Tayari-teri. The men, who were all adolescents when Lizot entered the field in 1968, had firsthand contact with Tayari-teri and with Lizot. I know all three—Jaime, Pablo Mejía, and Timoteo. Today they are Christian evangelicals, fluent and literate in Spanish, who live at the village of Koshirowa-teri on the Padamo River, near an unaffiliated Protestant mission. Originally, Pablo Mejía knew Lizot at Bisaasi-teri, but all of them visited Tayari-teri as teenagers. (Lizot also visited Koshirowa-teri.) Pablo and Timoteo currently play leadership roles at Koshirowa-teri, which numbers about four hundred individuals, most of them evangelical Christians.

Pablo Mejía: Remember one of my relatives that lived there [at Tayari-teri]? . . . He knows all the people at Tayari-teri because he lived there as a son-in-law. And that guy told me, "When I was asking Lizot for work, Lizot said, 'Yeah, I've got good work for you. Come over to my house.' So there were a lot of people at Lizot's house when we got there, and he made them all leave and he made me stay all by myself. Then he said, 'Now I'm going to give you your job.' He thought he was going to show him a job in the house. But Lizot just got in his hammock and he called the boy and he stood beside him. And he really didn't know what the guy [Lizot] was going to do. He hadn't heard any of the stories. And Lizot got all naked and got back in his hammock and said, 'Here take your hand and go up and down my penis.' And although he was really afraid he thought, 'Well, I sure want a lot of his stuff.' So he went ahead and grabbed hold of him anyway and began moving it up and down."

Jaime: Yes, I heard the story from the same guy. And I went to him and asked him personally, "Did you do that or did you just hear that from somebody else?" And the guy said, "No. You see my gun here? That's how I got it. And that's how all the young guys *[huyas]* who have guns in Tayari-teri got their guns. That's the way that we have to earn stuff *[madohe]*. And I also got a suitcase from him. And by doing that with his penis, I got a radio, too." So I asked him, "Didn't your hands get all dirty and slimy?" And the guy said, "Yes, my hands sure did get all slimy and ugly, but I kept doing it anyway because I wanted the stuff he would give. . . . Everybody did it because of the stuff they wanted."

Pablo: Yes, it's something you would normally never hear about. But over there, to hear them tell it, it's just so common. After I heard that boy tell it, I asked another, older man from Tayari-teri if that was true. And the old guy from Tayari-teri said, "Well, that's how all the young people at our village got all their trade goods. We older men never did that. We wouldn't do that. But we shared the trade goods that everybody got." And whenever he [Lizot] was in the urge, he would call all the young guys and he would say, "If you guys want work. . . ." And there was always somebody who wanted it. So he would say, "Who wants it this time?" And they would come.

Timoteo: I asked him [another individual], "How many times would you have to do it to get a gun?" And that guy, who had a 12-gauge shotgun, told me he did Lizot six times. And if he wanted a shirt or something else, it was just once or twice. He went to Caracas and then to France. He told me that when he asked Lizot for a shotgun, Lizot told him, "You'll have to do me six times." And he gave him a 12-gauge. And he had three 12-gauge shotguns, and I bought one of them from him.[61]

These accounts appear to confirm Lizot's claim to Good that he practiced some kind of masturbation with the Yanomami. The missionary Gary Dawson told me that one of the men interviewed had performed sexual favors for Lizot, but would not admit it on camera.[62] However, it was not very difficult to surmise which man that was, since he shifted from describing other boys' encounters with the naked Lizot to a conversation he himself had with a naked Lizot. And when he tried to explain how he got one of Lizot's shotguns, both he and Dawson started laughing.

Just as scientists and missionaries were—and are—reluctant to tell all they know about Lizot, the Yanomami fear him and his very real political power, based on his long influence with the government bureaucracy and the French embassy. His former partners may also be constrained by shame. Whatever homosexual practices the Yanomami had prior to Lizot's arrival, shotgun-driven prostitution is nothing to brag about in their culture. The attitude of other Yanomami toward Lizot's alleged sexual partners has ranged from sarcastic contempt to outright opposition. When one boy at the village of Koshirowa-teri obtained a beautiful watch from Lizot, he also got a new name from the other children: "One-Who-Strokes-A.H.'s Penis."[63]

Elsewhere, an armed alliance against Lizot's village of Tayari-teri was emerging. According to Ritchie's informants, the initial reaction was based on the Yanomami's abhorrence for the anthropologist's sex practices. One of the Bisaasi-teri's two headmen, Paruriwa, found that his son was involved with Lizot and drove him from the *shabono* at arrow point.[64] A coalition of village leaders allied with Bisaasi-teri then met to plan Lizot's murder. One of the strongest advocates for killing Lizot was Rerebawa, Chagnon's old guide and one of the three men to whom *The Fierce People* was dedicated. Paruriwa said, "I say we kill him!"[65] Paruriwa was unable to muster enough support for an assault against Lizot, because the other Yanomami were fearful of killing a *nabah* and because Lizot's trade goods were indispensable to Karohi-teri and other Tayari-teri neighbors. "He could never do this if we weren't so poor," Rerebawa lamented. "We're trapped. We're backed into a spot with no escape."[66]

The child-molesting controversy became part of the growing conflict between Tayari-teri and Bisaasi-teri for supremacy of the Upper Orinoco. Bisaasi-teri had not forgotten Chagnon (Shaki). In fact, Kaobawa, Bisaasi-teri's other headman, moved with his faction to a new *shabono* named Shakita, in honor of Chagnon.[67]

It took Chagnon ten years to visit his namesake. By 1976, Venzuelan scientists at the leading research institute, IVIC, were "not very enthusiastic in continuing sponsoring Chagnon's research," because they felt James Neel "had taken advantage of IVIC's facilities and resources without giving too much in return."[68] Still, Chagnon might have prevailed at IVIC, through the intervention of Marcel Roche, an influential doctor who arranged for Chagnon to give a presentation there in 1975. Unfortunately, at the end of the talk, a student asked Chagnon a question: What was he going to do to help the people he had been studying for so long?[69] Kenneth Good recalled, "When we gave our presentation to IVIC, he was very uppity, very arrogant, coming in with a quarter-million dollar grant. When someone asked him, 'What are the Yanomami going to get out of it?,' he answered,

'Well, they're going to get a hell of a lot of machetes and trade goods.' "[70] According to the anthropologist Leslie Sponsel, "All of them gave a talk at IVIC, but at the end a student asked, 'You've been working with them for ten years. What are you going to do for the Yanomami?' He said he couldn't interfere. He was a scientist. After the meeting, spontaneously a group of students and faculty met in the IVIC library upstairs, and all hell broke loose."[71] The Venezuelans received additional ammunition when Chagnon then offered an Andean archaeologist at IVIC, Alberta Zucchi, a $1,000 consulting fee for the Yanomamo project, even though she had no Amazonian experience and no interest in cultural anthropology. Since Zucchi's husband was the bureaucrat in ultimate control of all research permits at the Ministry of Justice, this was interpreted—perhaps mistakenly—as a clumsy bribe.[72] For ten years, from 1976 to 1985, Chagnon got no permits.[73]

His long absence hurt Kaobawa. Whereas Bisaasi-teri had acquired many young women from neighboring villages during Chagnon's tenure,[74] the terms of exchange reversed as Kaobawa ran out of steel. Far more women left the village than married into it.[75] Kaobawa was increasingly marginalized. "Chagnon no longer lived with them, taking away a major source of upper Bisaasi-teri's wealth, military security, and status."[76]

During a roughly corresponding period, the population exploded at Tayari-teri, not because of medical attention or lowered infant mortality but because of immigration—almost all of it male. Tayari-teri's population went from forty-one in 1974 to eighty-eight in 1979. Napoleon Chagnon was the first to point out a number of demographic anomalies: "Whence came the new residents? One suspicious thing is that the adults outnumber children by a wide margin, so the increase is unlikely to have been due to new births. Another suspicious item is the adult male/female sex-ratio of 143 to 100 [31 males, 21 females]—proportionately far more adult males than is normal for Yanomamo villages."[77]

At every other center of outside power, where trade goods purchased outside wives, the sex ratio shifted in favor of the Yanomami who resided with a *nabah*.[78] This was particularly evident whenever the Yanomami acquired shotguns, because the shotguns gave such a decisive advantage in hunting that a man could now support more than one wife.[79] But at Tayari-teri, in spite of unprecedented trade goods and shotguns, there was a curious dearth of women. And instead of an influx of women from other villages, a host of young boys were arriving.[80]

They came from all over,[81] including Bisaasi-teri's Shamatari ally, Momaribowei-teri,[82] from where Pablo Mejía's cousin emigrated. But most of the young immigrant males came from Karohi-teri, which had been Bisaasi-teri's closest ally.[83] Karohi-teri was Lizot's second residence. By 1969, just a year after Lizot had been active at Karohi-teri, the village had considerably more wealth in manufactured goods than even the richest Bisaasi-teri village.[84]

But Karohi-teri, for all its trade goods, was just the outer courtyard to Tayari-teri's fabulous inner sanctum of cigarettes, machetes, clothes, and shotguns. Lizot's narrative shows that boys were moving first to Karohi-teri and then on to Tayari-teri; according to Lizot, these boys were searching for brides.[85] However, Lizot's own account pointed to something else. For instance, a fourteen-year-old named Fama came down from the Parima Highlands to visit Karohi-teri.[86] It was a long trip, on foot, and his family stayed only one night, just long enough to drop Fama off. Fama decided to stay—even though in the Parima villages there was a *surplus* of women.[87] But, skewed as the male-female ratio was at Tayari-teri, it was even worse at Karohi-teri.[88] Whatever Fama was doing, he was not primarily searching for a wife.

And, if Lizot played a role as "a go-between" in the Yanomami love affairs he wrote about, he was also a keen observer-companion of new boys pilgrimaging from Karohi-teri to Tayari-tari. His book *Tales of the Yanomami* narrates the adventures of Hebewe, who was growing up at Karohi-teri. Hebewe had moved beyond masturbation and sodomy, though he still enjoyed tormenting the newly arrived boy Fama—by throwing him on the ground and untying his penis, the ultimate humiliation. "They laugh all the more and shout their pleasure."[89]

After this nastiness, Hebewe moved on to Tayari-teri. Lizot portrayed the Tayari-teri's great *shabono* as a timeless icon of Yanomami culture:

> There are many young people at Tayari, producing a merry and friendly activity. Hebewe . . . watches the children who are playing in the central plaza . . . do battle with long clubs cut out of soft wood that doesn't cut the skin but can raise welts. The blows are haphazard; those who are struck grit their teeth not to show their pain and try to trade blow for blow with their opponent. When they are finished, the children plant their weapons in two parallel rows.
>
> Nearby a group of youths is inhaling a hallucinogenic drug. As a joke,

> a tall devil of a fellow hails a young boy of about ten and orders him to participate, asserting that it would be cowardly to refuse. The child dares not back away; they blow into his nostrils several doses too strong for him and, stunned by the drug, he collapses, his head striking the ground.[90]

Shortly after his arrival at Tayari-teri, Hebewe fell in love with a young girl who was already married (to another immigrant from Karohi-teri). Rashly, Hebewe and two friends tried to ambush the older youth and shoot him with arrows. Their attempted murder failed; Hebewe was beaten and driven from the village for a short time. But the family of Hebewe's intended victim could not fathom the boy's motives for this unbelievable treachery. "They wanted to kill him for fun! For fun! There was no other reason."[91]

A third Karohi-teri émigré, Tohowe, became Hebewe's confederate in romantic intrigue. He was an orphan who had no relatives at Tayari-teri, so abandoned before being adopted in the village that he used to go around naked. "Adolescence refined his features and gave his body firmer, more harmonious lines, while the texture of his skin retained all its smoothness and beauty. . . . Tohowe was about twelve when he arrived at Tayari." Sadly, Tohowe was showing signs of a nervous breakdown. "His rages, however, are uncommonly violent; he breaks and tears anything that comes into his hands, first his own belongings and then other people's, for in his blind passion he no longer fears anyone. Afterward he retreats into absolute and persistent silence. . . ."[92]

Tayari-teri was full of troubled teenage boys who had just arrived.

Lizot's village was certainly a demographic and cultural anomaly. And the *shabono*'s "merry and friendly activity," as Lizot depicted it, is unfamiliar to me. Although I have visited over thirty Yanomami villages, including Karohi-teri, I have never seen any of the persistent, ritualized forms of cruelty Lizot described. In Lizot's extraordinary opening scene, the Tayari-teri *shabono* was a theater of pain: while a "devil" of a youth forced a drug overdose on a child that knocked the ten-year-old unconscious, other children beat each other with carefully chosen sticks to see who would scream first. The adults, including Lizot, did nothing. Elsewhere, Lizot's hero, Hebewe, incited a group of adolescents to a molestation of another boy, and then, without any remorse or serious consequence, tried to murder a newly married man.

The anthropologist Alcida Ramos observed, "What is especially disturbing is that Lizot's Yanomamo seem to be so very whimsical in their nastiness, inflicting pain on each other for no other reason than that it strikes their fancy."[93] Hebewe's attempted ambush was unique—as bizarre in the annals of Yanomami violence as having sex under the full sun in the *shabono.* Under normal circumstances, it would be self-destructively stupid for a fourteen-year-old newcomer to try and kill an older youth who was related to the village leaders at Tayari-teri. Stranger still, after the ambush turned into a fiasco, was Hebewe's decision to just hang around. He received only a mild beating and, after a brief timeout, was restored to the good graces of the community.

Lizot did not explain that Hebewe was his favorite. A nearby mission doctor nicknamed him El Príncipe because Hebewe would come downriver in the middle of a big, motorized dugout canoe wearing an immaculate white towel around his body, drenched in cologne. "The only thing he didn't have were two Yanomami fanning him. So everybody called him El Príncipe."[94]

Lizot probably distributed more clothes and shotguns than any other individual among the Yanomami. Yet he publicly opposed Western dress and firearms for the Indians, and even at Karohi-teri he would sometimes act like an angry Savonarola. On one occasion, he made all the Yanomami at Karohi-teri turn in their clothes, which he burned in a great bonfire of the vanities.

He could be very demanding of his boys. Women were not allowed at the bungalow at Tayari-teri. And for a long time he did not permit the boys to hang around with the mission Yanomami, whom he regarded as culturally impure.[95] "He also felt he was in competition with the missionaries," said Kenneth Good. "He didn't allow his guys to go near them, even though Tayari-teri was a fission group of Mahekoto-teri of Platanal. I remember he threatened them."[96]

For all these reasons, tension built between Tayari-teri and the poorer mission groups. This was dramatically illustrated in 1978, when Kaobawa, at the head of an unsuccessful hunting party, approached the village of Tayari-teri to ask for some food. He was pelted with mud balls, while children heaped abuse on him.[97] They laughed at Kaobawa, whom they called a poor, friendless nobody who had lost his *nabah.*

Publicly humiliated, Kaobawa retreated to Shakita. The war of words escalated into a war of arrows. Two Tayari-teri were killed first, followed by one Bisaasi-teri. A Salesian nun, María Eguillor García, who had lived in the area for three years and was completing her Ph.D. fieldwork, recorded what happened next: "[I]n their pride that they could not be defeated, they [Tayari-teri and its allies] carried out an amazing feat: They attack Klawaoitheri of the Upper Ocamo, burn their *shabono,* and capture five women. The news spread like fire all over the Upper Orinoco. This time it is too much. . . ."[98]

This time Kaobawa and Paruriwa were able to assemble the largest coalition in the annals of Yanomami warfare—comprising fourteen villages and 150 warriors—to attack Tayari-teri. Coming stealthily, by boat, they surprised the Tayari-teri, burned the village to the ground, killed seven people, most of them with shotguns, and seized "a bounty of cultural and manufactured goods," along with one woman.[99] After this, according to Eguillor García, "Tayari-teri disappeared from the Yanomami geographic map."[100]

When Napoleon Chagnon finally reentered Yanomami territory in 1985, he documented this war and presented his findings a year later, in Santa Fe, New Mexico, at an American research conference on the anthropology of war. "He gave a paper but no one was allowed to quote from it directly without the author's permission, something Chagnon often does," said Brian Ferguson, a professor at Rutgers University. "I can talk about what we discussed, however. The Yanomami had described this as a war between 'Chagnon's village' and 'Lizot's village.' The whole point was that 'Shaki's people' had 'exterminated' 'Lizot's people.' Chagnon's informants said that Lizot had been in his house [at Tayari-teri] when the fighting started and they had considered killing him. According to Chagnon's people, one man was shot with an arrow after he walked out of Lizot's house and Lizot was forced to flee to Karohi-teri." Although Chagnon's off-the-record paper undermined Lizot, it also raised questions about Chagnon's own influence in Yanomamiland. To Ferguson, it seemed amazing that two anthropologists would have villages named after themselves and that these unusual groups would wage war against each other for regional supremacy. Chagnon's account also confirmed what the author Mark Ritchie's Yanomami informants reported: that the Bisaasi-teri—"Chagnon's people"—had plotted to assassinate Lizot, but did not carry the plan to completion. They did, however, succeed in driving Lizot from the main course of the Orinoco.[101]

After Tayari-teri was burned down, at the end of 1979, Lizot left his bungalow and relocated to his older, secondary center of Karohi-teri, on the Manaviche River. "It was an amazing place," recalled Jesús Cardozo,[102] a Venezuelan anthropologist and former student of Chagnon's who helped his mentor return to Yanomamiland after his long exile. But, like Good, he soon broke with Chagnon. Cardozo became Timothy Asch's partner in Venezuela.[103]

"I remember Timothy Asch insisted that we meet, although Chagnon had forbidden me to meet Lizot," Cardozo said. "When I first went to his house, it was very far away and I thought I'd gotten lost because we just went on and on and on. Finally it was getting dark, and I was getting worried—I was with Tim—we see these Yanomami. So I asked where Lizot was, and they jumped into our dugout and took us a couple of miles upriver and I couldn't believe. Did you ever see *Apocalypse Now*? There was Richard Wagner music coming full blast out of the jungle. And there were these Yanomami with headsets on, listening to classical music. Then, later that night, Lizot explained to me that they had a passion for Mozart and for rock, for acid rock, that combination apparently hit them. It was very funny, that whole thing. He had dozens of boys working for him. Everybody seemed happy. Everybody was getting all sorts of gifts. And, oh, he had *lots* of things to give away. And, sure enough, some Yanomami had their own outboard motors and their dugout canoes, and some of them had shotguns. I mean they were rich people—all the Yanomami in the area that lived for him. We used to call it Boys Town. They all had perfume and whole necklaces and stereo equipment they'd listen to. It was wild. It was obvious what was going on. It was not something that you had to do a lot of research about. It was just out in the open.

"So I remember talking at times when I would just be sitting by myself talking to Bórtoli [head of the Salesian mission at Mavaca]. 'Well, Bórtoli what's your opinion? Don't you think this is a type of prostitution that's going on here? I mean sexual favors are being bought. I mean it's open.' People would even tell you how much—what exactly they had to do, that kind of stuff. . . . I mean they've actually described every single physical movement they do. They would actually act it out. . . . Kind of jerking people off and that kind of thing—sexual activities. To be quite honest, I would listen to the things but I felt kind of disgusted."[104]

Cardozo later went on, "I remember one night when I went to Karohi, and Lizot was not there, one of the men came out and sat down next to me. He seemed very sad. I asked him what was the matter. He said, 'Why must we keep doing this?' I asked what he meant—'Doing what?' He made the motions of masturbating with his hand. 'Why must we keep doing this for this man?' he asked. He was so disgusted with it. But it was also as though he considered it absurd, that this was the only way they could get trade goods, by masturbating an old white man."[105]

I asked Cardozo why he didn't do anything about this obvious abuse.

"Now, the problem is that Lizot has had enemies, he's got enemies, and they in fact have tried [to expel him]. The first time I remember in 1986 when I returned to Venezuela I met with this French couple Jean Chiappino [and Catherine Alès, both anthropologists who study the Yanomami and work for ORSTROM]. That night, I had just pretty much gotten off the plane, and they said, 'You should not meet Lizot.' And I said, 'Wait a second. That's what Chagnon says.' He said, 'No. The thing is that if you meet Lizot you will regret the day you were born.' I said, 'What's this?' He said, 'Lizot is the devil incarnate. He is evil itself.' I said, 'How can a person be this bad?' So that evening they told me the most incredible stories about Lizot—Lizot would chase them through the jungle and shoot at them, you know? Apparently they worked together, and they had their run-ins, so they parted ways. And in the midst of all these problems, on one occasion Lizot would chase them and shoot them through the jungle. So one of them, this guy Chiappino, returns to Caracas and files at the Dirección de Asuntos Indígenas a formal complaint against Lizot. That's what Chiappino told me. He filed an accusation of Lizot practicing sodomy with the Yanomami. Lizot comes back and hears about it and in turn accuses Chiappino of being a homosexual and having sex with the Yanomami. Apparently then what happened was that Lizot was supported by the French embassy, the school where he works, the School of High Studies in Social Science, where he's affiliated in France and a very important person in Venezuela . . . and that quieted things down. . . ."[106]

The Venezuelan Indian Agency's approach to this conflict was to investigate neither Lizot nor Chiappino. So Lizot continued his research. On one occasion, Good saw Lizot with a cast on his arm "after a club-fight with a Yanomami man." Another time, Lizot got into a boxing match. "I saw Lizot and a Belgian photographer fighting it out at the Ocamo Mission," Good said. "The Belgian was beating the shit out of him until Lizot picked up a burning log. Then the Belgian walked away."[107]

In Caracas, I went to visit Jean Chiappino's wife and research partner, Catherine Alès, at their spacious high-rise apartment overlooking the crowded valley of Caracas. We sat on her balcony as evening lights came on, and we could still see the 10,000-foot-high Avila Mountains. Alès, who has been doing research among the Yanomami for two decades, is a woman of striking elegance and long, red hair, and she spoke looking toward the mountains. I thought, when I mentioned the name Lizot, that a shadow of fear came across her face. She obviously did not want to talk about him. When I asked whether her husband had denounced Lizot for child molesting, she said, "Yes, but even if the Yanomami themselves denounce him, no one will believe them. People will say the Yanomami are liars. So you see—there is no proof."[108]

Jesús Cardozo became Lizot's friend—"sorts of friends," he said.[109] He remembered that when he next called on Lizot at Karohi-teri, he was welcomed by him effusively and then finding himself abandoned, alone in the forest, when Lizot disappeared without a word the next morning, taking the whole village with him. "He was surrounded by boys. I would say, they appeared to be around twelve years old." Cardozo asked another researcher, an archaeologist finishing a Ph.D. at American University who has also worked among the Yanomami, how old the boys were. She said the group included boys from around the ages of ten to twelve. "Yeah, ten to twelve years old," Jesús agreed. "They were walking with an effeminate swaying of the hips that, as you know, is not at all normal for Yanomami boys. And they would giggle and point to each other's asses. 'That's the place.' It was very strange to see. Lizot made no attempt to hide it from the Salesians. In fact, he would come downriver with a whole boatload of these boys, wearing their jewelry, all painted up. And, you know, [Father] Bórtoli and [Brother] Juan Finkers would be smiling and welcoming them as though it were no big deal. Lizot was really quite open about it. I attended a meeting with Lizot and the Salesians in which Lizot said, 'I think everyone who comes into contact with the Yanomami should be tested for AIDS.' There was some embarrassed coughing, but Lizot just continued, 'I always get tested for AIDS whenever I return to the area.' "[110]

The stories of Lizot's homosexual life among the Yanomami were widely rumored abroad. I heard about them in Boa Vista, Brazil,[111] and it was frequently alluded to in interviews with scholars from all camps of the Yanomami controversies.[112] These reports were so well known inside Venezuela that in 1986 a military commission was sent to investigate. Jesús Cardozo was at the Platanal airstrip when the military plane bearing the commission, headed by an army major, landed. He remembered having the following conversation with the major:

"Who are you?"

"I'm an anthropologist."

"Sergeant! Take note: We've just met the anthropologist of the Upper Orinoco. We're going to take this to Caracas for a report. Are you married?"

> "No, I'm not."
>
> "Are there other anthropologists in the area? Are they married?"
>
> "Well, I'm not really privy to their personals."
>
> "Ah, do they engage in any type of irregular sexual activity?" You know here we are next to the plane, and he's having this sergeant write all these things down. So he comes out finally and says, "Look man, is there a faggot here or not? I understand there's another guy who's French and he's a faggot."[113]

But Cardozo did not want to be the sex policeman of the Upper Orinoco. So he just gave the location of Lizot's village to the major, who, on learning that it was a two-day hike from Platanal, walked around the airstrip for ten minutes and flew away to Caracas and did nothing. "It's so bizarre," Cardozo admitted. "Everybody knows about the whole thing. But for some reason nobody ever did anything. And what Bórtoli told me on several occasions was that he felt that any type of exploitative attitude carried out by a *nabah,* a foreigner, against a Yanomami, be it heterosexual or homosexual, for him was equally censurable in his point of view. So if somebody would go and, say, flirt with a woman and pay her a machete or some rice or spaghetti or whatever in exchange for sexual favors, he thought that was just the same as Lizot or anybody paying a little boy to do something like that. That's how he felt. Morover, he felt that Lizot's impact upon the Yanomami had been minimal. And he actually used cases like Hepewe, who is in fact married and who seems a well-adjusted Yanomami. . . . I decided not to press the point. I just don't know. It's a complicated issue."[114]

Giovanni Saffirio, a Catholic priest who lived with the Brazilian Yanomami for twenty years and received a Ph.D. in anthropology from the University of Pittsburgh, confirmed Cardozo's encounter with Bórtoli. "I went with Dom Aldo [Mongiano, the bishop of Roraima, Brazil] on a trip to Venezuela in February 1987. We traveled up the Rio Negro and then crossed over to the Orinoco by the Casiquiare. Since I had heard so much about Lizot's homosexuality with the Yanomami, I asked the head of the mission, Bórtoli, about it. He told me it was true, that Lizot did have sex with boys, but that they didn't seem to suffer any permanent damage from it. They just grew out of it and got married like the other Yanomami boys. That was his explanation—that there was no long-term damage. I thought it was very strange because, for me, that kind of child molesting is a terrible thing."[115]

When I spoke with Father José Bórtoli at the Platanal mission, he was not as direct. Bórtoli is a tall, thin man worn to near transparency by repeated bouts of malaria (from which he was actually suffering when I spoke to him). When it came to the subject of Lizot, he parried my questions with questions. "Is there so much interest in Lizot's sexual life because it was with the Yanomami or because it was supposedly homosexual?" When I answered that the real problem, for me, was pederasty, Bórtoli said, "It is very hard to establish the ages of children at villages outside the missions. And if everyone who had sex with the Yanomami is going to be prosecuted . . ."[116] Here Bórtoli's voice trailed off, and he shrugged.

At any rate, Lizot worked for the Institute of Higher Studies in Paris, not the Salesian mission. The Indian Agency issued his permits, not Bórtoli. The Salesians had tried to remove Lizot and failed. Many people had tried to stop Lizot. No one had gained health or good fortune from it. Two (Claude Bourquelot and Gary Dawson) had tried to kill him; three (Juan González, Padre González, and Jean Chiappino) came to blows with him. No one had succeeded in halting his peculiar enculturation program. The best anyone had achieved was partial banishment—from Parima, Mavaca, Platanal, and Hasupuwe-teri, successively.

L'affaire Lizot forced everyone to confront the same dilemma: Lizot was publicly doing more good than anyone else while privately satisfying his appetites on Yanomami children. In structuralist terms, there was an absolute polarity—a perfect asymmetry—between Lizot's public and private personas. His Apollonian, above-ground activities as defender of the Indians stood in natural opposition to his Dionysian, chthonic rituals. He became everyone's favorite monster.

After living for twenty-five years among the Yanomami, Lizot finally returned to France in 1994. Today, he sees things through a long-distance literary lens. His preferred vehicle of expression is the first-person story, told exclusively by Yanomami narrators. There's no analysis in these lyrical, stream-of-consciousness histories. Lizot has achieved a tasteful, French antithesis of Napoleon Chagnon's cowboy Western. If Chagnon was the self-centered, heroic, gunslinger in the Amazon, Lizot was neither seen nor heard. Critics have praised Lizot's anthropological absence, his willingness to let the Yanomami speak for themselves.[117]

"I could of course have evoked my own experience of life among the Indians," Lizot noted in *Tales of the Yanomami,* published by Cambridge University Press, "but I wanted to speak of other things, for strictly personal reasons: I am not yet ready to speak of the terrible shock that this experience was for me, nor of the price I had to pay to become closely acquainted with a civilization so radically different from my own; perhaps I will never be able to speak of these experiences, for I would have to evoke so many harrowing things that touch my inner being."[118]

Chapter 9

That Charlie

Brewer Carías Incarnates Venezuela's closest approximation to Indiana Jones.—*ExcesO*[1]

The parachutist who sailed into Adulimawa-teri in June 1969 to disarm the mad French anthropologist was the amazing, and almost equally mad, Charles Brewer Carías. Brewer had become addicted to parachuting at the University of Michigan, when he was supposed to be studying genetics. "One month into his stay there," recalled James Neel, "he discovered sky-diving. The discovery was not good for his scientific training."[2] Although Brewer never completed his graduate work in genetics, he was uniquely equipped to rescue Lizot from the wrath of Claude Bourquelot, a rescue Brewer came to regret.

The introduction to one of Brewer's illustrated travel books says that he is "not only a botanist, a zoologist, an entomologist, a geologist, an astronomer and a naturalist. He is trained in all these fields of knowledge, and he has united them with a rare capacity for leadership and organization that has permitted him to be recognized all over the world today as one of the greatest explorers of all time."[3]

"That Charlie," said a British member of a Brewer expedition, "is well out of his tree. He is seriously bonkers. He ought to be put away."[4]

For thirty years Brewer Carías, a Venezuelan national, has symbolized La Conquista del Sur, the Conquest of the South, Venezuela's rain forest equivalent of manifest destiny. An ex-dentist, he has devoted himself to promoting tourism, colonization, and scientific exploration of Venezuela's southern frontiers, usually with the support of the ministries of tourism and defense. This has made him, and his undeniable survival skills, all but indispensable to a whole generation of First World scientists trying to study everything from soils to natives to insects in the Venezuelan Amazon. Brewer became one of the New York Botanical Garden's most productive research associates and most popular guest speakers.[5] A superb photographer, with seven picture books to his credit, Brewer starred in so many television specials that he became a trademark of sorts. He marketed Omega watches[6] and a six-inch steel blade with Rambo-like features called the Brewer Explorer Survival Knife, which converts into a harpoon.[7] He needed protection. "Everyone out there wants to kill me," he told the *Times Literary Supplement* editor Redmond O'Hanlon, explaining why he carried a huge Browning automatic handgun to a gym workout. "From the government of Guyana to the lowliest nut."[8]

"Have you read Redmond O'Hanlon's description of Charlie?" asked John Walden, a doctor from Marshall University in West Virginia who accompanied the explorer on one of his helicopter trips to the Siapa valley. "Now, if you didn't know Charlie, you would think, 'No. This is too much. This must be exaggerated.' But that's Charlie. He's really wild. I've been to his house in Caracas, and it's wild. In West Virginia, we keep moonshine in glass jars. Charlie keeps gold dust in glass jars. You have to love Charlie on some level; he's like an old pirate with a patch over his eye. I could tell you stories about Charlie."[9]

He wouldn't, though—scientists who have been Charlie's fellow travelers are fairly protective of him. But Rafael Salazar, a Venezuelan composer who was recently a Fulbright scholar at the University of Pittsburgh, was a bit more explicit about how wild things really were at Brewer's house. He and his wife, Teresa, a sociologist at the University of Caracas, had been invited over by Charlie to discuss a youth concert, during the time Brewer was minister of youth (1979–82). "He had a lot of power then," Salazar recalled.

"When Teresa and I entered, he told us, 'Come over here. I want to show you something.' He had a patio with some birds of prey, big ones, hawks or something, in a cage. He grabbed a chicken with a leather glove and gave it to his birds to eat. Then he waited there with a watch to see how long it would take them to kill it. When the birds didn't kill the chicken in the allotted time, he put his hand inside the cage to hurry them." This scene made quite an impression on Teresa, too. "I remember that he waved his hand around and shouted at the birds to kill the chicken more quickly," she said. "And I thought to myself, 'So this is our minister of youth.' "[10]

As minister of youth, Brewer left a colorful legacy of debacles and substantial achievements. He sponsored a concert near the Orinoco by one of Venezuela's most promising groups of musicians, Madera, whose eighteen members drowned after the event in a boat without life preservers. Brewer was not on board, but, as a former Olympic swimmer, he probably would not have sunk like the rest."[11] In spite of the tragedy, one of modern Venezuela's worst boating accidents, Brewer remained a quirky hero to many people in Caracas, where he turned downtown streets into Sunday pedestrian malls, humanizing the city. His ultimate solution for urban sprawl, however, was colonizing the rain forest, and it got him into trouble again and again.

Brewer released a film promoting La Conquista del Sur, which featured a new town near the border of Brazil meant to showcase the government's colonization efforts in the rain forest. In fact, the settlement was not new and the government did not build it. And it was really located five hundred miles from the Brazilian border, in the Orinoco Delta.[12] It was a larger version of what Brewer had done at Patanowa-teri for *The Feast.*

Brewer also led a national campaign to improve youth fitness, setting the example himself with public jogging sessions around Caracas. He successfully lobbied to ban cigarettes on buses and planes. He did not touch alcohol or coffee—"I am nothing but an ascetic"[13]—and rose before dawn every day to meditate. It seemed there was nothing Brewer was incapable of—except keeping a low profile. His mercurial meddling in Venezuela's foreign affairs overshadowed his real domestic accomplishments, which might have been the prelude to a major ministerial post. Brewer precipitated a crisis with Colombia by marching off on his own to Bogotá, where he unveiled a slogan that sounded like an ultimatum: Not one more centimeter for Colombia![14] Later, he compared Venezuela's border relations with Brazil to the Battle of Stalingrad.[15] He kept dreaming of battles with neighbors who had never fired a shot in history.

Brewer appeared to be a dynamo whose energy could hardly be contained. The final blow came when he began organizing young street toughs into a private paramilitary force. "They were sort of the Venezuelan Brown Shirts," said the anthropologist Terence Turner of the University of Chicago.[16] Brewer then led his gang on an invasion of the former British Guyana. Guyana mobilized its armed forces; Brewer withdrew and got fired. But he boasted that the Pentagon appreciated his videotaped evidence about Guyanese Marxists.[17]

After his dismissal, Brewer left his wife of twenty years and his comfortable home in Caracas, and made his way to El Dorado, a small mining camp in what was then virgin rain forest. There, in 1982, not far from mountains that he had explored for the first time, he began a second incarnation—as the founder of a mining company, Minas Guariche, with extensive capital, holdings, infrastructure, and legal problems. *ExcesO* magazine said it was as if Alexander von Humboldt "had decided to open a Manchester sweatshop at the dawn of the nineteenth century on the banks of the Casiquiare Canal."[18]

Here in this former rain forest existed another world, much like Conan Doyle's *Lost World,* a land marked by the sudden rise of *tepuis,* sheer mountain mesetas of black granite and ocher sandstone. The highest one reaches over nine thousand feet above sea level; the longest ones stretch for thirty miles. Sir Walter's fabled El Dorado, which he located near a "mountain of crystal," is now considered the first European description of a *tepui,* though to the Indians *tepuis* were always the homes of the mountain gods.[19]

The *tepuis* have a lunar look—the oldest mountains on earth, so isolated that each one is an eccentric island of evolution, a planetary archive three hundred million years old with hundreds of plant and insect species to itself. The area has been called a biological El Dorado because so many species adapted to the outlandish *tepui* terrain in novel ways, producing outrageous varieties of ferns, fungi, and insects. But these mountaintop riches appeared austere, beaten down by winds, and drenched by continual downpours that drowned all but the most resistant species in what amounted to a flooded desert of rocks and leached soil.

The Pemon Indians named this region Roraima, "the Mother of Waters," and here there is a profusion of waterfalls. Angel Falls, the world's highest, losing itself in gravityless mist of 3,000-foot freefall into the rain forest below, is one, and perhaps not the most impressive, of the three hundred waterfalls on the single meseta of Auyán-tepuí.

Brewer Carías was the first known man to explore several *tepuis,* and he described their fragile treasures in his book *Roraima: montaña de cristal.* That is why it was all the more surprising that he had opened a huge *garimpo* right at their feet—two sets of gold-mining concessions covering 25,000 devastated acres. He called them Triunfo II and Triunfo III. Manuel Nuñez Montano, head of the Conservation Society of Guyana, described Brewer's strip mines as "a desolate panorama—the common denominator is irreversible damage."[20] He said Brewer's mines flagrantly violated Venezuela's Law of the Environment, the Law of Forestry, Soils, and Water, and the Law of Mines, in addition to the terms of his concessions, which were questionable to start with. Brewer began mining six years before his permits went into effect, and he was ten years late in presenting an environmental-impact statement. When he did finally present the plan, it was simply based on the principle of "natural regeneration."[21]

It did not help that Brewer's strip mines were diverting the headwaters of the Cuyuni River, which were ecologically protected. The journalist Tania Vegas wrote, "Nothing stops Brewer Carías, not even the majesty of the imposing mesetas of the mountains of Supamo, on whose flanks you see the destructive effects of his machines. . . . Hundreds of water jets destroy our ancient natural heritage, but of all who do this, only Brewer Carías calls himself a conservationist."[22]

Until Brewer came along, there were two kinds of mining in the Amazon gold rush. The most common was done by mobile teams of four to six men with diesel pumps that powered hydrojets—very nasty, but small-scale. It was the work of poor men, the street people of the Amazon. The second type was big, open-pit excavating by North American companies, usually wildcat Canadian firms that traded on the Wild West Vancouver exchange and could ignore indigenous rights and environmental laws. But Brewer was carrying out industrial mining with water-jet technology, reducing large tracts of forest to mud soup, a designer dream come true for the anopheles mosquitoes, which carry malaria. It was the worst of all possible mining worlds.

Brewer personally has permission to move 133,600 liters of gasoline, diesel fuel, and other petroleum products each month to his mines in Bolívar State. That gives an idea of both the scope of his strip-mining activities and his personal involvement as director of operations.[23]

Brewer's Minas Guariche owns another 4,649 acres of concessions farther on, near a town called Kilometer 88, where the Pemon Indians' traditional farming and hunting lands have been completely overrun by gold miners.[24] Brewer has been repeatedly denounced by Pemon leaders, and these particular holdings, Vemeru I–VI, appear in Survival International's document *Venezuela: Violations of Indigenous Rights,* on a map called "The Mining Invasion of Eastern Bolívar State."[25]

As Brewer became one of Venezuela's biggest *garimpo* entrepreneurs, he also became the gold rush's most outspoken advocate—and the leader of a political movement that aimed to block recognition for indigenous land rights. The Venezuelan anthropologist Nelly Arvelo Jiménez, the senior social scientist at Venezuela's Institute of Scientific Research (IVIC), compared Brewer's role in orchestrating the Venezuelan gold rush to that of the notorious Ze Altino Machado, head of Brazil's Garimpeiro Union.[26]

Brewer was the bitterest antagonist of any recognition of Yanomami's right to land. Both IVIC and a Catholic think tank, Fundación La Salle (FLASA), had put forth proposals setting aside about thirty thousand square miles for the Yanomami, an area that both institutions designated as a "biosphere"—only to be dismissed by Brewer as agents of foreign powers. He made his case during a shouting match with Congressman Rafael Martínez, an encounter witnessed by several journalists and congressmen. Brewer charged that "the Yanomami reserves and the biosphere" were a leftist conspiracy to create "an indigenous nation . . . which arose out of a meeting between the sociologist Esteban Monsoyi and Libyan colonel Muhammad Gadhafi, on the occasion of publishing the Green Book of the Libyan Revolution in Venezuela."[27] Brewer claimed that Shining Path terrorists from Peru were also involved in this plot. These sensational charges created an uproar that sank the biosphere proposal in 1984.

This was also the year the spectacularly rich cassiterite (tin ore) deposits of the Upper Orinoco first came into play, with over 100,000 acres of concessions initially granted in Yanomami territory, before a popular reaction forced the government to cancel them. Brewer and the Salesian missionaries began their antagonism over this plan, which Brewer openly supported. In a May 10, 1987, article for *El Nacional* newspaper, he denounced the "supposed Indian experts who proclaim and suggest the creation of autonomous Indian territories within Venezuela, attacking our sovereignty." He attacked legislation prohibiting mining, which he called a "blessing" and an "obligation," particularly "in the case of the deposits of almost pure tin that are located on the slopes of the Sierra Parima," the very core of the Yanomami's ancestral homeland. "In this region there are only four Yanomami *shabonos,* at great distances from each other, with fewer than 200 individuals each. These indigenous people, seminomadic warriors who are accustomed to continually moving their villages, could be relocated without traumas to nearby territories, but far from the mining activity that would take place in the area; that is, if our scientists decide they shouldn't be included in the change that will occur there."[28]

In fact, the Sierra Parima is the most densely populated part of the Yanomami Reserve. Still, Brewer's invocation of "our scientists," had an authoritative ring to it. It also showed a genius for kaleidoscopic scapegoating, combining Cold War stereotypes of Communists, Colonel Gadhafi, and Shining Path terrorists to characterize his opponents, while using all the props and paraphernalia of science—from botany to archaeology—to promote his vision of the Conquest of the South.

Science was Brewer's ally in his mining ventures. He shuttled scientists from the Smithsonian, the American Museum of Natural History, and the Royal Geographical Society to Cerro Neblina, the highest mountain in the Amazon outside the Andes, where hundreds of new species of plants and animals were discovered. And he kept expanding his gold-mining activities, using the scientific expeditions as cover.

Venezuela's National Guard in Amazonas caught Brewer gold mining—in July 1984—along the Lower Ventuari River, near the Maquiritare village of Kanaripó, in a rain forest area where all commercial mining was prohibited. *El Diario de Caracas* reported that "the ex-minister . . . was arrested together with other people by the National Guard troops at Kanaripó, because he didn't have the necessary permits to travel in that area, where—in addition to gold—he was also commercializing and exporting Venezuelan fauna and other species without authorization."[29]

At that time, the naturalist was using unsalaried Maquiritare Indians as workers. "Brewer Carías destroys not only nature but also the men who work for him," said Sergio Milano, a police official and anthropologist who directed the investigation, referring to the loss of cultural identity among Indians who become *garimpeiros*.[30]

Milano sent me the police report.

It showed how Brewer did his gold mining while supposedly conducting research for the Venezuelan Foundation for the Development of Physical and Mathematical Sciences. But Brewer used the foundation-paid helicopter rides to drop American scientists off at Cerro Neblina and then backtracked to his gold mine. The real flight records—which the pilot nullified—were included in the police file, along with the pilot's testimony that the receipts were faked at Brewer's request. The fraud amounted to six hours of flights, worth some twenty thousand dollars. Receipt 1089 was voided and replaced by 1090.[31]

In his defense, Brewer charged that the army had framed him out of jealousy. He blamed Milano for having a personal grudge against him, and accused him of both gold mining and torturing witnesses. While Brewer admitted to buying gold and diamonds, he claimed that he had merely flown over the area of Kanaripó en route to his real destination—Cerro Neblina. This explanation was not accepted in the Venezuelan congress, however, since Kanaripó and Cerro Neblina were in opposite directions from Brewer's takeoff point.

Three congressional leaders went to Kanaripó, where they confirmed that the police had found "machines for the extraction of gold . . . whose owner was Brewer Carías."[32] They dismissed the accusations against Milano as "simply a smoke screen to distract the focus of attention and to disqualify the indictment." Their report concluded that Brewer had violated Law 2039 by "the exploration, exploitation, and marketing of gold in the Amazon Territory," and that "by secretly entering Indian lands he has violated Law 250." They asserted that he had contracted Indians to work clandestine mines for him: "He leaves them alone in the places of exploitation and then returns to pick up the minerals collected and pays them for their work."[33]

The list of scientists and journalists who took rides on Brewer's helicopters included prominent figures from the Smithsonian, the American Museum of Natural History, and the Royal Geographical Society. The twenty-five species of insects and birds named after Charlie were all tributes from grateful scientists. Hundreds of journalists hitched rides, too. They paid their tribute by burnishing Brewer's iconic image as the explorer-hero. Redmond O'Hanlon of the *Times Literary Supplement* called Charlie "the great explorer and photographer of Venezuela," and described his connection to the American Museum of Natural History, but kept Brewer's mining and politics out of their jungle adventures.[34]

Actually, Brewer invited me into the jungle, too, But this happened only after he discovered El Dorado. "I have an idea for you," Brewer said. "You know how there have always been legends about the Incas moving their capital to the Amazon? Well, the Brazilians have finally found Inca ruins on Cerro Neblina. I can give you an exclusive on this."[35]

Brewer did not want Venezuela to lose the race to the fabled Inca city. He wrote letters about the fantastic discovery to two government ministers who favored opening Indian lands to mining. "The fact is that the ceramic samples, stone mortars, axes and quartz fragments which I have excavated close to where the Brazilians have found Inca remains, permit me to suggest that we mount a multidisciplinary Venezuelan expedition which I would direct."[36] He was counting on the participation of various scientists" who had accompanied him before into the area, including Napoleon Chagnon.[37] He proposed contacting Yanomami groups, doing botanical studies, the whole panoply. He raised the tantalizing possibility of finding the city where the Incas had hidden most of their gold from Francisco Pizarro.

Brewer's ceramics shards were not, in fact, Incan. And the chances of finding Inca ruins at Cerro Neblina were outrageously remote. But that was not the point. The language of science had a magical quality, mystifying everything about the real historical nature of Brewer's mission. Instead of a cross, he and his scientist-missionaries carried cameras, computers, and blood-sampling equipment into the wilderness as a polite prologue to the engineers of destruction who always followed on their heels.

Like Sir Walter Raleigh's fables—in which natives without necks became exotic embellishments of imperial expansion—Brewer's hype about Inca cities entertained, distracted, and simultaneously justified the ongoing Conquista del Sur. I would not be surprised if I found myself reading accounts of an Amazonian Cuzco by the time this book comes to press—with a new scientific expedition acting as a Trojan horse for gold mining in the rain forest. Brewer's political strategy was already old by the early seventeenth century, when one of Simón Bolívar's ancestors in Caracas wrote, "Would to God El Dorado had never been discovered."[38]

Chapter 10

To Murder and to Multiply

Lethal raiding among the Yanomamo, it seems, gives the raiders genetic success.—*Richard Wrangham and Dale Peterson*[1]

In February 1988, as the gold rush was reaching its zenith and Brazil was locked in a national debate about the Yanomami's fate, Napoleon Chagnon published a major study in *Science,* "Life Histories, Blood Revenge, and Warfare in a Tribal Population."[2] He reported that 30 percent of all Yanomami males from his study group were killed in warfare, while 44 percent had murdered someone. But the real sociobiological bombshell was this: men who had killed had more than twice as many wives and three times as many offspring as nonkillers. It echoed and re-echoed in the press, and quickly became an issue in the Yanomami's survival.

Chagnon's findings seemed to confirm the most radical sociobiological belief about the "selfish gene"—that males will do almost anything, even the most antisocial things, to pass on the greatest number of genes.[3]

Sociobiologists praised Chagnon's *Science* story as a pathbreaking work.

Harvard's E. O. Wilson endorsed the piece, saying Chagnon had found a "powerful, potentially selective" link between violence and reproductive competition.[4] Favorable accounts of Chagnon's article soon appeared in *Scientific American* and other journals. It has since been reproduced hundreds of times, making it one of the most traveled social science studies of all time and a cornerstone in the edifice of sociobiology.[5]

But specialists on the Yanomami generally rejected the study. They attacked it in a dozen articles that set fire to anthropology journals, challenging Chagnon's ethics, statistical analysis, fieldwork methods, and interpretive bias.[6] While Chagnon repeatedly claimed he was taken completely by surprise with the "stunning" reproductive success of killers,[7] he glossed over the fact that his mentor James Neel had predicted such an outcome as early as 1962. Neel made documenting the sexual success of violent men "the number one priority," and a potential link to genetic dominance.[8] Although Neel never discovered his "leadership gene," Chagnon's critics felt that his disciple had finally come up with its cultural equivalent. In the Yanomami Eden, Cain acquired multiple wives and prolific progeny. Abel and other less aggressive men were banished from the gene pool, just as Neel had prophesied.[9]

The *Science* study provoked two debates—one over the political impact of Chagnon's study and the other over its scientific accuracy. The first was both more public and more personal. Jacques Lizot wrote, "Chagnon's theories have—with the author's collaboration—become the object of sensational publicity in the U.S. press. A grotesque and malevolent image of the Yanomami has been put forth in indisputably racist terms, the Indians being presented as bloodthirsty people obsessed by the desire for murder."[10]

The press's embrace of Chagnon's study was based on a central misconception that Chagnon cultivated. He wrote, "I demonstrated that Yanomamo men in my 25 year study who had participated in the killing of other men had approximately three times as many children and more than two times as many wives as men their own ages who had not."[11] This was false. Among mature men of the *same age group,* reproductive success of killers was not nearly as impressive—ranging between 40 and 67 percent, a fraction of the 208 percent advantage that Chagnon broadcast to the press.[12] Chagnon included a large sample of unmarried young men in his study, which hugely inflated the relative reproductive advantages of the *unokai,* almost all of whom were over the age of thirty. Had Chagnon's study included prepubescent boys and babies, the apparent advantages of *unokai* males would have been even more spectacular. "When I tell people that Chagnon is lying about his data," the anthropologist Brian Ferguson complained, "they say, 'So what?' "[13]

Chagnon told the *Los Angeles Times* that when the Yanomami were not collecting honey or hunting, they were killing each other.[14] *U.S. News & World Report* ran a piece titled "A Laboratory for Human Conflict." It began, "They are probably not the kind of people you would invite over for afternoon tea," and included Chagnon's claim that the Yanomami were "fighting like hell over women."[15] The *Washington Post,* like the *LA Times,* called the Yanomami one of the most violent societies on earth.[16]

Brazil's biggest newspapers quickly picked up the story. *O Globo* ran a piece, "Anthropologist Underscores Violence among Indians," along with a captioned picture: "An Indian educated to kill."[17] *O Estado de São Paulo* featured a similar story, "Violence, Mark of the Yanomami."[18] The political fallout was disastrous. Six months later, the Brazilian government formally split the Yanomami area into nineteen islands. Military Chief of Staff General Bayna Denys justified this drastically reduced space by explaining that the Yanomami were too violent and had to be separated in order to be civilized.

The Brazilian Anthropological Association (ABA) protested. The ABA's past president María Manuela Carneiro da Cunha accused Chagnon of doing violence to the Yanomami's chances of survival through his theories of violence. She said it was not the first time. In the late 1970s, Brazil's military junta had picked up on a *Time* magazine review of Chagnon's work—an article titled "Beastly or Manly?"—and had seized upon the apelike images of Yanomami warfare to postpone the demarcation of Indian lands.[19] It took Brazilian anthropologists, with the help of Survival International and the superb photographer Claudia Andujar, over a decade to create a hauntingly romantic image of the Yanomami, effective in garnering public support in both South America and Europe. But the U.S. office of Survival International closed; very little assistance was sent from the United States to the Yanomami. "People in the United States who'd read *The Fierce People* would ask why anyone would want to help such a horrible group," said the anthropologist Kenneth Taylor, who headed Survival International's U.S. office. "Chagnon has been told, over and over again, about the harm his work has done, but he just doesn't seem to get it. He keeps coming out with things that are more and more outrageous."[20]

Science finally ran a sequel to Chagnon's *unokai* thesis. Called "Warfare over Yanomamo Indians," the new article revealed a rift between scientists in the United States and those in South America. The president of the American Anthropological Association said, "There's nothing short of not publishing to keep this sort of thing from happening." The head of the Institute for Advanced Studies at Princeton dismissed the complaints by noting, "Compared to the enormous forces changing societies, anthropologists are rather small potatoes."[21]

At that time, that was also my impression. It was during this controversy that I first spoke over the phone with Chagnon in 1988.[22] He seemed genuinely anguished over the furor his study had set off. He told me he wanted to shift his priorities from pure research to a more humane involvement in relief and human rights. And he helped me in a simple but significant way—he gave me the phone number of Giovanni Saffirio, a priest and former Ph.D. student of his, who became my closest friend in the Brazilian Amazon.

My first stop in in Yanomami territory was Saffirio's small mission on the Catrimani River, where the gold rush was in full swing. In the general confusion, an old headman, Chico, had died, and his people, the Opik-teri, were very frightened at his departure. They blamed his death on a distant, uncontacted *shabono.* And their new headman said that if they didn't avenge Chico's death, his spirit, in the form of a huge jaguar, would keep roaring at night and making everyone sick. Moreover, they had to avenge four other brutal murders committed in recent years by the same enemies. They had impaled one Catrimani man on a pole; a second was chopped to death with a machete; a woman was drowned by having her head held in a bucket; an old man was killed, too. Later, the mission nurse told me all four had died of natural causes.

When I asked the local headman, Pedro, about the number of enemies he intended to kill on his planned raid, he said, "One," while holding up two fingers.[23]

In the end, the raiders never found their archenemies. This was fairly normal. Only about one Yanomami raid in four even reaches its destination, and no opprobrium is attached to returning without having fired an arrow.[24] But the excitement of the raiding party, with all the men marching around waving their weapons and shouting their lethal intentions—followed by a long, apparently pointless ten-day trek—served a key ritual purpose. Chico's irate ghost stopped roaring. People no longer heard him in their sleep. Calm returned to the *shabonos* of the Catrimani River.

I tried to find out whether Chico, who had sired more than thirty children, had killed anyone or not. At that time, I was quite enamored of

Chagnon's thesis that murderers had more offspring, and I had cited his *Science* study favorably in my own book on ritual murder in the Andes.[25] But I was accustomed to investigating killings where I could go to a site. I would normally inspect a body whose cause of death could be determined, usually surrounded by ritual paraphernalia and the like. Trying to reconstruct Chico's personal participation in warfare proved impossible for me, even with the help of two missionaries and an anthropologist who spoke the local Yanomami dialect. People's opinions about Chico were imbedded in intense local political rivalries. The University of Paris anthropologist Bruce Albert expressed one point of view when he said, "Everyone considered Chico a coward."[26] Father Guillerme Damioli expressed another when he said, "Chico was a great warrior."[27]

So how did Chico manage to achieve such reproductive success? He founded two small villages through his many sons. To begin with, Chico was unusually successful at getting other people—starting with rubber tappers and neighboring Indians and moving on to missionaries, Indian agents, highway workers, tourists, and, finally, gold miners—to give him wealth, protection, and medicine.

In fact, two of Chagnon's students—Giovanni Saffirio and Raymond Hames—did a study of the Catrimani Yanomami that focused on negative acculturation at Chico's *shabonos.* With the arrival of Brazilian highway workers in the 1970s, Chico's people became professional beggars. "They succeeded in receiving many Brazilian goods and became excited with the new way of life. They thought it would be easy from then on. It was only necessary to live along the highway and to beg tenaciously for goods."[28] By the early 1970s they owned shotguns,[29] and by the early 1980s they possessed a diverse hoard of steel goods with which to buy wives.[30] The last time I saw Chico's progeny, they were living on the banks of the Rio Branco with a gold miner about fifty miles outside the Yanomami Reserve—eating his food, drinking his beer, and helping him smuggle bananas out of Yanomamiland.[31]

Chico was many things to many people—headman, shaman, trader, traitor, regional politician, possible killer, and definitely the entrepreneur grandfather of a large, peripatetic group of male panhandlers. In the effort to evaluate Chagnon's thesis about prolific killers, Chico's case helps focus on the many variables that have to be related or excluded in considering homicide as a reproductive strategy. A survey like Chagnon's should include knowledge of all killings and all offspring, along with the ages of each killer and each father. Then it would be essential to ferret out spurious correlations, because many factors reportedly affect marriage alliances—including age, shamanic prowess, headman status, hunting skill, homicide, trade goods, and descent group. Finally, the risks of these alternative paths to conspicuous fertility must be weighed—in this case, whether killers are themselves killed.

The first question is, How do you count the dead in a culture that counts only to two?

In societies where status accrued with killing, you usually had to bring proof in the form of coups or trophy heads. There was a good reason for such requirements, according to Christopher Boehm, an anthropologist with USC's Jane Goodall Center who did his Ph.D. on the last head-hunting group in Europe, the Montenegrins of Serbia and Albania. "I asked them, 'Why did you take heads?' " he recalled. "The answer they gave me was, 'Listen, Painful One, anyone can say he killed somebody, but if you have a head that's the proof. Montenegrins are the biggest braggarts in the world. You have to make sure you bring back the head, or no one will believe you.' In a way the custom is driven by the very lack of veracity."[32]

And that raised another question: How did Chagnon count the dead? By counting the number of men who had undergone a ritual purification for murder, called *unokaimou.* It was a painful ordeal, involving fasting, celibacy, and immobility.[33] After completing this ritual, Yanomami say, they are *unokai.* It is Chagnon who calls them "killers." But many Yanomami *unokai* caused death by animal surrogates or magical substances or procedures like stealing a person's footprint. Others accompanied a raiding party and shot arrows in the dim light of dawn or twilight when most attacks took place and ran off without knowing what had really happened and joined in the group penance afterward. There were no referees to decide who had truly killed, and many times the Yanomami warriors themselves were not sure. "Recruitment to the *unokai* status is on a self-selective basis," as Chagnon pointed out.[34] He also noted, "I did not accompany raiding parties and did not witness the killings that occurred while I lived there."[35] There was no forensic evidence to buttress his statistics on Yanomami violence.

In fact, far more Yanomami men claimed to have "killed" on raids than the number of real victims.[36] Some men were what we would call accessories. Most of the *unokai* events (209 of 385) were claimed by large groups of men—8.3 on the average—who performed the ritual purification together after twenty-seven different raids.[37] At best, there was an uncertain relation between the Yanomami ritual category *unokai* and physical homicide, and Chagnon did not offer enough data to figure out what that relationship really was.

Nor was there anything but the most tenuous connection between killing, raiding, and the capture of women. The number of women captured in the warfare of the Yanomami is low, despite their reputation. Chagnon has never published any data on war abductions for the villages in question. When prodded by Albert, he stated that total "abductions" came to 17 percent, a much higher rate than that for any other Yanomami population but still a small number. It turned out this figure was a bit misleading, however: "I would like to make clear," Chagnon wrote, "that most abductions are not the consequence of raiding (making 'war' on) a distant village and 'capturing' women."[38] Most of these were women taken during feasts with allies, and such "abductions" worked only if the woman in question did not want to walk home again. Other anthropologists called these elopements. In a survey of four hundred marriages in the neighboring highlands, Lizot found that less than one percent of the women had been captured on raids.[39]

Yet the popular image of the Yanomami waging war for women persisted. Chagnon deftly created it by repeatedly claiming that men went on raids, captured women, and raped them at will afterward: "A captured woman is raped by all the men in the raiding party." The case was exactly opposite. On the rare occasions when women were captured in war, only *nonkillers* in the raiding party were permitted to have sex with female captives.[40] If a killer had sex without undergoing the long *unokaimou* purification, his life, as well as the woman's, was endangered.[41] So Yanomami warfare appeared to have other elements of negative selection apart from the real risks of being murdered in a raid—or having someone else make love with a warrior's wife while he was on the long paths of war.

In the academic wars over his *unokai* study, Chagnon presented himself as a "traditional scientific anthropologist" defending objectivity against the a new generation of "applied anthropologists" who promoted " 'politically correct' data."[42] Although there is, in fact, a good deal of politically correct whitewashing of native cultures these days, it was Chagnon's lack of hard data that aroused suspicions of similar distortions. Professional demographers tended to dismiss the study on the basis of paternity alone.[43] His charts on fertile killers looked good on paper, but there was no way to confirm or refute them. Not only were the "killers" anonymous, so were the twelve villages they came from. In the *American Ethnologist,* Jacques Lizot accused Chagnon of having created villages whose demographics were unlike any known communities, and whose exact location was "impossible to determine."[44]

Doshamosha-teri, Siapa Highlands, 1996

César Dimanawa, Chagnon's nemesis, Mavakita, 1996

Alberto Karakawe, Chagnon's ally,
Ocamo Mission, 1996

Girl at Mokarita-teri,
Siapa Highlands, 1997

Mother and baby,
Homoxi-teri, Parima
Highlands, 1996

Ashidowa-teri, Siapa Highlands, 1997

Piloto, suffering from rare skin disease, Platanal, 1996

Homoxi-teri, Parima Highlands, 1996

Girl at Cerrito, Upper Orinoco, 1996

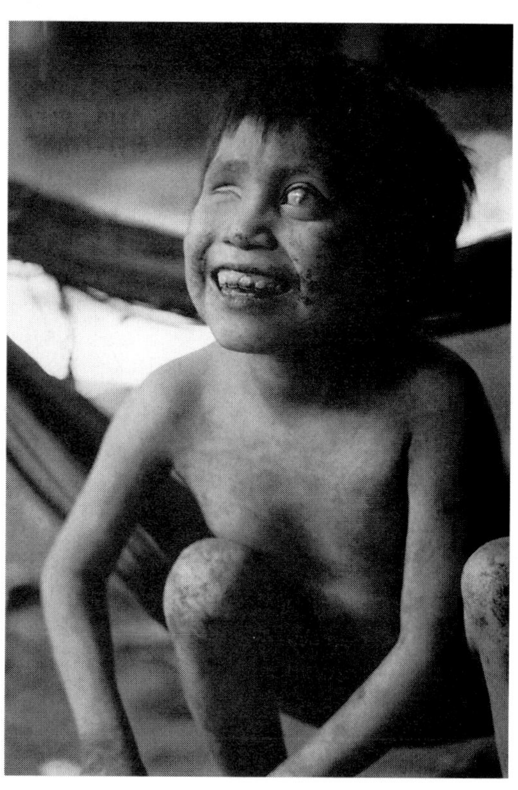

Blind boy with onchocirciassis,
Cerrito, 1996

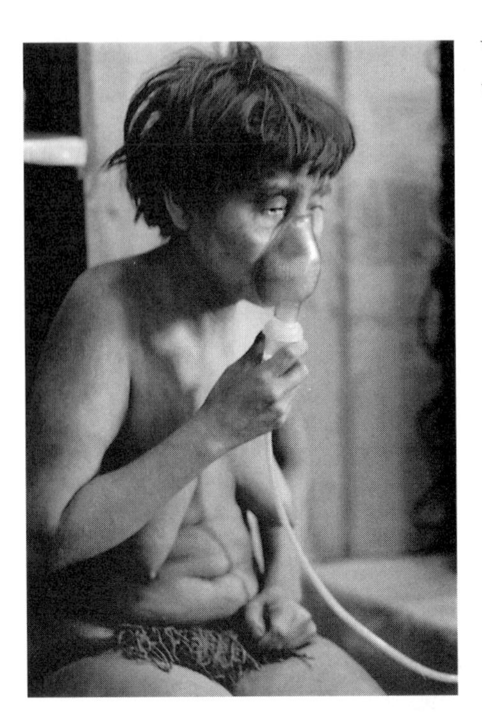

Woman with oxygen mask, Homoxiteri, 1996

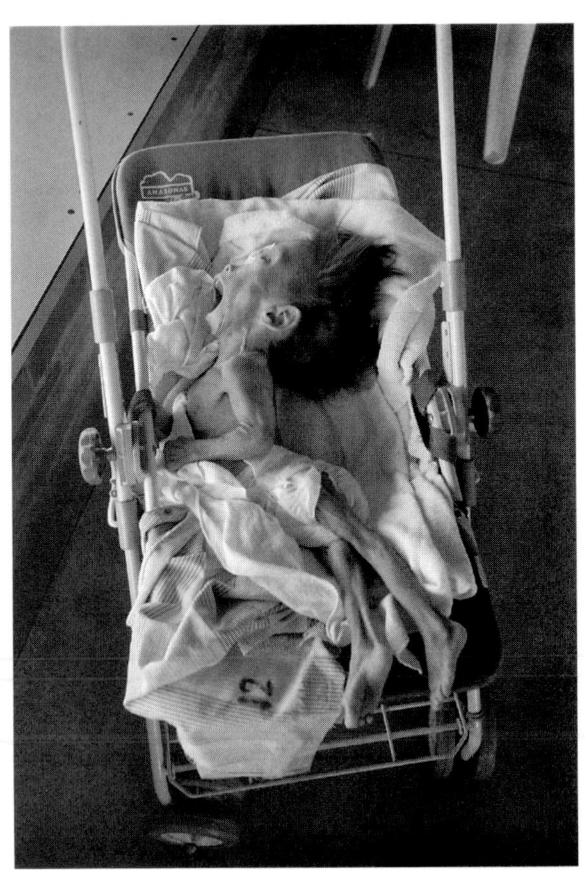

Renata, Xiriania Yanomami girl, Casa Hekura Clinic, 1996

Jacques Lizot, Platanal Mission, 1997

Monkey study, Irokai, Siapa foothills, 1996

Helena Valero, kidnapped by the Yanomami in 1932, Upper Orinoco, 1996

Girl at Demini, Brazil, 1995

Kaobawa, Chagnon's mentor, 1996

César Dimanawa, with bow and arrow, Mavaca River, 1996

Garden at Mokarita-teri, 1997

Men at Mokarita-teri, 1997

Survivors of the measles epidemic of 1968, Shashanawa-teri, 1996

Palisade at Homoxi-teri: the gold wars in the Parima Highlands, 1996

Yarima, who married Kenneth Goode and lived in New Jersey, 1996

Karohi-teri, who built a new village for a BBC/Nova special, 1996

César Dimanawa and his six wives, Mavaca River, 1996

Alfredo Aherowe, who ordered FUNDAFACI's cameras destroyed, Siapa trail, 1996

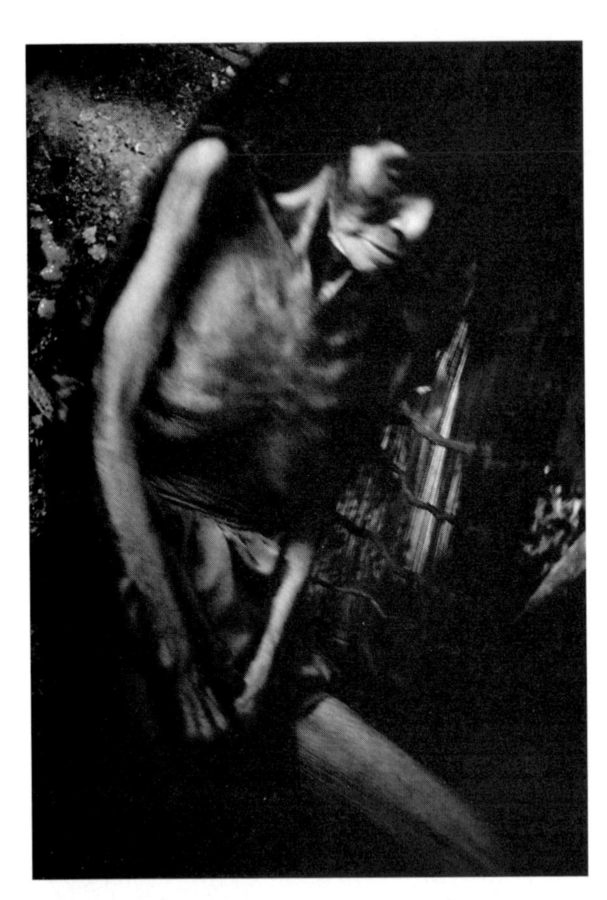

Yanomami woman dying at Toki, Marquiritare village, Padamo River, 1996

Orinoco expedition, 1996

Helena Valero, Ocama, 1996

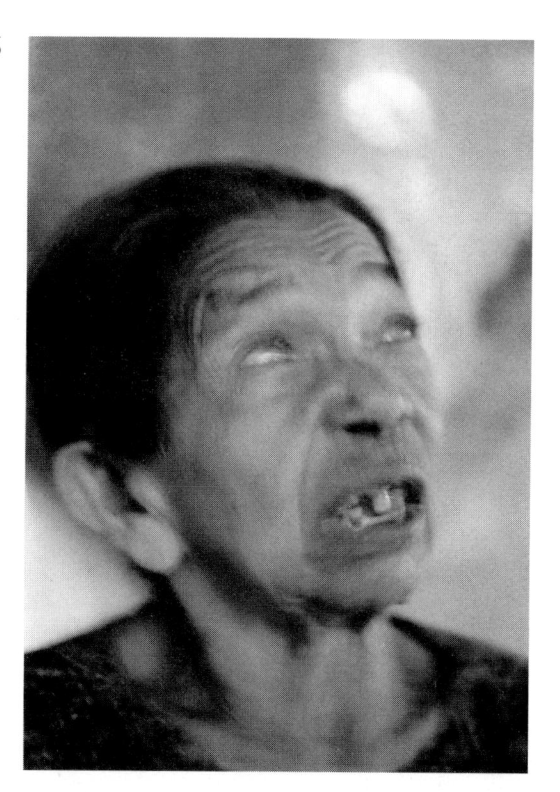

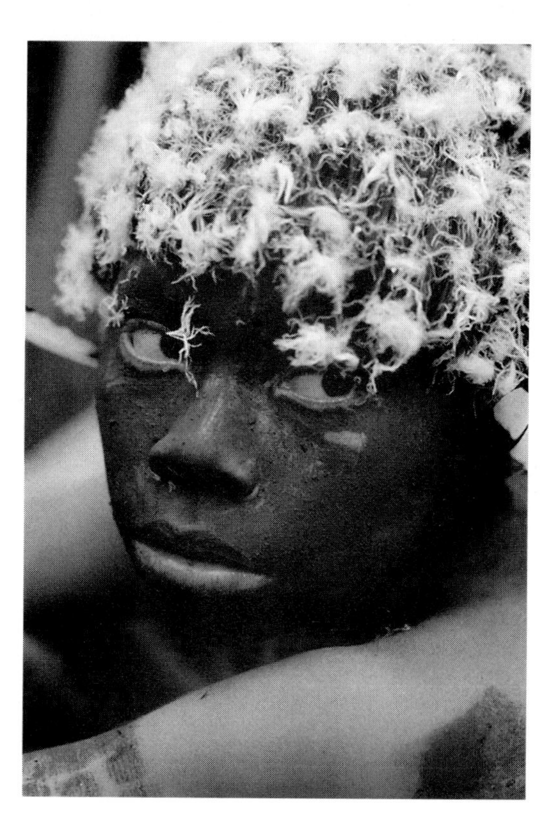

Boy with buzzard down,
Mokarita-teri, 1997

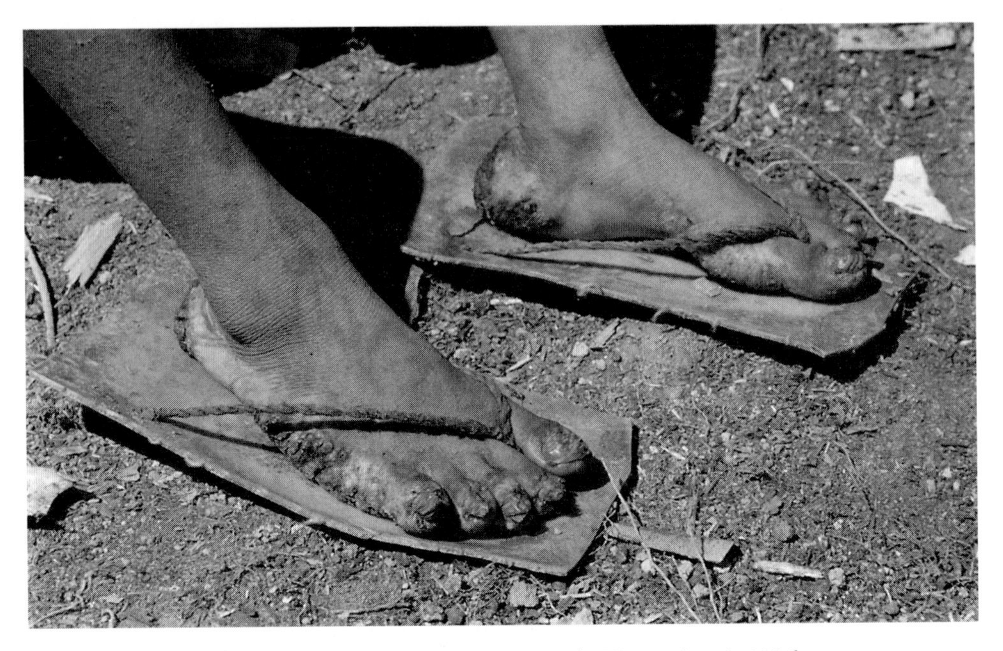

Feet infested with Amazonian jiggers (*Tunga penetrans*), Homoxi-teri, 1996

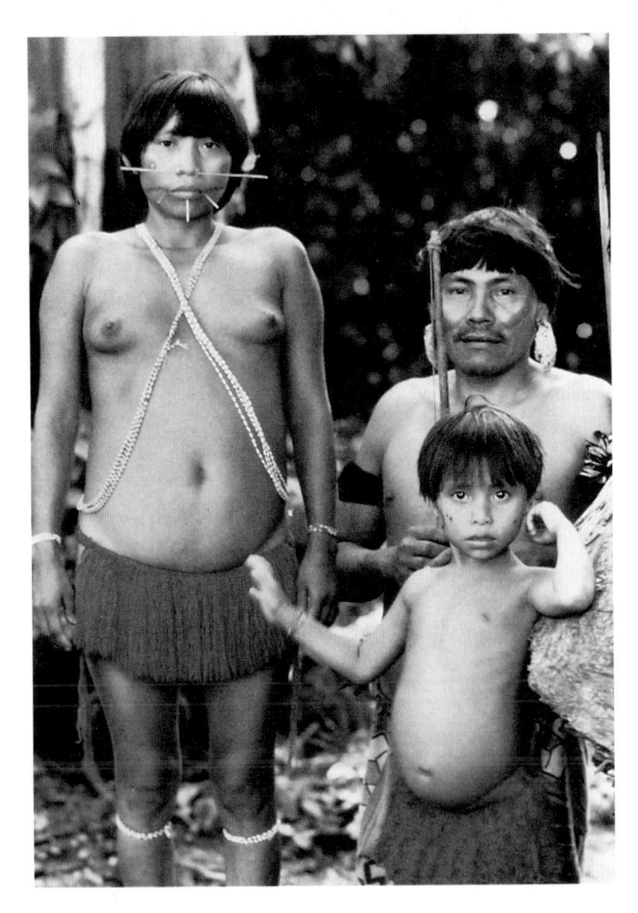

Isabela, the widow of Chagnon's guide, Ushubiriwa, 1996

Not impossible. It took me quite a while to penetrate Chagnon's data, but, by combining visits to the villages in the field with GPS locations and mortality statistics, I can identify nine of the twelve villages where all the murderers come from in his *Science* article.

Village No.	Real name	Place	Population	GPS
5	Bisaasi-teri	Boca Mavaca	188	2 31 65 10[45]
6	Bisaasi-teri	Boca Mavaca	121	
7	Bisaasi-teri	Boca Mavaca	105	
51	Kedebabowei	Mrakapiwei	164	2 02 65 09[46]
53	Dorita-teri	Shanishani River	136	2 09 64 15
84	Nasikibowei	Mavaca River	66	2 29 65 11[47]
90	Haoyabowei	Mavakita	55	2 11 65 08[48]
92	Mishimishi	Upper Mavaca/ Siapa	188	1 36 65 16[49]
93	Washewa-teri	Washewa River	105	2 02 65 04[50]

Chagnon did not invent the twelve villages for *Science,* as Lizot insinuated. Nor was his choice of villages arbitrary. These were the same *shabonos* where he had spent the great majority of his forty-five months on the Upper Orinoco. Villages 5, 6, and 7 were originally Upper Bisaasi-teri, Lower Bisaasi-teri, and Monou-teri.[51] All the others I have identified were offshoots of the Upper Mavaca villages of Mishimishimabowei-teri and Iwahikoroba-teri. Of the remaining four villages, three (52, 53, and 54) are offshoots of the Patanowa-teri: Dorita-teri, Sheroana-teri, and Shanishani-teri.[52] Altogether, these made up the cluster of villages that were most intensely studied and filmed by the AEC.

Chagnon opened his *Science* article by reporting, "Warfare has recently diminished in most regions due to the increasing influence of missionaries and government agents and is almost nonexistent in some villages." His villages, however, "were actively engaged in warfare during the course of [his] continuing field research."[53] In this undisturbed region, the Yanomami still fought with bows and arrows, carrying out traditional revenge raids, revenge that originated in disputes over women.[54] His study covered 153 war deaths "during approximately the past 35 years."[55]

That total was five times higher than the known war deaths for these villages during the years 1953–88.[56] The anthropologist Bruce Albert immediately questioned Chagnon's claim that his area of study was pristine or that his figures for war deaths were reliable. It was more likely, he wrote in *Current Anthropology,* that they were "calculated from genealogies covering three to five generations."[57]

In *Science,* Chagnon offered three concrete examples of warfare. One of them was the raiding between Monou-teri/Bisaasi-teri and Patanowa-teri that occurred between 1965 and 1975.[58] I showed in chapters 6 and 7 that two of the raids—one in 1965 and one in 1970—were filmed, transported, and coordinated by Chagnon. Moreover, the raiding was not, as Chagnon represented in *Science,* continuous. There were no killings after Chagnon left the field in 1966 until *The Feast* was filmed. Then Bisaasi-teri launched an attack and killed the Patanowa-teri headman, Komaiewa, who had starred in *The Feast.* The raiders blew Komaiewa's head off with a shotgun.[59] Chagnon brought seventy Mishimishimabowei-teri men down the Mavaca River with the help of the mission priest to join Bisaasi-teri in a raid on the Patanowa-teri. The Patanowa-teri, who did not have canoes or guns, were in no position to retaliate effectively against the Bisaasi-teri. The pattern of violence was completely outside the formula of bow-and-arrow, tit-for-tat exchanges presented in *Science.*

Chagnon's second example was the war between Bisaasi-teri and Tayari-teri, which I discussed in chapter 9. Chagnon set this strange conflict in the context of natural warfare and biological striving.[60] Yet, he had three years earlier briefed colleagues at a conference on Lizot's role in the war, presenting his rival as a feudal lord in the bush. Five years later, in 1992, he blamed this same war on the Salesian mission of Mavaca.[61] For *Science,* he left out the role of shotguns and motorized transport, along with the fact that it was "Chagnon's village" battling "Lizot's village."

The third reported instance of war was of a man who organized a raid on a distant village to assuage his wife's grief over the loss of a sister to raiders. "The role of female grief in Yanomamo warfare is an important factor in perpetuating revenge killings."[62] Elsewhere, however, Chagnon explained that this was another shotgun killing. The Yanomami themselves considered these shotgun killings to be *badao,* "without cause."[63] One of his informants told him, "When you give a fierce man a shotgun, he becomes even fiercer and wants to kill without cause."[64]

These three wars illustrated a new kind of violence, directed exclusively outward from western centers of power, where men with shotguns could attack remote villages. Under the influence of the missions, however, they did so only rarely. The missions themselves were safe, and their population boomed. Only three people were killed in war in all the Bisaasi-teri groups at Boca Mavaca from 1968 to 1983—and they all died in the hostilities with Lizot's village.[65] Another of Chagnon's twelve villages—No. 84, Nasikibowei-teri—had relocated a few minutes by motorboat from the mouth of the Mavaca. And, on the Upper Mavaca, two Mishimishimabowei-teri groups had come under mission influence. One village, Haoyabowei-teri, was at a small post manned part-time by the Salesian brother Juan Finkers. It had a school, as did the nearby village of Kedebabowei-teri, No. 51. Mishimishimabowei-teri proper, No. 92, had also made peace with its former enemies, and was no longer subject to attacks from the Orinoco. It was far more tranquil from 1977 onward, according to Chagnon.[66] And the sporadic raids by Bisaasi-teri against Patanowa-teri also ceased as the mission created a regional schooling system, with a big boat as a mobile school bus, which brought young men together from different communities. All the villages had some degree of mission contact, according to Chagnon. Many joined a trading cooperative.[67] From 1975 onward, during Chagnon's decade of exile, the only major disturbance was the war between Chagnon's village and Lizot's village. When that ended, in 1982, the region experienced general peace.[68] Traditional warfare had all but disappeared from the Upper Orinoco. The vagueness and the brevity of Chagnon's three anecdotes in *Science* were essential parts of their anonymous appeal, like the villages with no locations. Stripped of historical reality, devoid of context, the numbered villages and nameless war stories were elements in a struggle for reproduction; with their names added, these same deaths became weapons in a political struggle against Lizot, or missionaries.

The most intriguing village was No. 51. It possessed, as of 1987, the highest concentration of killers in Yanomamiland. Indeed, eight of the eleven Yanomami men who murdered more than ten people come from this village. These eight men participated in at least ninety-one killings. If Chagnon called the Bisaasi-teri "the Fierce People," and the Mishimishimabowei-teri were "the Fiercer People," then Village 51 could be crowned "the Fiercest People."

The village is actually a splinter group of Mishimishimabowei-teri that moved down the Mavaca toward the Salesian mission after the epidemics that killed 27 percent of the original *shabono.* It became known as Kedebabowei-teri, named after the river that empties into the Mavaca along a sandy beach, Mrakapiwei, "the place of sand." Although determining homicides after the fact with no physical proof is an inexact science at best, I submit that the actual number of people killed by the warriors of Kedebabowei-teri was insignificant for the period (1953–88). Kedebabowei-teri's grandmother village, Mowaraoba-teri, was an unusually aggressive force from about 1910 to 1935. It had obtained steel goods earlier than other Yanomami villages, and experienced explosive population growth coupled with territorial expansion.[69] Mowaraoba-teri captured Helena Valero in the mountains of Tapirapecó, in the Brazilian Amazon.[70]

But after the killing of their great war leader, Ruwahiwa, in about 1942, they became a very dispirited group. They launched a couple of desultory raids but turned back because of bad omens, according to Helena Valero, whose husband killed Ruwahiwa.[71] During this time, their mother village was driven from the Mavaca River drainage and into the Siapa River valley, where homicide levels were much lower than in the conflict-ridden Orinoco. As I showed in chapter 6, they had a pathetic war record. They failed to avenge not only Ruwahiwa's death but also a series of other killings through the 1960s. Their tally against Bisaasi-teri had become 0–11 by 1970.

But the faction that splintered to Kedebabowei-teri was the weaker subdivision of Mishimishimabowei-teri. These were the men who lost the ax fight and, after taking a beating, merely left the village (and the film crew's trade goods).

So how were they involved in more than a hundred killings?

Partly through rewriting. The single biggest battle in Yanomami warfare occurred on February 3, 1951.[72] It was a *nomohori,* a treacherous feast, given by Riakowa, the Iwahikoroba-teri headman, in which eleven to fifteen men from Kreibowei-tedi (the Bisaasi-teri's village at the time) were killed. This single event accounted for a high percentage of the total of thirty-one war deaths reported by Chagnon for all the Namowei Yanomani (Villages 5, 6, 7, 52, 53, and 54) from 1930 to 1966. He has given the following accounts of the killings:

> Shortly after arriving at the village, the Kreibowei-tedi men were attacked by Riakowa's men with sharpened staves and axes, and eleven of them were killed. Those that managed to break out of the palisades were shot from ambush by Sibariwa's followers [Mowaraoba-teri] and Hasabowa-tedi, resulting in the wounding of a large number of Kreibowei-tedi men. (Ph.D. thesis, 1966)[73]
>
> The men of Kaobawa's group [Kreibowei-tedi] danced both singly and

en masse and were invited into the homes of their hosts [Iwahikoroba-teri]. At this point their hosts set upon them with axes and staves, killing about a dozen of them before the visitors could break through the palisade and escape. . . . Once outside, they were shot from ambush by the Mowaraoba-teri, who managed to kill a few more and wound many others with arrows. (*Yanomamo,* 2d ed., 1977)[74]

The revenge of Ruwahiwa was collaboratively orchestrated by the Iwahikoroba-teri and the Mishimishimabowei-teri. Both had earlier been members of the same village but had fissioned with each other before Ruwahiwa was killed. The Iwahikoroba-teri invited Kaobawa's people to a feast. . . . But they also secretly invited the Mishimishimabowei-teri, Ruwahiwa's "closest" people. These unknown guests hid in wait and attacked them by surprise as they lay defenseless in their hosts' hammocks. (*Yanomamo,* 5th ed., 1997)[75]

When it came to figuring out who killed whom, there was, in the first place, a plus or minus factor of about 25 percent. (Similarly, in the battle at Tayari-teri, Chagnon, who arrived six years after the event, gave a figure of "8 or 10" killed;[76] María Eguillor García, just downstream and keeping tab of war deaths for her Ph.D. thesis, said seven;[77] Lizot, who lived there, six.)[78]

But it was the question of who actually killed whom that was the thorniest. In the first history of the *nomohori,* Chagnon gave eleven deaths, all at close range, inside the *shabono,* by the Iwahikoroba-teri; the Mishimishimabowei-teri and the Hasupuwe-teri played an ineffectual role, wounding but not killing any men outside the *shabono.* In the second version, the Mishimishimabowei-teri had "managed to kill a few more" and the Hasupuwe-teri had been dismissed from the scene. The final account performed a complete reversal: the Mishimishimabowei-teri now hid inside the *shabono* and killed all the men in Kaobawa's group as they lay in their hammocks.

Chagnon also rewrote the history of the two allied *shabonos*—Mishimishimabowei-teri and Iwahikoroba-teri—by making them originate from a single village just prior to Ruwahiwa's death.[79] Chagnon even redrew the map of their genesis.[80] Why all this extra work? Apparently, he had to make the massacre of 1951 match the sociobiological prediction that close kin must protect and avenge each other's deaths. The Mishimishimabowei-teri's continued failure to avenge their dead was not acceptable, at least not to

Chagnon. This was also the ideological key to Chagnon's reinterpretation of the *Ax Fight* described in chapter 7. Timothy Asch was right when he remarked that Chagnon was unconsciously venturing into postmodernism.

I think the same zeal that allowed Chagnon to see things on film that were not there enabled him to discover a farrago of serial killers at the downtrodden village of Kedebabowei-teri. He spent twenty-five years preparing the *unokai* study, finding misfits to give him the names of the dead, then hustling them off to mission posts where he could interview them without pressure from the village.[81] And always paying for the taboo information with steel.

That was the rub. Checkbook journalism is suspect—for good reason—and checkbook anthropology is no better. What happened at Kedebabowei-teri, Village 51, was this: Chagnon's principal informants conspired to rig the *unokai* data. They resented his coercive methods, including his ability to get the real names of the dead by checking their testimony with enemy villages. But they discovered a weakness in Chagnon's *unokai* research. They knew Chagnon would not stray far from the main rivers, and so they concocted "murders" in villages far into the Siapa Highlands.[82]

In the opening scene of *Studying the Yanomamo,* Chagnon provided a snapshot of his steel-for-murdered-people economy:

> The mood was volatile and a few of the men were becoming uncontrollable. I was concerned about what the headman's younger brother, Yahohoiwa, might do. Living in the shadow of the headman's renown, he had every reason to be concerned about his status and ferocity. He was, however, an unpredictable character and quite capable of violent expressions. Earlier in the day he expounded about his ferocity to me at considerable length and named the men he had killed on various raids—just before demanding a machete.[83]

Yahohoiwa's behavior defied the most sacred restriction of Yanomami society—naming the dead. Aside from the possibility of a magical summons to the discarnate entity, publicly naming one's victims invited reprisals; after real killings, the Yanomami send out spies to find out who the perpetrators were, so they can make effigies of them and attack them by physical and magical means.[84] "It's very hush hush," said the anthropologist Kenneth Good.[85] However, getting a new machete trumped all taboos. I consider this one of the most revealing encounters of Chagnon's career. Men were approaching him in the middle of the *shabono,* shouting out the names of their victims and demanding steel presents. Chagnon took it all in stride. Yahohoiwa was "violent," "unpredictable," "volatile," and "uncontrollable." According to Chagnon, Yahohoiwa's only compass was concern about "his status and ferocity." In reality, he needed a machete for his garden. I believe he had never killed anyone to this point in his life. If he had, he was another contrary case for the theory of reproductive violence. According to Chagnon's 1971 census, Yahohoiwa was only one of two men over the age of twenty-five who did not have his own wife.[86] Yahohoiwa was still sharing a wife with another man, a sign of subordinate status. Instead of killing, he should have spent more time in the garden, like the other men his age who could afford to feed a family.

Though Chagnon was superhumanly determined, he did not possess a subtle register. For instance, in the 1997 edition, the fifth, of his textbook, *Yanomamo,* Chagnon reported that Moawa, the headman of Mishimishimabowei-teri, single-handedly murdered twenty-two people.[87] In his Ph.D. thesis, Chagnon wrote that Yanomami raids required a group of at least ten men. In *Science,* he stated, "Lone raiders do not exist."[88] Yet, according to Chagnon's latest construction, Moawa alone killed more people than Billy the Kid, not by shooting curare-tipped arrows, but by sneaking up behind his victims and cutting their throats with a sharpened arrow point. Every Yanomami specialist I spoke to said this was bizarre. "I'm dumbfounded," responded Brian Ferguson. "Moawa killed all these people by himself without getting killed himself? This mysterious way of killing and the numbers are certainly unprecedented."[89] But Chagnon wrote that Moawa died without leaving any descendants.[90] In other words, the Yanomami's equivalent of Billy the Kid was a zero from the standpoint of sociobiology.

Once I had the names and real locations of villages that total 81 percent of Chagnon's sample, I could start unraveling some of the study's other puzzling features. First of all, Chagnon's villages had a big bachelor herd of unmarried men over the age of twenty—136 bachelors, 36 percent of the adult male cohort.[91] In sharp contrast to these bachelors, there was an unusual number of men with many wives, some with as many as six at one time.[92]

Chagnon's explanation of Yanomami warfare followed a classical sociobiological argument. "Some men succeed in obtaining two, three, and sometimes up to five or six wives. That means that some men will have no wives—and these men become a chronic threat to social cohesion because of their frequent attempts to seduce the wives of other men, or the occasional unmarried daughters of all men, even young girls who have not even reached pubescence."[93]

Of course, there are a few men in Yanomamiland who have an unusual number of wives. Chico had four or five, depending on how you counted his adulterous youngest wife, who was going off with younger men. But Chico was not a traditional kind of leader. In the more traditional Siapa Highland villages, Chagnon himself reported only one polygamous man per village, on the average.[94] That was what I also found, and none of these men had more than two wives.[95] The Salesian anthropologist María Eguillor García, who did a census among the Bisaasi-teri villages four years before Chagnon's survey of the same group, found that 17 percent of marriages were polygamous,[96] but that men with more than two wives were very rare. Drawing on all the mission populations, a group of more than a dozen *shabonos* from Ocamo to Platanal with a population of well over one thousand, she reported four men with three wives or more.

Man	**Place**	**Village**	**No. of Wives**
Asiawe	Platanal	Mahekoto-teri	3
Paruriwa	Mavaca	Bisaasi-teri	4
Yutuwe	Ocamo	Iyewei-teri	5
Kasiawe	Platanal	Mahekoto-teri	5

These men, like Chico, became all things to all people. They were headmen, shamans, and owners of shotguns who had lifelong alliances with non-Yanomami.[97] Asiawe was the gorgeously painted feast ambassador in *The Feast* and *A Multidisciplinary Study.* Kasiawe, the Mahekoto-teri headman, who negotiated his village's participation with the AEC,[98] was the leading purveyor of machetes for the Upper Orinoco by 1945.[99] When Father Luis Cocco first arrived at the confluence of the Ocamo and the Orinoco, in 1958, there was little polygamy. (According to a medical doctor who visited the area at this same time, there was only one polygamous marriage among all the Yanomami along the course of the river.)[100] Cocco observed that the increase in polygamy was directly related to an increase in shotguns, which improved hunting efficiencies, so men could feed larger families.[101]

There was no doubt that lowland communities had significantly more polygyny than highland villages. And all the villages in Chagnon's study of murders lay outside of the traditional highlands. Still, no one had a clue about the actual composition of the villages. Fortunately, the recently released CD *Yanomamo Interactive: The Ax Fight* offered a more detailed breakdown of Chagnon's data for the large village of Mishimishimabowei-teri, based on his 1971 census. Although taken fifteen years earlier, it was a profile of the largest Yanomami village in Chagnon's area, the mother of three communities included in the *Science* article. I opened the genealogies and found far more men with three wives than Eguillor García reported for the whole Orinoco.

Wadoshewa	3	Dedeheiwa	5	Nanokawa	8[102]
Ishiweiwa	3	Howashiwa	6		
Barahiwa	3	Moawa	6		

Thirty-four wives for seven men—4.8 wives each.[103] I could not believe it.[104] So I decided to take all the information about all the 271 individuals at Mishimishimabowei-teri that was contained in two long appendixes of Chagnon's book *Studying the Yanomamo* and put them in my own database.[105] It was a very tedious and time-consuming task. It took me a week to enter and analyze the information. But then something satisfying happened: I was inside the almost mythical village of Mishimishimabowei-teri.

What I found as I made my rounds of the *shabono* was intriguing. On the one hand, Mishimishimabowei-teri really *was* different from any Yanomami village I have ever seen—including populations around Boca Mavaca that composed 30 percent of his study.[106] Of fifty marriages, thirteen were polygamous; two other men had one wife and shared another with someone else. So 26 percent of the marriages were purely polygamous, a very high number. However, only two men out of the whole village actually had more than two wives. One had three; the other had six. The actual number of wives for the prolific men listed above was as follows:

Dedeheiwa	1[107]	Wadoshewa	2[110]	Nanokawa	6[113]
Ishiweiwa	1[108]	Barahiwa	2[111]		
Howashiwa	1[109]	Moawa	2[112]		

In reality, these seven men had fifteen wives (2.1 each). The other "wives" were dead or divorced. (Any man, or woman, who manages to live very long in a tribal population with life expectancy of twenty-one years will have several consecutive spouses.) And the only man with more than two wives was Nanokawa, headman of a faction of the village. Of fourteen men with five or more children—the average for the killers in *Science*—all were 35 and over except one, and he was 33. Of those with eight or more, all except Nanokawa were over forty.[114] The most prolific by far were the two oldest—Dedeheiwa, 60, with sixteen children, and his brother Ishiweiwa, 70, with thirteen children.

This breakdown offered an important insight into what Chagnon's overall findings are likely to be, if he ever reveals them. The Yanomami marriage system, which requires a bride price to parents in the form of meat and food in the highlands, or trade goods in the lowlands, favors older, established men, who essentially purchase a girl at birth or shortly afterward.[115] The Chicos of the Yanomami world are the men who survive to old age and continue to sire progeny into their sixties and seventies. The terribly tight correlation between age and reproduction suggested that, if Chagnon broke down his data more carefully, into a series of five-year brackets, the statistical advantages of killers would weaken.[116]

The picture of Mishimishimabowei-teri also suggested that around the central axis of age, shamanic prowess and headman status were secondary variables in marriage alliances. The two old shamans had almost a fifth of all the children (19 percent), but Dedeheiwa, with a regional reputation,[117] was more prolific than his older brother. Chagnon called him the spiritual authority of the village, and men like him can have more wives.[118]

Most headmen were also shamans.[119] Under normal circumstances, the most crucial determinant for headman status was age. Nanokawa and Moawa, the two polygamous headmen of Mishimishimabowei-teri, were both 35, a bit young to be leaders. But neither of them was a headman when Chagnon made his first contact at Mishimishimabowei-teri.[120] Nanokawa and Moawa both rose to power in the context of bitter struggle over Chagnon's trade goods. In fact, Moawa threatened to bury a hatchet in Chagnon's head[121] immediately after the anthropologist favored Nanokawa and his followers with fourteen machetes—machetes that Moawa demanded for his own faction.[122] Nanokawa went from being a person of little significance to being a leader with six wives, three of whom were not actually present at the village,[123] indicating how unusual his behavior had become. He was keeping two families at two different garden sites.

One of the weaknesses in Chagnon's *Science* article is that it did not take into account either shamanic attainment or headman status. That a headman accumulated more wives, in war or peace, was commonplace in the Amazon. James Neel originally wanted to discover how a man became a headman and, by virtue of having more wives, had more offspring.[124] Chagnon did a fine study of Yanomami headmen, and found that they had an average of 8.6

children[125]—many more than the average killer, who had 4.9. But in his *Science* piece all headmen were also included as "killers," a confusion of categories; when the headmen were factored out, the study's statistical significance in one of its major age categories collapsed, Chagnon admitted. He would not say which category it was. But no headman was ever reported to be under the age of 30.[126] And since most headmen are village elders,[127] it seemed likely that the age category that fell below statistical significance once the headmen were removed was the one with men over the age of 40. This was the group of older men with most of the killers (55 percent) and two-thirds of all the offspring (67 percent).[128] Again, Chagnon maintained a tenacious silence in the face of public challenge, this time by the anthropologist Brian Ferguson.[129]

In spite of the polygamous elders, there were no permanent bachelors at Mishimishimabowei-teri. Every man over the age of 27 was married. Of nine men aged 20–24, six were married and one was divorced, and they had a total of six children. Of nineteen teenagers, seven were married and three were divorced, and they had three children. The marriage and reproduction rates for the young men at Mishimishimabowei-teri at first contact, then, were completely different from the profile in *Science*. By Chagnon's 1987 census, there had been a remarkable decline in young male fertility and marriage rates. No teenager was married or had any children. And adults 20–24 had one-third as many wives and children.[130] Chagnon's data also diverged sharply from María Eguillor García's 1983 census at Mavaca—where 30 percent of the Yanomami from Chagnon's 1987 census lived. In contrast to Chagnon, she found that 31 percent of men had married by age 19 and 70 percent by age 24.[131] And, in all her years among the Yanomami, she had known only one lifelong bachelor—a refugee from the village of Patanowa-teri.[132]

No young man at Mishimishimabowei-teri, whether a murderer or not, had many offspring. (Only two men under 33 had even three children, and one of the two was in a polyandrous relationship, sharing a wife.)[133] At least not according to Chagnon's original census. So I was disconcerted to find in the new *Yanomamo Interactive* CD that two young men, Ruwamowa and Mohesiwa, aged 27 and 24, now had four and five children, respectively. Suddenly, Mohesiwa had nearly half the children for the 20–24 age group; and Ruwahiwa had three times as many as his 25–30 cohort. I thought I had made a mistake. And it occurred to me that there might, after all, be some correlation between violence and reproduction at Mishimishimabowei-teri.

Mohesiwa was the man who started the ax fight by beating his aunt, Sinabimi, with a club when she refused him some plantains. (Ruwamowa ignored the fight completely and went casually to his hammock; but Chagnon listed Ruwamowa as one of the many fighters.)[134] On checking Chagnon's original census, I discovered that Mohesiwa was listed as being born in 1938 and Ruwamowa in 1939[135]—making them 33 and 32, not 24 and 27. I found no other mistakes in the census transfer. And, if these were mistakes, they were statistically perfect ones. Mohesiwa had five times as many children as the other 20-year-olds—exactly like the young killers featured in *Science.*[136]

Minute manipulations in each age category could easily skewer all the results. For example, the spectacular superiority of killers for the entire study depended on a big bachelor herd under age 25 whose members were both peaceful and infertile. Only five men this age had killed, and all of them had wives. Of the seventy-eight who had not killed, only ten were married. This was the most stunning statistical correlation of the study.

But were the five killers really under 25? Chagnon expressed second thoughts in a rejoinder to Bruce Albert published by *Current Anthropology.* "Since the Yanomamo cannot specify their ages in years, the five men who were estimated by me to be younger than 25 years may, in fact, be 25 years or older."[137] Apparently, he was mistaken, because he removed these five murderers from the corrected version of his data. Now the fertility rates further plummeted for the young men in the study (to .18 children and .13 wives), while the statistical correlation for Chagnon's most impressive age category vanished altogether. There were no killers left under the age of 25.

And that was strange.

The strongest generalization to emerge from cross-cultural studies of killing is that murder rates rise abruptly in the midteens, peak in the early to midtwenties, and decline afterward. "Even though the actual rates are different among different societies, the age and sex of the murderers are the same the world over," wrote William Allman, the science writer for *U.S. News & World Report,* in his *Stone Age Present.*[138] Unmarried men committed the great majority of murders. And poor, low-status men were disproportionately involved. In the only study done in the United States, 73 percent of the killers were unmarried.[139]

The sociobiological argument for this pattern seemed unassailable: men were likely to become involved in violence at this age because they were at the peak of their physical strength and sexual drive, when competition for female respect would prompt high-risk strategies by poor men with little to lose. If this was still true in urban, industrialized societies, where guns and machine transport equalized physical power, it would seem to apply more forcefully to jungle fighting, which involved long treks and chases, in addition to hand-to-hand confrontation. The sociologist Margot Wilson, coauthor of *Homicide,* considered the definitive text on the subject, was one of many who cited the Yanomami as exemplars of urban gang warfare—motivated by sex, carried out by small bands of young males seeking status. According to Wilson, tribal violence followed the age/sex/status variables found in her cross-cultural study. "I think it's absolutely universal," she said. "I really think so. I would be really interested in places where it's the reverse."[140]

Chagnon's Yanomami were the reverse. No living teenager or, in the revised statistics, young man under 25 participated in a killing. The rates then increased in dramatic fashion, beyond cumulative effects, until 62 percent of those over 40 claimed to be killers.

But whom were they killing? Almost exclusively young men in their twenties, according to Chagnon.[141] And they did so without running serious risks themselves. "Also, recent wars in two other regions of the study area resulted in the deaths of approximately 15 additional individuals, many of whom were very young men who were unlikely to have been *unokais* [killers]."[142] There was no question, then, that the Yanomami had a unique gerontocracy of violence: old men were killing young men with impunity.

The tragic history of Yanomami war leaders was quite different, however. Most of them were killed. The few who survived lived in fear or exile. But they were not part of the study.

One of the least-noticed marginal fine print in the *Science* article was the unexplained removal of "living children whose fathers are dead."[143] In other words, all the warriors killed in battle were left out—that is like doing a study about the reproductive success of inner-city gang leaders without including those knocked off by rivals. Well-known killers among the Yanomami were targeted by their enemies, who made wooden effigies of them prior to raids. "Most of the men identified as war leaders were killed in war," Brian Ferguson wrote in *Yanomami Warfare.* The children of such leaders were sometimes killed, too. Ferguson suspected that, when all the dead *unokai* were taken into account, it was quite possible killers would have fewer offspring than nonkillers.[144]

Chagnon acknowledged that this was a serious question. He wrote, "Being *excessively* prone to lethal violence may not be an effective route to high reproductive success, but, statistically, men who engage in it with some moderation seem to do better reproductively than men who do not engage in it at all."[145] Of course, this was markedly different from the way the study of Yanomami killers was being promoted in the press. Now only a *moderate* degree of killing paid off. But the difficulty was that the graphs of killers in *Science* showed that nineteen men, who committed an average of 9.4 murders each, were responsible for the majority (52 percent) of all killings.[146] If such immoderate murder was *not* conducive to progeny, then most of the Yanomami killings were clearly counterproductive—from a sociobiological standpoint.

Historically, there was no known Yanomami man (except Moawa) with as many murders as any of these prolific killers; leaders with three or four killings were usually assassinated.[147] In modern society, men who murder with such abandon are called serial killers, and their social profile is that of weird loners. The situation is more complicated in tribal societies where ritual warfare could require taking many scalps or many heads. The Cheyenne, considered the most warlike tribe among the Plains Indians, certainly rewarded young war chiefs with high status, but a study of the reproductive success of Cheyenne leaders showed that peaceful leaders had 50 percent more offspring. "Most often the ambitious young war chief was brave, braver, and then dead."[148] Among the Waorani of the Ecuadorian Amazon, a tribe with the world's highest known rate of attrition in war, every known male had killed at least once. But warriors who killed more than twice were more than twice as likely to be killed themselves—and their wives were killed at three times the rate of other, more peaceful men. Most prolific killers lost their wives and had to remarry—which made it look as if they had more wives *if they survived.*[149] Among the Waorani's neighbors, the Jivaro, headhunting was a ritual obligation of all males and a required male initiation for teenagers. There, too, most men died in war. Among the Jivaro leaders, however, those who captured the most heads had the fewest wives, and those who had the most wives captured the fewest heads.[150]

"Chagnon's *Science* article on killers is anthropology's Piltdown Scandal," mused Kenneth Good. "But the Piltdown Scandal was pretty mundane—the guy who faked the fossil knew it. I don't think Chagnon does. He's a true believer. He really believes Moawa went out there all alone and murdered twenty-two people with a sharpened arrow point. *Jesus.* He's so deeply into his own reality, I don't think he's going to be able to come out of it."[151]

Even if all of Chagnon's data proved accurate, it revealed a series of startling anomalies. The Yanomami differed dramatically not only from *homo sapiens* but from the chimpanzees, to whom they were often compared. Even among the higher primates, where close ties developed, a dominant male could not murder other adults with impunity.[152] The Yanomami's homicidal and reproductive peculiarities, per Chagnon's data, included the following:

- teenagers who have no children
- 15- to 25-year-olds who do not kill
- adult males who do not target men who kill their own kin
- older men who become more violent with each passing year
- young men who are the principal victims of violence without inflicting violence

Chagnon's statistical structure, like the body of a medieval heretic on the rack, had to be stretched simultaneously in two directions. Older men were assigned all the murders and children; the youngest men got none. This deformed body of data stretched and stretched until it became a crooked parody of sociobiology. Or, to put it in evolutionary terms, Batesian mimicry. Chagnon's graphs and Darwinian language perfectly parroted the mating rituals of serious science.

And Chagnon drew his supporters closer by using below-the-intellect attacks on common enemies. In *Current Anthropology,* he claimed Albert considered Darwinian theory "repulsive."[153] In another reply for the *American Ethnologist,* Chagnon attributed strange statements to Brian Ferguson at a conference where they debated Yanomami violence. According to Chagnon, Ferguson had said, "I don't understand why 'you sociobiologists' keep bringing in reproduction. After all, if you have enough to eat, reproduction is more or less automatic."[154] He added a footnote, "His comments were both tape-recorded and heard by a dozen or so other participants in the symposium."[155] In fact, the debate *was* taped and the tape showed that Ferguson made no such statements.[156] Nor did the conference chairperson recall Ferguson's saying anything of the kind.[157]

Chagnon claimed his opponents were postmodernists and Marxists who hated objectivity and Darwin and motherhood and the flag. At the same time, Chagnon insisted that, somewhere in his computer, he had all the data necessary to vindicate his claims about murder and reproduction. He would publish them all, by and by.

Another twelve years have gone by, and none of Chagnon's closely held data has been published.

Missing Data

Data	Year Promised	Years Elapsed
Complete census of all villages	1974	26
Genetic paternity	1974	26
Unokais with headmen factored out	1989	11
Unokais with those killed factored in	1989	11
Female abduction	1974	26

Yet there was something familiar about Chagnon's strategy of secret lists combined with accusations against ubiquitous Marxists, something that traced back to his childhood in rural Michigan, when Joe McCarthy was king. Like the old Yanomami *unokais,* the former senator from Wisconsin was in no danger of death. Under the mantle of *Science,* Tailgunner Joe was still firing away—undefeated, undaunted, and blessed with a wealth of offspring, one of whom, a poor boy from Port Austin, had received a full portion of his spirit.

Chapter 11

A Kingdom of Their Own

It is better to let them die in their own country.—*Cecilia Matos*[1]

By the time I reached Caracas in February 1992, Napoleon Chagnon was closely allied with a political clique centered on the Venezuelan president's mistress, Cecilia Matos, and the naturalist Charles Brewer. The three planned to create the world's largest private biosphere in the heart of Yanomami territory.[2] They were the talk of the town, from the military high command, where I gave a briefing on Brazilian gold miners, to the presidential palace.

Outside the palace, the ruins of a failed military coup were still in evidence. The joke around the city was that the officers who attacked the palace picked the one place where President Andrés Pérez could never be found. He was more likely to be traveling in the city in disguise, meeting with Cecilia Matos.

In fact, Cecilia Matos was much more than the president's mysterious brunette mistress. She was a president herself—of a quasi-governmental foundation to help indigenous and peasant families (FUNDAFACI). The courts later indicted Cecilia Matos on numerous counts of fraud,[3] and she absconded before facing trial. This was not the first time that controversy had surrounded Matos; as early as twenty years before, she became known as El Bidón, the Funnel, during the first presidential term of Carlos Andrés Pérez.

Napoleon Chagnon wrote that he was "deeply indebted" to Matos for the "unprecedented support" provided to him during six field trips made in 1990 and 1991, "support that included transportation by both fixed-wing aircraft and helicopters."[4] But the lavish support that FUNDAFACI gave to Chagnon—millions of dollars in helicopter and plane transport for the most ambitious expeditions in the history of anthropology—became a key element in Cecilia Matos's downfall and Chagnon's own expulsion from Yanomami territory.

The question was, Why did America's most famous anthropologist involve himself with Matos and Brewer?

Chagnon and Brewer had a long association. It dated back to the University of Michigan, the filming of *The Feast* and their joint allegiance to James Neel and his brand of neo-Darwinism. They had a lot in common, including reputations as fighters and a fascination with violence. Chagnon used his students as decoys for attack dogs; Brewer measured the time it took his birds of prey to devour chickens.

But their aggressive personalities had frequently clashed during the AEC expeditions. "Keeping the peace between Napoleon and Charlie was not easy, let me tell you," James Neel recalled. "Those are two strong egos. They had a love-hate relationship."[5] For a while, they were estranged. Chagnon sharply criticized Brewer's *Conquista del Sur,* a colonization program for the rain forest.[6] It was a principled position to take, and it must have been a difficult one. For ten years, Chagnon could not get into Yanomamiland, and he made no attempt to rejoin Brewer, even though Charlie was flying dozens of scientists into the rain forest.

By 1990, however, Chagnon had become the most controversial anthropologist, certainly in South America, and probably in anthropological circles worldwide. According to his own bibliography, it had been many years since he had published a substantive article in any anthropology journal, although his supporters attributed this to a bias against sociobiology.[7]

In South America, Chagnon had become the face of an academic *conquista.* When word came to Boa Vista in November 1989 that Chagnon was due to arrive with a BBC film crew, a protest parade was organized to meet him at the airport. He canceled the trip. A similar protest took place in Caracas, at the Universidad Central. While Chagnon remained a star in the U.S. media, it was increasingly difficult for him to continue his field research.

Venezuelan law required him to have a collaborative project with Venezuelan scientists. But, as Chagnon wrote in 1990 to one of the Salesian missionaries, "The local anthropologists do not like me."[8] Instead, Chagnon suggested that the Yanomami indigenous organization, SUYAO—an economic cooperative—or the Salesian anthropologists should sponsor his visits. Later, Chagnon would dismiss SUYAO as a creation of the Salesians, with no real representation of the Yanomami.[9] But his letters and public statements about both the Salesians and SUYAO were flattering between 1988 and early 1990.[10] Chagnon tried to gain the support of the Salesians through promises of publicity and research assistance. At the same time, he held out the prospect of direct financial help to SUYAO through his newly created foundation, the Yanomamo Survival Fund.[11]

In spite of the economic incentives, however, Yanomami leaders of SUYAO were not willing to work with Chagnon. The academic opposition to Chagnon at Venezuelan universities and think tanks was matched by increasing resistance within the Yanomami communities. That became surprisingly clear at the mission post of Mavakita, his base of operations over the years, when he returned to Yanomamiland after a decade of banishment by Venezuelan authorities.

"I first went into the Yanomamiland with Chagnon in 1985," said Jesús Cardozo, president of the Venezuelan Foundation for Anthropology (FUNVENA) and a former graduate student of Chagnon's at the University of California at Santa Barbara. "We hadn't even gotten our boat moored to the shore at Mavakita, when Yanomami started coming out and shouting, 'Go away! Shaki brings *xawara!*' Shaki is their name for Chagnon. *Xawara* is the cause of illness. So they associated Chagnon's visits with sickness right from the start. Within our first twenty-four hours, three children died—two in the night and another in the morning. It was terrible to witness that and to hear the wailing that began. On our second night, half of the village fled into the forest in order to get away from us. And a group came from Nasikibowei [splinter group of the Iwahikoroba-teri, a village whose members had shot an effigy of Chagnon after the measles epidemic] who asked us please not to visit them. Now, I didn't speak Yanomami at this point, so this is what Chagnon told me. When the people from Nasikibowei asked him not to come, I said to Chagnon, 'You really don't seem to have a lot of friends among the Yanomami.' "[12]

"I was just surprised," Cardozo said another time. "An anthropologist is supposed to have a special rapport with the people he studies, but the Yanomami were kind of running away from us. They ran away. The Kedebabowei-teri, who were visiting Mavakita, left because of all this thing about Chagnon and *xawara.* He was really angry because all the people had left. Some of the local people had left along with the Ketibabowei-teri, so really we had an empty *shabono* with lots of space and very few people. So he said, 'I'm going to go and look for another group.' That's when he went to the Washewa River, looking for the Iwahikoroba-teri, the ones who had tried to kill him before [in 1971]. Chagnon came back and said, 'God, I really have found them.' This, I remember. That night we were lying in the hammocks, and he turned over to where my hammock was and he said, 'Jesús, get ready. Tomorrow you enter the Stone Age.' "[13]

"When we arrived at Iwahikoroba-teri, everybody was sick, throwing up, and moaning and lying down in their hammocks. I remember this little girl, Makiritama. She was vomiting blood. She was defecating blood, too. I remember her husband—she was very young, she was to be his future wife—showed me where she was spitting up and everything. And I went up to Chagnon and said, 'You know these people are really sick. Some of them could die. We've been in one village where three people died within twenty-four hours. Here people are spitting blood. I think we should go and get help.' There were doctors at the mission. There was medical help that could be gotten just a few hours away. And Chagnon just told me that I would never be a scientist. A scientist doesn't think about such things. A scientist just thinks of studying the people. That's what he told me. He didn't want to deal with it at all. It was death. Death was going around. But he said, 'No. No. That's not our problem. We didn't come to save the Indians. We came to study them.'

"So that night we discussed the book *The Mountain People,* by Colin Turnbull. Have you read it? It's about an anthropologist who's entered a situation of general starvation and famine. And he just locks himself up in his jeep at night and eats crackers, and in the morning he opens the door and sees all these corpses around the jeep, people who had died of hunger during the night. He hears people moaning. It makes it difficult for him to sleep to hear people agonizing from hunger, but he's got to do the work. He chose to take their pictures and record how many people died. He tallied them up every night and hid his crackers. So Chagnon's position was that that's what you've got to do—you've got to do your work. You know, it's okay if they die as long as we get the data."[14]

Cardozo also said, "He just kept working, writing letters on their chests with a magic marker. It just wasn't . . . I don't know . . . human."[15]

Chagnon's standard practice was, in fact, to number the Yanomami, as if they were going to a college mixer, "with indelible ink to make sure each person had only one name and identity."[16] It is clear from Cardozo's account that the health of the Yanomami on the Upper Mavaca, which had been relatively good when Chagnon made first contact in the 1960s and early 1970s, had deteriorated sharply. Chagnon's advice to Cardozo sounded like an echo of Neel's disregard for the Indians during the measles epidemic. "Anybody can walk into a village and treat people. This is *not* what we're here to do."[17]

Some of the Mishimishimabowei-teri were now living at the mission post of Mavakita, where politics and economics had profoundly changed since Chagnon's "first contact" in 1968.

"In 1989, we were having a feast at Mavakita when I told the Yanomami that Chagnon had written a letter saying he was coming," recalled the Salesian brother Juan Finkers. "I mentioned this because they were going on *wayumi* [foraging trek], and I wanted to know how long they would take since Chagnon wanted to visit them. They shouted, 'Why do you send us this man? We don't want him to come, because every time he comes afterward we have fights.' "[18]

The fighting that Chagnon provoked, according to Finkers, was related to Chagnon's genealogical research. "The thing is that Chagnon asks people for the names of the dead, and that creates conflict among them. One will accuse another, 'You've told him my father's name.' And that's how it starts. They blamed me. 'Why do you send this man here?' I said, 'I don't send anybody, but he's written a letter saying he wants to visit you.' They answered, 'Tell him not to come.' I told them I wasn't going to write the letter. They know how to write. So they had a meeting and delegated César Dimanawa to write a letter to Chagnon. That's when this whole thing started."[19]

Dimanawa's blunt, handwritten letter was published in *La Iglesia en Amazonas* in March 1990. "We the people of Mavakita, Washawa, and Kedebabowei-teri and Mishimishi-teri do not want you to return to the Upper Orinoco."[20] There were many reasons—almost as many as there were families on the Upper Mavaca. One group said it did not want Chagnon to come back, because he had threatened to burn down their village with his "fire weapon."[21] Some of the Mishimishimabowei-teri recalled that Chagnon had beaten two children with a belt—one had stolen some of his food and the other tripped him.[22] But these were all events from the early 1970s. Why had they become so important in 1989?

In the end, Chagnon's film and media projects were probably the biggest obstacle to reconciling his differences with SUYAO and to cementing his incipient alliance with the Salesian anthropologists. Initially, Chagnon had promised to "do something ambitious with the American and international press to dramatize the current situation among the Yanomamo. In addition, I take every opportunity to promulgate, in my conversations with the press, the extremely important role of the Salesian missionaries and their activities in helping the Yanomami."[23] It turned out, however, that Chagnon's idea of publicizing the Yanomami's plight was arranging for a BBC documentary to commemorate the twenty-fifth anniversary of his fieldwork, to be filmed on the Upper Orinoco in November 1989. Catholic nursing nuns asked that Chagnon's television project be prohibited, although they did not object to Chagnon's continuing his personal research. With so many Yanomami dying in the gold rush, the Salesians were not alone in doubting the value of a documentary celebrating the silver anniversary of Chagnon's research on Yanomami violence. CONIVE, a national Indian rights organization, also objected to Chagnon's television permits.[24]

I think Chagnon might have worked out his differences with the indigenous organizations, the Salesian medical staff, and the Indian Agency, but it would have meant restrictions on his freedom to bring big film crews into the field. He would have had to conduct research more like a scientist than a media star. But this was unacceptable.

By this time, Chagnon realized that power came from a high media profile, not academic journals, where his credibility was now questioned. With Matos and Brewer Carías for allies, Chagnon devised a characteristically bold media plan to permanently circumvent all the institutions that controlled the Yanomami Reserve. The three men would simply create their own, private reserve, a Yanomami park. At the same time, and in a dramatic reversal of Chagnon's previous statements, they began a fierce press campaign against the Salesians. Chagnon and Brewer accused the Catholic missionaries of arming the Yanomami with shotguns and callously neglecting their medical needs.

Chagnon helped Brewer get an honorary appointment to the University of California at Santa Barbara's anthropology department for 1991–92.[25] In a reciprocal appointment, Chagnon became a "technical adviser" of

FUNDAFACI and helped arrange funding for the FUNDAFACI trips from UCSB. Brewer and Chagnon—with Cecilia Matos occasionally on board—took individuals from *ABC News, Prime Time Live,* and other journalists on trips into the Siapa valley, where the last unacculturated Yanomami villages were located. Although Brewer and Chagnon greatly exaggerated when they told James Brooke of the *New York Times* they had "discovered" ten previously unknown villages with 3,500 Yanomami, the Siapa River valley was remarkably free of outside influence—until FUNDAFACI started promoting its Siapa River valley biosphere plan in 1990.[26]

Chagnon and Brewer knew how to attract the interest of the international press corps—by providing helicopter rides to virgin villages. Once there, Chagnon and Brewer displayed their laptop computers and GPS instruments for the Yanomami and for the cameras. The anthropologist Brian Ferguson said, "When I read James Brooke's articles about Chagnon going into the Siapa together with journalists, scientists and army officials, I kept wondering, How can this possibly be a safe thing to do with uncontacted Indians?"[27]

To American journalists like John Quiñones of ABC and James Brooke of the *Times,* Chagnon portrayed the Salesians as a group of aggressive proselytizers out to attract the Yanomami to Christianity by distributing trade goods.[28] But the Salesians had not baptized anyone since their missions had been reorganized in 1976, and their well-regarded schools did not even mention Christianity. Actually, three of the Salesians were anthropologists, and they had been doing demographic research together with Chagnon prior to the FUNDAFACI adventure.[29]

Within Venezuela, the Salesians' principal opponents were either evangelicals—who accused them of being anthropologists, not Christian missionaries—or proponents of the Conquest of the South. The most vocal of the latter were Charles Brewer and his gold-mining allies, who attacked the Salesians as radical Marxists promoting native nations, not Bible thumpers out to convert infidels.[30]

While Chagnon and Brewer would later take credit for helping legalize Venezuela's 32,000-square-mile Yanomami Reserve, both ABC's *Prime Time Live* footage and James Brooke's articles for the *New York Times* reveal that Chagnon and Brewer were promoting their own, independent FUNDAFACI reserve, where they planned to create a giant scientific research station. Their domain would have covered about 6,000 square miles, only a sixth the size of the biosphere as it exists now, and would have included a similar fraction of the Venezuelan Yanomami. In fact, the plan was an alternative to a much bigger indigenous biosphere that Brewer had been opposing as a terrorist plot since 1984. The Brewer-Chagnon plan would have opened up large areas of the Upper Orinoco to development—including the tin ore deposits that Brewer had been coveting for years—because the FUNDAFACI biosphere effectively denaturalized the "Mission Yanomami" and afforded neither them nor their lands any protection. At the same time, this private biosphere would have given Brewer and Chagnon a scientific monopoly over an area the size of Connecticut.

The FUNDAFACI plan was never approved. It did, however, have the salutary effect of finally concentrating the minds of Venezuelan anthropologists and Catholic missionaries, who saw the drastically reduced "biosphere" as a coup against the Yanomami territory.[31]

The filmmaker Timothy Asch was particularly bitter about Chagnon's renewed association with Charles Brewer. Chagnon dismissed Asch's objections: "As to the comments in your letter about what's going on in my relationships with the Salesians in Venezuela, you are very badly informed about the events lying behind all of this . . . and equally badly informed about Charles Brewer. The Salesians tried clandestinely to get my research privileges terminated in 1989—before I resumed contact and collaboration with Charles Brewer."[32] And when Asch sent two twenty-dollar checks to Chagnon's Yanomamo Survival Fund, Chagnon returned them.

> My decision to create the *Yanomamo Survival Fund* has led to the most astonishing and spiteful actions on the part of some of my colleagues. I have reluctantly decided to put things on the back-burner for the time being and am not accepting contributions until the air clears.
>
> I am amused that you remited [*sic*] these checks at this point in time. It strikes me as comparable to the Sinner who decides to repent on the deathbed. In your case, I am referring to the failure of Jesús Cardozo to pan out as Mr. Yanomamo. . . . It is possible that your shot at resuming filmwork among them might have gone down with his ship.[33]

Asch wrote back, "I think you are reading more into my reasons for sending the checks than was actually there. You said you were going to start a survival fund a long while ago, which would not have administrative expenses and where most of the money sent in would go directly to the Yanomamo . . . because a lot of other survival funds do spend a great deal of money on overhead, administrative expenses and on anthropological careers by anthropologists who write articles, etc."[34]

Of course, Asch was baiting his former partner. And, like many others who sent contributions to find out what would become of them, he suspected that Chagnon was using the Yanomamo Survival Fund for his own research. To some extent, Chagnon invited such scrutiny by a curious mission statement that appeared in the fourth edition of his *Yanomamo* textbook (but was removed from the fifth edition). He wrote that he would help doctors who are serving the Yanomami, not with desperately needed funds, medicine, or equipment, but with "a reliable base of the kinds of demographic data anthropologists like I normally collect for our own research purposes—census and genealogical data. . . . Monetary contributions to the Yanomamo Survival Fund or AFVI [American Friends of the Venezuelan Indians, another group Chagnon founded] will be used to support these and similar efforts."[35]

In a subsequent letter to Jesús Cardozo and the archaeologist Hortensia Caballero, Asch wrote, "I can't resist describing a letter of Napoleon Chagnon's that arrived when I returned. There isn't much to say about it, it speaks for itself, but it is a bit weird. I guess even if I clearly told him what we were trying to do together, ther[e] is probably no way he could conceive of the project because it is so far from the self-serving projects that he seems to get involved in."[36]

Asch was impressed with Cardozo's ability as a mediator in the hot political climate of Caracas.[37] They jointly organized a conference, between December 4 and 7, 1990, to promote a Yanomami biosphere much larger than the one proposed by Brewer and Chagnon. It resulted in a series of recommendations signed into law by President Carlos Andrés Pérez on August 1, 1991. While the conference marked a breakthrough for the Yanomami biosphere, it also led to a complete rupture between the two former filmmaking partners.

Neither Chagnon nor Brewer was invited to the meeting. According to Cardozo, "Brewer Carías came and said, 'I can give you a lot of money. How much do you need?' I mentioned some of the expenses for the convention, and he said, 'No. I'm talking about a *lot* of money. But there's one condition—FUNDAFACI has to enter as a cosponsor.' I said I couldn't do that, and he stormed out. Then Cecilia Matos called me and offered to sponsor it on the condition that Chagnon and Brewer Carías be speakers. I told her I couldn't, because SUYAO had held their annual meeting and there was a great deal of anger against Chagnon. SUYAO's new president, César Dimanawa, said he would kill Chagnon if he saw him in Yanomami territory. Cecilia Matos became angry and said, 'We will see what the consequences for this will be.' "[38]

The consequences seemed fairly dire to Cardozo when, four months later, at a British embassy party, Brewer came up to him and said, "If I see you on the street, I'm going to shoot you."[39]

In the Siapa Highlands, a different sort of party was under way. Scores of people fresh from international flights were coming and going into the Siapa wilderness without permits, as though visiting the least-contacted cluster of aboriginal villages in the world had become a huge FUNDAFACI fiesta.

By the 1990s, the principal Amazon countries had enacted legislation designed to avoid the tragedy of "first contact." Brazil created an entire department with its Indian Agency whose purpose was to *prevent* contact with the remaining isolated Indians. In Peru, a group of scientists led by the anthropologist Kim Hill of the University of New Mexico were denied permission to contact a tribe in Manu National Park. They were turned down even though they took quarantine precautions, promised to provide medical care for a full year, and refused to take any journalists. Hill said they turned down an offer of $90,000 from WQED Pittsburgh, which was doing *National Geographic* specials at the time.[40]

The Siapa experience worked differently. The cosponsors FUNDAFACI and the University of California at Santa Barbara shuttled at least sixty-three different individuals into this final tribal frontier. Most of them were flown in directly from Caracas, many right off international flights. Dozens of military personnel also accompanied them (four to a flight, at least twenty-six flights), but they were not counted in Brewer's log. Indigenous guides were not included either, and there were many of those. All in all, probably 100–120 people toured the most remote villages of Yanomamiland.[41]

By all accounts—including Brewer's own—these trips violated the most basic medical rules of first contact. Guidelines laid out in recent years include these three: (1) screen all expedition members during a quarantine period prior to contact; (2) have medical personnel present at all times to give initial health checkups and inoculations; and (3) maintain a permanent medical presence during the first year after contact to administer antibiotics as needed. "A failure to do so has always, in native Amazonian history, resulted in unacceptably high mortality," according to Kim Hill and a colleague in an article for *National Geographic Research* a year before the Siapa expeditions started.[42]

Hill's own research group followed all these rules when attempting to contact a group of twenty to thirty hunter-gatherers in Manu National Park, Peru. Some three thousand Yanomami lives were at stake in the Siapa region, and the expeditions took place while hundreds of Yanomami in Brazil were dying of uncontrolled contact with gold miners.

When I read James Brooke's articles in the *New York Times,* I was in Boa Vista, Brazil, struggling with outboard motors, malaria, and the day-to-day chaos of the gold rush. Although Brooke had repeated Chagnon's claim of having discovered 3,500 uncontacted Yanomami, I learned that anthropologists and human rights workers with long experience in the area did not agree. The French anthropologist Bruce Albert, whom I spoke with at the Yanomami village of Toototobi near the Venezuelan border, scathingly dismissed Chagnon's claim to "first-contact." I was surprised at Albert's vehemence and skeptical when he told me that Charles Brewer was a big gold miner.[43]

Yet, according to Josefa Camargo, Venezuela's assistant attorney-general for indigenous affairs, "All of these trips by Brewer and Chagnon into the Siapa region were illegal because there is no evidence they even submitted their plans to the DAI [Indian Agency] for approval."[44]

When I asked Chagnon's closest collaborator over the years, the anthropologist Raymond Hames of the University of Nebraska, what he thought of the Siapa expeditions, he readily answered, "Anybody who comes in a helicopter and goes quickly is putting remote villages at risk. It's pretty stupid and reckless to fly in from any urban center. It's crazy to fly big groups in from Caracas without any quarantine precautions. Chagnon made a stupid decision to work with Brewer."[45] Hames, to his credit, stayed away from the helicopters and the haze of publicity.

Kim Hill, who shares Chagnon's sociobiological approach and has defended Chagnon's right to do research in Venezuela, commented about the Siapa expeditions, "Getting involved with Charles Brewer Carías is probably the worst mistake of Chagnon's anthropological career. The reason I defend Chagnon is purely an issue of academic freedom. Chagnon got involved with Brewer Carías mainly because of academic repression. . . . Chagnon flipped out when they cut off access."[46]

Brewer's expedition records showed that the FUNDAFACI Siapa excursions started on August 8, 1990, yet no doctors accompanied Chagnon and Brewer until January 1991.[47] The doctor stayed for four days at a very remote village and then flew back to Caracas. By then, the FUNDAFACI camp at the Yanomami village of Washewa was caught up in an epidemic.[48] On January 26, the Fourth Air Force Group of Amazonia received an emergency summons to fly in a team of doctors, who stayed for less than an hour. The pilot of the emergency mission was surprised that the rescue team included Cecilia Matos and a European prince—but no first aid equipment. He was even more surprised when none of the dying Yanomami were evacuated as originally planned. Only Napoleon Chagnon and his son Darius were rescued, both ill with malaria and severe diarrhea. According to three air force colonels, who made declarations to the press and a Venezuelan court, "Before leaving, Mrs. Matos was asked about the sick Indians. She said they were in such critical condition that they couldn't be saved . . . *por lo cual era mejor dejarlos morir en su tierra."*

"That's why it was better to let them die in their own country."[49]

Brewer offered a defense of their medical decisions at Washewa. "The purpose of this flight," he wrote, "was to evacuate Dr. Napoleon A. Chagnon and Dr. Darius Chagnon. . . ."[50]

Carlos Botto, the head of Venezuela's tropical medical institute (CAICET), observed the FUNDAFACI expeditions firsthand. "I was at the Ocamo mission with the doctor Magda Magris, and we were trying to help a Yanomami woman whose fetus had died but which we could not remove. She was in danger of dying, and we wanted to take her out to Puerto Ayacucho. At that time, the helicopter of FUNDAFACI passed overhead all the time, but in spite of our emergency request by radio, it refused to land. After about five days, when the woman was by then almost unconscious, the enormous helicopter descended at Ocamo. Then some high-ranking military men came down the stairs and placed themselves at attention below as if they were waiting for an important dignitary. There, in the helicopter's exit, appeared a woman all dressed in white. She wore enormous boots and an immense white hat. Cecilia Matos. She descended. She was very worried because she had discovered in the Siapa Highlands that the Yanomami had fleas. She'd come because she wanted us to treat the Yanomami against fleas. We explained to her that this wasn't so terrible—that the Yanomami know how to remove them and they can even eat them—but that we did have a woman who was at the point of death and who would die if she weren't flown out. Matos went to see the woman and said that she couldn't fly her out. There was no room and the woman smelled bad. Of course the woman smelled bad—she was septic. We had to be very insistent and argue for a long time before Matos finally took the patient."[51]

While the international campaign to institute a new order in Yanomamiland was a success, local snags developed. Citizens at Puerto Ayacucho, where the helicopters refueled, told the pilots it was normal for Brewer Carías to conduct his gold-mining activities on government helicopters. That made the pilots feel like idiots. They soon detested ferrying Cecilia Matos, her friends, and press celebrities into the most remote reaches of Yanomami territory. If Matos wanted fuel, she just demanded it; if she wanted a helicopter, she did the same. No authorizations were offered. Squabbles were frequent—between the pilots and Brewer Carías (who wanted to fly the helicopters himself and keep the flight plans secret), between the pilots and Matos, and between the expedition and various Yanomami groups. The air force officers saw a dozen foreign scientists carrying out huge amounts of plant and animal samples, without approvals from the Indian Agency (DAI) as required by law. And they also saw Brewer Carías furtively collecting something in sealed metal boxes that they were not allowed to inspect—yet another violation of their regulations. Brewer told them it was "explosive material,"[52] which proved to be a political prophecy.

On November 27, 1992, a group of young air force officers decided to attack the presidential palace. The uprising failed, and the officers were either imprisoned or fled the country. But the aftermath of the coup brought allegations by the pilots against Matos, Brewer Carías, and Chagnon. Four of the air force officers who led the coup gave FUNDAFACI's Yanomami expeditions as a reason for the rebellion.[53]

"I accuse Brewer Carías and Cecilia Matos—they trafficked with the patrimony of our Indians in the Amazon," said Captain Luis Manuel Jatar, a Super Puma helicopter pilot who had flown FUNDAFACI around. "On those flights, I transported Mrs. Cecilia Matos and Mr. Brewer Carías. They were accompanied by various American scientists, led by a man named Napoleon Chagnon, who represented a prestigious North American university. . . ." At the FUNDAFACI camps in Yanomami territory, Captain Jatar saw large collections of rare birds like the guacamayas and pauxi—the latter an endangered species. "Generally, on each flight great numbers of these animals were carried out. But the most curious thing were some metal boxes hermetically sealed that couldn't be examined. Charles Brewer Carías jealously guarded those boxes."[54]

Brewer Carías said there was blood, not gold, in his sealed boxes. But as far as the Venezuelan public, congress, and press were concerned, it made little difference whether Brewer Carías, Matos, and Chagnon were profiting from the Yanomami's gold or their blood—or the endangered species of the rain forest. It was all against the law.

Cecilia Matos and FUNDAFACI became objects of a criminal investigation in the courts, a parallel inquiry in the congress, and a continuous scandal in the press. When the *Fiscalía*—the legal entity that is the nation's constitutional protector and legal controller—learned that President Andrés Pérez had been party to a huge diversion of defense funds into private bank accounts, impeachment procedures began that ended in his imprisonment. Cecilia Matos left the country in February 1993.

The ex-president's mistress was charged with seven counts of fraud and influence peddling. The Superior Court of Salvaguarda in Caracas ordered an inquiry into accusations that she smuggled gold and made unauthorized use of air force planes and helicopters.[55] A judge from the Sixth Penal Court requested her extradition from the United States. By this time, FUNDAFACI had vanished along with records of Chagnon's scientific expeditions, thought to have cost the Venezuelan people between five and ten million dollars.[56] A special prosecutor has spent six years pursuing Matos and her foreign bank accounts. These accounts hold more than half a million dollars and may, according to the special prosecutor, hold as much as twelve million dollars, although Matos's lawyer says this figure is exaggerated.[57] The Superior Court of Salvaguarda, Caracas, in April 1998 ordered Matos, whose whereabouts were unknown, to detention in a woman's prison until her case is resolved. That decision of arrest was ratified by the Venezuelan Supreme Court in August 1998.[58]

Although Brewer was not indicted, he had become, in the wake of this storm, one of the most unpopular and notorious people in Venezuela. But Chagnon and Brewer had never been excessively concerned about being sweet. True disciples of James Neel, they believed that dominant men could intimidate and impress lesser ones. They were about to test themselves, and their theory, against all the odds.

Chapter 12

The Massacre at Haximu

The massacre site was horrifying. But equally worrisome was the effort of the local authorities—the Roman Catholic Salesian missionaries—to derail our investigation and keep the plight of the Yanomamo hidden.—*Napoleon Chagnon*[1]

In July 1993, a heavily armed gang of Brazilian gold miners massacred twelve Yanomami Indians at a plantain garden located in the high, cool Parima Mountains. The miners killed everyone they could surround—mostly women and children—with shotguns, pistols, and machetes. They also burned down the victims' village, Haximu, a two-hour walk from the scene of the murders.[2]

The Haximu massacre dominated world headlines.[3] Initial reports were exaggerated—claiming seventy-three victims—because the first Brazilian official to visit Haximu came back and gave a press conference in which he described seeing many decapitated bodies.[4] The man wept as he gave the horrific and imaginary account. In fact, there were no bodies at the massacre site, because the Yanomami had cremated all of them. As a result of this hysterical story, the president of Brazil's Indian Agency was fired.[5] The official who gave the bizarre testimonial was allowed to keep his job at the National Health Foundation (FNS) of Brazil, though he never recovered. He committed suicide less than two years later.

The massacre seemed to bring out madness in everyone.

The next battle began over which reporters would fit into a Brazilian helicopter headed for Haximu. The press held a lottery for the three positions available, but the reporters from Brazil's two biggest news competitors, *A Fôlha de São Paulo* and *O Globo,* were not favored by chance. As the helicopter started to lift off, the last of the lottery winners—a photographer—jumped on board. But as he did so, the reporter from *O Globo* ran past the security guards and grabbed hold of the photographer's legs. Seeing that, his rival from *A Fôlha* also sprinted out and tackled the *Globo* man around the waist. All three were dangling as the helicopter went airborne. Their behavior amazed the Yanomami warriors standing by the airstrip, who were all painted black and armed with bows and arrows, preparing for a revenge raid against the miners.

"I had to rush out in front of the helicopter and wave the pilot back down," said Leda Martins, an FNS official and former journalist, who was coordinating press relations. "They were twelve feet off the ground, and the *Globo* guy was still hanging on to the *Folha* guy, while the photographer, who was barely holding on himself, was kicking his legs like mad to get rid of them both. It was really wild. When the helicopter landed, and we finally got them untangled, the two reporters from *O Globo* and *A Fôlha* started tearing each other apart. They completely ripped up each other's coats, ties, and shirts. The Yanomami warriors were mystified. They came up to me afterward and wanted to know what the fight was all about. But I wasn't able to explain the feud between *O Globo* and *A Fôlha* to them."[6]

In Venezuela, the anthropologist Napoleon Chagnon and explorer Charles Brewer Carías also hopped onto a helicopter going to Haximu. And, like the reporters, they found themselves in a war of their own making.

Even though President Andrés Pérez had rejected their Siapa valley biosphere proposal,[7] Chagnon and Brewer remained hopeful. Chagnon's writings from 1992 hinted that something was afoot. While noting that the 32,000-square-mile reserve for the Yanomami had been signed into law in 1991, he added, "Specific rules to carry out the law have yet to be established. A commission appointed by the president will write them, and as of now the composition of that important commission is not publicly known."[8]

Although President Andrés Pérez made many mistakes while in office, he never appointed the presidential commission Chagnon wanted. Luckily, the constitutional crisis that sent Pérez to jail coincided with the Haximu massacre, and this opened a new opportunity. After Pérez was impeached and his interim successor, Ramón Velásquez, was officially instated, Chagnon and Brewer were appointed to a presidential commission with sweeping powers over the new biosphere. Although Chagnon would repeatedly refer to this as a committee to investigate the Haximu tragedy, there was no mention of the massacre in the new law. Instead, the legislation—long prepared—reversed the previous rules on the Yanomami park and biosphere reserve. Brewer and Chagnon were authorized to make decisions about everything from scientific research to tourism to "the natural resources of the region," thus replacing the Ministry of the Environment and the park service as arbiters of the Yanomami's destiny. The presidential decree empowering the two men appeared on the page following the decree impeaching Andrés Pérez.[9]

The same week Chagnon and Brewer were given control of the Yanomami Reserve, Venezuela's First Congress of Amazon Indians opened in Puerto Ayacucho, Amazonas State. Three hundred representatives from nineteen Indian tribes rallied in the streets and chanted, "Brewer Carías and Chagnon out of Yanomami territory."[10] Similar protests took place in Caracas, led by university faculty, political parties, relief groups, and the Catholic Church. Antonio Guzmán, the elected president of the Yanomami trade cooperative, SUYAO, said, "We are tired of being constantly investigated, of them taking our blood and using us as little animals. And on top of everything these investigations are published abroad for the private benefit of Mr. Chagnon."[11] Another representative at the congress said that making Brewer the president of a massacre commission would be "like asking a wolf to count the sheep."[12] Survival International's anthropologist, Fiona Watson, who was present, said, "If the Yanomami themselves do not want these two persons to represent them, I believe that we have to respect that decision."[13]

Virtually every scientific organization within Venezuela denounced the decision to make a gold miner like Brewer Carías head of the commission to investigate gold miners. These included the National Council of Scientific and Technological Investigation (CONICIT),[14] the Venezuelan Institute of Scientific Investigation (IVIC),[15] the Central University of Venezuela (UCV), and the national association of anthropologists (CAV).[16] To these were added the Attorney General's Office,[17] members of congress,[18] and the federal, state, and private ecological organizations in the Venezuelan rain forest.[19] At a national meeting of governors with the interim president, the governor of Amazonas, Edgar Sayago, banged his fist on the table. "In Amazonas, we do not accept him [Brewer]!"[20]

The Congress of Amazonian Indians and the governors' conference also coincided with a meeting of Venezuela's National Conference of Catholic Bishops, whose members drafted a statement condemning the Haximu massacre while urging an investigation by "people who inspire confidence."[21] The presiding archbishop, Ovidio Pérez Morales, made it clear that Chagnon and Brewer inspired "anti-confidence."[22]

There was general amazement that Brewer and Chagnon had the audacity to show their faces after the indictment of their patroness, Cecilia Matos, and the impeachment of Andrés Pérez. Although a precise equivalent cannot be conjured, it would be as if discredited members of the Nixon White House had shown up a few months after their boss had flown off in disgrace and were suddenly, legally, made the sole administrators of one-fifth of all U.S. territory.

Different sectors of society had their own grievances and their own reasons for piling on. The universities and think tanks still remembered how Brewer had denounced the "Yanomami Reserve and the biosphere" as a plot by Libya's Colonel Gadhafi to create "an indigenous nation."[23] Yanomami leaders who disliked Chagnon came to Caracas and did the talk show circuit (Chagnon countered that they were alive thanks to his measles vaccination in 1968).[24] And, since the scandal erupted during the final months of a presidential election, two candidates—from the ruling party and the green party, respectively—called for Brewer's and Chagnon's removal from the Haximu investigation.[25]

In modern Venezuelan history, there had rarely been such a broad consensus of repudiation. Hundreds of articles condemned the duo.[26] Chagnon later wrote that "unprecedented press, radio and TV attacks on members of the Presidential Commission paralyzed the government."[27] The vacillating president withdrew all support for Brewer and Chagnon, gave them no funds or permits, and let the *Fiscalía*—a powerful independent omsbudsman that combined the functions of the Attorney General's Office with those of a national comptroller—take charge of the case.

In 1993, the *Fiscalía* was the most powerful institution in Venezuela. Its director, José Antonio Herrera, became a hero after the Fiscalía uncovered the million-dollar fraud that forced President Carlos Andrés Pérez from office. On September 15, the day after President Velásquez announced he was turning the massacre investigation over to the *Fiscalía*, Herrera invited Brewer and

Chagnon to his office. Afterward, Brewer said, "The *Fiscalía* will decide what commissions should be structured and the functions each should fulfill in the area, and we will collaborate with our expertise."[28]

Outside the *Fiscalía*, the press questioned Brewer about his gold mining. "That mining concession is in Bolívar State, a thousand kilometers from Amazonas State," Brewer answered. "My role as miner has nothing to do with my role as writer, ecologist, biologist, and naturalist. . . . It is another activity, but it cannot be seen as illegal, because it is being accepted and regulated by the Ministry of Mines."[29]

But the *Fiscalía* did, in fact, consider Brewer's mines on Pemon Indian lands in Bolívar State to be illegal. When I gave a talk to the *Fiscalía* staff about gold mining in May of 1992, the assistant fiscal for indigenous affairs, Josefa Camargo, maintained that the entire system for granting mining leases in Bolívar State was unconstitutional. At the same time, police investigators were informing the *Fiscalía* of Brewer's expanding role as a gold entrepreneur. On May 12, 1993, his mining company, Minas Guariche Limitado, changed its directory. The new vice president was named: Robert M. Friedland, Canadian passport No. EM269189.[30]

Like Brewer, Friedland had a colorful history. His romance with toxic chemicals began at age nineteen, when he was arrested with what was then the largest supply of LSD in the history of Maine.[31] After serving time in federal prison, he started his own mining company, Galactic, which acquired the Summitville Mine, at 12,500 feet in the Colorado Rockies. Using cyanide heap-leaching technology to concentrate gold, the mine also concentrated one of the largest deposits of toxic metals in mining history[32]—leaving the Environmental Protection Agency with a $150 million cleanup bill after Friedland abandoned Summitville to the U.S. government in 1990. Efforts by the U.S. government to seize Friedland's foreign assets failed. Friedland next turned to South America. His company, Golden Star, became a partner in Guyana's Omai gold mine—another cyanide-concentrating operation that, in an eerie rerun of Summitville, became South America's greatest mining disaster.[33]

In Venezuela, foreign mining companies had acquired 3.5 million acres of holdings, mostly on Indian lands and in supposedly protected areas of the rain forest, through the Corporación Venezolana de Guayana (CVG), an entity that literally owned the Pemon Indians and everything else in the state of Bolívar.[34] It worked like this: the CVG's vice president for mining, José Arata, and its president, Alfredo Gruber, both invested in a gold company,

Minera Cuyuni. They redrew the map of Canaima National Park, where Angel Falls is located, to give 221,000 acres of mining concessions inside the park. One of the chief beneficiaries of these holdings was their own Minera Cuyuni. Arata then sold his shares of Minera Cuyuni to Midas Gold, a wholly owned subsidiary of Green Forest, which was a wholly owned subsidiary of Bolívar Goldfields, whose principal owner was Vengold . . . which belonged to Robert Friedland.[35] A report for the Venezuelan congress's environment committee suggested that Vengold's labyrinthine alliances had a Venezuelan accomplice: "In this twisted web of purchases and associations, Charles Brewer Carías appears to be implicated."[36]

Because of this "twisted web," Brewer was also deeply suspect in the eyes of the police, who were watching every move he made. And his next move had a certain inevitability, given his bold personal history. In spite of all the opposition, and in spite of their promise to follow the *Fiscalía*'s lead in the Haximu investigation, Brewer and Chagnon managed to persuade the defense minister to let them reenter the Yanomami Reserve on a military helicopter. Only one other member of the original ten-man commission accompanied them—an ethnomusicologist with little experience among the Yanomami—along with Brewer's seventeen-year-old son. They were delayed for several days, according to Chagnon, because they "met with opposition, suspicion, sabotage, and deliberately imposed obstacles every step of the way—from military pilots, local military commanders, members of the Salesian missions, and members of the *Fiscalía.*"[37]

After three days of insults and delays, they were flown to Haximu on September 28.[38] The *Fiscalía*'s commission, headed by Judge Nilda Aguilera and Assistant Attorney General Josefa Camargo, arrived the next morning with Colonel Oscar Márquez and a Catholic bishop. According to a report by the *Guardia Nacional,* Judge Aguilera "asked Dr. Brewer Carías to leave the site because she was directing a crime summary for a tribunal that had been constituted *in situ.* Dr. Brewer Carías consulted with the colonel, who suggested he obey the judicial order in order to avoid inconveniences of a legal nature. . . ."[39] The judge was furious. She ordered Brewer and Chagnon to cease and desist or face arrest, which is why they never reached the actual garden where the massacre took place. Chagnon was escorted to Caracas by Colonel Márquez, who took his notes and urged him to leave the country immediately, which, in fact, Chagnon did.[40]

"We told them to leave because they had no business being there," Assistant Attorney General Camargo explained. "When the president denied them resources for their commission, their commission was effectively terminated, even though the presidential order was never formally revoked. The judge ruled that they had no competence in the investigation."[41]

"Chagnon and Brewer Carías tell a different story," I told her when we spoke.

"Yes," she answered. "And they will tell it until the day they die."[42]

Chagnon wrote an op-ed piece for the *New York Times* entitled "Covering Up the Yanomamo Massacre." It was concerned primarily with his own travails, not with the murders of the Haximu-teri. Chagnon decried a conspiracy against him by the Salesian missionaries, who were also supplying the Yanomami with guns that were being used on raids against "distant, defenseless villages." "Yanomami from the missions," he wrote, "kill the men with guns, abduct the women, and gang-rape them for days afterward. . . . It is likely that many more Yanomami die from mission policies than at the hand of garimpeiros."[43]

A similar, shorter article, entitled "Holy War in the Amazon," appeared in *Newsweek.* "The first independent investigation into last August's massacre of Yanomami Indians in the Venezuelan Amazon was nearly derailed by one of the institutions responsible for their welfare—the local Catholic church."[44] The text repeated Chagnon's assertion that the Salesians were distributing shotguns to "converts" and ended with the comment "the Venezuelan Amazon is the last theocracy in the Western Hemisphere."[45] Chagnon had a good relationship with *Newsweek,* having taken one of the magazine's writers to the Siapa village of Doshamosha-teri, subject of a *Newsweek* feature story named after Chagnon's book *Last Days of Eden.*[46]

Chagnon developed his *New York Times* editorial into a much longer piece for the *Times Literary Supplement* (the *TLS*'s editor, Redmond O'Hanlon, had known Brewer since receiving from him assistance in getting around in the rain forest in 1984 for his classic book, *In Trouble Again*). Chagnon's essay—published on December 24, 1993, called "Killed by Kindness?"—attacked missionaries, Yanomami indigenous leaders, Venezuela's "Attorney-General's office," "left-wing anthropologists," "left-wing politicians," and "survival groups."[47] While other observers thought Haximu was an example of the gold miners' culture gone beserk, Chagnon argued that the real villains were misguided do-gooders.

As with the piece in the *New York Times,* the principal targets of the *TLS* piece were Catholic missionaries. ("The theocracy in Amazonas extends to Caracas.")[48] But the article also savaged Survival International without actually naming it. Chagnon had threatened, two months earlier, to begin a campaign against Survival International if he did not receive a public apology from the organization.[49] He felt he deserved one following the anthropologist Fiona Watson's statements in support of the Yanomami's right to exclude him from their territory. Survival International never apologized.

Terence Turner, a University of Chicago anthropologist who headed the American Anthropological Association's Special Commission to Investigate the Situation of the Brazilian Yanomami, wrote a rejoinder to Chagnon in the *Anthropology Newsletter*. It was called "Truth and Consequences." Turner analyzed a series of statements in Chagnon's *New York Times* editorial. In the first place, Chagnon claimed to be representing the Venezuelan president at Haximu, when the president had named another commission to replace Brewer's. Second, Chagnon reported that the Salesians evicted him, when it was actually Judge Nilda Aguilera. Moreover, Chagnon said he had reached the site of the massacre; he was actually a several hours' walk away. Turner challenged Chagnon's statement that he had interviewed the Yanomami at Haximu, because they speak a different dialect. And, whereas Chagnon claimed the Salesians were orchestrating opposition to him, Turner showed that there had been a broad, national outcry against his and Brewer Carías's presence on the investigating team. Turner dismissed the Salesian "coverup," too, saying that "the real investigation had already been done by the Brazilians" and that Chagnon had already received a copy of it, to which he added very little except mistakes.[50]

Chagnon responded that Brewer's commission had not been abolished when they flew to Haximu. This was true. It was abolished later.[51]

I interviewed thirteen experts on the Yanomami about Chagnon's *New York Times* editorial; all condemned it. Two told me they had written letters to the *Times* in protest, but their letters were not published. (The Salesians were allotted a brief rejoinder more than two months later.)[52] Bruce Albert, a Yanomami expert from the University of Paris who conducted the official Brazilian inquiry into the killings, wrote that "Chagnon's text is a 'montage'—a fraud—mixing essentially a 'plagiat' of my report with some invented (and erroneous) details and very few things he could have gathered in loco."[53] The biologist Irenaus Eibl-Eibesfeldt, who had been studying the Yanomami since 1970, called Chagnon's version of events "grotesque" and "slanderous."[54] Eibl-Eibesfeldt also endorsed a rejoinder by Jacques Lizot, in which Lizot urged Chagnon, among other things, to find a psychiatrist. "Your state of paranoia disturbs me," Lizot wrote.[55]

Although the controversies surrounding Chagnon's *Science* article in 1988 had been touted as the most vicious in anthropological history, the aftermath of the Haximu massacre set a new standard for cruelty, one that spread around the world on the Internet. Kim Hill of the University of New Mexico, an expert on lowland South American Indians who shares Chagnon's sociobiological orientation, wrote, "For those of us who have met Lizot, his suggestion that Chagnon needs psychotherapy is supreme irony. . . ."[56]

The massacre at Haximu coincided with the apogee of Chagnon's standing in the world of sociobiology. The New York Academy of Sciences held a special session on Chagnon on September 27, 1993, to hear Brian Ferguson's analysis of Chagnon's lifework. Even critics still praised his "extraordinary devotion to anthropology as a science."[57] He was simultaneously elected president of the Human Behavior and Evolution Society, where his inaugural address became a vehicle for his defense. In his view, *l'affaire Haximu* was a kind of Scopes Monkey Trial for the 1990s. Chagnon's enemies were the enemies of Darwin.

> One of the frequently cited reasons why I must be removed from the Venezuelan Commission created to investigate the Yanomami massacre is the fact that I am a proponent of evolutionary theory and use this theory to guide my research among the Yanomamo. In short, I am a sociobiologist. . . . Venezuelan anthropologists have also convinced Salesian missionaries who work with the Yanomamo that evolutionary theory is hateful, racist, and widely denounced by responsible anthropologists.[58]

The strategy was successful. An eminent group of sociobiologists, including E. O. Wilson, Richard Dawkins, James Neel, and Robin Fox, only increased their support for Chagnon following his expulsion from Yanomami territory.[59] Interestingly, Chagnon's theories about the selective value of violence appeared to work well in the American media and academia. In these wars, Chagnon's small team of well-known sociobiologists and journalists was able to overcome superior numbers and the weight of factual evidence by aggressive and well-coordinated attacks.

The campaign involved sending press kits to key journalists, such as Matt Ridley, science writer at the *Economist.* Ridley has written two well-regarded scientific best-sellers, *Red Queen* and *The Biological Basis of Morality,* which expressed sociobiological principles in a moderate and intelligent way. His correspondence with Chagnon, however, revealed the kind of abject admiration many male journalists apparently felt for the great anthropologist. "I have written it in the way that the International Editor wanted, which means 'impartially.' (She is a bit PC, herself.) So you may find it less unambiguously sympathetic to you than you might have hoped, but it is about as far as I dare go, if they are to publish it at all. I think an impartial reader will still find it kinder to you than to your foes, and therefore it will bet [*sic*] better than nothing at all. I do hope you like it; Jacques Lizot's letter is the most offensive and dissembling document I have ever read!"[60] Ridley's letter and proposed article, which the *Economist* never published, were included as document No. 18 in Chagnon's large press kit.[61]

Lizot's letter was, of course, quite extreme. "Napoleon A. Chagnon wants to be original in everything. His inaugural address as president of the Human Behavior and Evolution Society is a heartless and repugnant settling of accounts, full of lies, slander and duplicity . . . I weigh my words when I state that Chagnon is a counterfeit and an intellectual accomplice of the *garimpeiros.*"[62]

Yet Lizot offered specific, detailed rebuttals to most of Chagnon's assertions about the massacre. How could Ridley know that Lizot, rather than Chagnon, was dissembling? Although Chagnon claimed to be a sociobiologist, he was not a biologist by training. Only one biologist has worked over a long period of time in Yanomami territory—Irenaus Eibl-Eibesfeldt. He had achieved international recognition for his collaboration with Konrad Lorenz in studying dominance behavior in geese and rats.[63] He then succeeded Lorenz as head of Human Ethology at the Max Planck Institute for Behavior Physiology. Throughout his career, he has not been at all squeamish about describing violence and its role in natural selection. If Chagnon's critics were "anti-scientific" and "anti-biological," as he claimed, it was odd that Eibl-Eibesfeldt and another researcher from his Max Planck Institute with experience among the Yanomami rebuked Chagnon—"whose behavior we find disgusting."[64] They wrote, "We are puzzled by Chagnon's articles. . . . He not only misrepresents the situation, but he got many important facts wrong."[65]

Any fact checker with a couple of hours to spare could have uncovered contradictions in Chagnon's accounts. In his *New York Times* editorial, Chagnon accused missionaries of giving shotguns to win converts and then looking the other way when their converts killed non-Christian Yanomami.[66] But for twenty years Chagnon had credited the missionaries with reducing

Yanomami violence.[67] And, in his turnaround, the only concrete examples of the Salesians giving the Yanomami shotguns were from the 1960s.[68] The villain was Father Luis Cocco, whom Chagnon once called "one of the great human beings of all time,"[69] and previously praised for his courageous role in halting warfare.[70] It was strange, then, that Chagnon had misrepresented missionaries in general, and Cocco in particular, for so long.

"The whole shotgun debate is ridiculous," said the anthropologist Kim Hill, one of Chagnon's defenders in the controversy. "A shotgun only lasts five to eight years down there, so if the Salesians gave out shotguns prior to 1985 it would be irrelevant now."[71] But since Chagnon's charges made the Salesians accessories to murder, they did not consider it irrelevant. After failing to get the American Anthropological Association to send a fact-finding team,[72] they turned to Frank Salamone, an anthropologist from Iona College who is an expert on missions and their impact on indigenous cultures.[73] He traveled to the Upper Orinoco and Mavaca rivers in November 1994 and did a shotgun census. "There are a couple of hundred shotguns, most of them made by the Brazilian brand Rossi," Salamone told me. "The Salesians didn't distribute shotguns. Chagnon has also criticized the SUYAO cooperative, which is independent, because they sold a few shotguns. The SUYAO cooperative sold exactly seven shotguns; I counted them. For Chagnon to make a fuss over SUYAO selling seven shotguns is a joke. Bórtoli [head of the mission] went in person and stopped the selling of shotguns, and the Yanomami are not happy about this. The Venezuelan law allows every citizen to have a shotgun. The Yanomami say SUYAO is their cooperative, and they want to sell shotguns. Now they get shotguns from other people—the military, *garimpeiros,* traders. Chagnon's accusations sound good in the press, but you can't have an Ethiopia in the Amazon—where people are chasing planes with spears. The world doesn't work that way. The *garimpeiros* and the military don't work that way. Charlie Brewer doesn't work that way."[74]

Chagnon did not work that way either—he always carried a shotgun with him in Yanomamiland.[75] Salamone also learned that he had sold one shotgun to a Yanomami man.[76]

In his *New York Times* op-ed piece, Chagnon wrote, "The Salesian policies include attracting remote Indian groups to their missions, where they die of diseases at four times the rate found in remote villages."[77] This apparent fourfold difference between mortality at missions and mortality at remote villages has become one of the most frequently quoted statements in the

Yanomami controversy (appearing most recently in a feature story called "Napoleon Chagnon's War of Discovery" in the *Los Angeles Times Sunday Magazine*).[78]

Yet all sources, government and private, Brazilian and Venezuelan, showed the long-term benefits of mission stations.[79] The myth that missions have mortality four times higher than that of "remote villages" is based on a study Chagnon conducted at seventeen Yanomami communities between 1987 and 1991. Chagnon divided the villages into three categories of mission contact: "remote," "intermediate," and "maximum." The villages with maximum contact had a 30 percent *lower* mortality rate than the remote groups, while the "intermediate" villages suffered four times as many deaths as the missions. Chagnon's data thus confirmed what all other researchers have found, but in the *New York Times* Chagnon converted the "intermediate" villages into "missions," which they are not.

The following graph shows mortality rates at these seventeen villages as Chagnon has presented them in his *Yanomamo* textbook.[80]

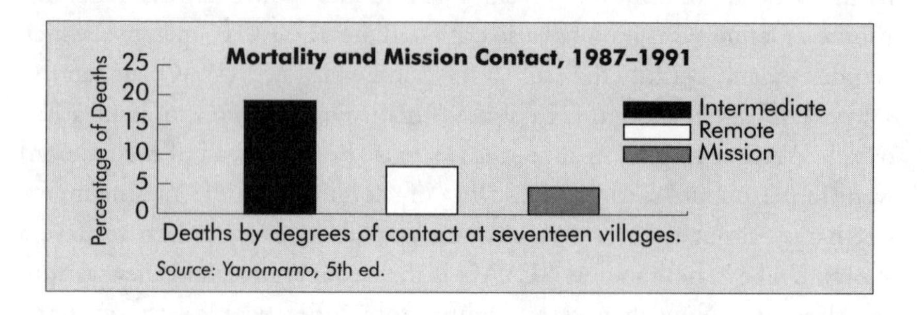

There is no doubt that the debate about mission mortality has been based on misinformation. But I believe that even Chagnon's study greatly understates mortality at the "remote" villages for the period 1987–91. He has assigned the village with the highest mortality, Dorita-teri, to the "intermediate" category, even though it is by far the most inaccessible village of the group.[81] The Dorita-teri have not been attracted toward a mission; in the 1960s they moved away from the missions into the heart of the Unturán Mountains, where their *shabono* can be reached by a three-day trek or by helicopter. (There is no record of the Salesians visiting Dorita-teri during the time of the study, though Chagnon landed there three times with large expeditions and also invited the villagers to a feast.)[82] Chagnon has also failed to count twenty-one deaths at the "remote" village of Kedebabowei-teri, which FUNDAFACI used as a helicopter supply base from August 1990 until January 1991. In fact, mortality at Kedebabowei-teri increased from less than two per year from 1987 to 1990 to almost two *per month* during sixteen months of intensive FUNDAFACI visits.[83]

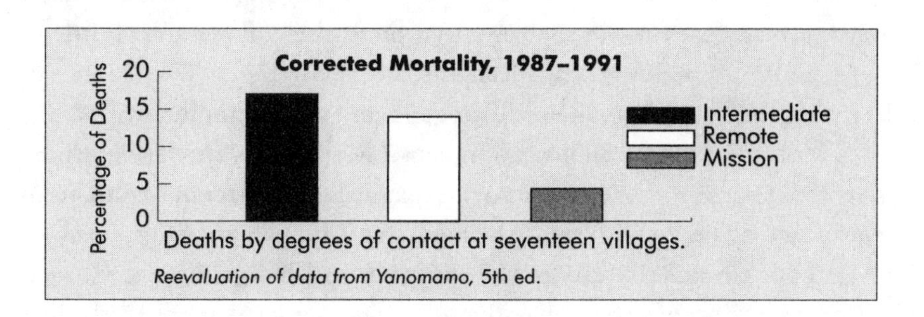

To explore this fully, I have included an appendix, which identifies all of the villages in Chagnon's study, shows their real geographic coordinates, and examines their history of contact with the missions and with Chagnon's expeditions (see the appendix: "Mortality at Yanomami Villages"). To summarize: actual mortality at both the "remote" and the "intermediate" villages was about the same from 1987 to 1991; both suffered a mortality rate about three times higher than that at the missions. A more meaningful distinction would be between villages that were visited by large FUNDAFACI expeditions and those that were not.

The next phase of the press war unfolded at the *Chronicle of Higher Education.* One of Chagnon's former students and longtime research associates, Raymond Hames, contacted the managing editor, sounding her out about doing a feature story on the controversy, which Hames portrayed as "a defamation campaign"[84] against Chagnon and the scientific method.[85] Chagnon then followed through, sending mountains of materials, while invoking the names of E. O. Wilson, Richard Dawkins, James Neel, Richard Alexander, and other sociobiologists who were supporting him.[86] Chagnon was apparently used to getting a good degree of editorial control in these cases. He suggested what the theme of the article should be "Church vs. State; Theology vs. Science; Science vs. Post Modernism; professional jealousy; and avaricious attempts by Industrial Strength Survival Groups to claim exclusive rights to 'represent' the Yanomamo now that they have become the most visible and famous tribe in the world . . . in large measure due to my research and publications on them. I happen to be in everybody's way."[87]

Chagnon's vast press kit was passed on to the journalist Peter Monaghan, who did what no editor at the *New York Times* or *Newsweek* or the London *Times* had done: he began contacting the anthropologists and missionaries whom Chagnon repeatedly denounced, to see what their side of the story was. The result was a piece in the *Chronicle of Higher Education* that revealed, for the first time in the American press, that Chagnon had little, if any, support from Yanomami specialists. Headed, "Bitter Warfare in Anthropology: Charges of academic distortion and grandstanding mark a dispute over the Amazon's Yanomamo Indians," Monaghan's article, while fully presenting Chagnon's side of the story, explained that Chagnon's conflicts in Venezuela originated with anthropologists and human rights groups, not the Catholic Church. He quoted the University of Chicago's Terence Turner: "He contends openly what many other Amazon researchers say privately: that Mr. Chagnon is 'using aspects of the Yanomamo tragedy to dramatize himself on the basis of patently false claims.' "[88] Monaghan also showed that Chagnon was at war with the Yanomami's most visible tribal leaders.

While this might seem like fairly elementary fact checking, it immediately deflated Chagnon's offensive. Within a month, Chagnon was looking for reconciliation with the Salesians, offering to drop his attacks if the Salesians agreed to sponsor his return to Yanomamiland.[89] But the Salesians balked at renewing any association with Chagnon unless he retracted his incendiary accusations, which Chagnon was unwilling to do. So the war went on.

It was at that point that I became involved. Peter Monaghan approached me because I was the only person in the United States who had actually been to the mining camp of Raimundo Nenem where the assassins had organized their assault against the Haximu-teri. I was present when the miners' clandestine airstrip at Raimundo Nenem was dynamited by the Brazilian army in November 1990. My principal informant among the Brazilian *garimpeiros,* Michelle Rodríguez Costa, lived nearby for three years, and I had obtained a long, harrowing oral history of the gold rush in this part of the Parima Mountains. No anthropologist had ever worked in there, and Chagnon's unfamiliarity with the terrain was obvious.

In the *New York Times,* Chagnon wrote that the miners hired "renegade Yanomamo trackers from a Brazilian mining site who guided them" to Haximu.[90] In the *Times Literary Supplement,* he went further and identified the Yanomami as being from "Paapiu, a well known mining camp in Brazil."[91]

Paapiu was on the Couto de Magalhães River, a tributary of the Mucajaí, where the Yanomami gold rush was kicked off. The Yanomami village of

Marashi-teri was nearby, and a quarter of its inhabitants died in the gold rush.[92] Their village was, in effect, occupied. It became a main hub of the rush, with close to five thousand miners, until December 1989, when it was torched and dynamited by Brazil's army. Unlike most of the gold camps, however, Paapiu was occupied by Brazil's Indian Agency (FUNAI) and by a French medical group, Doctors of the World. By 1993, there were no mining camps on the Mucajaí River and no way for the Marashi-teri, who lived there, to reach Raimundo Nenem airstrip, a seven-day walk. Even at the height of the gold rush, when air travel between the two mining hubs was possible, though unusual, I had never seen or heard of any Marashi-teri going to Raimundo Nenem. Different mining mafias controlled the two hubs. Nor would it have made any sense for the miners to hire the Marashi-teri as "trackers" *from* Raimundo Nenem, to trek for several days across a mountainous jungle that the Marashi-teri had never even visited, to attack another village from a different linguistic group whose inhabitants they had never met. The veteran miners still working around Raimundo Nenem were among the toughest jungle survivors I have ever seen, very familiar with the terrain, and they did not need or use Indian trackers. On a killing mission, they would not have wanted witnesses, especially Yanomami who spoke Portuguese, from a distant, acculturated village with resident government officials, Western doctors, and daily radio contact with the outside world.

Chagnon claimed this "new clue" came from an anonymous Indian whom he met walking around near "the massacre site."[93] (This mysterious group numbered seven in his first account,[94] then increased to eleven for the *Times Literary Supplement.*)[95] One difficulty with Chagnon's story was that he and Brewer never reached the real massacre site. A picture later published by a member of Chagnon's commission proved to be of the village several hours away from the actual place of the murders, from where Chagnon and Brewer were expelled by Judge Aguilera.[96]

And, however improbable Chagnon's account, it was by no means "new." A claim that the Yanomami from Paapiu participated in the killings had circulated for weeks in the mining town of Boa Vista, Brazil. It was the gold miners' version of the massacre. This was partly the fault of a FUNAI official, a grizzled frontiersman with good intentions, who had no experience at press conferences and was unwisely made a spokesperson during the massacre frenzy. When asked how the miners got to Haximu, he answered, "How should I know? Maybe some Yanomami from Paapiu helped them."[97]

The reason that Paapiu even entered the harebrained discourse of the massacre was that a band of four young Yanomami men had learned gold mining and made Paapiu their base from 1987 to 1992. Those of us who knew them nicknamed these boys "the Four Horsemen of the Apocalypse." Three of them were involved in robbing me at gunpoint in May 1990. In March 1992, however, a gold miner murdered their ringleader, Iaduce. Two of his remaining cohorts were then hired by the Brazilian government as translators, and the final member of the band was relocated to his original village, far down the Mucajaí River, and placed under the watchful eye of a very competent social worker. So when the disoriented FUNAI official invoked the Yanomami of Paapiu, he was visiting an Indian Agency nightmare on the world, but one that no longer had any basis in reality.

A government investigation found that no Yanomami from Paapiu accompanied the miners. By 1993, the Marashi-teri at Paapiu were beginning to recover from the devastation of the gold rush, which had brought thirty-seven deaths from disease and five murders. The shamans and French doctors were working together to restore morale, and the community had returned to its traditional activities. For anyone who knew the sad history of the Marashi-teri and their struggle with the gold rush,[98] it seemed cruel, and careless, to blame them for an impossibly remote tragedy.

And it had terrible consequences inside Brazil. Soon afterward, one of Brazil's most fanatical opponents of indigenous land rights adopted the rhetoric of Chagnon's massacre accounts in two long articles for South America's largest newspaper, *A Fôlha de São Paulo.* One article, a call for mining in the Yanomami Reserve and other native lands, took its title directly from Chagnon's favorite attack phrase—*Uma teocracia na Amazonia.*[99] A second feature, called "Yanomami Fraud," quoted from Chagnon at length to prove that the Yanomami themselves were responsible for the Haximu massacre.[100]

Chagnon never distanced himself from these articles. It was difficult to understand how he could stand by and let the gold rush lobby pummel the Yanomami with his writings. Or how he could so easily darken the reputation of a struggling group like the Marashi-teri—without consulting any of the scientists who had aided the village, including the anthropologist Bruce Albert. In response, Albert wrote, "This is a true scandal! . . . I believe this is the worst piece of Chagnon's manipulation: To give the impression that the Yanomami are partly responsible for the massacre."[101]

I might have agreed with Albert's assessment—except that I found Chagnon's accusations against the Protestant missionaries on the Mucajaí River just as objectionable as his inventions regarding the Mucajaí Yanomami.

He included the following incident in the fourth edition of his best-selling textbook:

> In 1967 while working in Brazil near the Mucajai mission, I discovered that men from the Protestant mission there had recently attacked and killed a group of men from a remote village with shotguns. When I brought this to the attention of the missionaries, they conceded that they, too, knew about this incident. They added that of the seven or so shotguns used, at least one or two had been provided by a Brazilian trader who had visited there. They weren't sure which guns were used in the killings and were not going to ask, adding: "If we ask and find out that our guns were used, we would have to confiscate them from the Yanomamo. They would then move away from the mission. You don't know how hard we worked to establish our mission here. My husband carried our kerosene refrigerator on his back from the top of that mountain over there where the cargo plane dropped it off."[102]

I was puzzled, for several reasons. The Mucajaí Yanomami were the subject of the most comprehensive demographic study ever done of the Yanomami, or any other Amazonian group.[103] It was based on continuous outside observation from their day of first contact, in 1958. Between 1958 to 1967, the Mucajaí people had launched exactly one raid—in January 1959—and it was driven by the fear of sorcery that accompanied their first encounter with outside respiratory disease.[104] There was no war in 1967.

And the quotation was from a group of missionaries, a sort of chorus. Yet "they" had a husband. And this was no ordinary husband. He was capable of carrying a refrigerator over a mountain. That was also a geographic oddity of no small order, because there is no mountain next to the Mucajaí mission. Nothing whatsoever separates the mission house from the airstrip. No mountain. No way. If you wanted to bring a refrigerator to the Mucajaí mission, you could either take it in by boat or fly it right next door to the mission house, where there has been a short airstrip since 1959. You could not, however, land a "cargo plane" at this airstrip unless you wanted to commit suicide. Even expert gold rush pilots had to be cautious when landing with their little Cessnas.

It sounded to me as if Chagnon was describing an evangelical mission at Surucucu, two hundred miles distant, long ago abandoned, which I also saw. In Surucucu there is, in fact, a mountain that separates the old mission site from the military base of Surucucu, where big cargo planes have landed for decades.

There were other contradictions. The only married woman at the Mucajaí mission in 1967 was Aurora Anderson. Yet I could not picture Aurora, an outstanding nurse, making the stupid, insensitive remark Chagnon anonymously attributed to the married woman at the Evangelical Mission of Amazonia (MEVA). I spoke to her in 1990 on the Mucajaí River. It was in the middle of a malaria epidemic brought by miners, who had also seized what had once been her house. Her husband was lying prostrate in his hammock with fever. "A lot of terrible things have been said about us in the last few years," she said. "People have accused us of trying to convert the Indians by bribing them . . . even though we've never done that. We've had only a handful of baptisms here. I don't remember exactly how many—maybe a dozen or so in all the years we've been here. The last one was about ten years ago. We know that even those who were baptized don't fully understand what Christianity is about. But we love the Indians, and they know we love them."[105]

Having seen Aurora Anderson's undeniable concern for the Yanomami, it was difficult to believe that she would have looked the other way if she had known about shotgun killings. During the gold rush, I had watched her making a valiant effort, in the midst of total chaos, to try and retrieve a revolver from one of the *shabonos,* after a young Yanomami man had stolen it from a *garimpeiro.* Her main concern was to prevent anyone from being killed—including me. Her alleged statement was totally out of character.

"I'm very suspicious of that quote," said the sociologist John Peters, who lived for eight years on the Mucajaí River. Peters had left the Mucajaí just a few months before Chagnon arrived for a brief stay in 1967. "I cannot imagine any missionary making a statement like that. Nor can I imagine them feeling threatened at MEVA on the Mucajaí River. The Yanomami on the Mucajaí never threatened to leave the mission if they didn't get shotguns. That incident did not occur on the Mucajaí River."[106]

What mission and what missionaries was Chagnon talking about?

I got my answer in Chagnon's unpublished article "The Guns of Mucajaí," which he sent to the *Chronicle of Higher Education.* Here the anonymous speaker had a husband named Bob. There was only one Bob working for MEVA in 1967, and he was married to another missionary. So when I contacted Bob and Gay Cable, I knew I had found the right people. They were familiar with Napoleon Chagnon, whom they called Nap. They had met him in 1967, all right—not on the Mucajaí River, but in the Yanomami heartland of Surucucu, which I suspected was the real setting for Chagnon's confused scene. When I read Bob the shotgun quotation, I thought he would be upset. Instead, he laughed and laughed.

"That's all malarkey. That's all false. We never lived on the Mucajaí River. We never sold guns to the Indians. I've never carried a refrigerator on my back. Hey, Gay, come over and listen to what Nap Chagnon has written about us."[107]

Gay listened, but she did not laugh.

"Nap was very friendly with us. He was with James Neel, the geneticist. They stayed with us for ten days or two weeks. He didn't seem to dislike us then. I don't know why he's attacking us now. He isn't quoting me, because the Indians in Surucucu were Indians of war and we gave them no guns or ammunition. I don't know who he could be quoting. Probably nobody. It makes no sense on the Mucajaí River to carry anything up a mountain, because the mission post there is right next to the river. There's no mountain. In Surucucu there's no river, but there are mountains and we did carry some things over the mountains. I remember the Indians helped us carry a washing machine over the mountains on poles. I don't know of any husbands who carry refrigerators on their backs. Whoever he's quoting, he's misquoting. It's like a bad newspaper article where different pieces of information from different people and different places are all pulled together and jumbled into the same quote."[108]

The British anthropologist Kenneth Taylor supplied the final piece to this composite puzzle. He lived with the Yanomami for three years, received his Ph.D. from the University of Wisconsin, then taught at the University of Brasília and accompanied an official commission to Surucucu in 1975. "That's when miners were beginning to come into Surucucu for the first time, searching for cassiterite [tin ore]. MEVA was quite innocent, really. They had a little trading post, where Yanomami could get things they needed in exchange for certain amounts of work. It sounds arbitrary, but it was actually perfectly practical. But the miners upset their economic system. The Indians got quite aggressive and began demanding the same things that the miners had—including shotguns. So MEVA acted responsibly and left. They decided to just shut down the mission at Surucucu rather than give the Yanomami shotguns."[109]

Chagnon's saga of the guns on the Mucajaí conflated two different Yanomami subgroups that lived in completely distinct topographical regions, along with two families of missionaries and two completely different sets of circumstances. The aggressive demand for shotguns occurred in Surucucu, around 1975, and during an invasion of tin-ore miners—not on the Mucajaí River in 1967. Bob and Gay Cable behaved responsibly by not supplying their villagers with guns, instead of callously looking the other way. But Chagnon's colorful story has the indestructible quality of myth—that of an Amazonian Sisyphus who is forever hauling a refrigerator on his back over a nonexistent mountain by the Mucajaí mission.

I have done an archaeological dig into this story because it is similar to many of Chagnon's startling, anonymous tales, including his improbable version of the Haximu massacre—and because he has been promoting this Mucajaí myth to substantiate his version of the Haximu killings. In the case of "The Guns of Mucajaí," there is no wiggle room. It is false, and it can be proven that it is false. Said Milton Camargo, the head of the MEVA mission, "It is obviously absurd."[110]

Chapter 13

Warriors of the Amazon

Outside diseases are decimating the Yanomami; ironically, now only outside help can save them.—*Nova, Warriors of the Amazon*[1]

The air freshened and the mosquitoes abated as we pushed upstream into the bony foothills of the Parima Mountains. It was pleasant traveling up the Manaviche River. The boat slowed down, and the doctor who accompanied me was sleeping with his feet dangling overboard. As the river narrowed, the gallery forest began to overlap, filtering dappled sunlight through a cathedral canopy. Gray porpoises surfaced in our wake, while capuchin monkeys picked fruit and then scrambled to escape the riot of our outboard motor. The tonsured capuchins, named for a Franciscan monastic order, suspected we were on a hunting expedition, and they were right.

I was hunting for a mysterious Yanonami *shabono* featured in a *Nova*/BBC special, *Warriors of the Amazon.* The documentary had received good reviews and strong audience response and has been shown three times nationally on

PBS since its U.S. premiere, in April 1996. One critic called it the finest film ever made about the Yanomami.[2]

Nova had continued the American tradition of portraying the Yanomami as a warrior society. "This is the world of the Yanomami; it is a world marked by aggression and revenge," the narration began. The film focused on "the threat of warfare that overshadows their lives," a constant menace that obliged men to "go off to fight two or three times a year."[3]

The plot was simple and familiar. Just as the film crew reached the field, an unidentified village decided to erect a giant new *shabono.* Why were they doing this, the film crew wondered? Then, revelation: the villagers suddenly invited their archenemies to a feast consecrating a new alliance, much to the film crew's surprise. All this supposedly reflected customs that had endured for thousands of years.

Shortly after the documentary's debut, I got a call from the anthropologist Brian Ferguson, author of *Yanomami Warfare.* Ferguson had never found a Yanomami village involved in any war lasting more than two years. So he was curious to know where this extremely bellicose group was actually located. And he made a prediction. "The village they filmed was obviously not at war, and I suspect it was built because the film crew paid them. The village is wide open. There are no palisades around the *shabono,* nothing to block the entrances. And they've left all the underbrush around the village uncut, so any enemy could just crawl up to within a few feet of the *shabono* and fire arrows into it. I also suspect they paid the Yanomami there to undress because it looks like a highly acculturated village. You can see Western haircuts, shotguns, and outboard motors. Even low-status individuals in this village have shotguns. As the film unfolds, the explanation is that the headman of the host village wants to see his sister who ran away some ten years or so ago, which is not a very good explanation. So the feast is happening because the film crew is there."[4]

When I reached the Upper Orinoco, I contacted a family of American missionaries, the Dawsons. I watched *Warriors of the Amazon* at two different "showings," with eight members of the Dawson clan. They had collectively spent over two hundred years with the Yanomami and had contact with every linguistic group except the Ninam of the Mucajaí. They were intrigued as *Nova*'s narrator gravely described how a Yanomami child was being initiated into the world of violence—his brother systematically hit him with a stick while his mother taught him to hit back, tit for tat. None of the Dawsons had ever seen this kind of ritual, except on film, where it has been a standard feature in Yanomami documentaries for decades. Several missionaries burst out laughing when Yanomami boys began beating burning embers together to make sparks fly over their bodies as pain conditioning. "The Yanomami are the biggest babies in the world," said Wilma Dawson. "I mean, I've been treating them for cuts and bruises all these years, and they react far more to a small cut than an American would. You'd think they were going to die. But we're used to seeing this sort of thing on the films."[5]

One of the missionaries immediately recognized the Yanomami in the film. "It's Lizot's village," Mike Dawson said, referring to the well-known French anthropologist. "It's Karohi-teri. Two or three raids a year doesn't sound at all familiar to me at Karohi-teri."[6] Lizot himself had criticized the notion that the Yanomami make incessant war.[7] In fact, during his first eight years at Karohi-teri, from 1968 to 1976, deaths from violence were negligible.[8] After that, Karohi-teri was involved in the wars between Chagnon's village (Bisaasi-teri) and Lizot's village (Tayari-teri).[9] When that peculiar conflict ended, Lizot observed a dramatic reduction in warfare among all the villages where he worked.[10] According to the nearby Salesian missionaries, it had been many years since the Karohi-teri had been engaged in an active war with the guests at *Nova*'s feast.

In the film, new trade goods could be seen all over the freshly constructed *shabono* (pots without smoke in them, shiny machetes and axes). The Yanomami also had wads of cash. On camera, Hisiwe, the headman, demanded cash and trade goods from the film crew so that he, in turn, could pay the guests at the feast. The actual Yanomami words suggested the Karohi-teri had worked very hard for the crew: *Fei yamaco nolableaabiyei.* Hisiwe demanded payment "for what we have suffered" for the film.[11]

Their suffering was also evident. The missionaries stopped laughing when the camera followed the progressive weakening and death of a mother and her newborn infant, which occurred over the weeks the film crew was in the village. "In most cases, death from fever is a very preventable death," said Mike Dawson, who has lived since birth with the Indians, over forty-five years. "With just a little bit of help, they could have pulled through. The film crew interfered in every other aspect of their lives. Let's be real. They're giving them machetes, cooking pots, but they can't give a dying woman aspirin to bring her fever down?"[12]

"I can't believe they actually stayed there and filmed her death," agreed Paul Griffiths, who has lived for fifteen years with the Yanomami, mostly in Brazil. "That's really weird."[13]

The woman's death turned out to be the highlight of the film—her cremation ritual was the most vivid ever captured. The director, Andy Jillings, expertly framed close-ups of the still-beautiful young mother being adorned and then consumed on a blazing pyre while the villagers wept and screamed to placate her unhappy ghost.

The Yanomami call filming "stealing the spirit." Even in the best of times, taking pictures among them can be problematic. But the spirit of a recently dead person was considered a dangerous force—*bore*—which haunted the living and threatened them with sickness unless the proper funeral ceremonies were respected. At this delicate juncture between worlds, the Yanomami were extraordinarily conservative and fiercely traditional. I had never known the Yanomami to allow snapshots during a cremation, much less filming. "What they paid for filming her funeral must have been enormous," said Mike Dawson. "I mean, they've never let us film a funeral, and we've been with them all these years. Believe me, a lot of lives could have been saved for what they paid to get that film."[14]

I showed *Warriors of the Amazon* to a small group of Yanomami at the Padamo mission, about thirty-five miles from Karohi-teri. "The film crew is not showing all the anger against them," observed Pablo Mejía. "I am very angry when I see this film."[15]

At Karohi-teri, there was also anger. Actually, "Karohi-teri" no longer existed, at least not the same community that hosted the *Nova*/BBC feast. The large *shabono* where the documentary was filmed was now completely empty. Its owners had split up as a result of the conflict over the filming of the young woman's cremation. One group had not allowed the filming; the other group, which accepted a huge payment in cash and trade goods, had. Three years had passed, and they were still not on good terms. No one blamed the filmmakers, who did not speak Yanomami and could not communicate directly with the Karohi-teri. "Lizot told us to build the *shabono* and have a feast for the *nabe* [foreigners]," said Renaldo, the leader of the faction that had opposed filming the woman's funeral. "Lizot wanted her *noreshi* [spirit] taken."[16]

Something was obviously wrong at Karohi-teri. The splintered groups were so small and dispirited that they could hardly be called a village. There was a young girl with severe burns on her face whom I took to the nearest hospital, in Puerto Ayacucho, where I found that she had a fractured skull from a traumatic blow to the head. There was a sullenness in the air that no amount of medical attention could heal. We all found the remains of Karohi-teri depressing.

I do not know how much of this was left over from the film and how much was due to the influence of Jacques Lizot, who had lived here for many years. When the anthropologist Jesús Cardozo visited Karohi-teri in 1985, he was reminded of *Apocalypse Now:* young boys doused in deodorant and dressed like queens were smoking imported cigarettes and playing Wagner from powerful stereo equipment. According to Renaldo, Lizot choreographed Karohi-teri's participation in *Warriors of the Amazon* down to the last details. Among other things, it involved taking off the Western clothes I saw them in, as Ferguson had guessed.

"Lizot has been doing that for years," said the anthropologist Kenneth Good. "When a Swedish theater company came in 1975, they wanted the Yanomami to see their performance. They thought their dancing was so powerful it evoked a universal response, and they wanted to prove this with the Yanomami. Lizot couldn't accompany the Swedes for some reason, so he wrote down instructions on an envelope and gave it [to] me to read to the Yanomami. When the Swedes did their performance, the Yanomami just sat there and giggled until I read Lizot's words: 'Stand up, dance, and play with the actors.' So the Yanomami got up and sort of danced. The Swedes were really happy. I thought, 'This is all bullshit.' "[17]

I was given another perspective on *Warriors of the Amazon* by Father José Bórtoli, the head of the Salesian mission of Mavaca, which is just an hour by motorboat from Karohi-teri. "The film was staged," he said. "But all the films I have seen in twenty years with the Yanomami have been staged. There is not one instance where the film was not staged, in my experience. In this case, the film crew really wanted to film a more distant group and couldn't get there, because the river was low. So they stayed at Karohi-teri and brought the more distant village to them. I believe Jillings was trying to do a good job. He worked very hard at it. Of course, no Yanomami village goes to war two or three times a year, and for a little group like Karohi-teri it would be obviously absurd. I don't think Jillings wanted to name the film *Warriors of the Amazon.* I think those kinds of exaggerations about Yanomami warfare were added by American TV to make it sound more like Chagnon's *Fierce People.*"[18]

The BBC's version of the documentary was called *Survivors of the Amazon,* and it showed more of the film crew's impact on Karohi-teri. As the feast was being organized, an ambassador from the guests said, "Now that I see you have all these trade goods from the film, I'll come to the feast."[19]

Although I do not know Andy Jillings, I believe he was trying to do an honest job of filmmaking and also treat the Yanomami humanely. Despite the fact that he would not discuss how the four Yanomami died during filming, I learned from the missionaries that Jillings made a special trip to bring milk for the infant that died. Everyone liked him, including the Yanomami. He wanted to do two things. He sincerely desired to act as a peacemaker between previously hostile groups and simultaneously to break new ground in Yanomami filmmaking. The problem was that the two hostile groups—Karohi-teri and Arimawu-teri—had simply avoided each other for a long time. By bringing them together, Lizot and Jillings acted like gods of peace. But theirs was not an original creation, nor was it one in which they could control the ultimate outcome. Artistically, they repeated the oldest cliché in the Yanomami film repertoire—a misleading film about a fake feast that co-generated a dangerous new military alliance. And it was all performed in the middle of a deadly outbreak of falciparum malaria.

That a good man like Jillings failed in his idealistic mission was due to his being caught in the filmmakers' trap: the need for a sexy, violent subject filmed within time constraints that made staging almost inevitable. Jillings was straightforward about his commercial dilemma. "I was very disturbed by the fact that Chagnon's work was being used in Brazil and Venezuela to justify genocide," he told me. "I was looking for a group that was fairly unacculturated and that was at war and suing for peace. So Jacques [Lizot] and I went out to another group that was at war, but they often weren't home much of the time. I wanted an unacculturated group because you can't make a film about the Yanomami if they're wearing Black Sabbath T-Shirts. We spoke to the more remote group but, basically, we were hijacked because the Karohi people said, 'Why don't you have the feast here?' They saw all our trade goods and they didn't want them going to the other group. The feast of reconciliation was in one sense a set-up. We might have facilitated it. But they wanted it."[20]

Jillings originally called the film *The Art of Speaking Well.* It was not considered a catchy title, and *Nova* opted for *Warriors of the Amazon.* "This wasn't what I intended," Jillings admitted. "But I found myself surrounded by death and having a responsibility to film it."[21]

Four Yanomami died during five weeks of filming at Karohi-teri. The headman died shortly after they left, making five deaths out of ninety people. Jillings would not answer questions about how the four Yanomami died when I spoke to him. Nor would he let me tape-record our conversation. And he did not respond to a letter with detailed questions about the Yanomami deaths. Nevertheless, he defended his film crew during our one talk. Everybody who went in there had medical checkups, he said. Later, he wrote, "We all went into the field 'clean' and came out sick."[22]

Jillings deserves credit for requiring medical checkups, something no other film crew I know of has done. But that still does not answer the question of what responsibilities went with the power to build a village and broker an alliance for a film event. Nor does it explain why a woman at such obvious risk was not removed by motorboat during the week of her slow death. They were an hour away from the Mavaca mission, where there was an infirmary with nursing nuns. A young government doctor came to see her at one point, but he came without proper diagnostic equipment or medicine, and he left without accomplishing anything. That was part of the daily reality on the Upper Orinoco, a reality too tragic, senseless, and prosaic for an exotic film to expose and, perhaps, help to remedy. At the heart of the incompetence, there was, I believe, a deep, unconscious conflict of interest between the film's need for a climax and the woman's need for emergency evacuation. When one of the film crew's cameras broke, a substitute was flown in from London in two days.

"The most distressing part of the film for me was watching the woman and her child die of fever," said the anthropologist Brian Ferguson. "I don't know if it was malaria or something the film crew brought in with them. Either way, they didn't seem to be doing anything about it. They seemed to treat it as an act of God."[23]

The Yanomami who saw the video at Kosh were equally nonplussed. "Are they crazy?" asked a man named Timoteo.[24]

The film crew did not introduce malaria to Karohi-teri. Malaria has been endemic at these lowland villages for years. Marinho De Souza, a Brazilian microscopist who accompanied me on my second visit to the village, in September 1996, did a malaria census and found that 10 percent of the people were infected. He was able to medicate all carriers within a few hours—at which point they stopped transmitting malaria. (At a nearby village, a woman with malaria gave birth to a child—also with malaria—and Marinho had them quickly evacuated to the Mavaca mission, where they survived.) Controlling malaria in a village does not take rocket science. It is actually rather easy. "The TV crew built the village," Marinho observed. "They could have stopped the epidemic."[25] And it would have cost a tiny fraction of what the crew spent on building a new village or sponsoring an alliance—or bringing another camera from London. By failing to deal with the disease, Jillings's crew members contracted malaria and became carriers themselves. Their trade goods attracted Arimawu-teri (the guests of the feast) down from the hills, who also entered the growing circle of contagion, which they spread still further when they returned to the highlands. By organizing a feast and moving populations during an epidemic, the film crew changed the nature of the outbreak. And by distributing thousands of dollars in trade goods, it facilitated a wide dissemination of the disease.

The content of *Warriors of the Amazon* followed the commercial requirements of television sequels—in this case, the never-ending need to obtain supplies of savage, remote Indians. It was the same drive that enticed Alfred Kroeber of the University of California to photograph Ishi to extinction while ignoring all the discomfiting details of his tribe's destruction. In *Warriors of the Amazon,* Jillings's exquisite skill with the camera made death memorable, beautiful, marketable, and . . . inevitable. The images of the Amazon Madonna and her child as they died and were offered to the flames became a self-fulfilling metaphor. Divorced from any historical moorings, the deaths at Karohi-teri confirmed a predisposition to see the Indians as losers in the Darwinian struggle. The camera, which has played such a crucial role in the high-stakes competition for Yanomamiland, seemed to seal their fate. The role of the film crew was kept invisible even as it made the ultimate decisions: to evacuate the sick or let them die.

Jacques Lizot was also invisible, but that was part of the film crew's deal with him. The crew agreed not to show him, his house, or his impact on the Karohi-teri in any way. *Warriors of the Amazon* distilled the illusion of timelessness from a group that had witnessed a generation of horrible sexual violation.

The crew knew about Lizot. That's why Jillings first turned to Kenneth Good to be the film's anthropologist. Good refused, feeling that he had been burned so badly in a *National Geographic Explorer* special the year before that he did not want to be involved in another production. "You don't have good choices," a member of the *Nova*/BBC crew said, on condition of anonymity. "I came to admire Lizot in a way. He was a man with a particular problem, who could never have been happy in a normal society, and he found a place where he could work that out and still do some good on the whole. I mean, three generations of Yanomami boys have been in and out of his hammock without any lasting harm."

Jillings was understandably upset about my investigation of his impact on the Yanomami. He accused me of having an agenda, which was true. I was testing Ferguson's hypothesis about *Nova*'s documentary. Still, I could not have guessed how badly things had really gone at Karohi-teri.

Warriors of the Amazon derived its drama from two tragedies. The first was the death of the young woman and her baby. The second was the new political alliance that the *Nova*/BBC feast created. Significantly, the narration itself warned that any Yanomami feast between adversaries was a potentially violent affair. In this case, both villages became embroiled in wars immediately afterward. Arimawu-teri attacked another *shabono* in the highlands. Karohi-teri fared worse.

"After the film, Karohi-teri split up," Renaldo said. "We fought among ourselves. Half of us fled down the Orinoco. The Bocarohi-teri stole some of our women; they took advantage of them and raped our women." These Karohi-teri fugitives joined another village, not far from the Ocamo mission. Then this new faction became embroiled in another war, in which two Karohi-teri men were killed. "The gun Lizot gave was used to kill," explained Renaldo, referring to the French anthropologist's generosity in giving out shotguns, one of which, through trade, had wound up in the hands of Karohi-teri's new enemies.[26] When I went to the spot where the shotgun deaths occurred, I found yet another abandoned *shabono,* overgrown with weeds and bad memories, on the banks of the Orinoco River.

It was all very sad and very complicated. "Yes, the reasons for such wars will always be very complicated," said Brian Ferguson. "But the likelihood of them occurring is much greater after a filming event like *Warriors of the Amazon.* It's happened on a number of occasions, and it happens again there. This is parallel to the 1968 film *The Feast.*"[27]

Part III

Ravages of El Dorado, 1996–1999

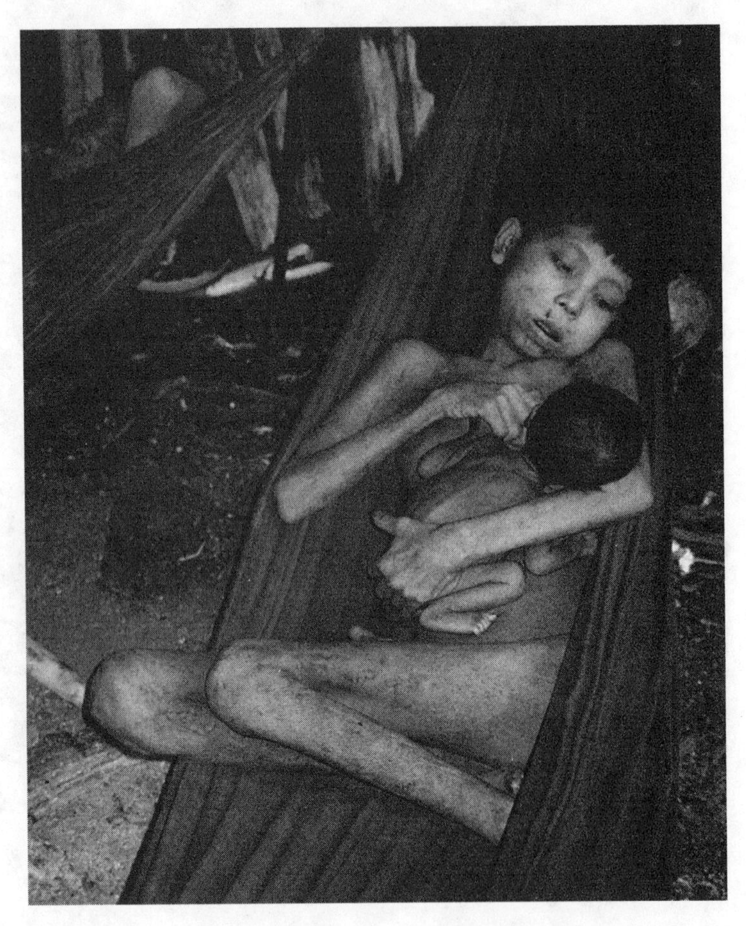

Mother and child at Ashidowa-teri, 1996 (photo by Patrick Tierney)

Chapter 14

Into the Vortex

The promise of many presents caused some Indians to collaborate with Mr. Chagnon. This caused a division between communities: those who were on Mr. Chagnon's side and those who were against his visit. It reached the point where they fought a war.—*Juan Finkers*[1]

Just before Christmas 1993, as Napoleon Chagnon was publishing "Killed by Kindness?" in the *Times Literary Supplement* (London), two of his guides were killed, not by kindness, but by enemies on the Upper Mavaca River, in one of the most unusual manhunts in the history of Yanomami warfare. These killings, strange as they were by the standards of Yanomami culture, were simply the sequel to a regional upheaval that pitted villages allied with SUYAO, the Yanomami trade cooperative, against those that joined Chagnon and Brewer's FUNDAFACI camp. The outbreak of the wars occurred at around the same time as Chagnon's entry into Yanomami territory, in the early summer of 1990; most of the killings (twenty-one) took place while FUNDAFACI's helicopter visits were most intense, between July 1990 and the end of August 1991.[2] The death total had reached at least thirty-one by September 1992—making it by far the deadliest war

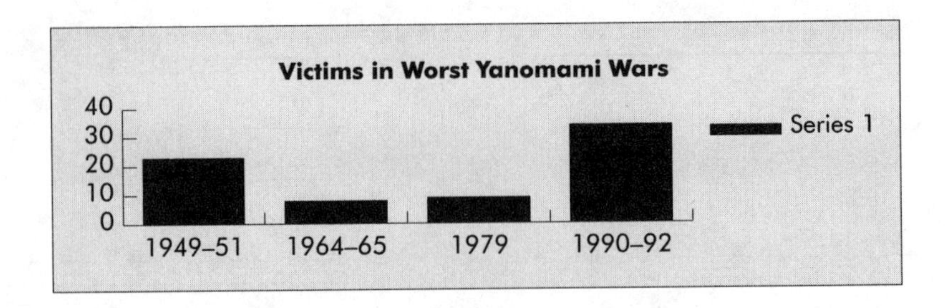

ever recorded among the Yanomami.[3] This violence surpassed that of all previous conflicts—when American missionaries first made contact (twenty-two deaths during a twenty-six month period, 1949–51), Chagnon's initial fieldwork (nine deaths during fifteen months, 1964–66), and the general uprising against Jacques Lizot's village (nine deaths, 1979).* When this war finally ended, in 1981, there was very little fighting in the region until 1990, as is attested by Chagnon's data, Lizot's writings, and mission records.[4]

Many anthropologists, familiar with the cultural and epidemiological risks of contact, wondered why Chagnon and FUNDAFACI had elected to descend upon the extremely isolated Siapa valley. What scientific gain could justify the inherent gamble of such large expeditions? Chagnon told the *New York Times* that he hoped "to study the impact of an agricultural revolution produced by the arrival of steel tools like machetes and axes."[5]

Arguably, that study had already been done. "In February 1990, I sent Chagnon a paper called 'Savage Encounter,' six months before he went into the Siapa valley," said the anthropologist Brian Ferguson of Rutgers University. "In that paper, I made a clear prediction that major Western intrusion would lead to destabilization and probably warfare in areas where it occurred. He was aware of that, and my feeling is that that is why he went into the Siapa valley and said that that was what he was going to study."[6]

If Chagnon wanted to disprove Ferguson's thesis, which linked the arrival of Western tools, guns, and germs to dramatic increases in tribal warfare, events derailed Chagnon's plans. Although he had once dismissed the notion that outside intrusion created tribal warfare as "the bad breath theory,"[7] his latest writings on raiding in the Siapa and Mavaca drainages sounded as though he had converted to Ferguson's persuasion. Instead of attributing Yanomami warfare to inherent ferocity, Chagnon now blamed fighting on

*For an account of the wars in 1949–51 and 1964–66, see chapter 3; for the wars involving Lizot's village of Tayari-teri, see chapter 8.

meddling missionaries and Yanomami leaders with guns and political ambitions. In fact, Chagnon wrote that SUYAO's warfare against Chagnon's allies was based on a desire to preserve the economic cooperative's monopoly of machetes, axes, fishhooks, and other items that Chagnon was dispensing without the control of former Yanomami middlemen at the missions.[8]

Chagnon brought these wars to the world's attention through his articles in the *Times Literary Supplement* and the *New York Times.* He demanded a full investigation—possibly by the United Nations—of the Salesian Catholic missionaries' role in the murders.[9]

The Salesians, for their part, invited Chagnon to accuse them formally before a Venezuelan tribunal and let a judge determine whether they were guilty of arming mission Yanomami or whether Chagnon was guilty of libel (a serious offense under Venezuelan law).[10] Nothing came of this challenge; neither Chagnon nor the Salesians took their grievances before the Venezuelan courts. Instead, the Salesians sent José Bórtoli to the American Anthropological Association's annual meeting, at Atlanta, in December 1994, to request a commission of inquiry. After getting some discouraging responses from his initial contacts, however, Bórtoli never formally requested an investigation by the AAA and none occurred.[11]

I knew nothing about the Salesians, and had never done research along the Upper Orinoco, until I reached Puerto Ayacucho, Venezuela, on June 2, 1996. I was dismayed to see the Orinoco's waters rising dangerously. The Maipure Rapids, just beyond the city, were roaring with record volume after weeks of monsoon-like storms. From the air, the whirlpools and eddy currents looked big enough to swallow houses.

I had arranged for a plane to fly me into Yanomamiland with the Missionary Aviation Fellowship (MAF), a Protestant group. On June 4, a sweltering day at the height of the rainy season, the MAF Cessna Skylane took off from Puerto Ayacucho. As the plane picked up altitude over the raging Orinoco, tabletop mountains and stiletto peaks began jutting out from the endless expanse of rain forest: from here to the Bolivian border, it was said an enterprising monkey could swing for two thousand miles, excepting rivers. The plane headed for a 9,000-foot-high granite mesa, Mount Marahuaca, with dozens of waterfalls parachuting off its vertical cliffs. The single-engine Skylane climbed steeply in the face of cascading waters to make the hop over Marahuaca, then began a circling descent to a Protestant mission post at the confluence of the Orinoco and Padamo rivers, the border of the Yanomami Reserve. The pilot was worried about the water accumulating on the dirt runway, and we lurched on landing, spraying water and mud in all directions. The river was already lapping the foundations of large American-style ranch houses, a sort of suburbia in the rain forest.

I headed there with Jodie Dawson, a quiet twenty-one-year-old missionary born on the Padamo River, who is married to a Yanomami woman. Jodie has translated for *National Geographic* and is as fluent in Yanomami as any outsider is ever likely to be. Still, the microdifferentiation between dialects is so daunting that he insisted on taking along Pablo Mejía, who was born at Momaribowa-teri, a Shamatari-Yanomami village from the Upper Mavaca. "I have to have Pablo with me to ask about the dead," Jodie explained.[12]

Since naming the dead is a fighting offense (a dangerous summons to the deceased spirit, in Yanomami perspective), it involved oblique, poetic use of idioms and metaphors that incorporated features of the local geography and history that were baffling to any outsider. And Pablo Mejilla had another valuable trait to recommend him. He was the cousin-brother of César Dimanawa, Chagnon's former guide and then nemesis.

In three hours, we reached the Mavaca mission, where we found children splashing inside the *shabonos,* swimming among the half-submerged banana trees. Even the mission school was a foot in water. It made river travel easier. We went up the Mavaca at full throttle and reached Mavakita in a couple of hours. It was an inviting place, a village built on a steep slope, a circle of thatched huts, connected, like row houses: a *shabono* on the hill.

There was an old shack on the northern fringe of the village, at the water's edge, where Juan Finkers, a seventy-year-old Salesian brother, lived. There were no crosses, no statues of saints. There was a school, but the bilingual books did not mention Christ or Christianity. An archaeologist going through the rains of Mavakita would find paltry evidence of Catholicism as such. Finkers, the only Catholic in the village, is a phlegmatic, white-haired man whose Dutch-accented Spanish is slow and ponderous. He has collected plants and written a book about his discoveries, while also raising bees and helping out with the school and health care.[13]

Yet, with his benign, monastic efficiency, Finkers has contributed to a revolution at Mavakita—and no revolution can be completely benign. There was a large *depósito* where handicrafts from half a dozen villages were collected each month, to be taken downriver and eventually sold in the Caracas tourist trade under the auspices of SUYAO, the United Shabonos of the Upper Orinoco. In June 1996, the month I was there, SUYAO sold over $1,000 in handicrafts from its *depósito* at Mavakita alone.

César Dimanawa was presiding over a meeting of three villages. Over 150 villagers were present, many adorned for a feast. Dimanawa was wearing toucan arm bracelets and a feather headdress. His body was painted a startling red with *onoto* vegetable dye (which the Yanomami prepare from the seeds and fruit of a cultivated plant, *Bixa orellana*). The bright pigment on his cheekbones highlighted his dark, intense eyes in a way that was a bit scary. I had heard and read a lot about Dimanawa—that he was a murderer,[14] that he was crazy[15]—but no one had told me that he was so imposing, so impressively authoritative, and, I can't think of any other word, so beautiful. He looked like one of the great Indian chiefs we do not have any more. Of course, we cannot be sure if the Yanomami had such chiefs either in their world without whites. What is known for certain is that such chiefs do not exist in the Parima Highlands, where all of today's Yanomami originated.[16]

Dimanawa was used to dealing with the press. He began by laying down the photo-op ground rules: no pictures of him in ceremonial paint and feathers. He wanted to be photographed in Western clothes, at school, or playing volleyball. He said the Yanomami were tired of being portrayed as *salvajes.* He was fluent in Spanish, and I needed no translation.

At the mention of Chagnon, Dimanawa exploded. "Chagnon!" he said, pronouncing the anthropologist's full name instead of his Yanomami nickname, Shaki. "We kill people like Chagnon. Talk to that man there—he went to work for Chagnon because Chagnon promised him an outboard motor."[17]

The man was Bokoramo, one of Chagnon's Siapa guides and a brother of the two other guides who were killed. He was obviously afraid of Dimanawa and angry at Chagnon. "I knew Shaki when I was little," he said, referring to Chagnon's visits during the late 1960s and early 1970s. "He left Mishimishi because he promised things, but wouldn't give them. So children began robbing his stuff. One boy stole a can of sardines. Shaki beat him and another boy with a belt. He hurt them badly. Afterward, the boys' parents were angry with Shaki, and he left quickly."[18]

Chagnon stayed away for ten years, from 1976 to 1985, because he could not obtain permits. When the anthropologist returned, from 1985 to 1987, Bokoramo's younger brother, a man in his early twenties named Ushubiriwa, became one of Chagnon's protégés. Chagnon preferred young informants for the task of prying into the secret names of the dead. In *Last Days of Eden,* Chagnon recounted how surprised he was to find that Ushubiriwa had learned to read the names of the dead on the computer screen.[19] César Dimanawa, in his late twenties at the time, was another one of Chagnon's river guides who also helped collect the names of the dead.

Both Ushubiriwa and Dimanawa differed from traditional Yanomami youth. Instead of focusing on hunting and gardening, the tasks necessary for bridal payments to future in-laws, they learned Spanish, became literate, and began earning a Yanomami fortune in trade goods from outsiders. With goods and money, they could purchase more wives than normal; with shotguns, they could also steal them. Ushubiriwa acquired six wives; Dimanawa also claimed six, although I saw him with eight. Both of them became leaders at least a decade before they would even have been eligible for headman status. Dimanawa was Chagnon's principal guide on the anthropologist's return in 1985 to the Upper Mavaca, where he needed César's strong assistance in calming the Yanomami at Mavakita who shouted at Chagnon to go away because he brought *xawara,* disease.* By 1989, however, the controversies around Chagnon had grown so intense that Dimanawa could no longer broker Chagnon's return. Dimanawa wrote a letter asking the Indian Agency to cancel Chagnon's permits, but Dimanawa's motives were complex. "We don't want to continue in the same way we were when Shaki first came," he said, explaining why he wrote. "You go on and you move upward. *Se sigue y se sube.*"There was a host of complicating issues, including the political liability of being labeled the Fierce People. "All the communities asked me to write the letter. He said that we are savages. That's why we became angry."[20] Diamanawa later touched on another issue: "The people from Washewa were afraid because Chagnon had shown them his revolver. They said, 'This guy wants to kill us.' Chagnon can't deny it; the people of Washewa came here and said they didn't want him to return. They said Shaki had a bomb, *shereka siwaka,* a fire arrow."[21]

Finkers explained that Chagnon reappeared at the end of June 1990, along the main course of the Orinoco River, where he established a supply center next to the Ocamo mission airstrip. "With the backing of FUNDAFACI and the president's mistress, Chagnon was able to come back. . . . Chagnon came with a lot of equipment, a lot of presents, all kinds of things for his camp that he wanted to ship upstream. But when he heard that César Dimanawa was also upriver at Mavakita—where he would have to pass because the river is so narrow—he didn't go. He didn't want to face Di-

*See chapter 11.

manawa. So he sent two Maquiritares [from a neighboring tribe] and one Brazilian *criollo* upriver with his things. He told them to speak with the Yanomami to prepare a camp for him farther up the Mavaca where they could land with their helicopter."[22]

Apparently, most of the Yanomami rejected the overture—probably because the payoff was too small and the risk too large. But Ushubiriwa and his brothers, who lived just above Mavakita at a *shabono* called Shamatha, decided to accept the commission. "Shaki offered my brothers and me the base of an outboard motor *(el pie de un motor)* to help him on his trips into Siapa," Bokoramo told me.[23] At Bokoramo's suggestion, we traveled toward the Mavaca headwaters, to Chagnon's helicopter site.

The river narrowed gradually, and the hills, off in the distance, were higher. Once we were beyond César's domain, Bokoramo relaxed. We reached a beach, called Mrakapiwei, "the Place of Sand," where two ruined *shabonos* were right next to each other. The larger one had belonged to the Kedebabowei-teri, one of the largest Yanomami villages. It had featured prominently in Chagnon's writings, films, and sociobiological theories. Men from this village had been the protagonists of *The Ax Fight,* and in his famous *Science* article about Yanomami killers Chagnon revealed that Kedebabowei-teri had more serial murderers than the rest of Yanomamiland put together. Eight of the ten men who averaged almost ten homicides apiece came from this village.[24]

Bokoramo sat down on a rusting gasoline drum and gave me a seminar on FUNDAFACI history at the old helicopter site. "We cleared the helicopter landing site—six days of work," he said. "Shaki promised us many things, and that's why the other communities were jealous and began to fight against us. Here there was a great *shabono* of the Kedebabowei-teri, but they left because they didn't want fights with César over Shaki. My brother [Ushubiriwa] came here with four wives and one child; I had my three wives and two children. We had twelve other men. We made this small *shabono* next to that of the Kedi. Everyone came to work for Shaki, but César came to attack us."[25]

FUNDAFACI's Maquiritare and Brazilian employees arrived with money and ammunition and shotguns. These emissaries, whether known to Chagnon or not, were and are considered among the toughest, most dangerous men on the Upper Orinoco. Thanks to Chagnon's well-armed emissaries, Ushubiriwa and his brothers were able to intimidate the much larger village of Kedebabowei-teri, which had four times as many warriors. Ushubiriwa and his brothers fought the Kedebabowei-teri with wooden staves. There were many bashed heads and even more seriously bruised egos before the Kedebabowei-teri angrily left for the hills.

"According to Chagnon's own theory, village fissioning precedes Yanomami warfare," said the anthropologist Jesús Cardozo. "At Mavakita, the village fissioned over the issue of Chagnon's visits. It was a violent fissioning, and there were several deaths as a result of it. Antonio's group broke away from the others at Mavakita and became known as Chagnon's people. In Chagnon's narrative, Antonio is called by his Yanomami name, Ushubiriwa. A number of other wars appear to follow the basic logic of Yanomami villages opposed to Chagnon attacking those villages that received him."[26]

Within a month of clearing the helicopter site and driving off the Kedebabowei-teri, Bokoramo led a raid against the inland village of Toobatotoi-teri. This was the most isolated village in the Siapa Highlands. Chagnon gave the following account of the raid:

> Bokoramo himself shot and killed the headman, Watemosiwa, with a shotgun he obtained from the SUYAO cooperative. Two other Toobatotoi-teri men were killed with shotguns on the same raid, one of them, a man named Rowahiwa, was shot dead while he was up in a tree, presumably collecting honey or retrieving an arrow, helpless. A third man named Kumadowa was also killed with a gun. The astonishing aspect of this incident was that Bokoramo's village had never had any contact in the past with the Toobatotoi-teri village and had no previous quarrels with them. The raid and killings were completely arbitrary by Yanomamo standards and seem to have been provoked only because Bokoramo's group had shotguns and the defenseless Toobatotoi-teri did not.[27]

In this rendering, Bokoramo was a "mission" hit man, armed by the SUYAO cooperative, not an employee of Chagnon's who had just acquired new ammunition and weapons with outside help.

Yet I believe his actions made sense, in two ways. First, Bokoramo told me that he hoped to build the biggest trading center on the Upper Mavaca. He sadly showed me an empty *depósito* of impressive proportions, which he had dreamed of filling with handicrafts manufactured by all the groups in the area. If trade monopoly was his goal, the site of Mrakapiwei was strategic in every sense. It was near the junction of two rivers, Keti-U and Washewa-U; above Mrakapiwei, the river narrowed sharply, making Bokoramo's outpost a bottleneck for all the Yanomami farther upriver. With FUNDAFACI's helicopters, and the plan of a permanent outpost to house fifteen to twenty scientists and their assistants year-round, Mrakapiwei promised to be the handicraft El Dorado.

But that was a threat to SUYAO's predominance and to César Dimanawa's political position. César's own *depósito* at Mavakita was serving the very function that Bokoramo wanted to supplant. So Bokoramo's bold raid on Toobatotoi-teri was also an announcement to everybody on the Upper Mavaca that the new owners of the strategic site of Mrakapiwei intended to contest SUYAO's trade turf. It meant that the interior villages would be punished if they tried to take handicrafts to Mavakita.

César Dimanawa responded to this challenge. "César came to Shamatha with many men and put an ax in the head of my younger-younger brother, and left my second-younger brother, Antonio [Ushubiriwa], also wounded, almost dead," said Bokoramo. "They took three shotguns and three women. When I got back, I found my brother Antonio almost dead. But he got better, and we came here to Mrakapiwei."[28]

Up to this point, October 1990, Dimanawa's responses were measured. He could have killed Bokoramo's two brothers, but he left them nearly unconscious instead. What he did take were three shotguns, in an attempt to restore the status quo ante, which FUNDAFACI's incursion dramatically altered.

A month later, in November 1990, after FUNDAFACI's expeditions had been going into the Siapa for three months, César Dimanawa held a large feast in Mavakita. Chagnon wrote that this feast was a *nomohori*—"a calculated act of treachery"—or ambush.[29] The intended victims, according to Chagnon, were the people of Hiomita-teri, a village contacted for the first time by Chagnon in 1987 and revisited by FUNDAFACI. Whatever Dimanawa's motive, the feast turned into a chase that revealed the new alliance patterns of the Yanomami reserve: SUYAO allies pursued FUNDAFACI villagers to the village of Washewa (another village that welcomed Chagnon), where the SUYAO men, with shotguns, killed two of the Hiomita-teri at close range. Later, while Chagnon and his expedition were in residence at Washewa, Dimanawa was rumored to be approaching for a raid. Chagnon credited his own arrival with thwarting Dimanawa's plan to attack Washewa: "He and his followers turned back, knowing I was there and probably armed. . . ."[30] By his own account, Chagnon, with his gun and trade goods, had become a factor in these wars.

The anthropologist Frank Salamone, an expert on missions who was asked by the Salesians to investigate Chagnon's charges against them after the AAA declined, questioned César Dimanawa about these battles: "When I spoke to César Dimanawa, he said, 'I have killed but I am not a killer.' I believe he has carried out raids. Chagnon believes that the reason Dimanawa attacked certain villages is because Chagnon was there. Dimanawa denies this and says it was for revenge—because those villagers had killed one of his people. Personally, I think it is possible Dimanawa was attacking because of Chagnon's presence."[31]

César laughed when I asked him about it. "It's a huge lie." He laughed harder when he heard that he was afraid to attack another village, Washewa, because he knew Chagnon was waiting for him armed with a shotgun. "I sent word to Chagnon for him to stop coming by helicopter and to try getting upriver by boat—then we would see who would win. But he didn't dare. He knew I would damage his boat. I was furious at him."[32]

"César Dimanawa has killed somewhere between four and six people," said Raymond Hames, an anthropologist at the University of Nebraska, a former student and longtime colleague of Chagnon. "He is a wily political leader whose real motives would be very difficult to determine from what he says. His actions should be interpreted in light of the fact that he aspires to be a Pan-Yanomami leader."[33]

I agree with Hames about Dimanawa's political intentions. By the end of 1990, SUYAO leaders understood that FUNDAFACI planned to create a new political entity that would take control of the Siapa region and leave the rest of Yanomamiland without protection. What had begun as a battle over turf and trade on the Upper Mavaca between two of Chagnon's protégés now changed into a regional conflict, one that began to have national implications. At the annual SUYAO meeting, held at the Ocamo mission in mid-November 1990, Dimanawa was elected president. In his inaugural address, he promised to kill Chagnon if he caught him inside Yanomami territory.[34]

César told me in 1996, "I am going to become the president of all the Yanomami."[35] With this goal, driving out FUNDAFACI would have been at the top of his—and SUYAO's—agenda. Everything suggests that the more acculturated villages closer to Western trade centers held political and military advantages over remote villages, which the new FUNDAFACI regime would have drastically altered.

Of course, some groups welcomed the new wealth. Others were too afraid of war or had their own strategic trade advantages to protect. All of them were apparently caught up in the violence. "Over the last year," Brewer told the *New York Times,* " 'mission Yanomami' had shot to death 21 'wild Yanomami.' Brewer-Carías blamed the deaths on a mission policy of handing out shotguns."[36]

Warfare between pro- and anti-Chagnon forces also spread down to the main course of the Orinoco. Chagnon's largest supply depot was the Ocamo airstrip, where video footage showed a cargo plane disgorging tons of FUNDAFACI equipment, which Brewer moved with a military truck. "I went to Ocamo [a mission on the Upper Orinoco] with *National Geographic* in January and February of 1992," recalled the anthropologist Kenneth Good. "That's when Chagnon was buzzing around all over in military planes and helicopters. When we landed in Ocamo I saw all these big crates lined up by the airstrip and kind of close down by the Guardia Nacional [Venezuelan army] building. I saw FUNDAFACI written all over them, and I knew it was Chagnon and Brewer's. They were such big crates, five by eight, and it looked like such a tremendous amount of equipment. I'd heard that Chagnon was going to get millions of dollars and bring in fifteen or sixteen scientists of all persuasions and build the largest anthropological field station in history in the Siapa area. It particularly impressed me because up until then I had been the sole person to study the Siapa. It was the last area to be contacted, so I knew Chagnon would go after it. I thought, 'My God, they're going to build a city there. This must be a power plant or something. It's incredible. It's going to be permanent.' "[37]

It was hard keeping track of all the equipment. During the Siapa expeditions, Chagnon experienced repeated thefts of his equipment, including two solar panels, a tape recorder, and many other valuables. Once he awoke at three in the morning to find a visiting Yanomami trying to rob his personal computer. "Who would ever suspect that a Yanomamo would want a computer?" he wondered.[38]

All these thefts occurred, Chagnon wrote, "despite the best efforts of my Yanomamo friends at the Ocamo Mission to protect my possessions."[39] Chagnon and Brewer's closest political associates, Alberto Karakawe and José Valero, were both from Ocamo. Karakawe and Valero traveled on the helicopters to Siapa and were included as advisers on Brewer's presidential commission.[40] Karakawe was also interviewed on Venezuelan television, posing with a stethoscope, as if he were a nurse—which he, and the doctors in the area, told me he was not.[41]

Obviously, Karakawe was being groomed for a leadership role in the Siapa biosphere project, and this did not go over well with César Dimanawa. They became enemies. And, perhaps by coincidence, the worst single Yanomami battle in forty years soon took place between the Ocamo Yanomami and a coalition led by César Dimanawa from Mavaca. It occurred in September 1992.

In the *Times Literary Supplement,* Chagnon described the war:

> Many of the recent killings are of men in extremely remote villages, with whom the Mission Yanomamo had no previous contact or quarrel. . . . Last year . . . a violent man named César Dimanawa, organized approximately 100 men who, with boats and motors borrowed from the mission, descended on a small village. They killed seven men and two women and gravely wounded several others. Dimanawa himself shot a woman in labour, killing her and the infant. They then abducted three young women, took them back to their mission redoubt, and systematically raped them over a week. No information about this incident reached the press. It does not differ radically in scale or brutality to what happened at Hashimo-teri, but this time the villains were Mission Yanomamo rather than *garimpeiros.*
>
> The above information is based on scientific research.[42]

The above information was, in fact, based on two interviews conducted not by Chagnon but by Charles Brewer Carías—in Caracas and with Alberto Karakawe, César Dimanawa's enemy.[43] Neither Chagnon nor Brewer has been to Lechoza, which is near the Ocamo mission, to investigate.

One would not, however, need any local knowledge to realize that Brewer had elicited a distorted account of the raid. According to Karakawe's tape-recorded statements to Brewer, César Dimanawa began the slaughter by trying to abduct a woman in labor, who refused to go with him: " 'Well, if you are so stingy [about having sex], I will kill you.' He shot her, 'Pah!' He shot her in the belly at close range. She died on the spot and the child then dropped from her womb and wiggled around on the ground." Chagnon added, "Brewer's transcript does not indicate if the newborn also died."[44]

Karakawe told me that César Dimanawa probably did not shoot the woman in question, but he blamed the story on Father Nelson. "I didn't go near that fight. I didn't see it. It was Father Nelson who told me about it. The Padre was all alone when the houses were burning—one dead here, another one burned there." For his part, Nelson said he received five different versions of the woman's death, and Karakawe selected the most improbable one. Nelson said, "It's asking a lot to try and reconstruct what happened on a confused battleground."[45]

According to eyewitnesses of the fighting, Dimanawa shot a man at the onset of the battle and then ran off in the confusion. By all accounts, he was not present when the pregnant woman was killed, by accident, after she moved from her hiding place behind a thatched wall. No one knew with scientific accuracy exactly who killed the woman, because several fired at her simultaneously, and several claimed to be *unokai* over her death.

In the *Times Literary Supplement*, Chagnon touted his reporting of this conflict as an example of "scientific research."[46] But this was not the case of "an extremely remote village" being attacked by "Mission Yanomamo." Both were mission groups. Of the two, César's Mavaca alliance was less acculturated than Ocamo's. There are many evangelical Christians at Ocamo, including Karakawe. Ocamo also had a military base, with its attendant problems of native prostitution and alcoholism. Chagnon defended the Ocamo people because they were his friends, not because they were from "an extremely remote village."

But while Chagnon's loyalty was understandable, it led him into serious mistakes in attributing homicides in this massacre. It was similar to Chagnon's easy acceptance of the Yanomami slander against his former enemy Moawa—who Chagnon believed killed twenty-two men, unassisted, by means of a sharpened arrow point.* Chagnon's lifework relied on accurate authorship of homicides. If his "scientific research" was based on thirdhand, second-rate horror stories like the one Karakawe told Charlie Brewer, then the *unokai* thesis was as dead as that baby that wiggled onto the ground. Actually, Karakawe never claimed that the baby died—Chagnon dispatched the infant for the *Times Literary Supplement.*[47]

That account enraged César Dimanawa, who proceeded to write a letter of protest. "How can you say I have killed so many people? Did you see the fight?"[48] He was also furious that Chagnon claimed Dimanawa could never have written, by himself, a letter asking for the anthropologist's permits to be canceled. To prove his point, Dimanawa wrote another letter in my notebook, in stronger language, saying that Chagnon could never come back to the Upper Mavaca.

One of the frustrating things about trying to defend the Yanomami's right to self-determination was their passivity in the face of outside aggressors—

*See chapter 10.

and the ease with which their fragile internal cohesion shattered at the offer of the slightest bribes. That's why gold miners kept slipping back inside Yanomami lands. That's how Chagnon collected tribal secrets against the communal will. Dimanawa's rejection of Chagnon and his violent pursuit of his former mentor were qualities that, perhaps, marked him as a "misfit" among the Yanomami. But those were the qualities Chagnon selected for in his distribution of trade goods. As one of the Salesians with a Ph.D. in anthropology, María Eguillor García, observed, Dimanawa "admirably stopped Brewer Carías and Napoleon Chagnon in their tracks."[49]

With this victory, Dimanawa turned his attention to settling accounts with Ushubiriwa.[50]

Dimanawa sent word to Ushubiriwa that he wanted to meet him for a collective club fight. This ritual required groups of men on both sides to face off in a series of exchanged blows to the head. Ushubiriwa sent a provocative response: "We want to fight with arrows instead."[51] It was war.

In December 1992, the men of Mavakita motored upriver and tried to surround Ushubiriwa's *shabono* by stealth, as they had done at Lechoza. But Ushubiriwa proved a superior strategist. He had prepared a trap. By then, he had already sent all the group's women and children upriver, and had moved, with eleven men, to the far bank, where they waited in dense undergrowth in ambush for César. Ushubiriwa saw a man armed with a shotgun creeping along on the open beach. He opened fire. His aim was true, and the man fell into the shallow water, where he died.

But it was another unfortunate victory for Ushubiriwa. The victim was one of Ushubiriwa's own cousin-brothers, named Bruno, who had been raised in the same small community of Shamatha-teri. Bruno was married to a woman from Mavakita and had to fight with his in-laws against his blood relations, as is often the case.

After killing Bruno, Bokoramo said, "To escape César we hid ourselves, just my brothers and I, farther upriver, near Mishimishi."[52]

Ushubiriwa and this small band of fugitives fled inland, to the mountains, and from there to the Siapa River. But they could not find refuge anywhere, because even the remote *shabonos* knew that Dimanawa was after them. Then Ushubiriwa antagonized his own followers by shouting out the names of their dead ancestors, bizarre behavior for any Yanomami.

Chagnon had used Ushubiriwa for years to betray the names of the dead. In fact, the anthropologist noted how Ushubiriwa's literacy led to a "reduction in his sensitivity to the cultural prohibition on using the names of deceased people publicly." Chagnon added, "In fact, he seemed gleefully pleased that he and I shared a mysterious skill."[53]

Ushubiriwa's own followers apparently did not share this glee. Bokoramo returned to his father-in-law's village, Washewa, angered at his brother's insulting behavior. According to Juan Finkers, he told the Mishimishimabowei-teri that he did not care if they killed his brother.[54] A war party from Mishimishimabowei-teri killed Ushubiriwa with shotguns and his brother Mariano with arrows and machetes the week before Christmas 1993.

Another party from Mavakita, also searching for Ushubiriwa, met the Mishimishimabowei-teri, who thought that Dimanawa's group would be pleased at the killings. Instead, Dimanawa's allies were angry. "He was ours to kill!" one of the men from Mavakita exclaimed.[55] I had never heard of a manhunt so intense—or an animosity so deeply felt—toward anybody in the Yanomami world.

Yet Bokoramo said he would work with Chagnon again—as long as he was finally paid an outboard motor. *"Quiero mi motor,"* Bokoramo said, holding a shotgun and standing next to one of the badly rusted gasoline barrels inside the Kedebabowei-teri's old *shabono.* "I want my motor. Because of the war I am alone now, but I am not the only one. I have made others lonely, too."[56]

Back at Mavakita, one of Ushubiriwa's captured wives, Isabelita, looked lonely, too. She was about twenty-five years old, lovely even with the stick adornments Yanomami women prize jutting out of her cheeks and nose. She was pregnant now by another man, who had received her as a prize from César. It was to be her first child; Ushubiriwa had six wives but only one child, a daughter. Like most of the other *unokais* I have known, Ushubiriwa did not leave a large number of offspring.

Isabelita seemed terrified when I asked to speak with her, even with her husband's permission. She looked straight ahead and said nothing until Bokoramo started prompting her, "Talk about the motor! Tell him about the motor!"[57]

"Chagnon, I want my outboard motor," she said. "For the work my husband did, you were supposed to give us a motor, because we worked very hard. I am still waiting. I want my motor. That is why I am still waiting. As I woman, I am now sad because I have been left with nothing."[58] Her final sentence was evocative and rang true. She really had nothing—not even the freedom to say what she truly felt about her life as a captive.

By now it was getting dark. Bokoramo begged me not to ask any more questions about Ushubiriwa's death. He was afraid César would "remember" that Bokoramo was the brother of his village's enemy. For the Yanomami, words and forms are inseparable. There is no past tense or future tense. When they talk about something, they are *in* the experience. That is why the mention of the dead brought them such grief, and why the mention of these wars opened gaping wounds. Pablo Mejía pulled me toward the boat.

"Come on. Let's get out of here."

"We're supposed to spend the night here, and I want to do some more interviews."

"I'm not spending the night here," Pablo said. "César is my brother, but I'm afraid he might kill us."[59]

It was already dark, but a full moon was rising as we raced downriver toward the Mavaca mission. We camped at a *shabono* near the Orinoco juncture, and saw the next day that the mission was under three feet of water. We kept traveling, and passed Lechoza, the village where the massacre had taken place, near the Ocamo River. Water covered the whole village up to the thatched roofs. But we still were not prepared for the disaster that waited at the Padamo mission.

There was no Padamo mission to speak of. The white church was underwater with the cross still visible, like a periscope protruding from the waters. The power plant was powerless. The ranch house where I had slept on a waterbed had been ripped open—the wall facing the river was gone. It was the worst flood in half a century of record keeping. The Dawsons were all crowded into the second floor of a reinforced concrete building, their hammocks swinging from the rafters, jostling each other like bats.

We puttered around the watery ghost town in our boat, as Jodie and Pablo searched through the flotsam and jetsam, looking nostalgically for lost pieces of their lives. The strong current was sweeping everything away, picking the houses clean. In one of the eddies that backed up around a tree submerged to its branches, I thought I saw one of my socks, but it quickly disappeared, caught in the vortex like the rest of us.

"The sudden warfare on the Upper Mavaca and the Siapa coinciding with the FUNDAFACI expeditions is perfectly in accord with my model," said the anthropologist Brian Ferguson. "I suppose I should be happy about that, but it's all so terrible."[60]

Chapter 15

In Helena's Footsteps

We were thinking of jaguars all the time.—*Helena Valero*[1]

The Siapa River, which the Yanomami called "the River of the Parakeets," was our ultimate destination. It ran through one of the world's most isolated highland valleys—far beyond the snaking, chocolate-colored waters of the Orinoco. On August 27, 1996, as our plane passed over the hazy Amazon-Orinoco continental divide marked by the mountains of Unturan, the pilot set his course for a solitary peak that stood apart from the others like a perfect pyramid. We skimmed it, at 3,400 feet, and then saw in the distance another range, the mountains of Tapirapecó, almost twice as high, bordering Brazil. We began a sharply banking descent toward the village of Doshamosha-teri.

The thatched round house was built on ground of red clay, making the central plaza, where children were playing, stand out like an apple surrounded by an upright rind. The village was located by rapids on the Siapa named *Shukumena ka bora*—"the waterfall on the river of the Parakeets." The pilot circled, while inside the *shabono* pandemonium broke out. Children ran and pointed toward the plane. Adults extend both hands skyward. The control tower had given us permission to land.

But we had no intention of landing, even if it had been possible. Having confirmed the location of the villages we intended to visit, we returned to the Orinoco, where a microscopist with Brazil's National Health Foundation, Marinho De Souza, and I spent a week under medical observation. At the end of this time, we received a clean bill of health from the government doctors, and we began our journey upriver, stopping several times on the main course of the Orinoco to conduct malaria censuses.

Near the Ocamo mission, Marinho was surprised to find an old woman, completely blind, who spoke Portuguese. It was Helena Valero, who has lived with the Yanomami for over fifty years. The founder of the Ocamo mission, Father Luis Cocco, compared her to Homer's Helen,[2] but Valero's Odyssey took her much farther from her homeland than a ship ride to Troy. Kidnapped at the age of twelve on a tributary of the Rio Negro, she covered thousands of miles on foot, crossed the highest mountains in the Amazon, and became the first white person to explore many regions, rivers, and mountain ranges. She speaks Yanomami better than any other outsider. She has also written an incomparable book about her tragic and heroic life called *Yo soy Napeyoma,* "I am the White Woman."

She sat in a tiny hammock behind a torn mosquito net fully penetrated by gnats and flies, looking like a desiccated, emaciated mummy caught in an ancient spiderweb. "I am blind," she said. "I am always hungry here in this darkness."[3] Marinho took a flashlight to examine her cataracts. The artificial light harshly illumined the surroundings, revealing a reliquary of rusty objects collected over decades. There were tin cans with torn covers and a crucifix with Christ askew. Tears rolled down her eyes as she spoke to me in a mixture of Portuguese and Spanish. "When I was twelve years old, my suffering began, when the Indians captured me in Brazil. *Me lembro, me lembro de todo.*"[4]

Yet she remembered everything.

No one disputed that. Her uncanny recall of Yanomami genealogies has been praised by James Neel,[5] an exacting master of the art, and by the Italian medical researcher Ettore Biocca.[6] Yanomami lore, repeated at countless campfires over the decades, left indelible imprints in Valero's memory—of everything from hundreds of family trees to the chants for the dead and to the infrequent wars that took her beloved husband's life.

She said, "I arrived at the headwaters of the Shukumena-U, the river of the Parakeets, with the Kohoroshi-teri. I was just a little girl. I didn't have breasts. I was not a real woman yet. They called me White Woman, the woman from the moon. The men didn't want to take me. 'Leave her. She's too skinny.' But a woman said, 'I'm going to take her just like she is.' So she carried me."[7]

She recalled the confused scenes of her first arrival among the Yanomami. There was a flurry of excitement as word of a white girl's capture spread throughout the region. The Indians hoped she would show them how to make axes and machetes, objects so prized that men slept with them on their chests and sat on top of them when they rested in the forest. Villages fought over her, then rejected her. Some suspected her of withholding the secrets of metals; others blamed her for causing conflicts. There were doubts about whether she was a human being or a ghost. After she was accused of poisoning a child, an angry man shot her with an arrow dipped in curare poison. "He shot me with an arrow here," Helena said, pointing to her leg. "Then I escaped into the forest alone. I was just a child. I didn't have my period yet, and my chest had just begun to grow. It's incredible that a little girl could have lived alone like that, with the jaguar coming every day, sniffing for me, going *Huuuh Huuuh.* You should have seen the size of his teeth! I survived only because that Old Man up there was looking out for me. After seven months, the Patanowa-teri found me, and I married my husband. He was the only one who really loved me. His mother and his whole family loved me well, too. But the Indians killed him and wanted to kill my children."[8]

Valero successfully ran off several times while still in her early teens, but by then she was hundreds of miles away from her homestead. Some of the saddest moments in her story are her lonely wanderings through the forest, climbing mountains and trees to try and spot her home, shouting in Portuguese to the endless horizons of green and blue . . . and hearing only monkeys calling in reply. As she remembered these scenes from her long exile, Valero cried. Finally, she shook her head and smiled. "Well, you can't complain, but I'd like to die with my stomach full of food, not air. And these mosquitoes!"[9]

Of course, if Helena Valero's work and life had achieved more recognition, she might not be suffering from hunger in an insect- and malaria-plagued hut. But when the filmmaker Timothy Asch tried to make a documentary about Valero's life, Chagnon objected. He did not want to exploit her, he wrote.[10]

Yet he had already exploited her. Every geographic discovery and anthropological first contact in Chagnon's career had really been achieved by Helena Valero. It is a remarkable fact and a remarkable theft. Every single place—the Upper Mavaca River, the Siapa valley—and every single village—Patanowa-teri, Mishimishimabowei-teri, Iwahikoroba-teri, and a dozen others in the Siapa—that Chagnon has touted as his discovery, was intimately known and visited by Helena Valero.

In the history of discovery, accuracy mattered. In the history of Amazonian discovery, Chagnon misrepresented Valero's life. Students of the first three editions of *The Fierce People,* read only one sentence about her. "Helena Valero, a Brazilian woman who was captured as a child by a Yanomamo raiding party, was married for many years to a Yanomamo headman before she discovered what his name was."[11]

Chagnon's latest edition of *Yanomamo* incorrectly states that Valero lived her adult life with the Kohoroshi-teri.[12] She lived with the Kohoroshi-teri only for about six weeks,[13] and none of them as an adult.[14] Helena was kidnapped by the Kohoroshi-teri in Brazil, but quickly crossed over the Tapirapecó Mountains into Venezuela, where she was captured within a few months by Ruwahiwa,[15] the fabled Mishimishimabowei-teri war leader who played an archetypal role in *The Fierce People.*[16] Valero remained in Venezuela twenty-four years, living with exactly the same groups Chagnon studied. She intimately knew all the Mishimishimabowei-teri, Patanowa-teri, and Bisaasi-teri—the protagonists of *The Fierce People* and the book's accompanying films. It was true that she did not learn the real name of her husband, Fusiwe, a Patanowa-teri leader, for a time. But she eventually became familiar with the whole genealogical structure of the Namowei Yanomami. James Neel could not find a single error in her accounts. In other words, the Atomic Energy Commission could have spared its scientists, and the Yanomami, great suffering, if, instead of coercing the names of the dead through trade goods, it had just interviewed Valero.

But Chagnon never did that. Although he met Valero, he wrote that he never once asked her about her life among the Yanomami.[17] Again, he didn't want to exploit her. Chagnon's lack of curiosity about Valero's history was consistent with ten cardinal rules that guided his research. Rule number 5 was "Men are better informants for genealogical relationships than women."[18] Chagnon thus never approached Valero to help him clear up the discrepancies between his reconstructed histories of Yanomami warfare and her eyewitness accounts.

For example, regarding Valero's capture, Chagnon wrote, "The raiders attacked the foreigners at dawn but were driven off by gunfire; a number of Indians were killed, including a man called Hokohenawa, and some of the foreigners were allegedly hit with arrows."[19] Not so, according to Valero. Together with her mother, father, and baby brother, she encountered the Indians at a little-used homestead while reconnoitering in their canoe. The Yanomami did not raid the family at dawn. Her father had left his shotgun behind; had he brought it, Helena would almost certainly not have been captured. "A number of Indians" were not killed. Nobody was killed. Helena was not "allegedly hit with arrows;" she was pierced with three curare-tipped points, which sent her into a black swoon. She bears the scars to prove it.[20]

The Fierce People opened with a dramatic prologue, "The Killing of Ruwahiwa." This murder, in Chagnon's narrative, was a seminal event that set off interminable wars, which were still going on when the anthropologist entered the field in 1964. Chagnon's chronology of Ruwahiwa's death has varied. He first placed Ruwahiwa's killing around 1938;[21] then he decided it happened around 1950.[22] Valero has been more consistent. She said that Ruwahiwa, who was a friend of hers, died about 1943, and his murder, which she tried to prevent, had little, if anything, to do with the subsequent wars—which did not break out until 1949.[23]

The scientific prestige of *The Fierce People,* as well as the famous thesis that Yanomami murderers have more offspring than nonmurderers, depends on the reliability of oral histories. Chagnon maintained, from his Ph.D. thesis[24] to his *Science* article on killers, that he obtained "remarkable consistency in . . . reports on violent deaths."[25]

Yet Chagnon's reporting diverged remarkably from Valero's. This divergence began with motives and dates, but, most crucially, it included the actual number of victims and their specific killers. For example, Chagnon had Ruwahiwa die alone at the hands of a single, ax-wielding assailant, Mamikininiwa.[26] But Valero saw Ruwahiwa attacked by a different man, and the initial ax blow did not kill him. In her account, Ruwahiwa and five others were riddled with arrows.[27] It was a horrifying, unforgettable scene, as the victims, still in ceremonial regalia, were taken away:

> Then those who had killed him took lianas and tied the dead body. So they dragged it out; you could hear the arrows which went tr, tr, trr as they scraped along the ground. The *tushaua* [leader], sitting in his hammock, watched. The designs on that bloody body were still perfect; the ornaments had stayed in the ears and on the arms. No feather had fallen on account of the arrow-shots.[28]

Much later, Valero witnessed her husband ambushed alone.[29] According to Chagnon, he died in a "raid" along with two other men.[30] Since small variations like this are crucial in a population whose homicide total was thirty-one individuals over thirty-five years, the difference in numbers alone suggested that, at the very least, Chagnon should have tried to check his data with the only person in the culture who could count beyond two.

Apart from the issue of numerical accuracy, there was the deeper question of what these deaths meant in Yanomami history. Chagnon situated the murders of Ruwahiwa and Fusiwe within the context of biological competition between dominant males, and he enlisted their blood relatives to avenge their deaths. According to Valero, Ruwahiwa was killed at the request of his own brothers; and Fusiwe's relatives did nothing to prevent his ambush, though they knew it was coming. Both were killed with the complicity of their close kin—who were simply sick of them because they had both started too much trouble. Moreover, after his death, Fusiwe's children and his wives were targeted by his enemies, a fact at odds with Chagnon's theory of killers' reproductive success. Fusiwe's dying words to Valero were these:

> "Napeyoma: go away with these children. Don't stay any more with these people, because what happened to me will happen to you. Find your path. Go far away with these children, far from here. . . . I am sorry that I am going to die and abandon them. Go and raise them among your people." On saying this he squeezed the hands of the children tighter, looked at them, sighed deeply, closed his eyes, and was dead.[31]

But I think the main principle at stake was that, if Chagnon—or Lizot[32] or any of the media members who claimed first contacts in areas where Valero had lived—had given Valero, and a woman at that, due credit, it would have detracted from their own mystique. The charm, and market value, of first contact lay in its perfect virginity. You cannot write, "I was the first white to ever explore the Mavaca River with my outboard motor and make contact with the almost mythical village of Mishimishimabowei-teri, except for a twelve-year-old girl, who, after being shot with poisoned arrows, escaped, traveled five hundred miles barefoot through the jungle, and lived in the area for decades." After that admission, it would be hard to hold the reader. *Who cares about your outboard motor? What happened to that girl?*

Compared with Helena Valero's courageous adventure, everybody else's jungle narrative looked mighty small. And it must be admitted that Chagnon's series editors in *Case Studies in Anthropology,* including a past president of the American Anthropological Association, George Spindler, have collaborated for thirty years in the theft of Helena Valero's singular achievements. So did the National Geographic Society, which gave Chagnon a generous forum for his views.[33] The society was usually quite knowledgeable about the history of discovery. But Valero was not mentioned in *National Geographic*'s articles about the villages where she had lived.[34] When allocating credit for first contact, anthropologists, journalists, and media outlets were either ignorant of a life that was already legendary throughout the Amazon, or they did not consider Helena Valero a person.

Marinho and I continued up the Orinoco easily for another seventy miles, in a big dugout canoe, until we reached the Guaharibo Rapids. The water level was still high from the flood, and the wave action was fairly intense—the biggest waves were maybe two and half feet. By technical white-water standards, these waves were not too serious—maybe Class Two on a scale of five. But white-water recreation crafts are state-of-the-art creations, with expertly designed fiberglass hulls, and are loaded lightly. Most important, they normally go *down* the rapids. Driving a heavy, square-bottomed, hardwood canoe upstream through Class Two rapids takes careful planning. In Brazil, the Mucajaí Yanomami knew how to ferry their way upstream through rapids like this one. They came up behind big rocks where there were "holes" in the white water—eddy currents flowing upstream in reaction to the water exploding around the rock. Then they would catch these upstream countercurrents to the edge of the rock, inch the boat out on the safer side, and, holding it at a forty-five-degree angle, ferry across the white water to another eddy current. Ferrying seemed to defy common sense because the crashing waters looked as if they would capsize the boat or drive it downstream. Yet if you kept the throttle open just enough to hold the boat's forty-five-degree angle, the rough water worked in your favor—you rode the crest of one wave to the next. It required exquisite control, and the Mucajaí Yanomami had learned it from the gold miners.

My *motorista,* a Yanomami man who went by the name of Antonio, had a different approach. He just opened up the throttle and tried to tough his way through the highest wave action. We stalled, turned, and were lucky not to capsize as we were finally pushed out of the main channel and got stuck on one of the enormous granite rocks strewn in our way. We had to off-load and portage our way.

It was a historic spot. According to Alexander von Humboldt, this was where the Yanomami world began. In 1800, the Yanomami, whom Humboldt called White Guaharibos, marked their territory with a vine-and-pole bridge across the rapids. Just beyond it, Venezuelan troops massacred a group of sleeping Yanomami in the 1820s, and, in the same place in 1920, Hamilton Rice machine-gunned an unknown number of Yanomami. We walked past an island where the 1951 Franco-German expedition had camped. As we continued upriver, the banks, completely overgrown with vegetation, rose more steeply as the channel narrowed. About an hour farther, we came to the village of Patahama-teri. A new *shabono* was up, its thatching still green and fresh. Lying in our hammocks, looking at the stars and a dozen fires around the compound, we felt as if in another world, one that stirred deep memories in the Christianized Yanomami from downriver. "I used to live my whole life like this—it's lovely," Pablo Mejía said. But he decided not to travel with us to the distant villages. "There are too many wars going on now," he said, as he shoved off the next day.[35]

Alfredo Aherowe, a SUYAO leader from Mahekoto-teri, on the Orinoco, and his nephew Marco accompanied Marinho and me. Six boys from Patahama-teri also came, far more than I wanted to help carry our food and medical supplies. When we got up in the morning, we all gave blood for malaria slides (a routine Marinho made us practice each day) and were checked by Marinho for cold symptoms. Then we left the humid banks of the Orinoco and entered the gentle, filtered light of the forest. The terrain was easy, almost Appalachian. We followed winding streams of reddish waters, a product of sedimentation and concentrated tannin, leached from the riot of foliage, to Irokai-teri, an old village that Helena Valero knew well. Her husband, Fusiwe, was ambushed near here on his way to a feast in 1950.

As we approached the village, we heard a cacophony of shouting, and many of the women fled the *shabono.* It was noon, and, when we emerged from the cover of the forest, the central plaza of the round house was oppressively bright and hot. Three men were chanting and blowing hallucinogenic snuff up each other's noses, invoking healing spirits for a dozen people lying ill in their hammocks. Why had the women run away?

"We thought you were Kenny Good," one of the shamans said.

They were afraid Good had come to reclaim his wife, Yarima. While doing his Ph.D. research here, Good married a teenager who had been born at Irokai-teri. Good has told his love story in the book *Into the Heart.* He entered Yanomami society in a unique way, and his affection for Yarima translated into an enjoyment and high regard for Yanomami culture.

Nobody from the outside world had seen Yarima since the end of 1993. At the Padamo mission, she was rumored dead; elsewhere, it was rumored she was hiding out in the hills.

Still, I was startled, while helping Marinho take the malaria census, when someone tugged on my sleeve, and said in perfectly good English, "Hello. My name is Yarima. What is your name?" She was nursing a baby and looked radiantly healthy, one of the few people at Irokai-teri who did. She said her new husband was treating her well. Her eyes clouded over for an instant when she asked me about her three kids in New Jersey. Then she added, "Here good. Jersey bad."[36]

The basis for Yarima's discontent in New Jersey was easy to understand. For someone who had grown up in a communal round house, without privacy or loneliness, the life of a housewife inside a small apartment was alien in the extreme. She missed fishing and collecting fruit with her sisters and mother. Although Yarima was initially delighted with running water and electricity and the marvelous convenience of life without gardening or collecting firewood, she gradually became disenchanted. She observed that most Americans were more obsessed with shopping and watching television than with spending time with their families. "They love TV and malls, that's all," she told me.[37]

Yarima was also less impressed with Good outside the jungle. "They told me that I would be less enamored of my little Indian girl once I was in the city," Good said. "Yarima thought I'd become a real milquetoast in the United States. No shotgun. If a policeman told me to stop, I'd just stop. Once I bumped a little lady's car, and she got out and yelled and called me an idiot. Later, she [Yarima] said to her brothers, 'He's not so fierce. Downriver people laugh in his face, and he doesn't do anything.' I don't think she understood that."[38]

Good did not prettify the Yanomami. But he did not confuse their rhetoric and posturing with actual warfare, of which he saw almost none. "A *shabono* is an amazing model of cooperative living," he said. "I'm not saying it's all peace and flowers. No. There are problems, fights, insults. But you've got eighty or a hundred people living in this rather close quarters, without partitions or privacy. I don't think any Western society, group, or extended family could do it. They'd be at each other's throats in a very short period of time. The Yanomami have culturally adapted to live that way. And they should be admired for the ability to do that, not given the image of people who are always fighting."[39]

Like Helena Valero, Good observed that what Yanomami violence existed was directed mostly at women, not at other men in war. He honestly described his horror at a gang rape by Yanomami teenagers—and the equal horror at his own failure to stop it. Confused, "shaking with anger," Good asked himself, "Come on, Ken, what's wrong with you? Are you going to stand around with your notebook in your hand and observe a gang rape in the name of anthropological science."[40]

Good has also been forthright about the conflicts his marriage caused—to himself, Yarima, and the village of Hasupuwe-teri. When Good went downstream on business, the Hasupuwe-teri decided he was dead, and Yarima was raped by a number of men in the village. Good was angry. He publicly humiliated one of the rapists, his brother-in-law (from a Yanomami perspective, it was normal for Yarima to have sex with her sister's husband). Another time when Good was gone, Yarima was beaten and her ear partly ripped off. In the end, Good was on strained terms with many people in the village over their failure to protect Yarima during his business trips. "I am pissed off at all the people," he wrote. ". . . And at her brother, too. Maybe her brother especially. His basic attitude toward this is, So what? What do you care? It's just a *naka,* just a vagina. What's the big deal?"[41]

In the end, the Yanomami, who believe that every man who has sex with a woman contributes to the strength of the fetus, and who have no word to distinguish uncle and father, have different concepts of fertility and fidelity. To avoid the difficulties engendered by his periodic absence, Good finally took Yarima to New Jersey, where he was able to get a teaching position. Yarima gave birth to two boys in New Jersey and a girl on a return trip to the rainforest.

Yarima's impression of Western culture was fascinating. "At that time I did not understand what police were. It seemed to me they must live everywhere. . . . I asked Kenny where their village was, and he said they didn't have a village. But that didn't make sense. . . . I also saw the wives of the police. They wore the same clothes, and they carried guns, too. I wondered if their children also wore the same clothes and carried guns, but I never saw any."[42]

When I asked Good what life was like in New Jersey with Yarima, he said

Yarima had fun until they had kids. After that, "Yarima was lonely and bored. She didn't want to be here. Neither did I. But I didn't have any choice—we had three kids.[43] Yarima lived the life a lot of middle-class women live. She decided, 'Screw it. This is not the way humans are supposed to live.' "[44]

They both wanted to return to Yanomamiland, at least for a visit. Good's professorial salary did not give them that freedom. But a way presented itself when Yarima became the subject of a *National Geographic Explorer* special, *Yanomami Homecoming.* Basically, *Yanomami Homecoming* was a variation of the classic "fish out of water" story. In this case, it was the "fish out of water goes back to water" story. *National Geographic Explorer* came with three large boatloads of people to film the return of Yarima to the Upper Orinoco. As it happened, the Yanomami did not make much of a fuss over her return. The real story was the arrival of all these *nabah* with their trade goods.

But the Yanomami captured part of this story—the Yanomami history of the film crew—on video. They had acquired camcorders through the help of Timothy Asch. By the late 1980s, Asch had turned away from traditional documentary approaches, the ones driven by big budgets and big crews that arrived in the field with a predetermined agenda (like the one he and Chagnon implemented for *The Feast*). He wanted to empower the Yanomami, so he donated 8mm camcorders to SUYAO, helped train a few young men, and let them film whatever they wanted.[45] One of the Yanomami cinematographers, José Seripino, wanted to film the *National Geographic Explorer* crew.

The result was a small but important footnote in the history of ethnographic filming. It could be titled "National Geographic Caught in the Act." The anthropologist Jesús Cardozo, Asch's partner in the project, described Seripino's study in a report he sent Asch in March 1992, right after *National Geographic* left the field:

> It is a short but very interesting film with some very good takes that make me think that José has a special feel for filming. So while the National Geographic crew staged and filmed "the reencounter between the American-anthropologist-who-married-an-Indian-woman-of-a-primitive-Amazonian-tribe-and-now-returns-from-the-States-with-three-Indian/American-kids-who-are-fans-of-the-Ninja Turtles and the Catholic padre at the mission," José shot a much more interesting film of the whole "encounter-show." I don't think he was trying to make any point in particular or a "critical" film in any way, but the fact is that for

> him the interesting thing was the whole scene of Ken and Yarima's arrival with all the *nape* who came with huge cameras and loads of *matohi* and what they did in Mavaca. By the way, there is a funny scene in José's film in which we see a bewildered Ken looking at the camera (i.e., at José) and asking Bórtoli something like: "What is this guy doing? Why is he filming me? How did the Yanomami get film cameras?"[46]

For years people have suspected that *National Geographic* did this kind of thing. But it took a Yanomami filmmaker, a Stone Age artist, to prove it.

The film crew's distribution of trade goods also generated antagonism between Hasupuwe-teri and the neighboring village of Patahama-teri. (The night I slept near the Orinoco, the Patahama-teri headman delivered a long harangue, as rain poured down, denouncing the fact that Hasupuwe-teri received the lion's share of the film's largesse.) But I think it reflected the poor choices film crews face nowadays. To get a million-dollar budget for a 45-minute production, they needed a story set in steel before they even arrived in the field. But they stubbornly stuck to the "fish out of water back in water" at a time when they could have done a great deal of good by simply filming what was going on around them. While they staged the Yarima show, in February 1992, Cecilia Matos, dressed in her White Witch outfit, was buzzing overhead every day in giant helicopters with her millionaire friends. And, on the Upper Mavaca, Napoleon was at war with Caesar in the most bloody contests ever recorded in Yanomami culture. Transport planes were dropping off dozens of multiton crates to construct the biggest tropical research station ever built.

Kenneth Good was also unhappy with the film crew's behavior in the field. He complained that *National Geographic Explorer* reneged on a promise to bring a doctor on its expedition. "That's bullshit," the assistant producer Amy Wray said. "We would never bring a doctor in with us."[47] Most film crews think like that. I personally found Amy Wray to be an outstandingly gracious person. I really enjoyed meeting her. Yarima also enjoyed her company during the filming—at the mention of her name, Yarima's face broke into a huge smile. The problem is not personal. Extremely nice people, as well as not-so-nice people, have been staging extremely destructive films on the Upper Orinoco for over thirty years now. Even while the films preached about the dangers of outside contact and the urgent need for medical help to save Yanomami lives, the filmmakers excluded themselves from the picture. *National Geographic Research* magazine had warned for years about contacting remote groups, and how 30 percent could die within a short time unless permanent medical attention was provided.[48] But there was a total disconnect between the exploring society's scholarly articles and the video-sales and ratings-driven behavior of its commercial television unit.

Good felt that *Yanomami Homecoming* helped break up his home. He said members of the expedition pressed Yarima to criticize him, and insisted on interviewing her alone, with the help of Gary Dawson, for several days, a situation that strained their relationship. There is a hint of this on film, when Yarima responds to a question of why she married Good. "The *nabah* always ask me the same questions and now I am tired of answering. I married him because I loved him."[49]

A more difficult question was why Yarima decided to leave Good. That did not happen for another year, at the Platanal airstrip, when Yarima ran from the plane and became, for a time, a pawn of politicians in Caracas, who paraded her around to the talk shows. Good blamed one of the men whom *National Geographic* initially hired, Timenes, for influencing Yarima. Timenes was one of the heavily armed men Chagnon sent to the Upper Mavaca to purchase Ushubiriwa's allegiance and to secure a helicopter site.

But the real conflict between Good and Yarima ran deeper than Timenes or *National Geographic.* Just before the *National Geographic* crew flew to the Upper Orinoco, Good received word, in Caracas, that his father had died. Under the circumstances, he decided to stay and meet his commitments to the film crew. Yarima was horrified. For the Yanomami, mourning dead relatives is the most imperative obligation. According to Good, Yarima lost all respect for American culture from this point onward. On returning to Hasupuwe-teri, Yarima learned that her own mother had died. She burst into tears and was disconsolate. A chasm separated the emotional worlds that Yarima and Good inhabited.

I went for a long walk with Yarima and her new friend. We climbed up a limpid, meandering creek with shallow pools, where we swam and then sat on warm rocks sunning ourselves at the foot of a cascade. Just being in the forest was so healing and quieting; except for malaria, the dangers are overstated. But I could not resist the civilization game. I asked Yarima if she remembered how to count. "Yes," she said. "One, five, three. . . ." She laughed. "I forgot."[50] She sounded glad to have forgotten.

Yarima and Helena Valero, the two people who experienced Yanomami culture and the outside world, both chose to remain among the Yanomami. In colonial America, similar choices prevailed. Benjamin Franklin noted that whites kidnapped by the Indians almost never readjusted to civilized life, whereas Indians taken from the tribes looked for the first opportunity to run off even after years of education. There must be something very superior in the nature of their social bond, Franklin concluded.

Or there must be something very inferior in ours. Our "discovery" of Yarima turned out to be a Stone Age sensation. A Brazilian reporter learned about our encounter with her, and a series of stories broke out about Yarima's rebellion against modern life. *National Geographic* sent a photographer to find her; Yarima fled and he failed. (Instead, it put the Orinoco on the cover of its April 1998 issue, with a gorgeous two-page inside shot of Patahama-teri.)[51] Brazilian papers went berserk. *O Estado de São Paulo,* the country's second biggest, ran the headline *YARIMA, CINDERELA REBELDE.*[52] The *Times* of London ran three stories about a Stone Age bride who had exchanged her paradise in New Jersey for the rain forest.[53] According to the *Times,* a new expedition was being organized to relocate Yarima. To lure her out of the jungle, searchers were going to play tape-recorded messages, in Yanomami, from Yarima's American children in New Jersey—all begging "mummy" to come home. Apparently, a lot was at stake in this experiment. Would Yarima obey the pull of her maternal instinct and the attraction of her comfortable New Jersey home? Or would the call of the wild be so strong that no prerecorded announcement could lure her out of the primeval forest?

At least Yarima was far enough away that *National Geographic* could no longer find her, and the *Times* of London could only speculate about her choices.

But, though Yarima was enjoying her renewed life, Irokai-teri did not look so "good" from my perspective. Fifteen of thirty-seven people had falciparum malaria.

Inland, things were worse, the men told us. Epidemics and wars were both raging. They warned us not to go. In long, singsong narratives accompanied by side-slapping and gesturing toward the south, they told us about the dangers of visiting those remote, warlike groups that were currently raiding each other. I suspected that this was simply self-interested propaganda designed to keep rich visitors from taking their medicine farther inland. These days, medicine was gold. Several days later, I discovered the rumored wars were real.

Chapter 16

Gardens of Hunger, Dogs of War

"Mamá, when are we going to eat?" Miramawe asked.

I said, "This is what your father wanted for us when he decided he had to kill. Our hunger is the price of his bravery."—*Helena Valero*[1]

The village of Mokarita-teri sat at the bottom of a secluded valley where steep mountain slopes converged upon a neatly circular *shabono.* To one side, there was a narrow ravine, so densely overgrown that it was invisible from above. Every day, the Mokarita-teri climbed down this canyon to a trickling spring that was their only source of water.

As Marinho De Souza, the microscopist, and I worked our way down sharp embankments of reddish clay, we found ourselves surrounded by plantain and banana trees growing with abandon despite mediocre soil and rough terrain. We were eating what we gathered at the gardens we encountered, as we trekked through the Siapa Highlands in September 1996. There were hundreds of gardens—from neat, well-tended little fruit farms to semi-abandoned, riotous patches being reclaimed by the forest. But they were all predominantly banana and plantain groves, with fruit ranging from green, finger-size sticks—destined to be sweet, miniature bananas—to yellow, loutish plantains the length of a man's forearm, which one had to cook to eat. Nothing was easier to grow than a banana tree. A cutting was made and planted in the ground, and, in just three or four months, bananas sprouted. The trees themselves resembled giant celery, with extravagant purple stamens—drooping, foot-long decaying leaves hung languidly around the periphery, like tattered rags. But the prodigious growth was deceptive. Plantain production was often uncertain and plantain consumption often unsatisfying.[2]

We wearied of plantains long before we tired of walking.

Nike. That was the word I heard over and over again during our seventeen-day trek through the Siapa Highlands in September 1996: *Nike, nike.* My teenage guides from Hasupuwe-teri were complaining of "meat hunger," not asking for state-of-the-art footwear. Food became a problem, particularly since the Hasupuwe-teri detested my dried-milk supplement. Although armed with their six-foot poisoned arrows, and always on the alert for game, they killed nothing larger than a snake the whole trip. Fishing in the highland streams was also a waste of time: the biggest catch weighed a few ounces. The boys from Hasupuwe-teri were noticeably thinner by the time they went home. So was I. I lost seventeen pounds.

Marinho estimated about a quarter of the Siapa Yanomami were malnourished. Many of the children showed protruding ribs and stomachs inflated like little balloons.[3]

At night, however, the emaciated children disappeared from view, and our own hunger was also forgotten. In every village, we heard long ritual chanting around the fires. Sometimes the shamans were performing their healing rituals, invoking the spirits of animal familiars to drive away disease, particularly malaria and colds that were now endemic. But much of the chanting had to do with us, as it turned out, and other visitors—both real and mythological. The Yanomami are a singing people. Their normal discourse, accompanied by graceful gestures and pantomime, is far more rhythmic and varied than ours. It is as much a chant as speech, and its purpose is not to quickly convey some information but to repeat a delicious sound, to enjoy its resonance and assonance. The Yanomami are sybarites of sound. And they seem to enjoy staying half awake—keeping an eye on the fire, keeping the healing and trading chants going, warding off evil spirits and imagined animals through the long tropical night.

Helena Valero began to feel at home with the Yanomami one night when the Mahekoto-teri and Patanowa-teri asked her to sing at a great feast. They told her to sing in her own language, and they would accompany her. So she introduced the Yanomami to a mournful Brazilian love ballad that also aptly summed up her own situation and intentions:

Eu sou prisoneira
Escrava desse grande amor
Deixa disso, deixa disso
Eu vou te abandonar.
Estou cansada de sofrer
Por te adorar, por te agradar.[4]

I am a prisoner
A slave of that great love
Leave it, leave it
I am going to abandon you.
I am tired of suffering
To adore you, to please you.

The Yanomami began chanting along with Helena, and soon they had a wild, foot-stomping improvisation going. Prior to that night, she had run off several times. But she never again left the Patanowa-teri until her husband, Fusiwe, was killed.

During our final day at the village of Mokarita-teri, the headman began chanting around one in the morning. He was performing a *wayamou,* a trade chant, a kind of musical negotiation with Alfredo Aherowe, the SUYAO leader. We all listened, in the half sleep of hammocks, bathing in the sound, which grew particularly haunting after Alfredo joined in around four o'clock. They sang back and forth, Alfredo responding an octave higher than the deeper-voiced headman, until the sun rose. Trade chants employ archaic language, and the only thing I could understand was some references to "the outsider," *nabah,* and "trade goods," *madohe.* Alfredo told me that the central refrain, which the headman sang over and over again, was the following:

The sun is rising
The night is ending
It's time, it's time.
Tell this rich foreigner
To hurry up and give me some fishhooks.

By this time, we had treated ten people for malaria, and I felt that was sufficient payment to the village. They did not offer us any food, and, at the end, when Marinho asked for some bananas, the headman refused. "We do not want to give you bananas." Marinho, who spoke Yanomami and had spent four years treating the Indians in Brazil, was disconcerted. He said that the Yanomami had always fed him gladly. But we were operating in the wake of FUNDAFACI and the *National Geographic* crew, and expectations were high. The Mokarita-teri had walked three days to another village in order to ask Shaki, Chagnon, for presents, which he generously gave them.[5] I proved to be a disappointment.

The six boys from Hasupuwe-teri were supposed to go home, but, when Alfredo returned to the main course of the Orinoco, his nephew Marco was not willing to continue the trek unless we took all of the boys with us. Marco had spent his whole life along the Orinoco, and he regarded the people of the interior villages as "wild Indians." For their part, the Mokarita-teri refused to provide any guides to the next village, Hokomapiwe-teri. They were at war.

The path to Hokomapiwe-teri was blocked by broken branches tied together, clearly showing the hostilities between villages. We stopped several times to let the Hasupuwe-teri guides hunt with their bows and arrows, but they killed nothing. They wanted to know why I had not brought a shotgun, like other *nabah* they had known.

The scarcity of game was surprising to me. I had in 1992 traveled to lowland swamps on the eastern fringes of the Yanomami area in Brazil—about two hundred miles from the Siapa—where animals were so unaccustomed to hunters that monkeys and curassows came to inspect us docilely. The threats in those swamps were jaguars, also unafraid of humans, and snakes of many varieties. After killing a howler monkey, the miners I was with were frightened into letting go part of their prize when a big jaguar approached to within twenty feet and growled. One night a deadly fer-de-lance, attracted by my body heat, climbed a tree trunk next to my hammock and began swaying rhythmically just inches from my foot, before an Indian guide smashed it to jelly by a series of expertly timed club blows.

Here the night held less terror, but the days were marked by a greater monotony of both flora and fauna. On our first day out of Mokarita-teri, we saw a wild pig—the only one we saw the whole trip—that bounded off too quickly into the underbrush for anyone to even fire an errant arrow.

Whereas orchids and bromeliads blossomed profusely in some marsh areas, color was rare here. The Amazon has great disparities in its species wealth. In some places, the tenacious adaptation of insect predators has forced trees of the same species to grow wide distances apart. A single acre can have a hundred different varieties. In these highlands, we found groves of palm as well as hardwood species not unlike those of an Appalachian forest. The beauty of it was the ease of travel and the relative absence of heavy undergrowth—at least for the first couple of days, in the lower foothills. Here the streams cascaded into sunlit pools where we stopped to rest and fish and swim and look for game. Piranhas and caimans were not a problem. We did not see any crocodilians. We did not catch any fish larger than minnows.

So we kept eating bananas—morning, noon, and night. Europeans had introduced them to Amazonia in the sixteenth century. No one knows when the Yanomami began cultivating plantains, or *musa.* But there is no doubt the banana had a revolutionary impact on the diet of the Yanomami, and it may have been a decisive factor in their territorial expansion and in their population increase, which were both phenomenal between 1800 and 1960, just before the arrival of Western anthropologists.

What were the Yanomami like before they obtained steel cutting tools and plantains? An archaeologist excavating a Yanomami village would be struck by two facts. First, all the crucial agricultural tools were imported; second, plantains made up over 70 percent of the Yanomami diet.[6] Very few tribes anywhere in the tropical world rely on plantains as a staple. I know of no other tribe that gets such a high percentage of its diet from these starchy, sweet fruits. According to a study of Yanomami gardens in the Parima Highlands, up to 98 percent of their cultivated-food calories came from musaceous plants.[7] There was no fresh meat at the first four *shabonos* we visited.

It was a pillar of the sociobiological theory of Yanomami warfare that the Yanomami have more than enough protein, easily obtained with a few hours daily work. "Whatever else befalls them, all members are relatively well-fed by Western standards," noted Edward O. Wilson. "Yet they are so fiercely combative and territorial that the level of their aggression within and between villages is legendary; they are indeed 'the fierce people.' What, then, is the essential resource, if any, for which they contend? . . . In a protein-rich environment, reproduction by males is constrained primarily by control of access to women. . . ."[8]

For the Yanomami to be truly fierce, they had to be truly well fed. Otherwise Yanomami warfare would fit the most common denominator of tribal conflict: resource scarcity. A worldwide archaeological survey of tribal wars, based on forensic analysis of the victims and killers and the circumstances of their deaths, has concluded that the great majority were provoked by tribes undergoing population explosion and territorial expansion. The wars were over hunting grounds and trading routes and agricultural sites—not caused by biological striving.[9] As Chagnon has recently acknowledged, "The archaeological explanation of preference for causes of prehistoric violence appears to be resources shortages that provoke intergroup conflicts. . . ."[10]

But Chagnon argued that the Yanomami did not fit this worldwide pattern. He invoked James Neel and the AEC studies: "For most of my early research on the Yanomamo, I collaborated with a large team of competent and distinguished medical researchers whose thousands of biomedical, epidemiological, and serological observations on several thousands of Yanomamo led them to conclude that the Yanomamo were one of the best nourished populations thus far described in the anthropological/medical literature."[11] The authority Chagnon cited for this assertion was an article by James Neel, published in *Science* in 1970, which had no data about Yanomami diet or nutrition.[12] Elsewhere, Neel said his impression of the Yanomami's health was simply a snapshot, and his enthusiasm was based on an observation of adult men, who danced and sang throughout the night during his visits.[13] In recent years, the tendency to generalize about the health of Amazonian Indians on the basis of spot observations of men engaged in ritual activities has been criticized by female scientists, who have found the health of women, children, and elders to be far from robust.[14]

Less impressionistic methods of evaluating tribal health have been available for decades. The World Health Organization relies on measurements of height and weight. So did Neel. He and the doctors funded by the AEC measured and weighed hundreds of Yanomami men, women, and children, as shown by the tapes of his 1968 expedition, which are now at the National Archives.[15] Yet neither Neel nor any other scientist on the AEC-funded expeditions actually published a study of Yanomami height or body weight.[16]

A forensic archaeologist who examined the size and weight of the Yanomami would conclude they were among the most poorly nourished people on earth. When Humboldt encountered the Yanomami in 1801, he called them "midget Indians."[17] The only study in this century to focus on the diet and health of Yanomami in the Parima Mountains, where all existing groups originated, found that adult men averaged four feet nine inches (145 cm); women averaged four feet five and a half inches (136 cm).[18] The Yanomami at the Parima villages were, on the average, four feet seven inches tall.

Being short in stature did not, by itself, mean the highland Yanomami were suffering from malnutrition. The World Health Organization, however, regards stunting—short stature—as a strong indicator of environmental stress, usually including malnutrition.[19] Neel interpreted Yanomami size differently. He acknowledged that the Parima Yanomami were "pygmoid," meaning under five feet tall, but attributed this to genetic microdifferentiation.[20] Other experts have been struck by the Yanomami's stature, even in a part of the world noted for short tribesmen. "Indeed, they are the shortest people in Amazonia."[21]

Many observers in the Yanomami highlands have reported bloated bellies, emaciated limbs, and other telltale signs of chronic hunger.[22] The first ethnographers to leave recorded impressions of the highland Yanomami were two Germans, who briefly visited them in the 1840s and the early 1900s, respectively. They both noted that the Indians relied on small gardens with irregular crops, which left them exposed to shortages.[23] In 1924–25, the geographer Hamilton Rice penetrated to the eastern flanks of the Parima Massif, on Brazil's Parima River, where he noted obvious malnutrition. Rice concluded that these remote Yanomami "are not the fierce and intractable people that legend ascribes them to be, but for the most part poor, undersized, inoffensive creatures who eke out a miserable existence."[24] In 1951, a scientist on the Franco-German expedition that reached the source of the Orinoco was shocked to see newly contacted Yanomami so hungry that they were eating *mosquitoes.*[25] In two Parima Highlands villages, studied in 1980, a researcher, Darna Dufour, at the University of Colorado in Boulder, noted that while young adults looked good, the rest of the community was in poor health. Most of the children were "moderately to severely malnourished."[26] Similar conditions—swollen stomachs, painfully thin limbs—can be seen in Napoleon Chagnon's 1971 ID photos of Mishimishimabowei-teri's women and children.[27]

The anthropologist Kim Hill of the University of New Mexico told me the Yanomami have one of the lowest age-weight ratios he has measured anywhere in the world.[28] This is not surprising, since not only are the Parima Yanomami the smallest people in the Amazon; they are smaller than any other extant New World group.[29] And the second-smallest group in the Americas is also Yanomami—the nearby Shiriana Yanomami of the Upper Ventuari River. In fact, the Parima and Shiriana Yanomami are both smaller than several Old World pygmy populations.[30] It may be that only the Ituri pygmies of Central Africa, generally considered the world's tiniest people, are smaller than the Parima Yanomami. Even this is uncertain, however. The data of the anthropologist Colin Turnbull, the Ituri's best-known chronicler, shows that the Parima Yanomami are actually smaller, when male and female differences are averaged out, though by a microscopic measure (140.75 cm for the Parima Yanomami compared with 140.80 cm for the Ituri pygmies).[31]

Many scientists believe that the African pygmies possess a genetic limit on size that remains basically unaffected by calories, fat, and protein. But it is doubtful whether this is true of Amerindians in general or of the Yanomami in particular. Geographic location is the best overall predictor of size among American Indians. Tribes in the northwestern Amazon are generally smaller than others, while tribes in the Venezuelan highlands (Yanomami, Yupa, and Ayamanes) are the smallest of all. Moreover, anthropometrical measurements, rather than genetic traits, are the most accurate predictors of Yanomami village and subgroup membership.[32]

In the early 1970s, Father Luis Cocco measured 287 Yanomami at the Ocamo mission, who had been raised on the main course of the Orinoco with a more varied diet of fish, manioc, and rice. He found men averaged five feet one inch (156 cm) tall and women four feet ten inches (148 cm).[33] The men at Ocamo today are 4.3 inches taller than those in the Parima heartland. Nevertheless, they are tied for shortest in the Amazon when compared with the men of ten other tribes. The following chart shows the disparity:[34]

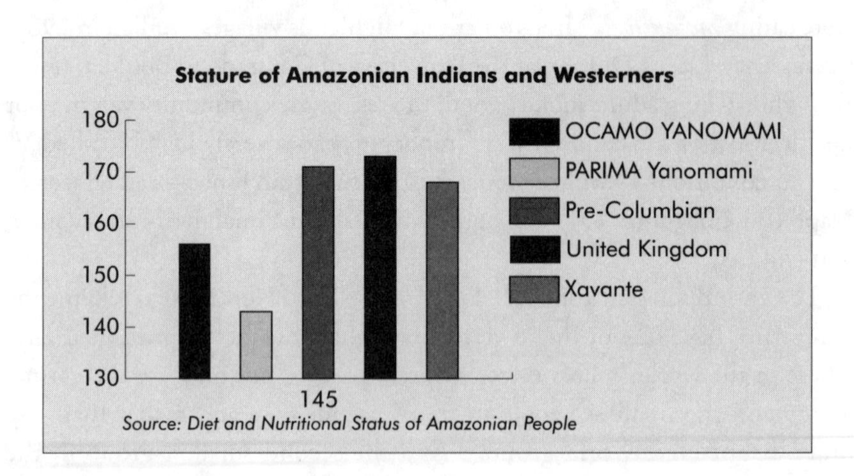

Source: *Diet and Nutritional Status of Amazonian People*

These facts have to be taken together with a broad spectrum of information suggesting that the Amazonian Indians have suffered a traumatic collapse in health since the Conquest. Anna Roosevelt, an archaeologist who has spent decades arriving at a new synthesis of Amazonian history, has shown that Amazonian Indians have a much longer—and varied—past than previously imagined. Their occupation of the region goes back to 10,000–12,000 years before present. Previous orthodoxy held that Amerindian groups were just crossing the Bering Strait 10,000–12,000 years ago and did not reach the Amazon until thousands of years later. Remarkably enough, these tribal groups a hundred centuries ago were "technologically and aesthetically complex," manufacturing fine stone points and leaving a corpus of rock art. By 6,000 years ago, sophisticated ceramics appeared in the Amazon—"decorated with curvilinear and punctate incised designs"—before they were apparently known in the Andes, suggesting that the flow of cultural influence, in this instance, went from the rain forest to the highlands, the reverse of what we were taught until recently. About 2,000 years ago, large settlements, some of them veritable cities, arose across a wide region of the rain forest, among the most impressive sites being the mouth of the Amazon and the Amazon regions of Bolivia and Ecuador. These societies relied upon intensive cultivation of maize, and other crops largely abandoned today, and were organized by political and military elites unlike those of existing tribes. People living in these complex chieftainships were not only taller than any Amazonian Indians today; they were taller than Europeans of the same period. In fact, the precontact Amazonians, the available skeletal evidence suggests, were equal in height to people in the United States today. They also showed fewer bone pathologies than did Indians after European contact. Similarly, lower-class Amazonian descendants of European colonists today are, on the average, as small as the local Indians. "Thus, comparable aspects of Amazonian Indians' development and physique may well be physiological adaptations to their present ecological and economic marginality rather than genetically determined adaptations to their environment."[35]

Strangely enough, none of these arguments from forensic medicine or archaeology entered into what became known as "the Great Protein Debate."[36] It was fought without fear or pity, starting in the mid-1970s, in a series of campus encounters between Napoleon Chagnon and the prominent anthropologist Marvin Harris, who was then at Columbia University. It was a great road show, and, if it occasionally sank to the intellectual equivalent of mud wrestling, the Great Protein Debate also raised some classic questions about war and society. In his Ph.D. thesis, Chagnon presented Yanomami warfare as a triangle of forces. First, female infanticide created a shortage of women, which then intensified competition over mates, sparking warfare that obli-

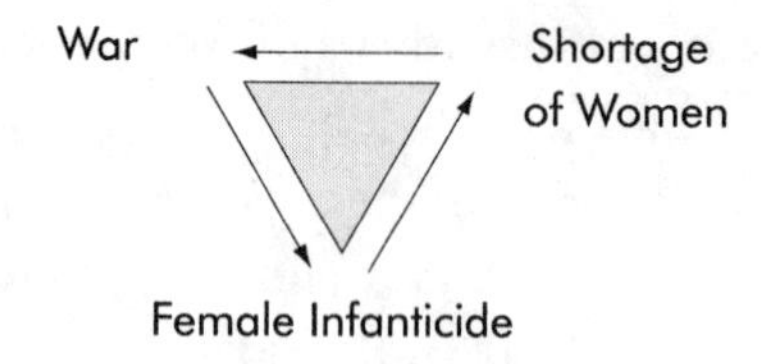

gated villages to raise as many male warriors as possible—thus reinforcing the practice of female infanticide.[37]

But why would the Yanomami kill off their baby girls when women were the very resource men then risked their lives for? Chagnon claimed that Yanomami parents preferred to kill off infants in order to avoid postpartum celibacy, a cultural requirement. Harris cited Chagnon's own history of Yanomami military-demographic expansion—based on an agricultural revolution brought on by plantain cultivation and secondhand steel axes—to situate the Yanomami within a well-known pattern of warfare over new territory.[38] According to Harris, the expansion had reached its limits, and the fighting that Chagnon witnessed was the result of exhausted hunting territories. "As meat becomes scarcer, internal conflicts increase (including those involving women); villages fission; the factions become enemies, spread apart, and create no-man's lands where game finds refuge. Coupled with the practice of female infanticide, Yanomami warfare may be viewed as a part of a system that slows population growth and protects game supplies."[39]

What passed almost unnoticed was that Harris's position was actually more in accord with Darwinian theory than Chagnon's was. For the first ten years of his Yanomami scholarship, Chagnon maintained that parents murdered their baby daughters simply to circumvent lactation taboos that prohibited sex for several years after birth.[40] This was the most *anti*-Darwinian interpretation of infanticide that anyone had ever proposed, as two of Chagnon's fellow sociobiologists, Martin Daly and Margot Wilson, pointed out. "In this one remarkable case, Yanomamo parents reputedly sacrifice the realized fitness of a child for the hedonic goal of mere copulation. It is ironic that this 'least sociobiological' rationale should have been recorded by Napoleon Chagnon. . . ."[41] Daly and Wilson, authors of *Homicide,* also agreed with Harris's basic argument that, if the Yanomami were killing their female babies—and thereby sabotaging their own reproductive expansion—there had to be an environmental constraint.[42] Darwin himself thought infanticide was a common tribal response to scarce resources.[43]

Maybe Chagnon's eccentric theory had some merit. Yet, as soon as he recognized the contradictions his own findings implied, he shifted the terms of the debate. Infanticide disappeared from his explanations of Yanomami conflict. By 1979, he was claiming that the female-male disparities he observed in Yanomami villages were nothing but a natural imbalance in the birth ratio.[44] This last suggestion contradicted James Neel's report that "20–25% of female infants are killed, simply because of their femaleness."[45] Among the Bisaasi-teri, near the Mavaca mission, the incidence of female infanticide was much lower, 10 percent.[46] Among the Mucajaí Yanomami, it was hard to quantify for the precontact period, but an undeniable reality.[47] Pushed from contradiction to contradiction, Chagnon finally closed down the Great Protein Debate by announcing that he could no longer reveal his statistics about infanticide, because he feared that the Yanomami would be prosecuted by the Venezuelan authorities for murder. This has never happened.[48]

For me, the puzzling thing was that Chagnon's theory of infanticide, like his theory of male killers' genetic success, sounded Darwinian. But, just as his *unokai* argument actually turned the fundamentals of sociobiology upside down—with perfectly celibate young men who committed no murders, and older men who became more murderous with each passing decade—so did his link between infanticide and warfare. His proposed equilibrium between copulation and violence invented a mechanical, almost hydraulic model of warfare driven by the frequency of orgasm. This was much closer to the quirky ideas of the psychologist Wilhelm Reich, who believed that the repression of orgasms gave rise to German militarism, than to Darwin.[49]

For both Charles Darwin and Alfred Russell Wallace, food scarcity was a universal constraint that drove the struggle for evolution. Wallace conceived the idea of the survival of the strongest while reading Thomas Malthus's *An Essay on the Principle of Population,* a treatise that proposed the grim principle that population increases geometrically but food production only arithmetically. Malthus's book inspired Darwin's 1838 essay, *Species and Speciation,* and influenced his *Origin of Species.* "Darwin's theory of evolution by natural selection is essentially that, because of the food-scarcity problem described by Malthus, the young born to any species intensely compete for survival."[50]

A tribe with no food constraints and no territorial conflicts over resources that also practiced female infanticide would be something of a wonder, at least from a Darwinian perspective; or from the perspective of resource competition as a universal, "territorial instinct," in Wilson's words. Yet Chagnon has been successful in dismissing Marvin Harris's argument as an arbitrary prejudice of cultural anthropologists. This is how the debate has been represented, from Matt Ridley's scientific best-seller, *The Red Queen:* "For a long time anthropologists insisted that war was fought over scarce material resources, in particular protein, which was often in short supply. Chagnon now believes that the conventional wisdom—people only fight over scarce resources—misses the point. . . . 'Why bother,' he says, 'to fight for mangango nuts when the only point of having mangango nuts is so that you can have women? Why not fight over women?' "[51]

Harris did not dispute that the Yanomami were "fighting over women." He was unique among Chagnon's critics because he accepted what Chagnon said about female infanticide, shortage of females, and warfare to remedy the sex-ratio imbalance. His contribution was to seek a different explanation for this state of war. In doing so, Harris also disputed whether the Yanomami could be considered models for our prehistoric relatives. Pleistocene man lived for eons in small, stable populations where technological change was achingly slow. The Yanomami, as Chagnon described them, were undergoing a revolution: "The Yanomami represent one of the last opportunities to explore the process of tribalization . . . when human populations made a transition from hunting and gathering economy to agriculture, and how the stability and productivity of the new economic order gave rise to a sharp increase in the rate of population growth and, as a consequence, expansion of groups into regions previously unexploited by agricultural techniques."[52]

Between 1800 and 1950, the Yanomami population increased from about 3,000 to 25,000, doubling every half century.[53] By contrast, Pleistocene populations doubled every fifteen thousand years.[54] Evolutionary biologists of Harris's generation believed that prolonged stability in prehistory was achieved through infanticide.[55] Today the biologist Susan Blaffer Hrdy attributes populations stasis among the !Kung of the Kalahari Desert to delayed sexual fertility, which is caused by food shortages. In either case—infanticide or "adolescent sub-fertility" forced by meager rations—the relation of food to fertility, territorial expansion, and warfare cannot be dismissed.[56]

No group can continue doubling its population every fifty years for long. If the Yanomami went on like this, they would soon overwhelm the earth. By 2550, there would be one hundred million Yanomami.

Chagnon cited the population explosion of the Yanomami as proof they were well fed. In a Malthusian model, that would be true. But Herbert Spencer, the greatest Social Darwinist of the nineteenth century, was the first to detect the fallacy of Malthusian assumptions about food, status, and

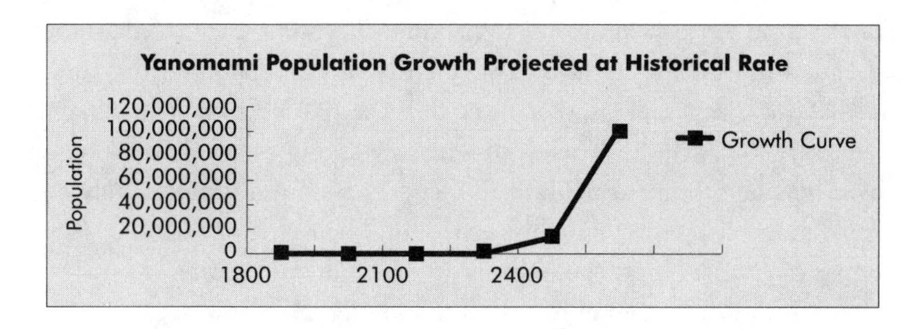

reproduction. Spencer noticed that the Victorian poor reproduced at a far higher rate than the richer, better-fed classes.[57] That has proven to be true all over the world, as with the tripling of India's population in less than three generations. "At first glance, such a finding seems completely antithetical to predictions from evolutionary theory," Blaffer Hrdy noted. "Around the world, there is a tendency for people who are better off to have a lower birthrate."[58] The popular jingle "The rich get richer and the poor get babies" summarizes existing data about birth rates better than Malthusian theory.[59]

But this paradox is nowhere illustrated so clearly as in the case of the Yanomami. (See the appendix: "Mortality at Yanomami villages.") One of my professors at UCLA, the anthropologist Johannes Wilbert, briefly studied the Sanema Yanomami several years before Chagnon arrived in Venezuela. In his book *Survivors of El Dorado,* Wilbert marveled that the Yanomami, with the simplest subsistence agriculture, and in the absence of hierarchical leadership, had nevertheless spread over a vast area of the jungle at the expense of other, better organized, more technically accomplished tribes.[60] The Yanomami's demographic and territorial expansion between 1800 and 1960 was one of the most astonishing in Amerindian history. Although the pre-conquista Yanomami occupied a very extensive region, all of the existing Yanomami are descendants of groups from a relatively small area of the Parima Highlands. In 1800, they, too, seemed destined for extincton.

The Yanomami's subsequent territorial and demographic success underscored one of the principles of coalition theory: weakness can be strength, and strength can be weakness. If the Yanomami had been a highly organized, well-armed military group led by a charismatic chief, they would have provoked immediate opposition and destruction. The last charismatic Amerindian chief in the area, Ajuricaba, proclaimed himself the king of Gran Manoa in 1720. The Portuguese promptly sent an army that crushed his coalition and carried Ajuricaba off in chains.[61] By contrast, the Yanomami expanded in all directions without a central authority. Everywhere they went, they asked for food, steel goods, medicine, clothes.[62] Two airstrips in Yanomami territory, one at Boca Mavaca, Venezuela, and one near Surucucu, in Brazil, were nicknamed "Give me."[63] The Indians' habit of endlessly asking for steel and food annoyed outsiders, but it also disarmed them. These tiny, technologically poor people multiplied eightfold while expanding their territory tenfold. Father Luis Cocco half jokingly spoke about "the Yanomami Empire."[64]

If so, it was an empire of mendicants and migrant workers.

Whole villages uprooted themselves and became indentured servants for neighboring tribes, like the Maquiritare Indians. Today some six thousand Sanema Yanomami live in subordinate relationships with the Maquiritare. Their peculiar symbiosis was explained in 1970 by Chagnon and Neel in the *American Journal of Physical Anthropology:* "They work for them in order to obtain the necessary and extremely desirable steel goods that make their agricultural economy more efficient. . . . Makiritare men (in mixed villages) demand and usually obtain sexual access to Yanomama women."[65]

If the Yanomami could not obtain the goods by barter, work, or giving away women, they sometimes went to war. An Italian explorer, Alfonso Vinci, who visited the Sanema Yanomami in the early 1950s found that the mere rumor of another village's steel wealth—"The tribe on the river so-and-so possesses a machete or a knife"—was enough to trigger an expedition to try and rob it. "It is not certain that a battle will yield the desired utensil or object, but then, it is never certain that a hunt will fill the pot. . . ."[66] A naturalist who visited the same Yanomami groups in the 1960s told of a war that broke out after a migrant worker, who had previously sent annual installments of axes and machetes to his home village, suddenly died of a snakebite. The steel-poor relatives of the dead man then attacked the more prosperous tribe, pilfering "all of the valuable objects and utensils, but not touching the women and children."[67]

The Yanomami were primarily hunting for steel, not game—or women. (Interestingly, the Siapa Yanomami referred to a machete as a "white woman.")[68] Marvin Harris did not fully appreciate the importance of outside wealth in Yanomami migration, but he recognized that their demographic and territorial expansion was not typical of our ancestors and could not be explained by native ferocity. Using Chagnon's own evolutionary logic, he raised questions that Chagnon has never been able to answer.

Yet Chagnon deftly recovered by challenging Harris to a test: they would measure the Yanomami protein consumption at remote, traditional villages in the highlands. If the Yanomami had adequate supplies of protein, Harris would "eat his hat."[69]

Kenneth Good was chosen for the study. He was then a friend and graduate student of Chagnon's. "I was supposed to do the study on animal protein to refute Marvin Harris, but then I found there were huge problems with protein deficiency in the highlands, once you moved away from the big rivers," Good recalled.[70] In his book *Into the Heart,* Good described how Chagnon ordered him not to speak to Harris. Finally, when it became clear that Good and Chagnon had irreconcilable differences, Chagnon said, "But, Ken, tell me, what are you going to do with yourself, go to work in your brother's dental lab? Because you're not going to get into any other anthropology department. I'll see to that."[71]

When Good defected to Harris's camp, Chagnon selected another student, Raymond Hames, to do a protein study at a Maquiritare village named Toki, where two Yanomami families were living. In *Science,* Chagnon and Hames reported that the Yanomami at Toki consumed an average of fifty-two grams of protein per day, with two-thirds of it coming from meat, one-third from fish.[72] It was more than enough protein for survival, a great coup. "My data will not change," Chagnon wrote. "He [Harris] refuses to change his protein 'theory'—at least not in one dramatic step. That is called 'eating Crow,' a high-protein diet eschewed by any concerned scholar." He concluded that Harris's resource school "is not science. It is not only 'vulgar Marxism,' it is blind Marxism."[73]

The *Science* article did not convince Harris, because he knew that Toki was not a Yanomami village. Nor had the Yanomami lived on the Padamo for two centuries.[74]

Hames himself told a different story in his Ph.D. thesis. He admitted that the Yanomami had moved into the Padamo Basin only in the 1920s, when they came to rob steel goods from the Maquiritare Indians. But those moves and those wars took place more than a hundred miles upriver from Toki. The Maquiritare did not migrate downriver to Toki until 1969, when a village relocated below major rapids to a navigable stretch of the Padamo to supply Protestant missions with manioc flour and manual labor.[75]

When Hames arrived, in 1976, Toki already had a Christian church, a government co-op, a school, a dispensary, a volleyball court, and a single household of Yanomami who worked for the Maquiritare as servants. Another household of Yanomami, from a different community, lived a kilometer away in the forest. This odd, newly created village of Toki, and its tiny, separate, equally eccentric Yanomami satellites, were what Chagnon and Hames boldly presented as a typical Yanomami village for the readers of *Science.*[76]

The Lower Padamo River differed almost as much from the Yanomami's mountain habitat as the game-rich swamps I traversed near the mountains of Youth. It was not surprising that the Yanomami servants at Toki ate well in the mid-1970s. Lowland areas have more game, and the Padamo River had abundant fish, so abundant that men went out at night, with flashlights and their outboard motors, and killed them with machetes. In the highlands, fish are a negligible source of protein. But things have changed at Toki as well. Today, fish are much harder to catch. And the big game has been hunted out. In fact, today the missionaries on the Padamo are trying to open a game reserve to restock the area.

Presenting the Padamo as a normal protein environment was as far removed from reality as presenting Toki as a normal Yanomami community. Kenneth Good called Toki "a bogus village." He said, "Lizot and I wrote a letter together to *Science.* We don't agree on too much—Lizot is a student of Claude Lévi-Strauss, and I studied with Marvin Harris—but we agreed that this was just outrageous. I got on the phone to the editor and begged her, 'Please. For the sake of science, you have to publish this.' But they didn't publish it. And Lizot and I were the only two anthropologists who were working in the same area as Chagnon at that time—and we'd both lived there for a lot longer than Chagnon had. So if we couldn't get a word in edgewise, who could know what the hell Chagnon was up to?"[77]

The most celebrated moment of Yanomami violence, captured in *The Ax Fight,* was a dispute that started over a few bananas. The most celebrated war, the one in which Helena Valero's husband was ultimately killed, originated in an argument over garden produce.[78] And the most bizarre incident, the destruction of Lizot's village, was also triggered by a refusal to feed Kaobawa, the Bisaasi-teri headman.[79] Food mattered.

Food was also a principal problem for the Atomic Energy Commission's earliest expeditions in the 1960s. Timothy Asch nearly starved to death on his eleven-day trek into the Siapa Highlands in 1968.[80] And Chagnon, in spite of his shotguns, was unable to provide sufficient meat for *The Feast*—"His hunters have done so poorly that he must make the meat go further than it should."[81] (The same failure to provide food led to bad feelings at an aborted film at Reyabowe-teri.)[82] When the geneticists left Patanowa-teri, Chagnon secured a supply of cheese instead of relying on local protein sources. On the unedited tape, Charles Brewer said, "Well, you know I went to do some hunting for them because they were really hungry."[83]

Kenneth Good stubbornly remained in the Siapa Highlands from 1975 to 1983 and did the protein study Chagnon had originally requested. He found that the protein intake of the Siapa Yanomami fluctuated widely, from village to village, and from season to season. When villages got too large, they quickly hunted an area out, even with bows and arrows. The only way to kill enough game was to keep on the move, spending about half the time trekking through the forest, living in *tapiris,* temporary shelters. (At Toki, the two Yanomami families had given up trekking altogether.) Good joined the Yanomami in this nomadic existence. He loved the freedom. But hunger could be a real hardship. By weighing the kills of 1,857 hunts by Hasubuweteri men over an eight-year period (1975–83), Good found an average daily consumption of 24.25 grams of animal protein,[84] half of what the Yanomami at Toki ate and less than a third of what the Maquiritare Indians ate.[85] While Hames conducted a brief study of a hybrid village, Good conducted his protein research over eight years among the relatively undisturbed Siapa villages.[86] I doubt anyone will ever be able to do such a study again, anywhere in the world.

In fact, Harris really won his protein wager with Chagnon. But the results were near the borderline of where Chagnon and Harris placed the protein cutoff. Protein was only part of the picture. Fat was just as important, particularly for children, and some Amazon tribes were critically short of fat.[87] Stomach parasites were also a key factor in any equation of nutrition. Marinho found that every Siapa Yanomami he tested had intestinal parasites; many had seven different types of intestinal parasites. Figuring out how many calories were being consumed by the Yanomami and how many were being shared by their parasitic comensuals was beyond any current calculation.

Chagnon's own calculations on Yanomami food sufficiency have fluctuated, like his data on infanticide. In his Ph.D. thesis, Chagnon related the bellicose nature of the groups he studied to their early participation in an agricultural revolution. He also saw a tight synergy between Yanomami warfare and garden sites because villages without strategic, inaccessible garden retreats could starve during wars. "They acknowledge the relationship between chronic warfare and village relocation by being constantly on the lookout for new lands to settle. Men are always discussing the agricultural merits of the forested areas in which they hunt. . . . Old gardens are desirable sites because they continue to produce peach palm long after they are deserted."[88] He described fights over these old gardens, particularly over peach palm fruit, a source of protein.[89] During his debates with Harris, Chagnon dismissed the possibility of warfare over food resources. After his Siapa expeditions, however, he reversed himself again. He acknowledged that game was scarcer and gardens more difficult to cultivate in the highlands.[90] And, in a recorded talk at the American Anthropological Association on December 2, 1994, Chagnon admitted it was necessary to "supplement or provide to the Yanomami ways of supplementing their nutritional needs, particularly in the form of protein and I hope you don't tell Marvin Harris I said that."[91] The moderator of the panel, the anthropologist Frank Salamone, noted, "Such a position reverses Chagnon's earlier one regarding sociobiological sources of warfare and places him in Marvin Harris's camp. . . ."[92]

Charles Brewer has been in Harris's camp for a long time. Brewer, who has spent more time than anyone else collecting plant and animal species in the Yanomami highlands, told the journalist Redmond O'Hanlon, "There's not much food in these forests. So when times are hard they kill the new-born girls; so there are never enough women to go around; so they fight over them."[93]

Food *was* scarcer in the highlands, as Brewer said (and as Chagnon eventually recognized).[94] Female infanticide was, in fact, practiced at many villages. But so was male infanticide, and the difference between them was, overall, apparently much smaller than the 6:1 ratio (25–30 percent vs. 5 percent) reported by Neel. A sex-ratio imbalance did, however, exist—in *some* places.[95] But there was actually a surplus of women in the Parima Highlands, where the Yanomami had lived the longest and where all present groups originated. The Parima Yanomami were also the tiniest and, presumably, hungriest of all Yanomami groups.[96] If Harris's theory had true predictive power, the Parima villages should have had the most severe sex imbalance and the highest rates of violence. They had neither. Violence was actually much higher at the lowland villages, whose members were bigger and apparently better fed.

The same contradiction held in the Siapa Mountains. Chagnon was initially astonished at how peaceful the Siapa Yanomami were compared with lowland groups. "The most startling difference is the degree to which violence and warfare—and the consequences of these—distinguish highland and lowland groups from each other. Warfare is much more highly developed and chronic in the lowlands."[97] And the higher up in the mountains

Chagnon went, the more peaceful the villages were. Here is the data from five villages in the heart of the Unturan Mountains:[98]

Mountain Villages	Population	% Abducted Women	% *Unokai* Men
Village 69	53	14.3	7.7
Village 59	71	0.0	21.4
Village 57	81	15	15.4
Village 68	54	14.3	0.0
Village 67	70	0.0	12.5
Total:	329	8.7	11.4

The *shabonos* at the highest elevations had only a fourth as many killers as the lowland groups and half as many abducted women.[99] Moreover, each *unokai* participated in fewer killing events—.96 in the highlands compared with 1.13 in the lowlands. *Unokai* events, however, were far more numerous than actual, physical deaths, as I showed in chapter 10. Again, there is no way to translate these statistics into actual deaths until Chagnon divulges more of his data. It is clear, however, that the Siapa mountain villages differ so much from "the Fierce People" that they appear to belong to a separate tribe altogether. This is what the geographer William Smole argued a generation ago when he compared the Parima Yanomami to the Bisaasi-teri.[100]

In Yanomami history, war has always worsened the problem of hunger. During war, hunters stayed closer to home and women worked less frequently in their gardens—only when men protected them. There was a telling moment in the life of Fusiwe, Helena Valero's husband, when his father advised him to avoid war.

> Oh my son, you must not shoot. You have two male children; one is growing up, the other has only recently appeared. Why do you think of killing? Do you think that killing is a joke? If you kill today, tomorrow your sons will be alone and abandoned. When a man kills, he often has to flee far off, leaving his children behind him, who weep for hunger. Do you not yet know this? I know it, because I am old. When we used to live on the other side of the great river, we used to fight with the Kunatateri, who had killed one of us. We fled, carrying with us the bones of the man who had been killed; on the journey we found no bananas and so much time passed before we could prepare the banana pap for the ashes. We ate only *inajá,* the

> fruit of the *balata:* sometimes not even that. The children wept and I wept for the pain of seeing my sons go hungry. Your father wept with me and you yourself, who were than a child, used to weep for hunger. So would you now wish to do the same?[101]

These were not words of advice from someone living in protein heaven. But they were also not words suggesting that hunger drove people toward war. Chagnon is correct when he says there is an inverse relationship between food resources and war among the Yanomami. For his part, Harris has admitted that, if he had known about the impact of steel tools on Yanomami warfare back in the 1970s, he never would have proposed his protein thesis. Today, he believes that most Yanomami wars, including the ones Chagnon witnessed, were caused by competition for sources of steel goods. Ferguson's research convinced Harris that the Yanomami were fighting over access to Westerners, particularly anthropologists.[102]

All researchers agreed that the Siapa Yanomami were more peaceful than the emigrés along the banks of the Orinoco and Mavaca. "Here we found only fear, timidity, discretion, even in circumstances where the proportion of forces was manifestly unfavorable toward us," Lizot observed on his visit to the Siapa valley in 1973. "And the reserved attitude toward us was to prove constant: we never felt threatened. Rather, we were the ones who inspired fear."[103]

So it was unusual that the Mokarita-teri described a war with their neighbors in which three people were killed. In fact, the Mokarita-teri had a strict border with Hokomapiwe-teri, something that Kenneth Good had never experienced in his years of trekking between villages.

Marco wanted to turn back on September 9, five days into our trek, because of these rumors of wars, and agreed to continue only when I quintupled his salary. After leaving Mokarita-teri's beguiling valley, we followed a path along a creek strewn with granite bolders. Although we walked with the sun at its zenith, we were in a blue-green darkness all afternoon. (The bluish pigment of the foliage helped capture yellow light for photosynthesis.) We rested at a *tapiri,* a temporary shelter of palm leaves hastily thrown up by Hokomapiwe-teri warriors. They had slept here, I was told, before raiding Mokarita-teri. According to the Mokarita-teri, the war began when the Hokomapiwe-teri refused to honor an obligation to give dogs to the Mokarita-teri. Angered, the Mokarita-teri attacked Hokomapiwe-teri, killing one man with arrows. The Hokomapiwe-teri, who knew the terrain better than the raiders, gave chase, cut them off, and killed one of the raiders, also with arrows. "This could have happened a thousand years ago," Marco said, after translating.[104]

A dog is an invaluable aid in hunting and one of the most desirable trade items throughout the Yanomami world. Prior to FUNDAFACI's arrival, villages with easy access to metal goods typically traded old machetes to remote communities in exchange for dogs, a pattern documented by Kenneth Good in the Siapa Highlands. "A machete acquired for a dog is used for a while and then traded to another village for another dog. That village will in turn use the machete for an additional period of time and re-trade it to a more distant village and replace the dog it had given."[105] But once FUNDAFACI was supplying these remote villages with armloads of new machetes, they had no need of giving up valuable dogs to groups with old tools. Chagnon reported that one of the reasons for the new wars was remote villages' "failing to deliver dogs as promised in an earlier trade."[106] It made sense that the Mokarita-teri, who had been promised dogs from a more interior village, would feel cheated when their trading partners found better terms of exchange after the FUNDAFACI windfall.

Similar disputes erupted throughout the region, one of them between Narimobowei-teri and Doshamosha-teri, the most distant villages we planned to visit. I learned that their dispute started when FUNDAFACI moved its camp from Doshamosha-teri to Narimobowei-teri, bestowing goods on the latter and angering the Doshamosha-teri, who raided them.[107] Until this time, all the visitors to Narimobowei-teri—Helena Valero in the 1930s and 1940s,[108] Jacques Lizot in 1973,[109] and Kenneth Good from 1977 to 1986[110]—had reported friendly encounters with the Narimobowei-teri themselves and peaceful relations between neighboring villages. We were not sure what to expect.

Marinho moved to the head of our party and began to lead the march through streams and swamps, where we were immersed in water up to our arms. It was a tough slog in places, because the granite or sandstone footing was especially tricky underwater. Fortunately, our teenage guides directed us to an idyllic camp spot for our first night—a small island above a waterfall. We dived into a deep pool just below the falls and swam under a clear sky, so clear we did not bother with setting up a shelter. We just swung our hammocks before the sun quickly went out. Marinho kept his fishing lines going under starlight, but he caught only shadows.

Marinho, who was born on the Rio Negro in 1968, of mixed Indian,

African, and Portuguese descent, was just an infant when his mother left home. His father raised him. After finishing high school, he rose to the rank of sergeant in the army, where he won an award for saving a boy who had gotten lost in the jungle. After leaving the service, he moved to the booming town of Boa Vista, in the northernmost Brazilian Amazon. And, after years without any contact, Marinho found his mother, and two teenage half sisters, in the same improvised suburb where he set up a sort of house.

"There's nothing I'd rather do than wander through the forests with the Yanomami," he said.[111]

We reached Hokomapiwe-teri after two days of hiking. It was empty, and there were signs of a recent cremation ceremony in the central plaza. Hokomapiwe-teri means "the place where the oca grows." We rummaged for this root crop, which grew in a large, producing garden just beyond the *shabono* walls. We quickly cooked and ravenously ate the starchy oca, which proved to be about as satisfying as plantains. While eating, I made the mistake of taking off my shoes and socks, and discovered that the sandy soil of Hokomapiwe-teri was an ideal breeding ground for large sand fleas, called *hiho.* Although I had my shoes off only for a few minutes, the fleas had migrated under my toenails. We all had them, and they had to be removed, one by one, with a needle. It was painful, but not as painful as allowing the fleas to hatch a brood of babies under your toenails, which invariably brought infection and made walking excruciating. Marinho recalled that, in the Parima Mountains, he had treated a Yanomami orphan boy who had so many fleas that he had become partly crippled; they just kept reproducing and expanding their territory until they had literally eaten away part of his foot.

From Hokomapiwe-teri—a village not mentioned in any of the Siapa literature—we hiked over a 3,000-foot pass in the Unturan Mountains. From the summits, we glimpsed Narimobowe-teri, a *shabono* much larger than any we had visited so far, sitting in a broad valley next to a small stream. On our way down, we passed FUNDAFACI's helicopter landing spot, and were spotted ourselves by an old woman in a plantain garden, who ran ahead to sound the alarm. Soon we heard a confusion of whistles and shouts, and the boys from Hasupuwe-teri insisted that I enter the village first.

There was a small problem. The Narimobowe-teri claimed that the Hasupuwe-teri owed them some dogs; the Hasupuwe-teri said it was the other way around.

The young Hasupuwe-teri, the *huyas,* were all excited. They were nervous and scared and proud that they were on this adventure so far from home.

They delicately painted each other's faces with red *onato* dye and bits of charcoal they had brought for the occasion. They reviewed each other's artistry, standing under rays of gold coming down through chinks in the leaves. With their red and black faces (and useless bows and arrows), they struck warrior poses—as though they were primping for the warrior prom. They also expressed disappointment in my appearance. Unlike Chagnon, I did not paint up, and they said I looked *shami,* dirty. In fact, I looked filthy. My clothes were torn and soaked, and hung limply on me like a vagabond's. So I went ahead alone, feeling extremely tired and a little vulnerable amid all the shouting and choreographing, wishing my guides had settled their possible deficit in the dog department.

After I crawled into their *shabono,* I found myself facing men with drawn arrows. It was a new experience for me—and it did not look as if most of the Narimobowei-teri had much practice at it either. They were still at war with Doshamosha-teri. When they saw I was not a Yanomami warrior, they immediately let their arrows go slack and gave huge, audible sighs of relief. *Oh, it's only you.* Even without the benefit of a translator, I understood that they had been hoping I would arrive ever since they had spotted our plane overhead. They had postponed a foraging trek on the chance of our coming.

Following this fire drill, the women came back to the *shabono,* and we were all addressed as "brother-in-law," a sure sign they wanted trade goods from us. One man came over to me, embraced me like a long-lost friend, and, as he kept patting my back, requested my watch, backpack, and even my shirt. Finally, he tugged gently at my beard and said, "I need everything you own, including your beard."[112]

We had arrived.

Chapter 17

Machines That Make Black Magic

For in the beginning the Indians regarded the Spaniards as angels from Heaven. . . . And they committed . . . acts of force and violence and oppression which made the Indians realize these men had not come from Heaven.—*Fray Bartolomé de las Casas*[1]

In 1941, the Venezuelan Yanomami saw their first airplane.[2] It was sent by the Venezuelan government to chart the Amazon-Orinoco divide. At first, the Yanomami thought the flying machine was Omawe, the Creator. Warriors raised their bows and arrows and chanted, *Omawe! Omawe!* Others feared it was a flying ghost. Shamans invoked the spirit of the wind—to blow it away—and the giant anaconda—to tie it up—while raising their arms heavenward to keep the sky from falling. Women put out the fires where they were cooking yams and *ocumo* roots. But when the Yanomami saw the plane disgorge a trail of smoke, they lost hope. "Look! It is dropping smoke from its ass!" They intuited that an epidemic was about to descend upon them. Some retired to their hammocks, where they soon developed fevers and expected death. "Many people wept," Helena Valero recalled. "For those Indians it was the end of the world."[3]

Actually, it was the beginning of the end of the Yanomami's world. However, the planes and their passengers would bring more than simply technological wonders, pathogens, and arms. They also introduced a cosmological dilemma that the Yanomami, as masters of the spirit world, could not avoid and have never solved. Where did these new creatures come from? Who created them? Were they men, after all? And, if so, how had they come to possess such astonishing powers . . . and spread such terrifying diseases?

Although their shamanism is concerned primarily with illness, the Yanomami do not have a conception for infectious disease. The same was true throughout the New World. A Cree Indian who observed a devastating smallpox plague sweep the Canadian wilderness in the eighteenth century remarked, "We had no belief that one man would give it to another, any more than a wounded man could give his wound to another."[4]

The Yanomami have associated epidemics with white people for over half a century. Fusiwe, Helena Valero's first husband, introduced respiratory epidemics visiting camps left by U.S. Army engineers. After many of his relatives had died, he said, "Last night I dreamed of so many white men, all clothed and with a cloak over them; when they shook the hood, smoke came out, and that smoke entered into us. When the whites undress, they leave the illness in their clothes. . . . White men cause illnesses; if the whites had never existed, diseases would never have existed either." Helena, for her part, commented, "Perhaps he was partly right. I have lived with them so long; I almost died under their arrows, but I never had any illness, never a headache. I have known illnesses only after I returned to the white men."[5]

Fusiwe's vision of deadly smoke coming from the personal possessions of white people was reexperienced and elaborated upon by other shamans during the Brazilian gold rush. In 1990, I interviewed the University of Paris anthropologist Bruce Albert at a Yanomami village in Brazil on the southern flanks of the Tapirapecó Mountains, a high range visible across the Siapa valley. Albert had written an article, "The Smoke of Metals," about a new religious movement among the Brazilian Yanomami.[6] Shamans began rallying communities against the gold rush, inspired by an apocalyptic perspective on gold mining created at the village of Paapiu. "What we're seeing is embryonic prophecy, the way new myths are made," he explained. "I visited Paapiu in 1989 and saw the devastation there, and how the few Yanomami survivors tried to account for what had happened. For them, the two most striking things about the *garimpeiros* were their tools—including machines and airplanes—and their illnesses. So the shamans began asking themselves, 'Who are these white people, with such powerful, fantastic, tools, and such terrible sicknesses?'

"At Paapiu, the Yanomami developed their own interpretation of the mining machinery—you know, the pumps and dredges that proliferate like infestations everywhere. They started calling these machines *maquinari a ne wakeshibi.* This means 'the machines that make sorcery.' They believe that the sicknesses, which killed almost all the Indians at Paapiu, arose from the smoke of the machines. They now perceive a web of destruction—from gold in the *garimpo* to metal in factories—a huge conspiracy of black magic made by whites to transform objects into venomous, taboo substances. It's the end of the world by smoke. The Yanomami call this smoke *xawara*—fumes which rise, expand to the sky, and cause the sky to fall, destroying the whole earth. This is a new development, a way their shamanism is responding to the crisis brought by the *garimpo,* similar, in some ways, to the Ghost Dance revival of the late nineteenth century by the Indians on America's Great Plains."[7]

In the Siapa Highlands, there were no gold miners, but the FUNDAFACI expeditions led to a similar crisis, in which wars and epidemics spread from uncontrolled contact by outsiders who possessed marvelous, terrible machinery. To start with, four villages had their roofs blown off in the tornado-like descents of the expedition helicopters.[8]

Narimobowei-teri had its roof blown away in August 1990. Napoleon Chagnon described this experience for *Santa Barbara Magazine.*

> A few feet away from landing, we aborted when we saw the leaves of their roofs being blown away by the chopper's downblast. We saw people fleeing in terror and men throwing sticks and stones at us as we retreated up and away. . . . This visit was the dream-come-true of my professional life—an expedition funded by the University of California, Santa Barbara, and FUNDAFACI, a non-profit organization influential in the affairs of the Venezuelan government, into the heart of the remote, untouched, region of the River of Parakeets.[9]

The Narimobowei-teri headman, Waupuruwe, recalled how, after Chagnon had moved on to the nearby *shabono* of Waborawa-teri, some of the Narimobowei-teri followed him there, asking for gifts. It was after this trek to FUNDAFACI's Waborawa-teri camp that, on returning home, many of the Narimabowei-teri became sick and died. Opinions about the cause varied. Waupuruwe said, "Shaki made *xawara* to kill us. We do not want him to return."[10] Waupuruwe lost a brother-in-law, a brother, and a small child.

The account of Chagnon cursing a village with death was part of a tribal interpretation that went back long before Chagnon, but had come to include him as a personification of outside contagion. I heard a variation of the headman's story about Chagnon's curse from a Narimobowei-teri boy while we were camping at a temporary shelter in the forest, far from any *shabono,* on September 16, 1996. In the twilight, I caught the words Shaki and *xawara*—Chagnon and disease—and asked Marco, the Mahekoto-teri translator, what it was about. We got the following tale, from a young man in his twenties named Hetoyaw: "When Shaki arrived, we didn't yet have *xawara.* After he came, the serious illnesses [*hariri*] started. Shaki made a *hehohi* [revolver], and then he shot the pistol into the air to frighten us. He said, 'This pistol brings *xawara,* and now you will have to wash that child.' I saw this."[11]

Were there any facts that could be salvaged from this incipient legend? To me, it sounded like a mixed-up regurgitation of Chagnon's by now regional reputation for shooting his gun off and spreading disease, but adapted to a local and largely fictional setting crafted to explain an increase in childhood illness. At a nearby village, a young shaman, Yanowe, gave a different account: "When I was in Narimobowei-teri, I saw that those of Narimobowei-teri were at war with those of Doshamosha-teri. So those of Narimobowei-teri asked Shaki to fire off his pistol to frighten those of Doshamosha-teri. Then he fired into the air."[12]

As far as I could see, these stories contained two kernels of truth. One was the estrangement between Doshamosha-teri and Narimobowei-teri following the FUNDAFACI expeditions, an estrangement caused partly by the competition for trade goods. The other was the epidemic that followed soon after the Narimobowei-teri trekked to Chagnon's camp at Waborawa-teri.

Brewer's expedition log, which covered all of FUNDAFACI's helicopter trips from August 1990 to September 1991, confirmed one part of Waupuruwe's story. Chagnon did invite the Narimobowei-teri to the nearby village of Waborawa-teri, between February 18 and March 6, 1991.[13] And it was a regular practice of FUNDAFACI to use one *shabono* as a base of operations to which whole villages from the nearby region were invited to come and be studied. Since the epidemic profile of groups varies so greatly, and the sudden influx of goods is so unsettling, mixing up entire villages has historically enhanced both the risk of disease and the risk of war.

Narimobowei-teri and Doshamosha-teri ended their trading and feasting relations, which had been steady for several decades. A fence surrounded the Narimobowei-teri village, and the only "door" was a tiny opening that you had to crawl under. At night, the opening was closed, and people relieved themselves in the center of the *shabono,* a sign of how nervous they were about enemy raiders.

The morning after we arrived, Waupuruwe sang a speech warning about the "Caraca-teri"—the people of Caracas—who were known to be prowling around the village at night, arriving by invisible helicopters, coming to bring diseases and steal the Yanomami's goods. He called these Carace-teri *waikas,* literally "killers," a term traditionally applied only to other Yanomami. Curiously, he imitated the Caraca-teri *waikas'* speech, sounding like a gringo trying to speak Spanish.

Marco and the boys from Hasupuwe-teri were mystified by all this. They had never heard of *waikas* from Caracas. As the song and story unfolded, it turned out that the invisible helicopters contained a motley crew. *Americanos* and *garimpeiros* accompanied the Caraca-teri, plotting dastardly attacks together. Since no gold miners or Americans or people from Caracas have actually been to the area since Chagnon's trips, it sounded like a garbled recollection of FUNDAFACI's descent from the sky.

This new form of shamanism was also noted by the Spanish anthropologist Javier Carrera, who spent four years among the Yanomami and revisited Narimobowei-teri in February 1998, seventeen months after our trek. "When they speak of the *waikas,* they speak in Yanomami, but it's as if they were trying to imitate our style of speech," he said. "This is something that is unique to the Siapa area, where there is a lot of it. Everybody is talking about it. It seems to me that it is a mythological reinterpretation of the arrival of the *nabah* [outsiders] among the isolated communities that had the very special contact of Chagnon with his helicopters."[14]

The Narimobowei-teri refused to accompany us to Doshamosha-teri, and they persuaded my guide Marco to go on strike, too. Then came word that an ally of Narimobowei-teri, Toobatotoi-teri, had been hit by an epidemic and needed emergency assistance. So Marinho, Marco, and I decided to trek into higher country to the northwest, where the Toobatotoi-teri lived in a mountaintop hideaway.

The headman, Waupuruwe, ran after us. "If you know someone who has a helicopter," he said in parting, "tell him I want to see him."[15]

We followed the stream as it meandered northwest through marshes and small palm forests. Birds were more plentiful, and we saw deer tracks. At dusk, we reached a lovely *shabono* in a clearing close to 2,500 feet above sea level, where there were signs of a recent shamanic initiation—bits of jaguar skin and a feather armband were wrapped around a pole where the initiation took place. "No one will touch these," Marco said.[16]

We were once again straddling the Amazon-Orinoco divide. All the streams to our north drained toward the Orinoco; but we were camped next to a small creek that emptied into the Siapa River, an Amazon tributary. On FUNDAFACI's map of the highlands, this *shabono* was labeled "Unknown."[17]

It was cold that night, but there was a good feeling in the air. Shamanic initiations are epic events among the Yanomami. If there is one art at which the Yanomami excel any people I have met, it is in the realm of the spirit. When a man becomes a *shaburi,* he fasts on a liquid plantain soup, while participating in marathon chanting and hallucinogenic drug taking, all under the direction of older healers. But the final initiation is a communal one. Everyone in the village directs his or her energy, which the Yanomami perceive radiating out of their hands and toes, toward the initiation pole, where animal skins represent spirits that the young shaman will welcome into his chest. The jaguar is a powerful *noreshi,* alter ego, which many shamans incorporate. Some shamans have more than one such spirit. Fusiwe, Helena Valero's husband, had a jaguar spirit and a howler-monkey spirit. In a mysterious way, the Yanomami consider themselves to be animal spirits *(noreshi),* nature sprites *(hekura),* and humans, all at once. For them, this tripartite being is evident on the level of perception. They sing until they actually see the bright spirits coming to them like showers of sparks, or clouds of butterflies, which take up residence in their hearts. The Yanomami are very confident of their knowledge of the spirit world, as confident as they are of our ignorance in that realm.

Even for someone who does not participate in the shamanic rituals, the Yanomami spirit world can be overwhelming. One night, Marinho woke me up to say that he could see the spirits of Yanomami around us, lying in hammocks, laughing at us.

After a chilly night around the initiation pole, we met three men from Toobatotoi-teri's main *shabono* who had come to gather plantains from the garden. They led us down another stream, into an abrupt, rocky valley. On our way down, a deer bounded across our path. Shortly afterward, we entered a small, recently cleared opening in the forest. I thought it was the initial stage of a garden, but learned that it was actually a landing pad for the *waikas.*

"*Americanos* land here—*boom.* Caraca-teri land—*boom.* Brazileros—*boom,*" one of the men explained. These people from America, Caracas, and Brazil were all coming down from the sky, at night. The clearing was far too small for a real heliport. Actually, the Toobatotoi-teri had opened this place up themselves in hopes of attracting the *Americanos,* Caraca-teri, and their compatriots. The shamans were calling to the helicopter owners, inviting them to set down.

It was a trap. Their strategy was to attract the *waikas* out of the sky and then to scare them after they had unloaded their precious cargoes of goods—but before they could spread diseases and confusion. It was a cargo cult, like the famous Fijian ritual airstrips built by Pacific islanders to attract departed warplanes after World War II.[18] But the Siapa cult had a distinctive twist: the Yanomami wanted the cargo without its owners.

One of the shamans invoking the *waikas* was the newly initiated young man Yanowe. He was dressed in a shaman's monkey cap, with a macaw-feather armband and large parakeet-feather earings. He painted red serpents all over his body before asking an older shaman to blow hallucinogenic snuff up his nose with a long tube. As the drug began to change his awareness, he started humming and chanting, rocking slowly, vomiting slightly, then blowing large bubbles through his vomit. Now in the spirit world, he began a graceful dance, gliding like a hawk, in circles, as he sang to the owners of the helicopters.

Yanowe told us about the mysterious night flyers. "We invoke these spirits, and they land in the middle of the *shabono*—Venezuelans, *garimpeiros,* and dead Yanomami also. Even the Yanomami dead come dressed in foreign clothing, and if they give this clothing away, we get skin diseases—*chivakoa.* The spirits say, 'We are spirits but neither Yanomami nor *nabah.*' They have Yanomami ornaments but *criollo* clothing; they dance and sing like Venezuelans. 'We are Venezuelan,' they say. 'If you want to listen to us sing, then listen.' These *waikas* appeared only after Shaki visited everyone. Before Shaki, there weren't any. The Yanomami at Braorewa-teri [a village south of the Siapa River visited by FUNDAFACI between January 8 and 12, 1991] saw these *waikas* first; afterward, everyone saw them."[19]

Beyond the *waikas* landing pad, where strange encounters of a fourth kind were taking place, we went up a steep ravine, where we found a *shabono* perched on a rock ledge—at once the safest and most precarious place I had ever seen for a Yanomami round house. It took us twenty minutes to negotiate the cliff to what Napoleon Chagnon described as "the last uncontacted group in this region."[20]

It was a maximum-security site because the Toobatotoi-teri were at war with the Mavaca villages of Washewa-teri and Hiomita-teri, which had both received FUNDAFACI expeditions. Originally, Bokoramo, Chagnon's Siapa guide, attacked Toobatotoi-teri with allies from Washewa, angry over the usual failure to deliver dogs.[21]

Later, the Toobatotoi-teri also fell out with the Hiomita-teri. The headman of Toobatotoi-teri said, "The dispute began when we went to visit Shaki at Hiomita and the Hiomita-teri began yelling at us, 'Why have you come here to get Shaki's presents? Go away.' Later the Hiomita-teri warned them they would kill any of us from Toobatotoi-teri who tried to trade directly with SUYAO. That's why I am living here so far away."[22]

Brewer's record showed that the Toobatotoi-teri were invited to visit Hiomita-teri between February 8 and 15, 1991. Brewer explained that FUNDAFACI created a camp there "in order to study the inhabitants of the area and also the Shapono of Toobatotoi-teri."[23] According to Alberto Karakawe, Chagnon's closest collaborator, the *shabono* at Hiomita-teri was also accidentally destroyed when Chagnon decided to land there against Karakawe's advice. "When he landed they wanted to blow up the helicopter," Karakawe told the anthropologist Frank Salamone. ". . . The Yanomami were prohibiting landing because they think strangers will bring illness—malaria, measles."[24] In fact, a double epidemic of malaria and severe dysentery struck at the time.[25] The FUNDAFACI scientists were able to smooth over fears of disease with gifts, but they were not able to control the competition for those gifts.

Our only problem at Toobatotoi-teri was leaving. One man demanded that I open my backpack and give him everything inside it. "The foreigners from other places gave us everything," he said, recalling Chagnon's generosity.[26] It could have gotten ugly, but fortunately the elders at Toobatotoi-teri told him to leave us alone. He would not give up completely, though. Even as we crawled out the small opening in the fence, he kept repeating that the other villages near the Orinoco did not need my things as much as they did. That hurt because it was true, but I needed my clothes and backpack, too.

The headman waved good-bye affectionately and said, rather wistfully, that if I had been a sky-riding *waika* they would have been justified in stealing all my stuff.

We wanted to get back to the Orinoco, which was twenty-seven miles north of us.[27] We walked a hundred and fifty miles to get there. The Yanomami are not heavily invested in straight lines. There was no point in going straight when you could zigzag, or make an enormous circle, or make an enormous circle *while* zigzagging—which is what we ended up doing.

Villages are the main geographic boundaries on the Yanomami horizon. When I asked if we could go directly to Patanowa-teri, our final destination by the Orinoco, I was told, "You can't get there from here." We had to follow paths that led from one ally to the next, beyond which villages did not trespass except at the risk of confrontation.

We began by trekking for two days to the village of Ashidowa-teri, southeast of Toobatotoi-teri. We were zigzagging through the mountains, leaving streams that emptied into the Siapa River and reaching the headwaters of the Shanishani River, which joins the Orinoco. The Shanishani started off as a series of waterfalls pouring over giant granite boulders in the mountains. There were places where the torrent rushed through sandstone channels, creating delicious natural water slides, but we had no time to enjoy them. Our guides urged us on, into a marshy valley where we camped for a night under palm-thatch lean-tos that we repaired from a previous camp. The Shanishani broadened and deepened, and we began winding through a plethora of old gardens, which made the going miserable. At noon, we came across the place where FUNDAFACI had originally visited Ashidowa-teri, a now abandoned village and garden site that was densely overgrown with thornbushes.[28] We crawled on hands and feet under the cruel spines for two hours to advance a few hundred yards.

We kept going, from garden to garden and, after the sun had set, finally to a circle of campfires where the Ashidowa-teri were living.

We needed to rest. Marinho had been bitten by a "twenty-four-hour ant," so named because its excruciatingly painful bite lasted twenty-four hours or more. His leg had ballooned up, and he was limping badly. All of us had foot infections, mostly from the carnivorous fleas, but also from various cuts and scrapes that had gotten dirty and never healed. My feet were bleeding through my socks.

Still, we were in excellent shape compared with the Ashidowa-teri. They were the most sickly, unhappy Yanomami group I had seen since the height of the gold rush. As soon as we entered the camp, a man grabbed me, crying, *"Hariri,"* sickness, and placed my hand on his sweating forehead. A teenage girl looked tubercular and her baby skeletal. The girl, who called herself Nape, had eight different kinds of intestinal parasites. Marinho prepared milk for Nape's infant, who was crying continuously because its emaciated mother had none.

The tests in the first two cases of suspected malaria both turned out positive—falciparum. Many people were painted black in mourning, and the children looked severely malnourished. Even some of the adults were so thin that they resembled stick figures.

But at night, under the clear, starry sky, in the light of the circled hearths, the pain seemed to recede. The village looked lovely, romantic—prehistoric. Then, around ten, the singing began. The energy and harmony, the inspired improvisations, were moving, especially in light of the Ashidowa-teri's suffering. Just closing my eyes, I could imagine they were passing on tribal lore, creation stories, from time out of time.

In fact, they were also obsessed with the *waikas* and their spiritual helicopters. They chanted one story about a helicopter that had just came down to a neighboring village. The *nabah* gave clothes for presents—but a woman and girl who received the clothes died of *xawara.* Basically, all these stories were the same story.

Marco translated, "They are singing against the Americans, the *garimpeiros,* and people from Caracas who descend from the sky at night. They say these spirits speak just like you do, and if the Yanomami catch them, they can capture their hammocks and things. I am very worried about this."[29]

People here also seemed to remember Shaki's visit as a kind of giant feast where they were fed, got lots of presents, and did not have to work. The downside was the *xawara*—the smoky, evanescent stuff of disease—which accompanied the white people. At Toobatotoi-teri, the headman said that Chagnon's helicopter had emitted fumes that caused everyone to become sick *(tho shaki keyeha a ha kerini, a ha hayuikuherini pe shawaramou nomatahaa),* much as the Yanomami in the 1940s had done.[30] At Narimobowei-teri, it was Chagnon's smoking pistol that covered the babies with disease-bearing *xawara.*

James Brooke of the *New York Times* offered this account of Chagnon's Ashidowa-teri landing: "After the helicopter descended in a cloud of dust on one garden, a group of 20 excited Yanomami men and boys emerged from the forest. One explained that they had never received non-Yanomami visitors at Ashidowa-teri. . . . [S]everal were startled to discover that there are people in the world who do not speak Yanomami."[31]

In fact, Ashidowa-teri was the closest ally of Helena Valero's village, Patanowa-teri, one that she and her husband visited frequently during the 1930s and 1940s.[32] Later, so did Kenneth Good, who wrote a letter, never published, to the editor of the *New York Times:* "This is now the second article where first contact has been claimed at a village where I have been visiting and researching since 1976."[33] (Previously, he had protested when the *Times* repeated Chagnon's assertions about "first contact" at Narimobowei-teri, a village that Good had visited three times for a total of two months.)[34] However, with his second protest Good sent a copy directly to James Brooke and received a prompt response. "Brooke called me and said, 'How the hell am I supposed to know? He [Chagnon] tells me we're the first people to get there."[35]

Carlos Botto, director of Venezuela's Center for the Investigation and Control of Tropical Disease, traveled to Ashidowa-teri on FUNDAFACI's final Siapa excursion. Botto told me it was one of the strangest experiences of his life. "I went to Ashidowa-teri to conduct a study of the Yanomami during two weeks when Brewer was there. When I got there, Brewer was giving a speech to the anthropologists and the others—about how he was an anthropologist and a naturalist and I don't know what else. He spoke for over an hour about himself; I'd never heard anyone talk so much about himself. He also said that the Yanomami at another village had tried to assassinate him by order of the Platanal mission but that he had knocked one of the Yanomami to the ground. He said he was going to write a book about this. There was a singer with Chagnon's group at Ashidowa-teri . . . a Hollywood star. He was a madman [*un loco*]. He dressed up like a Yanomami and painted himself and went around acting like a witch doctor. It was astonishing."[36]

Botto witnessed yet another helicopter mishap. "I was present at Ashidowa-teri when Chagnon landed in his helicopter and collapsed part of the *shabono.* When the poles of the roof fell, a number of Yanomami were injured and we had to treat them. We had to rescue the people who were buried under the poles and roofing of the *shabono.* It was a serious situation. The shamans and the elders began to practice their chanting because of the collapse of the *shabono.* The expedition left a tragic scar."[37]

One of the headmen at Ashidowa-teri, Mirapewe, recalled, "Shaki destroyed our other *shabono* with his helicopter. A pole ran through my wife's leg, and my brother and sister-in-law were injured on their heads. Shaki promised to return in one month to pay for the *shabono* and for the injury to my wife. I still wonder why Shaki hasn't returned."[38]

The Venezuelan anthropologist America Perdamo was inside the helicopter that partially destroyed Ashidowa-teri. She said several journalists were also aboard.[39] Perdamo headed the Malaria Department for Amazonas State. "The helicopter blew off the roof of the *shabono* when we tried to land," she said. "People came out running, and, as they fled, some of the broken poles fell on top of them. They were struck by the falling poles, and about five were injured—on the head and legs. Some of them had to be stitched up. We spent an hour and a half cleaning their wounds, until the helicopter pilots told us we had to leave in a hurry. The Yanomami objected to the damage done, and the wounded were all crying. It was sad. Of course, I knew this had happened at other places, too. From Ashidowa-teri, they carried botanical samples to Dorita-teri. It was a huge amount, the size of a small room, three square meters. And here in this country we don't know where those samples went."[40]

According to the University of California report on Ashidowa-teri, "The collectors Brian Boom and Charles Brewer-Carías participated on behalf of the New York Botanical Garden, achieving an important collection of ethnobotanical samples."[41] Boom told James Brooke, "For example, there is the toxic Andean nightshade in the gardens here. What is it doing with a people who were supposedly isolated for thousands of years?"[42]

The real question was: What were these scientists doing among a people who had supposedly been isolated for thousands of years? There was a large contingent from the anthropology department of the University of California at Santa Barbara, even though the University of California at Berkeley had been a pioneer institution in documenting the decimation of Native American Indians.[43] And, while the scientists were photographed by the press, Chagnon had the Ashidowa-teri posing with axes and arrows. "Chagnon was getting the Indians to assume very aggressive stances with their weapons, so that he could photograph them in a staged way," said Botto.[44]

Decades earlier, in 1974, Chagnon had concluded, "The Yanomami, like all tribesmen, are doomed."[45] And while many Yanomami remained fatally attracted by free trade goods, they had also evolved a mythology about their relationship with these goods and their owners that amounted to a doomsday prophecy. In the new myths, shamans portrayed Chagnon's scientific legacy as the machinery of black magic. Just as the Brazilian Yanomami defined the gold rush technology as *maquinari a ne wakeshibi*—machines that make sorcery—the Venezuelan Yanomami perceived Chagnon's scientific technology as a huge factory of *xawara,* the deadly smoke of disease. His cameras, guns, helicopters, blood-collecting equipment, and purchase of the names of the dead were ultimately interpreted as magical smoke screens for mass murder.

That became clear to everyone, including Chagnon and Brewer, when they landed at this village, Dorita-teri, on May 17, 1991. This was the scene of the ax fight between the Yanomami headman and Chagnon, filmed, but not broadcast, by Venevisión. Chagnon described what happened when he landed with Brewer and the six-man television crew:

> Harokoiwa began rocking rhythmically sideways, violently denouncing me for killing their babies and causing epidemics among them, slapping his thighs as he rocked from side to side to emphasize his hostile declarations. . . . At this point, some of them, inflamed by Harokoiwa's violence and vitriolic accusations, surged toward me and Charles, but we held our ground: one of them grabbed at Charles' arm, but on noting its strength and our apparent fearless resolve, backed away. . . .[46]

What actually happened was something more dramatic. Another member of the Siapa expeditions, Isaam Madi, described the attack in more detail:

> When Shaki, as he is generally known, entered the *shabono,* he was bitterly abused by the headman of the community, who denounced him for poisoning their spring water and making them sick with his Polaroid photographs, which stole their spirits. In the middle of the argument, the chief's son hurled himself against Chagnon with an ax to strike him on the head. He would have succeeded if it had not been for the timely intervention of Brewer, who, with great agility, intercepted the murderous weapon with his left hand, while knocking down the Yanomamo man with his right.[47]

Brewer simply noted, "Because of our photographs [the Yanomami believed] we had caused all of the illness and were responsible for the deaths of all the Yanomami over the past 20 years!"[48]

It is important to remember that the Dorita-teri are a Patanowa-teri splinter group. They have had a unique relationship with Chagnon. He financed, transported, and filmed a war party that attacked their old village in 1965. Those wars prompted the Patanowa-teri, after three deaths, to abandon their *shabono* near the Orinoco and relocate inland, not far from where Dorita-teri is today. This was where Timothy Asch, acting as a messenger for the AEC expedition, met them and paid them in steel pots to relocate once again, *back* to the Orinoco, for *The Feast, A Multidisciplinary Study,* and the 1968 measles epidemic.* Afterward, the survivors fled again to the inland, where they joined another village in making an effigy of Chagnon and shooting it full of arrows.[49] Chagnon then mobilized the enormous village of Mishimishimabowi-teri to form an alliance with Bisaasi-teri, leading to another film of another feast ending in another joint raid—*against* his former friends at Patanowa-teri.† In 1985, when Chagnon was allowed to return to Yanomamiland after a ten-year absence, Dorita-teri's allies sent a special delegation imploring Chagnon not to visit them.‡ And this was before Chagnon declared war on SUYAO in association with Brewer.

Chagnon and Brewer blamed the ax incident on Alfredo Aherowe, the elected SUYAO leader who accompanied me to Mokarita-teri. Chagnon called Alfredo "a representative of a Salesian mission"[50] because the Mahekoto-teri live at the Platanal mission. But there might have been no link between the Patanowa-teri and the Platanal Yanomami if the AEC had not brokered their alliance for its filmmaking. Alfredo's father, Asinwe, played a prominent role in *The Feast.* Alfredo, who is literate and fluent in Spanish, became a leader of the village through his new skills and his family's support. Unlike César Dimanawa, he is widely liked throughout the region. As the biologist Eibl-Eibesfeldt put it, "It is clear that Chagnon makes things up. Alfredo Aherowe is not a 'Salesian representative,' whatever that may be. . . . He is a respected man among the Yanomami of Mahekoto-teri, those of Patanowa-teri including Dorita, those of Hasupuwe-teri, and those of all the small villages located on the Upper Orinoco. By 1991 he had already been elected representative of the SUYAO . . . a cooperative which does not belong to the mission, but to the Yanomami themselves. Chagnon should know that and not maintain otherwise."[51]

Alfredo readily admitted he had asked the Dorita-teri headman to confront Chagnon, but not to kill him. "I told the captain to destroy Shaki's

*See chapter 6.
†See chapter 7.
‡See chapter 11.

cameras. When I was a boy, Shaki filmed us with the Patanowa-teri, and many of my people died of measles afterward. I was afraid if he filmed us again, we would get measles again." At Platanal, Aherowe also denied Venevisión access to the Mahekoto-teri *shabono.* Alfredo led a shoving match against the television crew and the president's mistress. "Cecilia Matos wanted to film here," said Aherowe. "I told her she could not. We put our hands over the cameras."[52]

Although the headman who accosted Chagnon was not present when I visited Dorita-teri in September 1996, one of the elders who had also confronted Chagnon, Kayopewe, spoke to me about the incident. "Shaki landed on the outskirts of the *shabono.* We didn't allow him to enter, because we told him he was with *xawara* and we didn't want to get sick again. We were afraid to let him in because we were afraid he would cast another spell on us and we would all die off again. That's why we wouldn't allow him to enter the *shabono.* The captain was going to chop the cameras with his ax. He almost chopped Shaki, too, when Shaki tried to enter the village."[53]

Of course, the Yanomami's shamanic vision of Chagnon as the smoking gun of disease and destruction also served to rally communities against him, just as similar myths rallied the Indians against the gold miners. One of Chagnon's Siapa guides, Enrique Lucho, who was also on the helicopter that went to Dorita-teri, explained the ax confrontation in this way: "Shaki wanted to command all of the Yanomami of the Upper Orinoco. Shaki Napoleon built a camp in order to become the chief of all the Yanomami. Since he can speak Yanomami, he wanted to be the chief. We, the Yanomami, got together to defend our territory. We rejected all that plan." According to Lucho, one of the many reasons the Orinoco villages banded together against Chagnon was that the anthropologist planned to control all the government funds destined for the Yanomami.[54]

When Chagnon was first sent to the field, he hoped to create a revolution in anthropology. But a strange thing happened along the way to buying thousands of vials of blood and thousands of names of the Yanomami dead. Chagnon became what the vials of blood promised but never produced: Napoleon himself incarnated the leadership gene. With his feathers and loincloth, his guns and machines, his shamanic drug taking and ritual dancing, Chagnon was, for a while, "the be-all and the end-all."[55]

But his magical aura wore off. César Dimanawa noted, "At first I thought Shaki was good. Then we realized he wasn't good."[56] In the end, the Yanomami concluded Chagnon was simply out to rip them off. He wanted complete control of the films and the blood and the budget, and he intended to give them only crumbs from his rich table. The man who had once incorporated the fearsome Vulture Spirit now landed his helicopter in the middle of *shabonos* in the company of Venezuela's leading *garimpeiro.* Chagnon had turned into the most banal of revolutionaries.

The wilderness had found him out.

Chagnon sought power over the people he studied. "This anthropology is, then, not an anthropology at all but a deformed social science in the service of the engineering sciences of destruction," wrote the anthropologist Jeffrey Rifkin about Chagnon in his article "Ethnography and Ethnocide." "In this context, photography, the numbering of people, and hundreds of Yanomami blood samples become scientifically ritualized inscriptions of the conqueror on the body and society of the subjugated primitive. This is not the study of a murderous people, but social science as rationalized murder."[57]

Chapter 18

Human Products and the Isotope Men

I must tell you that Nature always intends and strives to the perfection of Gold; but many accidents, coming between, change the metals.—*Roger Bacon*[1]

El Dorado was a man before it became a myth. In 1534, Spanish soldiers near Quito, Ecuador, intercepted an "itinerant Indian chief, called El Dorado," who, under questioning, revealed that he lived a twelve days' march to the north. This wandering chief had come south to seek the Inca's help; he was so far on the fringes of empire that he had not received news about the Spanish conquest. The soldiers let him go home, but he was never exactly forgotten. How "El Dorado" became a city, and the city acquired gold-paved streets—then expanded into the capital of an Amazonian empire that lured tens of thousands to their deaths—is a history as fantastic as the transformation of a boy from a poor family in rural Ohio into the director of a worldwide genetics empire.[2]

James Neel, who became known as the Mutation Man, would prove that some human populations—small breeding groups—could mutate much more quickly than conventional theory predicted.[3] That discovery, which came from Neel's inventory of Yanomami genes, was only part of his legacy.[4] A doctor from the Rockefeller Foundation wrote that Neel "has led what amounts to a revolution in medical genetics, whose influence stretches from the clinics of Ann Arbor through the consanguineous villages of Japan, the fava bean fields of the golden crescent, the malarial coasts of Africa, the jungles of the Orinoco, and back again to the amniocentesis centers of the United States."[5]

A decade after I first ventured into the Amazon, I still did not know much about James Neel, apart from the brief sketch of a Horatio Alger life described in his autobiography. I felt that Neel was the key to comprehending the Yanomami tragedy. Without Neel's support, and the prestige of his scientific achievements, Napoleon Chagnon could never have penetrated the bastions of cultural anthropology and popular culture.

But how did Neel choose the Yanomami for this endeavor? And why did the Venezuelans allow him such a free hand for such a long time? I had the Department of Energy's explanation, obtained through the Freedom of Information Act, that the Yanomami had been used as a control for the atomic bomb radiation studies, in order to determine mutation rates. It was logical, in a way—just as it was unethical. But, as each month brought new revelations about the Atomic Energy Commission and its peculiar subworld of human radiation experiments, such simple logic appeared increasingly improbable to me.

On January 2, 2000, the *Los Angeles Times* featured a long review, entitled "Die Hard," of *The Plutonium Files*—a thirteen-year investigation of the AEC and its predecessor, the Manhattan Project, by the Pulitzer Prize–winning journalist Eileen Welsome. She began her odyssey in radiation when she learned, in 1987, that human beings had been injected with plutonium. Plutonium, element 94 in the periodic table, was created in 1941 at the University of California at Berkeley when uranium was bombarded until it fissioned. The new substance was "fiendishly toxic, even in small amounts," according to the scientist who discovered it.[6] As the war reached its conclusion, scientists at Los Alamos were desperate to know exactly how toxic plutonium really was. The experiments devised to answer that question turned out, Welsome found, to be more toxic than anyone could have imagined.

Most of the injections took place at Strong Memorial Hospital. This municipal hospital was under the same roof as the University of Rochester Medical School; both were connected by tunnel to a Manhattan Project annex that employed 350 people in 1945. Stafford Warren, the man whose name has become synonymous with human radiation experiments, started off as chairman of the university's radiology department. In March 1943, he switched to the Rochester Manhattan Project as a consultant. Five months later, in November 1943, Warren was commissioned colonel and promoted to the position of chief of the Manhattan Project Medical Section for the entire United States. The Medical Section sought clear radiation tolerance thresholds by exposing animals to lethal doses of radiation and comparing them to controls. Between 1943 and 1945, animal tests were carried out on an industrial scale. At Rochester alone, scientists sacrificed 100 hamsters, 200 monkeys, 675 dogs, 1,200 rabbits, 20,000 rats, 277,400 mice, and 50 million fruit flies.[7]

But it was not enough. The radiation thresholds remained elusive when Warren was ordered to accompany the military to Japan immediately after V-J day. His team of scientists feverishly prepared to record human and ambient radiation with all available Geiger counters, many of them built at Rochester. For scientists whose subject was radiation biology, the bomb presented a frightful opportunity, like an earthquake vibrating off the Richter scale for seismologists. Warren's successor would refer to the Hiroshima and Nagasaki explosions as giant experiments.[8]

Before leaving for Japan, on September 5, 1945, Warren called a meeting in which he outlined a bold plan for human experimentation. His protocol called for the Manhattan Project to take complete control of a small hospital ward, the Metabolic Unit at Strong Memorial, and use it to inject patients with plutonium, polonium, or uranium. "The purpose of the study is to establish, on a statistically significant number of subjects, the metabolic behavior of the hazardous materials. . . ."[9]

Each patient was given a number: HP 1, HP 2, and so on. HP stood for "human product." Only two people could be experimented on at a time at the Metabolic Unit, where, over a twelve-month period, eleven patients were injected with five micrograms of plutonium each. Five micrograms was five times the dosage the scientists considered harmful. Initially, the doctors had debated doing experiments on their own bodies, but thought the better of it: "plutonium is considered to be sufficiently potentially dangerous to discourage our doing absorption experiments on ourselves." One victim, who was already moribund before injection, died six days after receiving his five micrograms. Surprised at the death, one doctor then suggested increasing the next injection to fifty micrograms, on the suspicion that another Manhattan

Project lab, Chicago's Argonne facility, was getting ahead of Rochester on dosage. "I will see what can be done," Samuel Bassett, the doctor in charge, replied.[10] It appears, however, that such a large dose was never given.

Many of the human products were considered terminal cases, but this was not a requirement. One of the scientists organizing the human experiments commented, "Undoubtedly the selection of subjects will be greatly influenced by what is available."[11] On Christmas Day 1945, it was hard to find patients to fill the Metabolic Unit, but Dr. Bassett did his best. "No one seems to want to be in the hospital on that particular day. I will do what I can, however, to keep the production line going."[12]

According to one of the doctors, a "deliberate decision was made not to inform the patients of the product that was injected."[13] Nevertheless, the patients had to receive "indoctrination" in collecting their own feces and urine, since these samples were scrupulously gathered each twenty-four-hour period.[14] The amount of urine and fecal matter overwhelmed the Los Alamos researchers; they requested less prolific patients. At the same time, the radioactivity of the excreta meant that special precautions had to be taken. All the urine was gathered in mason jars, and each jar was packed inside an individually constructed wooden crate for shipment from the Manhattan Project annex. Warren was so concerned about the details of the blood, stool, and urine packaging that each of the Manhattan Project's ten department heads—eight of whom were affiliated with the university—were instructed to help. It was truly a production line, and it required extensive teamwork to be kept going.

In the original protocol, "Col. Warren proposed Lt. Valentine as the one to do the injections." Since Colonel Warren was the ultimate authority for all medical research surrounding the atomic bomb, his "suggestion" was converted into a working plan that allotted responsibilities. "In charge of wards, Dr. William McKann and Dr. Sam Bassett. Injection, Lt. Valentine."[15]

Lieutenant William Valentine was James Neel's research partner from 1942 to 1945, during which time the two men carried out an acclaimed study of inherited anemia, thalassemia. "We met and made the obvious decision to combine forces," Neel wrote.[16] Valentine, a member of the National Academy of Sciences since 1977, has denied that he ever injected any patients with plutonium.[17] "It's an absolute and total surprise," he said. "I never gave injections to anyone."[18]

The journalist Corydon Ireland, of the *Rochester Democrat & Chronicle,* a Gannett newspaper, wrote a series of articles about the radiation experiments at Strong Memorial Hospital. He said that Valentine was the chief medical resident during the time when the plutonium injections were administered and that he was a protégé of both Samuel Bassett and Stafford Warren.[19] In the end, however, Ireland felt it was impossible to ascertain who actually gave the plutonium injections, because the relevant hospital records at the University of Rochester could not be found. ("Files on UR Radiation Injections Reported Lost" was the title of one of Ireland's stories.) The Atomic Energy Commission had assigned the task of keeping records of the study to the university, so the AEC had only partial accounts of the human experiments. One document stated that Dr. Bassett injected the first person. Valentine claimed that he was engaged primarily in animal radiation experiments "in a little lock-and-key lab on the third floor of the hospital." In his interviews with Ireland, Valentine defended Samuel Bassett, who "ran a superb metabolism unit."[20] (According to Eileen Welsome, the Metabolic Unit existed exclusively for the sake of radiation experiments from September 1945 until the end of 1946.)[21] Valentine acknowledged that he was "extremely close" to both Bassett and Stafford Warren, and accompanied them to UCLA in 1948, where the AEC established a center initially staffed entirely by former Manhattan Project scientists.[22]

The only surviving radiation victim, Mary Jeanne Connel, was twenty-four years old at the time of her injection. One of six Rochester patients who received uranium instead of plutonium, she was in good health, but underweight, and had been sent to the metabolic ward for rest. Connell recalled that many doctors were present when she was injected. She found herself strapped down to a hospital gurney and had no idea what was going on, though she soon felt nausea and burning throughout her body. In the days ahead, numerous members of the medical staff inspected her. On two occasions, she was taken to visit animal laboratories, an experience that terrified her—especially when one of the doctors mentioned that work on the animals was not progressing as rapidly as hoped. A member of the hospital staff told her she was the most famous patient in the building, which does not suggest that the study was enveloped in much secrecy.[23]

One of the mysteries of the Rochester experiments is how people were assigned to the Metabolic Unit in the first place. Although the patients selected for injection suffered from a variety of disorders, including anemia and hemophilia, they all wound up in the Metabolic Unit. Janet Stadt, a forty-one-year-old woman with a chronic skin ailment, became HP 8. She received the highest dose of radiation, but she lived for twenty-nine years, experiencing bizarre illnesses that could not be diagnosed. She finally died an agonizing death of cancer. Her son said, "My mother went in for scleroderma, which is a skin disorder, and a duodenal ulcer, and somehow she got pushed over into this lab where these monsters were."[24]

It was a hospital where patients were rerouted to the Metabolic Unit for intense doses of toxic substances and where the chief medical resident, Lieutenant Valentine, was up on the third floor, in a locked lab, irradiating animals. His research partner, James Neel, also served as "company commander" of the medical students and residents who ran much of the hospital.[25]

In 1946, Stafford Warren made a decision that changed Neel's life beyond recognition. On October 29, Neel was suddenly elevated from a routine assignment as a lieutenant doctor to acting director of the atomic bomb radiation studies in Japan. "In the military scheme of things, there is nobody much more superfluous, unloved, and unwanted than a first lieutenant on temporary assignment in a large military hospital," Neel wrote in his autobiography. "I could not help but observe how my status changed when orders came to report directly to the Pentagon. At such times it is wise not to be too explicit. Murmuring only something about a 'special assignment, can't talk,' I left the post for the railroad station in the colonel's car."[26]

Neel said he received his appointment through the intervention of another Rochester doctor, Captain Joe Howland, an adjutant to Colonel Stafford Warren.[27] Howland distinguished himself in the human radiation scandal by being one of the very few scientists to actually admit wrongdoing. He injected patients with plutonium because he was told to so, he said.[28] As Thomas Powers, author of *Heisenberg's War,* observed in the *Los Angeles Times,* "Far more common were researchers who told Welsome or official investigators that they had lost their files, couldn't remember who gave the orders, were out of the room when the injections were given, didn't know if consent had been obtained, believed subjects were 'hopelessly' or 'terminally' ill and insisted that all those 'premature deaths' would have happened anyway."[29]

Neel considered his own ascent to the post of acting head of the atomic bomb studies an incredible, inexplicable fluke—the whim of Warren and Howland. But Warren was not given to flukes. And those who entered the Manhattan Project (which then mutated into the AEC) were evaluated for loyalty to the goal of manufacturing the atomic bomb. Whatever the reason for his promotion, Neel suddenly became a representative of one of the most powerful and secretive institutions in the world. Eileen Welsome has observed, "The Atomic Energy Commission would indeed have godlike powers, controlling virtually every aspect of the nuclear weapons program for the next three decades."[30]

The AEC's genetic division became known as the Atomic Bomb Casualty Commission (ABCC). It was headquartered in Hiroshima, where Neel worked with two civilian scientists, the biologist Paul Henshaw and the physician Austin Brues, two senior researchers of the Manhattan Project. Henshaw later helped organize radiation experiments with prisoners in the United States, test that subjected their testicles to crippling doses of X-rays. Brues succeeded in tracking down one of the Rochester radiation subjects, a black man named Elmer Allen, and persuading him, on false pretexts, to return to one of the AEC's labs for further study. Welsome has described Brues as "bright, articulate—and deceitful."[31]

James Neel is not mentioned in *The Plutonium Files,* but Welsome knew about his career in the AEC. "They're all tied together," she recently said. "This is a very small world. Austin Brues was with Neel at the ABCC, and then he was trying to get Elmer Austin back to the Argonne Lab. Brues is a constant presence in my book. These were incredible people."[32]

Junior officers like Neel became lords in Japan. "MacArthur was the indisputable overlord of occupied Japan, and his underlings functioned as petty viceroys," according to the historian John Dower, author of *Embracing Defeat: Japan in the Wake of World War II,* which won a 1999 National Book Award. "Even mid-level staff offered advice or suggestions to Japanese functionaries that, though technically not orders, effectively operated as such."[33]

The authority of MacArthur and his "viceroys" extended, at times, to the National Academy of Sciences (NAS), which was theoretically a full partner in Neel's atomic bomb studies. In 1950, the NAS sent to Japan two representatives who informed Neel of the academy's decision to cancel the genetics research program. Neel, who had by then moved on to found the Department of Human Genetics at the University of Michigan, nevertheless remained on the ABCC. He expressed "total horror and indignation" to the NAS representatives in Japan. A few days later, MacArthur "strongly urged" the NAS to continue the genetics research, on the grounds that "discontinuation of the program would create a scientific vacuum into which investigators of uncertain credibility would be drawn."[34] The program continued, with genetic research being centered, in 1954, at the University of Michigan under Neel's direction.[35]

The military, the AEC, and the nuclear power industry have never since had reason to doubt Neel's credibility. His data on radiation toxicity has been consistently optimistic—with thresholds set four times higher than United Nations safety guidelines.[36] The University of Michigan's medical department received $6.8 million in funding from the Department of Energy and its predecessor, the AEC, for genetic research under Neel between 1965 and 1980, the only period for which I have information. The money endowed two professorships, supported worldwide population studies, and created a radiation research laboratory at Ann Arbor.[37] Scientists who reached less sanguine conclusions about radiation risk were discredited and sometimes driven out of government labs and academic institutions.[38]

In the aftermath of the Hiroshima and Nagasaki bombings, the military and the AEC managed to censor the scientific data and control the public perception of atomic bombs. But this required a concerted public-relations campaign. What the manufacturers of atomic weapons feared most was that images of mass destruction, particularly of civilian losses, would create such revulsion in the United States that no further nuclear weapons could be deployed. The bare fact that the United States had won a war by murdering 150,000 civilians was so appalling that a popular reaction seemed quite possible. After all, one occurred in Japan, where the atomic bombs became symbols of the ultimate evil and led to a new pacifism. "Defeat, victimization, an overwhelming sense of powerlessness in the face of undreamed-of weapons of destruction soon coalesced to become the basis of a new kind of antimilitary nationalism."[39]

In the United States, the atomic bomb had the opposite effect. It breathed life into a new, technologically triumphant, global American militarism. Nothing better exemplified the cheerful new perspective on atomic destruction than the testimony of General Leslie Groves, the overall head of the Manhattan Project, when he spoke to Congress in November 1945. Groves claimed that radiation brought death "rather soon, and as I understand it from the doctors, without undue suffering. In fact, they say it is a very pleasant way to die."[40]

In fact, General Groves's chief scientific officer, Colonel Stafford Warren, was privately telling Groves and the military high command just the opposite. In a series of talks where reporters were not permitted, he warned that one thousand atomic explosions might destroy all human life on the planet. Both Warren and Howland, Neel's superiors, saw alarming evidence around Hiroshima and Nagasaki of testicular atrophy and lower sperm counts in men who were far from the blasts. The men whose testicles were shriveled miles from the atomic epicenter were kept out of sight, as was the scientific debate about the genetic implications of radiation. "Stafford Warren was a key figure in establishing this policy of denial," Eileen Welsome said. "Stafford Warren and the AEC had a lot of data on the genetic effects of radiation very early on that were systematically withheld from the public. It's not unlike the tobacco scandal, where scientists withheld documents for years, except this is worse. Because in this case we were paying the salaries of those doing the concealing. And not just concealing, but the skewing of the scientific studies."[41]

Studying the AEC's inner world is now possible through a remarkable Energy Department Web site called Human Radiation Experiments: http://hrex.dis.anl.gov. It offers a powerful search engine to explore 250,000 documents, many of them recently declassified. One of these documents shows that in December 1949 Neel was appointed to a five-person "ad hoc committee" to advise Stafford Warren's successor, Shields Warren (no relation), "on the genetic effects of radiation and to interpret the data in terms of human populations. Such information will provide a basis for properly informing the public on this subject."[42]

In January 1951, the AEC began a series of atomic tests in Nevada that continued, on and off, until 1955. Recently declassified documents have shown that Shields Warren "had grave concerns about the health risks from fallout." Privately, he opposed the military's plan to station troops as close as possible to the atomic blasts. Publicly, however, he supported the tests and the dangerous troop deployments. As Eileen Welsome put it, "he had become firmly committed to the idea that whatever the risks, the tests were necessary to keep the United States safe from the Soviet Union and a world dominated by Communism."[43] The citizens of Nevada were not even advised to go indoors—for fear of alarming them—while hundreds of thousands of troops received dangerous doses of radiation. Some soldiers were literally knocked over and momentarily blinded by nuclear sunbursts so intense that all shadows were obliterated.

Yet the effects of the new atomic bomb tests proved subtler, more insidious, more enduring, and more far-reaching than methods of measuring could easily determine. In early 1954, Dr. Lester Van Middlesworth was working at a biology laboratory in Tennessee, when his Geiger counter was set ticking by the thyroid of a local steer. It turned out that *all* thyroids, from all over the world, showed traces of iodine 131 in the wake of each atomic explosion. "No one believed you could contaminate the world from one spot," he said. "It was like Columbus when no one believed the world was round."[44]

What did it mean? For Willard Libby, a physicist who passionately supported the hydrogen bomb tests and later won the Nobel Prize for his work on radiocarbon dating, it meant that the AEC would be forced to stop detonating—unless its scientists could find out what was happening. Libby was particularly worried about the accumulation of strontium 90, a by-product of plutonium fissioning. But to measure levels of this strontium isotope scientists needed human bodies in large numbers. Libby consulted a lawyer to look into the legal ramifications of body snatching, and decided it was too serious a crime in the United States. "I don't know how to snatch bodies," he said. "If anybody knows how to do a good job of body snatching, they will really be serving their country."[45] Some of his subordinates figured out exactly how to snatch bodies, along with thousands of skeletons, fetuses, and bone samples, the world over. It became known as Project Sunshine. Willard considered it the most crucial security program in the United States, after weapons.

As Project Sunshine was born, the Eisenhower administration created an international outreach called Atoms for Peace. It was a diplomatic offensive, promoting an atomic utopia. Nuclear-powered airplanes and ships would travel forever on handfuls of fuel. Radioisotopes would cure many diseases. Whereas Project Sunshine was a take-away operation, Atoms for Peace was based on the appealing principle of giving away radioactive isotopes to scientists from friendly, or potentially friendly, countries.

The apostle for isotopes was a biologist named Paul Aebersold. He had been in radiation research for a long time, starting as an assistant at the University of San Francisco, where from 1939 to 1941 he participated in a new therapy regime: bombarding 128 patients, in varying degrees of health, with phenomenal doses of neutrons. Almost half died within six months—in extremely unpleasant, unusual ways. Aebersold became known as Mr. Isotope within the AEC. "He committed suicide, which is probably the extreme of his nuttiness," another scientist, Merril Eisenbud, recalled. "But he was very fanatic about radioisotopes and what it was going to do for you."[46] One scientist compared the function of radioisotopes inside the human body to spies who enter by stealth and search for new information, thus providing diagnostic insights.

Aebersold functioned very much like a stealthy isotope tracer inside the world's body politic. On June 3, 1954, he set off to promote isotopes on a five-country tour of South America.[47] His first stop was Venezuela. And at the very top of his Venezuelan list, carefully marked by two asterisks, was the following name:

Marcel Roche, Instituto de Investigaciones Médicas
Plaza Morelos, Caracas, Venezuela

Marcel Roche was a member of James Neel's 1968 expedition, who, along with Napoleon Chagnon, administered the Edmonston B live vaccine to the Iyewei-teri at the Ocamo mission, from where the worst epidemic in Yanomami history spread.

Within a year of Aebersold's trip, Roche enrolled in two seminars in the Atoms for Peace program, designed to teach doctors from all over the world to work with radioisotopes. There is no doubt that radioisotope technology eventually improved medical diagnosis in many areas, but, because of the AEC's own primary interest in radiation pathology, there was often a double agenda behind the studies the agency sponsored. Atoms for Peace inducted scientists into a select society, with its own myths and rituals, where the ethical ambiguities of these studies were not questioned. Scientists watched films that promised universal progress through atomic power. "It's a huge fraternity, this order of the mushroom, and it's growing all the time," declared one of the AEC's documentaries about the Nevada atomic tests.[48] The training was presented as "President Eisenhower's plan for sharing the peaceful atom with other countries."[49]

Part of this peaceful plan involved experiments with trace amounts of radioactive material on human beings. The risks were uncertain, but that was the justification for the experiments. Declassified documents show that, as early as 1947, the AEC decided that tracer research was fundamental to toxicity studies despite the "Moral, ethical, and medico-legal objections to the administration of radioactive materials without the patient's knowledge or consent."[50]

It was this research world of tracer studies that Marcel Roche entered. At a conference on radioisotopes, which is not dated in the Department of Energy archives, Roche presented a paper on the metabolism of radioactive iodine. "Marcel Roche (Venezuela) described a mass population study of thyroid uptake in the lowlands of Venezuela and high in the Andes, showing 2 slides"[51]

Beginning in 1958, Roche carried out a long, difficult study of iodine metabolism, by administering radioactive iodine to the Yanomami and the Maquiritare Indians. It was a genetic study, with no known benefit for the Yanomami, who do not suffer from goiters. It was really an investigation into why this interesting population, without any real source of iodine, did *not* suffer from goiters, compared with highland populations that did. The discovery that Yanomami have astonishing capacities to concentrate radioactive iodine 131 in their thyroid glands—71 percent of the dose was absorbed in twenty-four hours—gave Roche a fine academic article.[52]

Today, the official position of the Department of Energy is that these "tracer studies" were unethical, because they violated the principle of informed consent and did no good to the subjects involved, but that they did not cause any real harm. "I don't think the tracer experiments benefited the patients, but they increased our knowledge," said Jay Stannard, who spent half a century in radiation biology and wrote a three-volume history of the radiation studies. "They [the doctors] were also under a great deal of pressure to find out how these new materials worked. At that time you have to remember the doctor was omnipotent."[53]

Some scientists take a more critical view of the tracer studies. "Iodine 131 is an extremely toxic substance, and it had to be harmful to the Yanomami—the question is how harmful," said Terrence Collins, professor of chemistry at Carnegie Mellon University and winner of a Presidential Green Chemistry Award in 1999. "It is completely outrageous for scientists to administer trace amounts of radioactive substances to remote tribes in the Amazon for the sake of genetic studies."[54]

Other Venezuelan scientists collaborated with the AEC to inject Venezuelan natives with radioactive iron Fe-59, as a way of testing iron metabolism. This experiment was relevant to the study of bomb survivors because intense radiation often resulted in iron loss and anemia. In fact, the University of Rochester AEC lab was anxiously searching for anemic patients after one of its subjects moved away. ("The patient was lost to us when he returned to his home after graduation.") The AEC Biology and Medicine Division was attracted to the Venezuelan Indians because they offered "a large group of such severely anemic patients . . . found in tropical areas where hookworm flourishes."[55] Berkeley's Donner Laboratory provided the iron isotopes for the study.[56] The results were published, and photos from the radioactive iron study were displayed in a traveling exhibit promoting isotopes in Latin America.[57] A live cobalt 2000-curie reactor accompanied the exhibit, apparently built by an engineer who was a refugee from the nuclear airplane project—a two-billion-dollar AEC brainchild that, mercifully, never flew.[58]

The AEC was also searching for human bones from the Amazon because "the primitive populations [live] in areas with soil of very low calcium content." Strontium 90 intake was highest in areas in which the soil was low in calcium. As a result, the AEC hoped its Amazonian outreach would "find the maximum concentration of strontium 90 in humans and thus estimate the high end of the distribution curve for strontium 90 in the world population."[59] Obtaining enough bone for strontium separation meant performing an autopsy, which was followed by a complex lab procedure called ashing.

Somehow, the AEC established a Venezuelan "bone sampling center," a laboratory of some sort. It processed far more bodies than any other South American Project Sunshine center. For example, between June 1955 and March 1956, 122 "bone samples" were collected, meaning there was an autopsy almost every other day. The Project Sunshine inventory of bones was organized by age, starting with an infant only four months old. Here is a typical entry:[60]

Sample No.	**Sex**	**Date of Death**	**Age**
B-423	female	7/12/55	30 years
B-996	female	11/5/55	30 years
B-609	male	8/4/55	30 years
B-643	female	10/27/55	33 years
B-997	male	12/25/55	34 years

In December 1956, Dr. Laurence Kulp, a Project Sunshine leader, went on a South American inspection tour "to improve communications and understanding at bone-sampling centers."[61] Venezuela was on his list. Kulp had been the first to see that "body snatching" was easier among the poor and in southern climes, starting in Texas. "Down in Houston they don't have all these rules," he observed. "They intend to get virtually every death in the age range we are interested in that occurs in the city of Houston. They have a lot of poverty cases and so on."[62]

Questions remain. Whose bodies were being autopsied in Venezuela and with whose permission? Was the AEC paying a finder's fee to morgues or hospitals? Who was performing all these autopsies cum plutonium separation? Did the fact that bodies could be sold for "bone samples" in any way affect care given to impoverished patients in their final hours of life? And, since the Project Sunshine scientists were particularly determined to measure strontium 90 in "primitive populations," did they pay a premium for Amerindian corpses? Venezuela was receiving more radioactive isotopes from the United States than any other Latin American country except Brazil. Was there an isotope-for-bodies exchange going on?[63]

The AEC's traffic in native blood and bones gave new meaning to the title of the historian Eduardo Galeano's book *Open Veins of Latin America: Five Centuries of the Pillage of a Continent.*[64]

The subjects of the AEC's experiments, from the Rochester plutonium injections to the administration of radioactive iodine to pregnant mothers in Tennessee, were predominantly poor. Stafford Warren did not sympathize with the indigent. He had an obsessive fear of government welfare. He saw the undeserving underclass as the true threat to his atomic legacy. On April 1, 1947, he told an audience at Yale, "Particularly if we have a depression, we will have a lot of people who will want to coast along on the government charities, and who will come up with the story that they worked for the Manhattan District somewhere, or they passed down the street to windward of the plant, and therefore they deserve compensation. It is going to be a very serious problem."[65]

It has turned out to be a very serious problem—especially in light of the National Academy of Sciences' latest, most comprehensive study, showing that many people did, in fact, die of radiation exposure at the bomb plants. On January 29, 2000, the front page of the *New York Times* ran the headline "U.S. Acknowledges Radiation Killed Weapons Workers."[66] It was a stunning reversal, according to Energy Secretary Bill Richardson. The government had spent tens of millions of dollars fighting lawsuits for decades and refuting charges of radiation hazards at bomb plants and elsewhere. "They've finally admitted what they denied for fifty years—that people exposed to radiation at the plants got cancer," said Eileen Welsome, when I spoke with her on February 3, five days after the Department of Energy announced the reversal. "They did everything they could to discredit everyone who claimed harmful effects of radiation. John Goffman of UC Berkeley was reviled because he said we were going to have an epidemic of cancer patients—until this week, when the government finally agreed."[67]

Stafford Warren and his successors believed that the human radiation experiments would protect the government from damage claims. The experiments not only violated the Nuremberg Code and federal guidelines; they also became rituals of self-absolution, the AEC's version of compulsive hand washing. As Welsome put it, "They were not just immoral science, they were bad science."[68]

The Fierce People paradigm can be understood only in terms of the AEC's own culture, which countenanced injecting people with plutonium to protect itself against lawsuits and spending billions of dollars in an effort to build a nuclear airplane that would have weighed as much as an oceangoing vessel. These men had some of the best minds in the world, but they had gone temporarily unhinged under the stress of the Cold War, which went on and on. They feared Communists abroad and welfare recipients at home. But once they gained entrance into "the brotherhood of the mushroom," they never had to worry about work again. They were enabled by a system of interlocking government sinecures, which allowed their temporary insanity to become institutionalized at unbelievable taxpayer expense. The world was their asylum. Moreover, as Welsome has shown in alarming detail, they reproduced prolifically within the federal bureaucracy, creating lineages of like-minded scientists—men obsessed with secrecy, who saw the American public as the real enemy.

The Fierce People were the great-grandchildren of Colonel Stafford Warren and the Manhattan Project. This is their genealogy:

Lineage of the Fierce People

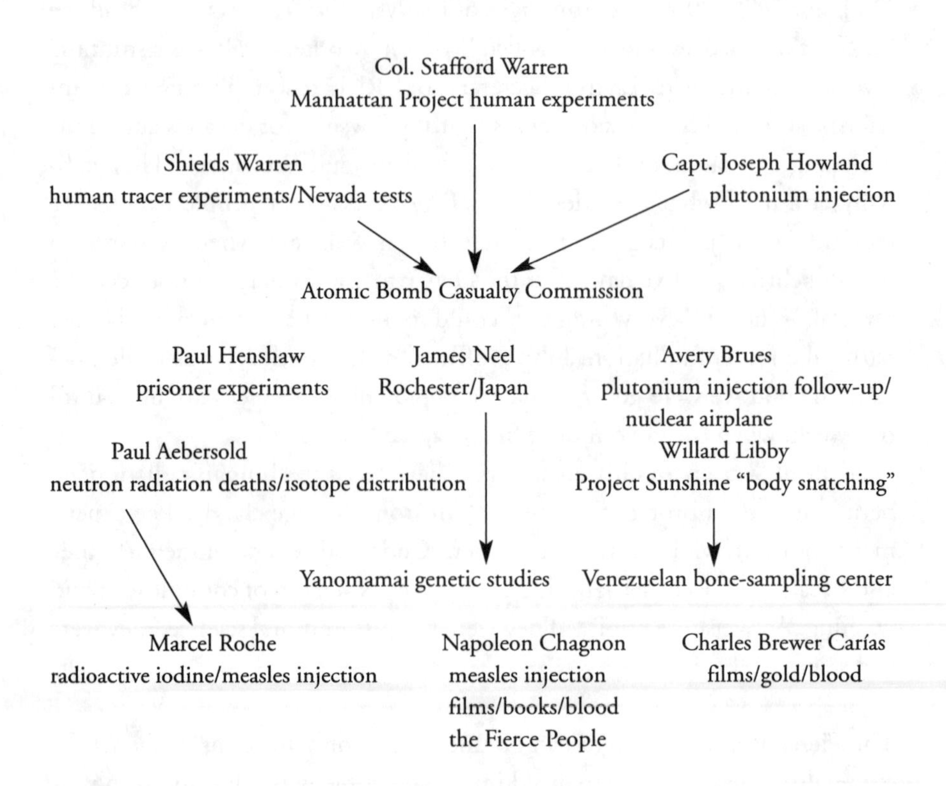

Each generation in this lineage initiated the subsequent one into violence. Stafford Warren, on his way to measure the effects of the atomic blasts on the Japanese, specifically asked that Lieutenant Valentine, his brilliant protégé, give patients doses of plutonium.[69] Why would Warren want do that to a young doctor? Warren seemed, out of all his concerns, to make a particular point of forcing Valentine to violate the Hippocratic oath: "I will give no deadly medicine to any one if asked. . . ."

There are societies other than crime syndicates where the passage to adulthood requires an act of violence by young men—one that is forced upon them by older ones. Greek boys, at about the age of fifteen, participated in a nocturnal ritual in which they engaged in symbolic murder and cannibalism as an initiation into a male fraternity, as the great classical scholar Walter Burkert described in his book *Homo Necans.* In these nocturnal mountaintop initiations, the domestic order of society was inverted: what was forbidden in normal, waking society became obligatory for all participants. The "unspeakable act" became the entrance price that transformed the participants. The boys became killer-outcasts, who went out to the forest to live together in a military band, and blood brothers in the clan of Zeus Lykos, Zeus the Wolf.[70]

Stafford Warren's closest aide and adjutant, Captain Joseph Howland, injected a black man named Ebb Cade (HP 12) with plutonium at the Manhattan Project's Oak Ridge Army Hospital. Howland said, "I injected a five-microcurie dose of plutonium into a human and studied his clinical experience. (I objected, but in the Army, an order is an order.)" Howland called this his "command performance," and there was a ritual element to it.[71]

In 1968, James Neel gave the Venezuelan doctor Marcel Roche and Napoleon Chagnon the Edmonston B measles virus and told them to vaccinate the Iyewei-teri Yanomami, a group that suffered from severe anemia.[72] On that ground alone, it was absolutely "contraindicated" to give the Iyewei-teri an antiquated vaccine like the Edmonston B, even if Roche had administered the gamma globulin that was supposed to accompany it.[73] Roche told me that he did not even ask what kind of vaccine it was and did not know that gamma globulin was recommended.[74] This sounds incredible by the standards of normal medical practice, but not by the standards of AEC practice. Roche had been administering radioactive iodine to the Yanomami and neighboring Maquiritare for ten years, and he was concerned primarily with this demanding protocol in 1968.[75] His Instituto de Investigaciones

Médicas had been highlighted with two asterisks on Paul Aebersold's itinerary and cannot be excluded as a possible site for the AEC's autopsies. The Instituto de Investigaciones Médicas mutated into Venezuela's most prestigious think tank, the Instituto Venezolano de Investigación Científica, whose library is named in Roche's honor. He is considered one of the most influential scientists in the history of his country. Still, he had no authority to administer a vaccine without permission from Venezuela's Ministry of Health. By 1968, he was an AEC veteran of thirteen years. "I gave out what they gave me," Roche said, echoing Howland's soldier reflex. "I injected what they gave me."[76]

The crucial moment in the 1968 expedition came just after noon on February 18, at the AEC's Mavaca base camp, when the scientists suspected that a full-blown epidemic had erupted among the vaccinated Yanomami. Rousseau, the radio technician, was in the process of calling Caracas to request penicillin *por los efectos de la vacuna*—"because of the effects of the vaccine"—and alerting the government doctors that the vaccine might be causing *brotes de sarampión*—"outbreaks of measles." And this was when Neel shocked Chagnon, telling him that, yes, the sick Yanomami were going to infect everyone they met. But they had to continue their research and filmmaking activities upriver and let the Yanomami fend for themselves until some other doctors arrived.[77] (This, too, was an AEC heuristic: When in doubt, make a film.) Upriver, among the Patanowa-teri, where Timothy Asch filmed *A Multidisciplinary Study* and *The Feast,* Neel mocked one of his doctors who was treating a sick man—"the physician ministering to his flock"—and ordered Asch not to film the scene. "Anybody can walk into a village and treat people. This is *not* what we're here to do. Now, I don't know how I can be more definite about it."[78]

Neel and the other members of the expedition had all rushed off by the time the full vaccine reaction set in among the Patanowa-teri. When I showed several Patanowa-teri elders *The Feast,* at the Platanal mission in September 1996, they said the nurse Juan Gonzalez helped them in their distress. "The Patanowa-teri all got sick," González remembered. "I accompanied them." So many of the Patanowa-teri died and so many survivors were suffering from measles that the customary cremation ceremonies were postponed. In fact, González had to assist them in hanging the bodies of the dead from trees. "They hung the children in baskets from the trees—the cadavers were placed inside the baskets, all rolled up tightly in leaves, like a metallic foil. The women were more loosely wrapped in leaves, and they were left hanging in their hammocks out in the wild among the trees. They tied the men up on poles, higher up in the branches. Look, all the way from below the mission at Platanal to where the Shashanawe-teri live, what a stench there was. Nothing but dead Yanomami, on both banks, here, there, everywhere. Who knows? Well, after a while. . . . Who was going to hang up the dead? Everybody was infected."[79]

Juan González estimated that about twenty-five Patanowa-teri died, but this was a vague tally based on different family groups whose bodies he had found hanging in the jungle.

The Patanowa-teri elders gathered together after the film and dictated the names of twenty-six adults, from Patanowa-teri alone, who died of measles after the filming of the *The Feast.* They wanted me to tell the owners of the film that they should not lie about what happened to these people. One of them repeated what I have often heard said about the measles epidemic: "Shaki cast a spell on us." Another said, "He stole our spirits, and we have never been the same."

Chagnon was part of an efficient team that stole their spirits—the names of their dead, the twelve thousand vials of their blood, and the pictures of their warriors—to cast a spell on the whole world of anthropology. And it has never been the same. Neel and his eugenic missionaries engineered a bold creation myth, a ferocious Garden of Eden, where the healthy, well-fed Yanomami fought for the fun of it and killed their infant daughters for sexual pleasure. Chagnon was the chief scribe, but not the inventor, of this social gospel for Americans in the midst of the Cold War. It was not the Yanomami but Chagnon's fellow Americans who belonged, in reality, to one of the best-fed, healthiest societies in history. America enjoyed abundance so delirious that it seemed, for a short time in the 1960s, that its citizens would not agree to the stress of world combat against Communism—or to the continual sacrifices of economic competition. At that critical moment, *The Fierce People* and its film sequels came to reverse a dangerous complacency, proof that the battle is never won, that the fight can never be abandoned.

According to the foreword to the 1977 edition of *The Fierce People,* "Yanomamo culture, in its major focus, reverses the meanings of 'good' and 'desirable' as phrased in the ideal postulates of the Judaic-Christian tradition."[80] In fact, it was the geneticist James Neel's eugenic creed that reversed not only the idealized postulates of the West's religious traditions but, more important, the ideals of liberal democracy and the ethics of medical practice. Dazzled by the elegance of Mendelian genetics and seduced by the power of atomic physics, Neel aspired to the role of physician to the gene pool. It was a cosmic conceit. But what Neel meant by "physician" had little in common with the tradition of Hippocrates. He wanted legal limits on reproduction and cited, with approval, plans to limit health care to the Third World's burgeoning population.[81] He had a profoundly pessimistic view of ordinary people, particularly of democratic gene pools where any misfit could marry. By contrast, the murderous Yanomami war leaders—the *unokais* who killed their rivals and took their wives—held the secret to history and to the conservation of "innate ability."*

Colonel Stafford Warren and his kin were the real *unokais,* who had killed 150,000 people, the great majority unarmed civilians, in Hiroshima and Nagasaki. But, though these nuclear *unokais* were richly rewarded in government grants and scientific status, the bomb proved a letdown. The nuclear age marked the end of both chivalry and treachery, as they had been known and loved by many men. In a world where strontium 90 was being mysteriously absorbed by everyone's bones, threatening to bomb the enemy into the Stone Age had become little more than a territorial display.

Radiation did not respect territory. The frontier was over. But the rituals of conquest went on, for a little while. America's atomic *unokais* built thousands of nuclear weapons, irradiated millions of laboratory animals and dreamed of Stone Age warriors in the Amazon. Was it nostalgia? Was it fear that civilization would be reduced to small tribal societies again? The search for the flawless, golden man draws everyone in some way, and Neel thought he could locate the precious genes of male dominance. Personally, however, I believe that James Neel wanted, at some level, a respite from the logic of scientific production, with its endless cycles of blood and urine collection and animal sacrifice. Maybe he hoped, away from the laboratories where he became the AEC's authority on rat and mice experiments, that the Yanomami would show him something, something we would all like to know.[82] Yes, he fantasized about aggressive headmen and their huge harems. But perhaps he just wanted to breathe again.

Sadly, he took both his beliefs and his experiments with him into the rain forest. Neel and his eugenic disciples imbued the impersonal nature of evolution with a personal animus: natural selection became selfish, mur-

*See chapter 4.

derous, cruel, and deceitful. Doctors trained by the AEC gave the Yanomami a radioactive tracer and a vaccine that was potentially fatal for immune-compromised people. Scientists kept on filming and collecting blood in the midst of epidemics. These brave men took a long walk on the dark side, but, in the artificial brilliance of ground zero, they could see no shadows.

Appendix

Mortality at Yanomami Villages

The Salesian policies include attracting remote Indian groups to their missions, where they die of diseases at four times the rate found in remote villages.—*Napoleon Chagnon*[1]

Now we read in Chagnon's text that mortality rates due to new diseases are four times higher [at the mission] than in the remote villages. This cannot be true. From where does he get his data? Close to the missions, according to our own observations, mortality is lower.—*Irenaus Eibl-Eibesfeldt and Gabriele Herzog-Schroder*[2]

No one knows exactly why the Yanomami survived and many neighboring tribes died out. Most of the Yanomami's precontact neighbors had bigger villages, greater social cohesion, and more regular access to steel goods. But those assets could be liabilities during epidemics because the trading routes were also the paths of disease, and sedentary populations without immunity were perfect hosts for the invading organisms. Mortality rates of 80–90 percent were not unusual.[3] Many tribes, like the once numerous Mavaca Indians, left only their names behind.

In the Parima Mountains, the Yanomami were less frequently exposed to outside infection. Their highland isolation was reinforced by suspicions that made trade and communication difficult between villages. Yanomami culture was characterized by relatively infrequent warfare but simmering suspicion of witchcraft.[4] During epidemics, groups avoided each other. This began to change from the 1950s on. For example, the AEC's massive distribution of trade goods and film orchestration broke down the distrust between villages that had maintained a cordon sanitaire. This occurred between the Patanowa-teri and the Mahekoto-teri, immediately before both villages were devastated by measles, and later between Mishimishimabowei-teri and Bisaasi-teri, immediately before Mishimishimabowei-teri was devastated by malaria and hepatitis.

The Yanomami also survived because they did not have malaria in the highlands. According to the medical historian Francis Black of Yale, no Amerindian tribe has survived falciparum malaria without outside medicine.[5] The protozoa parasites of falciparum attack blood cells, clogging routes to the brain, where they cause death. Falciparum malaria kills between one and two million people each year worldwide. And it is killing the Yanomami by the score because it is the predominant strain on the Upper Orinoco. When Helena Valero lived in these lower reaches of the Siapa Highlands, the Yanomami did not have malaria. Now, at every stream below two thousand feet, the Brazilian microscopist Marinho De Souza found infestations of anopheles mosquitoes, the carriers of malaria plasmodium. There was no way to run from the mosquitoes. And the disease marches were not effective against a recidivist disease—like the vivax malaria strain, which accounts for about 40 percent of cases—because the carriers of the next wave already harbored plasmodium in their livers, where the parasites could survive for months.

Today the population of the Yanomami is about as large as it was forty years ago, but its distribution and its pattern of growth have changed abruptly. Whereas the total population had been increasing at approximately 1–2 percent annually for 150 years,[6] from 1960 onward it suddenly shifted. In the traditional population centers outside the immediate mission sphere—the Ocamo River basin,[7] the Upper Orinoco above the Guaharibo Rapids,[8] the Parima Highlands,[9] and the Catrimani River headwaters[10]—population has either collapsed or stagnated. But population growth has simultaneously increased at the missions, where it has averaged 2–4 percent, about twice the historical rate.[11] That does not mean that groups at the missions are healthy or well nourished. They are not. The missionaries who serve them are often sick, too.

Today the single biggest reason why mission groups can increase in spite of malaria and heavy parasite loads, and in spite of falciparum malaria's resistance to chloroquine treatment, is that missionaries dispense mefloquine.

The Salesian missions of Venezuela arguably occupy the most difficult and crucial niche in the Yanomami world because their four posts are located along the main course of the Orinoco and the lower Mavaca, where malaria is considered *inabordable,* "uncontrollable," by Venezuelan health authorities.[12] Some three thousand Yanomami live near Salesian missions, and 100 percent of the Yanomami in the area have been recently infected with malaria. A third of them have pathological hardening of the spleen, caused by repeated onslaughts of the disease.[13] It is amazing that the Yanomami at these missions are experiencing a population explosion.

Nevertheless, Napoleon Chagnon maintained that Salesian mission policies were responsible for hundreds of deaths on the Upper Orinoco River—responsible for more deaths than the Brazilian gold miners.[14] The television and print journalists who repeated these charges probably did not realize that, when Chagnon said mission Yanomami were dying at four times the rate of remote groups, he meant something else.

He meant that mortality rates at the missions were lower than those at remote villages.

It took me a while to understand the nuances of his argument. Prior to his break with the Salesians in 1990, Chagnon had credited the missions with reducing disease. "The health problems that are beginning to emerge among the Yanomamo are far more serious in the remoter villages than they are at the mission stations, for at least at the missions there are radios with which assistance from Caracas can be solicited in emergencies, or stores of antibiotics and other medical supplies to treat the more common and more frequent illnesses," he wrote in the 1983 edition of *The Fierce People.* "And most of the local missionaries are patient and good people who are sensitive to the medical problems that arise within their groups, working indefatigably at times to nurse the children and adults through a lingering malady that no shaman on earth could cure. . . ."[15]

An altogether different picture of mortality around the missions emerged, for the first time, in a 1991 television program in which the Venezuelan reporter Marta Rodríguez Miranda showed graphs, based on Chagnon's data, revealing that mortality was two and a half times greater at the missions than at isolated Siapa villages.[16] By October 1993, the mission mortality had risen from two and a half to four times that of the remote villages, according to Chagnon's op-ed article for the *New York Times,* "Covering Up the Yanomamo Massacre."[17]

Three months later, in December 1993, Chagnon offered what appeared, at first glance, to be a rebuttal to his previous claims. Now villages nearest the Salesian missions had the same low mortality rate as the most distant ones. "Both those with minimum and those with maximum contact suffered relatively low rates of mortality. . . .—about 5 per cent in both cases. Those with maximum contact lived at or near the Mavaca mission, where fairly reliable medical attention was available on demand."[18]

Both the *New York Times* editorial and the *Times Literary Supplement* commentary, however, were based on a single study Chagnon conducted between 1987 and 1991 at seventeen unnamed villages. And he had already published these findings in the 1992 edition of his *Yanomamo* textbook. His own graphs clearly showed that mortality was 30 percent lower at the Salesian mission than at the apparently healthy remote villages.[19]

I will begin by identifying the seventeen villages in Chagnon's study, along with the category of mission contact assigned to each one. Chagnon has not divulged such information so far, even though the biologist Eibl-Eibesfeldt and the anthropologist Lizot have publicly challenged the authenticity of his data. The dispute over sources is similar to that over Chagnon's famous *Science* article about killers.[20] In chapter 10, I gave the names and locations of nine of the twelve villages that were the subject of a *Science* article about the reproductive success of Yanomami *unokai.* Interestingly, those twelve villages were touted as models of natural Yanomami warfare, unlike communities that had given up warfare under the influence of missionaries and government agents.[21] The total population for the twelve villages in *Science* was 1,394. The seventeen villages in Chagnon's study on the negative impact of missions have a nearly identical population, 1,407. In fact, they are exactly the same villages. And their populations are based on the same 1987 census. This is my summary:

Yanomami Mortality: 17 Villages, 1987–1991[22]

Minimum Contact

No.	*Real name*	*Place*	*Population*	*Mortality %*	*Deaths*
92	Mishimishi	Upper Mavaca	188	4.3	8
50	(Karohi-teri?)	(Manaviche)	63	2.7	2
51	Kedebabowei	Mrakapiwei	164	9.0	15
54	Shanishani-teri	Shanishani River	109	11.0	12
		Total:	524	7.0	37

Intermediate Contact

No.	Real name	Place	Population	Mortality %	Deaths
53	Dorita-teri	Shani Shani River	136	24.3	33
52	Sheroana-teri	Shani Shani River	94	22.3	21
93	Washewa-teri	Washewa River	105	14.3	15
90	Haoyabowe-teri	Mavakita	55	10.9	6
		Total:	390	19.2	75

Maximum Contact

No.	Real name	Place	Population	Mortality %	Deaths
88	Bisaasi-teri	Boca Mavaca		0.0	
89	Bisaasi-teri	Boca Mavaca		0.0	
94	Bisaasi-teri	Boca Mavaca		3.3	
5	Bisaasi-teri	Boca Mavaca		4.4	
87	Bisaasi-teri	Boca Mavaca		4.5	
83	Bisaasi-teri	Boca Mavaca		5.3	
86	Bisaasi-teri	Boca Mavaca		7.0	
85	Bisaasi-teri	Boca Mavaca		10.0	
		Subtotal:	414		
84	Nasikibowei-teri	Mavaca River	66	7.5	
		Total:	480	4.7	23

For the first two categories of contact—"minimum" and "intermediate"—all of the villages correspond perfectly to those cited in *Science.* Moreover, since Chagnon gave their exact populations, though not their names or locations, there should be no ambiguity about the mortality rates at the villages, when the populations are multiplied by the mortality figures given in the second study. There is, however, a tiny discrepancy. Chagnon's "minimum contact" group numbers 531, not 524; his "intermediate contact" group numbers 392, not 390.

Similarly, the population of the combined Bisaasi-teri and Nasikibowei-teri villages, which constitute the group of maximum mission contact ("Boca Mavaca only"), came to 480 in Chagnon's *Science* article, but appear as 484 in the second study. Because of these minute changes in population (0.93 percent), it is not immediately apparent that the villages from the *unokai* study and the study on mission impact are identical. In addition, Chagnon took two of the Bisaasi-teri communities at Boca Mavaca (formerly numbered 6, 7) and turned them into six or seven anonymous villages (Nos. 83, 85, 86, 87, 88, 89, 94). Here I cannot be certain whether Village 94 is actually a Bisaasi-teri splinter or a splinter of one of the groups on the Upper Mavaca (Nos. 90, 92, 93), and I cannot decipher exactly which Bisaasi-teri *shabonos* are ciphered by any of the new numbers.[23] But I am as certain about the overall population of the Boca Mavaca groups as I am about the more distant *shabonos* and what the real, physical villages are and were. When Chagnon conducted his 1987 census, Bisaasi-teri actually consisted of eight different *shabonos,* not three huge communities, as the numbers in *Science* suggested.[24] That was why Lizot complained that the populations listed in *Science* did not correspond to real *shabonos.*

The most serious problem arises from the confusion between intermediate, maximum, and minimum degrees of contact. According to Chagnon's text, "intermediate contact describes villages that are easily reachable by motorized boats, sometimes followed by relatively short walks, from Salesian missions."[25] Yet the community with the highest mortality, Dorita-teri (No. 53), was actually in the heart of the Siapa Highlands. In 1968, Timothy Asch got lost for eleven days when searching for the parent village of the Dorita-teri, Patanowa-teri. And that was at a time when the Patanowa-teri were living much closer to the nearest mission than the Dorita-teri are today.[26] A Venezuelan television crew, led by Marta Rodríguez Miranda, accompanied Chagnon and Brewer to Dorita-teri in May 1991. They apparently believed it was a remote village. Chagnon's presentation lent itself to that interpretation.[27] Miranda was understandably confused by the mortality statistics. As far as she was concerned, Dorita-teri, separated from any missions by a vast, unbroken expanse of forest that she flew over in a military helicopter, was one of the "pure" groups that her editors had sent her to film.[28] If Dorita-teri had a 25 percent mortality rate over four years, and the missions suffered rates two and a half times higher, then the mission mortality should have been 60 percent. Consequently, Miranda's documentary included a graph showing that 60 percent of all the Yanomami at the Salesian missions had just died.

The most peculiar village in the mortality study was No. 51, Kedebabowei-teri, with a population of 164. Chagnon's graph displayed mortality at Village 51 as 8.0 percent, and the village was grouped with those of minimum contact. There is an error in Chagnon's accounting for Village 51. In fact, Village 51, according to Chagnon's own data, lost 9.1 percent of its population between 1987 and 1991 (15 deaths in a village of 164).[29] This alone changed the mortality of the "minimum contact" group from 6.5 to 7.0 percent. The real difference in mortality between the Mavaca mission Yanomami and the most remote villages was 40–50 percent, not 30 percent. Yet Chagnon's text further contradicted his data. Elsewhere, he presented Kedebabowei-teri as a case of mission impact that suffered a nearly 20 percent mortality rate from 1987 to 1991.[30] And this did not include an epidemic at the end of 1991, which took an additional 21 lives. As in the overall study of Yanomami mortality, here there were kaleidoscopic and conflicting realities.

The real figure for mortality at Village 51 was 22.6 percent—36 people out of 164 died. Yet this bare figure did not do justice to the sad history of Kedebabowei-teri.

Kedebabowei-teri was FUNDAFACI's helicopter site, located at the juncture of the Kedebabowei River and the Mavaca. This was where Ushubiriwa and his brothers had built their rebel base in hopes of obtaining an outboard motor. After a club fight, Ushubiriwa and his brothers occupied the strategic juncture, driving the inhabitants of the much larger village of Kedebabowei-teri southward, into the hill country, where they had a second *shabono.* Later, some of them came back from the hills, asking for presents. The village was caught up in a conflict with César Dimanawa's alliance, which meant that it was isolated from mission influence after the summer of 1990. And it was precisely at this time that Kedebabowei-teri's mortality suddenly surged. Prior to FUNDAFACI's arrival, Kedebabowei-teri suffered five deaths between 1987 and 1990, normal for a tribal population but unusually low for the Yanomami during these peak gold rush years. Then, after Chagnon's three guides cleared a heliport for landing and refueling, the number of deaths shot up to ten within the following twelve months. Chagnon concluded, "Exposure was affecting their health."[31]

"Who caused the deaths at Kedebabowei-teri?" asked Juan Finkers, the Salesian brother and naturalist who lives twenty miles downriver from the village. "That is the question. I am not saying that Chagnon personally brought the epidemics, but it was his big FUNDAFACI expeditions coming in from Caracas with all those people. Look, there were so many people I couldn't keep track of them all. They had come to see the Indians from here, there, and everywhere. In the last chapter of Chagnon's book *Last Days of Eden,* it says that I knew the people of Mrakapiwei [Kedebabowei-teri] were sick and I didn't want to give them medicine. How can he describe what I did? He wasn't here. I was working day and night to get them medicine."[32]

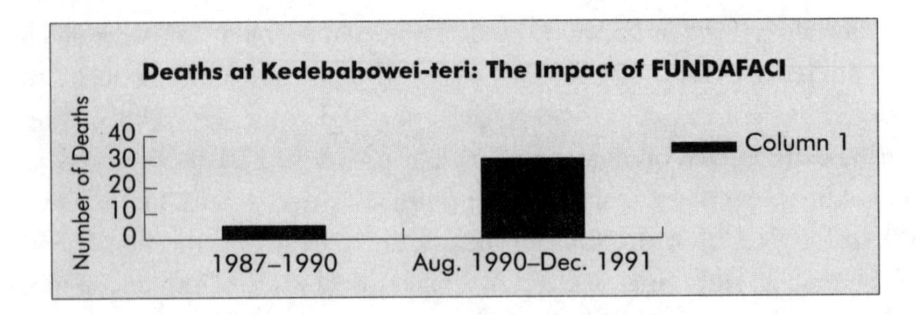

But a small window on mortality rates over just four years did not fully reveal the impact that uncontrolled contact had on the Kedebabowei-teri. This was one of the groups, known as Ironasi-teri, which had once been part of the enormous village of Mishimishimabowei-teri.[33] I showed, in chapter 7, how 27 percent of the Mishimishimabowei-teri died following the filming of *The Ax Fight.* But the faction of the Ironasi-teri was hardest hit, losing 40 percent of its people.[34] Their population began to recover until 1985, when they moved from the hill country to be near a German expedition that established itself at the juncture of the Kedebabowei River and the Mavaca. From this time onward, the Ironasi-teri became known as the Kedebabowei-teri. The only data I have for the impact of this move comes from Chagnon's listing of fifty adults, the principals in *The Ax Fight.* Remarkably, eleven of them died between 1985 and 1987, years of intense scientific activity, when both Chagnon and German researchers were in the field.[35] Many children also lost their lives during this period. In 1985, the anthropologist Jesús Cardozo witnessed two children die on his first night on the Upper Mavaca with Chagnon; Cardozo described how his mentor kept collecting data, writing numbers on the sick Yanomami, and refusing to send for help at the mission. Mortality returned to normal from 1987 to 1990, when the scientists (and filmmakers) had gone, then spiked upward to claim 22 percent of the village during the fifteen months after Chagnon set up a helicopter camp at the village.[36]

Another village that received FUNDAFACI expeditions was Washewa-teri. Called Village 93, it lost 15 percent of its population. This was the *shabono* from where the Venezuelan air force sent its Super Puma helicopter, presumably to rescue the sick Yanomani. Instead, Chagnon and his son were evacuated and the Indians were left behind.[37] Charles Brewer defended the decision to do so, explaining that only one Yanomami man was dying and that the doctors decided not to evacuate him "out of respect for the

Yanomami funeral rites because of his imminent death." Two research doctors from IVIC, Venezuela's leading think tank, had come on that air force flight. They brought a box of liquid nitrogen and vacutainers in order to collect blood samples for a viral study, but realized they had already harvested blood at Washewa-teri, and insisted on returning immediately to Caracas. The helicopter spent only a few minutes at the stricken village.[38]

"Chagnon has been saying that these remote villages are healthier, and that is not real," said Carlos Botto, a doctor who directs CAICET, Venezuela's tropical medicine center. Botto actually accompanied FUNDAFACI to Ashidowa-teri, one of the remote Siapa villages. "Only one patient in Ashidowa-teri didn't have onchocerciasis," he said, referring to the debilitating eye infection. "Sevenly percent had hepatitis antibodies. Eighty percent of the young adults had malaria. Forty percent of everyone had malaria, mostly falciparum."[39] Another member of Botto's team was America Perdamo, an anthropologist who headed the Malaria Service for Amazonas State. "That visit to Ashidowa-teri and Dorita-teri really changed my life, because I had never seen health conditions like that," she said.[40] It did not, however, change the perspective of the reporters from the *New York Times* and the Associated Press who reported nothing amiss at Ashidowa-teri.[41]

The fable about healthy remote villages created a convenient cover for government bureaucrats who did not want to spend money on the Yanomami.[42] "Brazil has twenty-six health posts for ten thousand Yanomami; Venezuela has five posts for fifteen thousand Yanomami," said Botto. "In Brazil you have one health worker for each three hundred Yanomami; in Venezuela, it is one per three thousand. In Brazil 75 percent of the Yanomami are covered by regular medical visits; in Venezuela 20 percent are."[43] Total gasoline funds, covering the whole Upper Orinoco, came to less than a thousand dollars per year. Nor a single government outboard motor was working when I was there. The doctors, all four of them, earned about one hundred dollars a month each.

John Walden, who heads Marshall University's international health department, accompanied Botto to Ashidowa-teri. "If we continue to just study the Yanomami, they're going to die," he said. "Hundreds of Yanomami are dying quietly every year. There is rampant falciparum. Villages are scattering, and they're dying right and left. You check anyone's spleen above the age of twenty, and it's as hard as a rock from the repeated insults of malaria. What are your options? Doctors are stars; they aren't going to put up with the mosquitoes and malaria. Neither are the anthropologists, at least not to provide medical assistance. My experience of twenty years in Latin American indigenous groups is that if you rely on doctors you will fail. Only health programs that incorporate missionaries will work for the Indians. You can tear your hair out, you can grind your teeth, but, in the end, the only ones who are going to stick it out in the jungle and take care of the Indians are the missionaries. The critical thing is the involvement of the evangelical missionaries and the Salesians. In my opinion, the Yanomami are the number one priority for South American emergency relief right now."[44]

When Marinho treated the people at Ashidowa-teri, a sixth of them had malaria, average for our census of the Siapa Highlands. "That is a critical situation," said Marinho.[45] It meant there was an ongoing epidemic affecting the whole region.

However revealing, a malaria census is just a snapshot. Only 22 percent of the Ashidowa-teri were under the age of ten,[46] a far smaller percentage than normal.[47] The headman Mirapewe said, "If you could count the dead, you would see how many of us there were."[48]

Notes

Abbreviations

AAA American Anthropological Association
AJDC *American Journal of Diseases of Children*
BWHO *Bulletin of the World Health Organization*
DEO U.S. Department of Energy
FOIA Freedom of Information Act
IVIC Venezuelan Institute of Scientific Investigation
JAMA *Journal of the American Medical Association*
NAA National Anthropological Archives, Smithsonian Institution, Washington, D.C.
NYT *New York Times*
TLS *Times Literary Supplement* (London)

Introduction

1. Edward O. Wilson, quoted in Eurípides Alcántara, "Indio também é gente," *Veja,* Dec. 6, 1995, p. 7.

2. Maria Manuela Carneiro da Cunha, letter to the editor, *Anthropology Newsletter,* Jan. 1989, p. 3.

3. This was a visit planned to the Catrimani Mission for the fall of 1989 in the company of a BBC documentary film crew. Davi Kopenawa and the Committee for the Creation of the Yanomami Park

threatened to stage a protest at the Boa Vista airport if the visit occurred and Chagnon's former student, Giovanni Saffirio, advised him that the visit was too problematic.

4. Pedro José Romero Farias, General de División Comandante de la Guardia Nacional, Nota informativa al Vicealmirante Ministro de la Defensa, Oct. 7, 1993, pp. 1–2.

5. Visita de carácter jornalístico documental sobre os contatos preliminares do Prof. Napoleon Chagnon com os índios Yanomami no Brasil e produção de material fotográfico para comercialização." Ministério da Justiça, Fundação Nacional do Índio, Autorização para ingresso em área indígena No. 059/CGED/95. Sept. 1, 1995.

6. Napoleon Chagnon, "Killed by Kindness?," pp. 11–12.

7. Napoleon Chagnon, "Covering Up the Yanomamo Massacre," *New York Times,* Oct. 23, 1993, Op-Ed, A-21.

8. Linda Rabben, *Unnatural Selection: The Yanomami, the Kayapó and the Onslaught of Civilization* (London: Pluto Press, 1998), p. 16.

9. He told me he did not have any money to pay for the plane. Later, I found out he had sold millions of books and CDs, that he is really quite rich, but he still had the nerve to ask for a free ride. At that time I just said the FUNAI was not in the business of giving free flights to photographers—or their assistants. He left, but then one of the New Tribes missionaries came over. He was worried because Chagnon had contracted with their pilots to fly into the jungle, but he thought there was something strange. Chagnon had struck a deal with the Rio Grande do Sul Genetics Department. Chagnon was going to collect blood samples for them. Suami Percílio Dos Santos, President of FUNAI-Roraima, Boa Vista, Brazil, July 16, 1996.

10. These expeditions, financed by the Atomic Energy Commission, collected thousands of blood samples for radiation studies. There were at least two Brazilian blood-collecting expeditions financed by the AEC in Yanomami territory—one in 1967 and the other in the early 1970s. They were both led by the geneticist James Neel. Giovanni Saffirio, who was then a missionary on the Catrimani River when Need arrived in around 1971, recalled that "Neel flew in on military planes with military personnel. This was the early period of the military dictatorship. You couldn't ask him what he was doing or why he wanted the blood." Giovanni Saffirio, Riverside, California, Jan. 1, 2000.

11. Antonio Mari, Letter to Patrick Tierney, May 12, 2000.

12. "Solicitamos dar conhecimento ao Professor Napoleon A. Chagnon quanto aos problemas que serão criados com a práctica da coletsa de amostras de sangue dos Índios Yanomami, conforme pedido feìto por pesquisador da UFRGS, podendo acarretar, até mesmo, a interrupção das actividades autorizadas por esta Fundação, acarretando a sua retida da Terra Indígena Yanomami." Otília María C. E. Nogueira, Coordenadora Geral de Estudos e Pesquisas, FUNAI, Memo No. 239/CGEP/95. "Lembramos a V.S.; que o Prof. Napoleon Chagnon obteve autorização de ingresso em área indígena Yanomami com outros objetivos e conforme normas vigentes não encaminhou a esta Coordenação qualquer projeto voltado a proposta em questâo. Otília María C. Ne. Nogueira, Coordenadora Geral de Estudos e Pesquisas, FUNAI, Memo No. 240/CGEP/95.

13. Antonio Mari, Letter to Patrick Tierney, May 10, 2000, p. 2.

14. Leda Martins, *Chagnon: O Dossier,* IX. Manuscript. Sept. 1995.

15. O Conselho Indígena de Roraima—CIR, entidade civil sem fins lucrativos, destinada a defesa dos diereitos a interesses indigenas neste estado, tomou conhecimento, com sumpresa, da autorização concedida por este órgao ao antropólogo norte americano Napoleon Chagnon, para adentrar aldeias do territorio yanomami. Chagnon tem uma longa ficha de problemas causados ao povo yanomami. Seu trabalho academico é contestado por pesquisadores de renome que discordam de suas teses e da maneira como são utilizadas para prejudicar esse povo. Segue, em anexo, relatório produzido pela jornalista Leda Martins, em que são descritos com mais detalhes os danos causados pelo antropólogo aos yanomami, principalmente os que vivem na Venezuela. O CIR vem protestar contra a Funai por ter concedido autorização de visita a terra yanomami, a pessoa danosa aos interesses e direitos deste povo, sem consultar os principais interessados, seus representantes, reinstaurando práctica que pensávamos extinta." Nelson Galé, Coordination of the Indigenous Council of Roraima, Letter to the President of FUNAI, Márcio Santilli, September 27, 1995.

16. Eurípides Alcántara, "Indio também é gente," *Veja,* De. 6, 1995, p. 8.

17. Ibid.

18. Patrick Tierney, *The Highest Altar: The Story of Human Sacrifice* (New York: Viking, 1989), pp. 412–13.

19. "One possibility [for representing the Yanomami] is Davi Kopenawa, a Spanish-speaking [*sic*] Yanomamo who has emerged as a tribal spokesman. Yet Mr. Chagnon and his supporters dismiss him as a parrot of human-rights groups and say he does not speak for the tribe. The charge has outraged Mr. Albert and others. 'It is very bad for the Yanomami that a well-known American anthropologist like

Chagnon should say this,' says Mr. Albert, who is helping David Kopenawa write his autobiography." Peter Monaghan, "Bitter Warfare in Anthropology," *Chronicle of Higher Education,* Oct. 26, 1994, A19. "Tension was never far below the surface at the meeting here. At one point, speaking from the audience, Terence Turner, a professor of anthropology at the University of Chicago, heatedly rebutted Mr. Chagnon's description of Yanomami spokesmen as 'parrots.' " "Parties in Bitter Dispute over Amazonian Indians Reach a Fragile Peace," ibid., Dec. 14, 1994, A18.

20. Diana Jean Schemo, "In Brazil, Indians Call on Spirits to Save Land," *NYT,* July 21, 1996.

21. Monaghan, "Bitter Warfare in Anthropology."

22. William Booth, "Warfare over Yanomamo Indians," *Science* 243 (1989): 1138–40.

23. Leda Martins and Patrick Tierney, "El Dorado: Lost Again?" *NYT,* April 7, 1995.

24. Patrick Tierney, letter to Gare Smith, Principal Deputy, Assistant Secretary, Bureau of Democracy, Human Rights, and Labor, Aug. 28, 1998.

25. See chapter 11 for an account of the events that led to Cecilia Matos's involvement with Napoleon Chagnon and her subsequent flight from the Venezuelan Superior Court of Salvaguarda and the Sixth Penal Court. So far, until the end of 1999, she has successfully fought attempts of the Venezuelan government to bring her back to Venezuela. In February 1998, the U.S. Department of Justice communicated to the Venezuelan government that Cecilia Matos had four bank accounts and that she had purchased a New York City apartment for $450,000 and had "a fixed rate account for $559,000." Patrick J. O'Donoghue, "Life of Riley for 'Lady Friend' Cecilia Matos Drops ex-Prez Carlos Andres Perez in Hot Soup." Vheadline.com, Feb. 9, 1998. On April 15, 1998, Judge Pedro Osmán Maldonados, president of the Superior Court of Salvaguarda, Caracas, issued an order of arrest against both Cecilia Matos and former president Carlos Andrés Pérez. The order was based on the evidence of "illegal enrichment" found in the unexplained foreign bank accounts. The judges ordered Matos to serve her arrest at a female penitentiary, should she be apprehended. Reuter, April 15, 1998. www.lanacion.com.ar. This double order of arrest by the Superior Court of Salvaguarda was confirmed by Venezuela's Supreme Court of Justice on Aug. 12, 1998. Judge Ivan Rincon Urdaneta said "the amount of money kept in the accounts far exceeded their earnings as public officials." Patrick J. O'Donoghue, "Disgraced Ex-President Carlos Andres Pérez Angered over CSJ Ruling Ratifying His House Arrest," Vheadline.com, Aug. 12, 1998. The legal process was frozen because Pérez was elected, while under house arrest, to the Venezuelan senate, in Jan. 1999. Senators enjoy immunity from prosecution in Venezuela. However, the Venezuelan Supreme Court stripped Pérez of his immunity on Oct. 21, 1999, and lawyers expected the illegal-enrichment charges to once again begin their long march through the Venezuelan justice system. CNN en Español, "Corte Suprema de Justicia reactiva juicio contra dos ex-presidentes de Venezuela," Oct. 21, 1999. The precise whereabouts of Cecilia Matos are uncertain. Her lawyer would not divulge them, and the Venezuelan special prosecutor assigned to the case, Romulo Villalba, stated, "I can't be certain that Cecilia Matos is actually in New York right now. She could be on her way to Panama or Costa Rica where it'll be more difficult to bring her to court on extradition charges. I mean, she's well-off and will have no problem establishing herself in any part of the world." Patrick J. O'Donoghue, "Special Prosecutor Still on the Trail of New Joint Bank Account Leads in Ex-Prez CAP Prosecution," Vheadline.com, April 16, 1998.

26. Napoleon Chagnon took Jaime Turón to an Open Forum meeting at the U.S. Department of State in Washington, D.C., in June 1998. Turón was elected to a municipality, Alto Orinoco, created by the state of Amazonas and whose existence was declared "null, absolutely null" by the Venezuelan Supreme Court. Corte Suprema de Justicia, Dec. 5, 1996, no. 43. Nevertheless, Turón functioned as the "mayor," despite the supreme court ruling, until Oct. 27, 1997, when another court ordered his imprisonment on charges of embezzlement of public funds: "se acuerda la Detención Judicial en el Comando de la policia local: . . . Jaime Turón, de nacionalidad Venezolano, natural de Cunucunuma, alto Orinoco . . . por la comisión del delito de peculado. . . ." Carmen Victoria Jordan, Secretario Titular de Tribunal de Primera Instancia en lo Penal y Salvaguarda del Patrimonio Público del Estado Amazonas, "Expediente Penal No. 97-5945/5982/5983." Among other things, Turón was accused of opening "private bank accounts with public money," including account no. 457-955976-9 in the Bank of Venezuela, and of making "extremely expensive improvements" on his private house with municipal funds. Turón claimed to have spent some $30,000 (17,500,000 bolivares) evacuating patients from the airstrip of Esmeralda between June and August 1996, months during which "the airstrip was totally flooded and it was impossible to land airplanes there." Copies of a check spent for personal reasons and other corroborating evidence were submitted. Melicia Velasquez, Denunciante, Juzgado de Primera Instancia en lo Penal y Salvaguarda del Patrimonio Público del Estado Amazonas, Puerto Ayacucho, Nov. 6, 1997. After this order of arrest, Turón became a fugitive from justice until lawyers in Caracas appealed the local court decision. In March 1998, he was released from the sentence of arrest pending "further investigation." Carmen Victoria Jordan, Secretaria Titular de Tribunal de Primera Instancia en lo Penal y Salvaguarda del Patrimonio Público del Es-

tado Amazonas, "No. 98-0354, exp. No. 6628-98," March 9, 1998. This means the indictment is still open and unresolved, like so many cases in the Venezuelan justice system.

27. Napoleon Chagnon, interview, Univ. of California at Santa Barbara, Oct. 2, 1995.

28. Napoleon Chagnon, interview, Univ. of California at Santa Barbara, Oct. 3, 1995.

Chapter 1: Savage Encounters

1. Charles Brewer Carías, quoted in John Quiñones, "A Window on the Past," *Prime Time Live,* July 26, 1991.

2. Issam Madi, *Conspiración al sur del Orinoco* (Caracas: self-published, 1998), pp. 71–72.

3. Marta Rodríguez, Venevisión, Caracas, Fundación Cultural Venevisión, "Los Yanomami," July 24, 1991. Charles Brewer Carías stated, "Los grupos humanos más puros que existen son estos." Padre Nelson, camcorder interview with Marta Miranda, Ocamo, May 13, 1991.

4. Napoleon Chagnon, *Yanomamo: The Fierce People,* 3d ed. (New York: Holt, Rinehart and Winston, 1983), p. 214.

5. James Brooke, "In an Almost Untouched Jungle, Gold Miners Threaten Indian Ways," *NYT,* Sept. 19, 1990.

6. James Brooke, "Reserve for Primitive Tribe Promised in 6 Months," *NYT,* Sept. 25, 1990.

7. James Brooke, "Stone Age Villages Found: Venezuela to Protect Yanomami Indians," *Gazette* (Canada), Sept. 27, 1990.

8. "Their primitive state and considerable isolation have made them ideal for studies from a biological and genetic standpoint." David Atkins and Timothy Asch, *Yanomamo: A Multidisciplinary Study, Field Notes* (Somerville, Mass.: DER, 1975), p. 2.

9. Quiñones, "A Window on the Past."

10. Spencer Reiss, "The Last Days of Eden: The Yanomamo Indians Will Have to Adapt to the 20th Century—or Die," *Newsweek,* Dec. 3, 1990, pp. 40–42.

11. Charles Brewer Carías, Napoleon Chagnon, and Brian Boom, "Forest and Man" (MS; Caracas: Fundación Explora, 1993), pp. 10–19.

12. "Gentilmente aceptaron el aterrizaje del helicóptero en el centro del *shabono* a pesar de que todos sus techos caerían con el viento." Marta Miranda, Venevisión.

13. Napoleon Chagnon, "To Save the Fierce People," *Santa Barbara* magazine, Jan.–Feb. 1991, p. 36.

14. Ameríca Perdamo, head of Malariologia, interview, Puerto Ayacucho, Venezuela, June 16, 1996.

15. Quiñones, "A Window on the Past."

16. "Because Napoleon's book is so popular, the films are used with the book in almost every University and College in the United States and also in Japan, Australia, England, Italy, France, etc." Timothy Asch, personal correspondence, Jan. 17, 1991, NAA.

17. James Neel, Timothy Asch, and Napoleon Chagnon, *Yanomama: A Multidisciplinary Study,* 43 min. (DOE, 1971).

18. Timothy Asch and Napoleon Chagnon, *The Feast,* 29 min. (DOE, 1970).

19. "Harokoiwa began rocking rhythmically sideways, violently denouncing me for killing their babies and causing epidemics among them, slapping his thighs as he rocked from side to side. . . . All the mysterious deaths in their village since I began visiting them were due to my ID photographs. They wanted to kill me to avenge all these deaths." Chagnon, *Yanomamo,* 4th ed. (Fort Worth: Harcourt Brace, 1992), pp. 236–237. See also Madi, *Conspiración al sur del Orinoco,* p. 71.

20. "The members of the Mahekototeri who feasted with the Patanowateri the next week were not inoculated and many of the members of this village died." Atkins and Asch, *Yanomamo,* p. 14. "Doritateri 15 de mayo de 1991: . . . que por causa de nuestras fotografías . . . éramos los causantes de enfermedades y culpables de la muerte de todos los Yanomami en las últimas dos décadas." Brewer Carías, Chagnon, and Boom, "Forest and Man," p. 15.

21. Napoleon Chagnon, *Studying the Yanomamo* (New York: Holt, Rinehart and Winston, 1974), p. 172.

22. Chagnon, *Yanomamo,* 4th ed., p. 237.

23. ". . . Charles Brewer-Carías, the man who has accompanied Chagnon for some years now on his trips into Yanomami territory, is indeed the owner of goldmines in Venezuela. It is more than understandable that a Yanomami, who is informed about the threats of goldmining on his own people's territory be alarmed about somebody like Brewer-Carías. . . ." Irenaus Eibl-Eibesfeldt and Gabriele Herzog-Schroder, "In Defense of the Mission" (paper, Forschungsstelle für Humanetheologie in der Max-Planck Gesellschaft, Feb. 28, 1994), pp. 6–7.

24. Madi, *Conspiración al sur del Orinoco,* p. 72.

25. Brewer Carías, "Teocracia y soberanía de Amazonas."

Chapter 2: At Play in the Field

1. Davi Kopenawa, interview, Boa Vista, Brazil, Nov. 3, 1990.

2. Napoleon Chagnon, *Yanomamo: The Fierce People,* 3d ed., (New York: Holt, Rinehart and Winston, 1983), p. 10.

3. Timothy Asch, "Bias in Ethnographic Reporting" (MS), p. 4.

4. "Between 3 and 4 million students of anthropology, just in the United States, have read one of my books about the Yanomami." Chagnon, quoted in Eurípides Alcántara, "Indio também é gente," *Veja,* Dec. 6, 1995, p. 8.

5. Peter Monaghan, "Bitter Warfare in Anthropology," *Chronicle of Higher Education,* Oct. 26, 1994.

6. David Cleary, *The Anatomy of the Amazon Gold Rush* (Oxford: Macmillan and St. Antony, 1990), p. 1.

7. Rubens Esposito, *Yanomami: Um povo ameacado de extinção* (Rio de Janeiro: Qualitymark Editora, 1998), pp. 1–35.

8. Brian Ferguson, *Yanomami Warfare: A Political History* (Santa Fe: School of American Research Press, 1995), p. 374.

9. Darna L. Dufour, "Diet and Nutritional Status of Amazonian People," in *Amazonian Indians,* ed. Anna Roosevelt (Tucson: Univ. of Arizona Press, 1994), pp. 151–76; William J. Smole, *The Yanomama Indians: A Cultural Geography* (Austin: Univ. of Texas Press, 1976), p. 19.

10. In the Parima Mountains, the Yanomami's most densely populated heartland, men were 145 mm; women, 136 mm. R. Holmes, "Nutritional Status and Cultural Change in Venezuela's Amazon Territory," in *Change in the Amazon Basin* ed. J. Hemming (Manchester: Univ. of Manchester, 1985), p. 251.

11. Kim Hill, phone interview, Jan. 17, 1995.

12. Napoleon Chagnon, *Yanomamo,* 4th ed., (Fort Worth: Harcourt Brace, 1992), pp. 189–90.

13. John D. Early and John F. Peters, *The Population Dynamics of the Mucajaí Yanomama* (San Diego: Academic Press, 1990), pp. 67–76.

14. Chagnon, *Yanomamo,* 4th ed., p. 220.

15. Napoleon Chagnon, "The Guns of Mucajaí: The Immorality of Self-deception" (MS, Sept. 1992, part of Chagnon's press kit), pp. 1–3.

16. He had also completely altered the geography of their warfare and multiplied it by using a single incident in three different contexts. See pp. 24–26, 210–14; Yanomamo Warfare, Social Organization Village Alliance, pp. 196–99.

17. Jacques Lizot, "On Warfare: An Answer to N. A. Chagnon," trans. Sarah Dart, *American Ethnologist* 21 (1994): 845–62.

18. Jeffrey Rifkin, "Ethnography and Ethnocide," *Dialectical Anthropology* 19 (1994): 295–327; Bruce Albert and Alcida Rita Ramos, "O exterminio academico dos Yanomami," *Humanidades* (Brasília), 18 (1988): 85–89; Chris J. Van Vuner, "To Fight for Women and Lose Your Lands: Violence in Anthropological Writings and the Yanomami of Amazonia," *Unisa Largen* 10, no. 2 (July 1994): 10–20; Asch, "Bias in Ethnographic Reporting."

19. The first article, on protein consumption—Napoleon Chagnon and Raymond Hames, "Protein Deficiency and Tribal Warfare in Amazonia: New Data," *Science,* 203 (1979): 910–13—is covered in chapter 16. The second, on homicide—Napoleon Chagnon, "Life Histories, Blood Revenge, and Warfare in a Tribal Population," *Science* 239 (1988): 985–92—is the subject of chapter 10. Two Yanomami experts from Chagnon's field research area, Kenneth Good and Jacques Lizot, objected to Chagnon's protein article on the grounds that it falsely presented a Maquiritare Indian village, Toki, as a traditional Yanomami community. See Kenneth Good and Jacques Lizot, letter to *Science,* appended to Marvin Harris, "Culture Materialist Theory of Band and Village Warfare," in *Warfare, Culture, and Environment,* ed. R. B. Ferguson (Orlando, Fla.: Academic Press, 1984), pp. 111–40. Their letter was never published. Nor were they, or any other expert from Chagnon's Yanomami subgroup—the Yanomami of the Upper Orinoco—consulted about the publication of the controversial homicide study. At the time of the publication of the protein study, Good and Lizot were the only other anthropologists in Chagnon's immediate research area. At the time of the second publication, there was a Salesian anthropologist, María Eguillor García, and a Venezuelan, Jesús Cardozo. None of these individuals was a reviewer for *Science.*

20. William Booth, "Warfare over Yanomamo Indians," *Science* 243 (1989): 238–40.

21. See chapter 9 for a full account of the police, congressional, and NGO reports on Brewer's illegal mining operations. Mayor Sergio Rafael Milano (Jefe), Teniente Luis Alberto Godoy y Geraldi Antonio Villaroel (Secretario), Expediente de la Comisión de la Guardia Nacional, Fuerzas Armadas de Cooperación, Comando Regional 6, Destacamento de Frontera No. 61, Puerto Ayacucho, April 18, 1984.

22. E. S., "Charles Brewer Carías: Inventario de supervivencia," *ExcesO,* April 1990, p. 65.

23. James V. Neel, *Physician to the Gene Pool: Genetic Lessons and Other Stories* (New York: John Wiley, 1994), p. 408, n. 8, 134–200, 310.

24. Redmond O'Hanlon, *In Trouble Again* (London: Hamish Hamilton, 1988), p. 15.

25. There is a two-page photograph of Brewer smiling and swimming with David Rockefeller and two of Venezuela's richest men in the shadow of a tabletop mountain, in E.S., "Charles Brewer Carías," p. 65.

26. Colonel Sergio Milano, IVIC, phone interview, Dec. 12, 1994. Milano, an anthropologist who worked on frontier security from 1984 to 1993, before taking a job at IVIC, described Brewer's mountaintop parties to me: "El hace un turismo muy exclusivo. Ultimamente ha estado pidiendo permiso para ir con helicóptero y comida muy fina arriba de los tepuis—Auyantepui, Roraima—para hacer fiestas, cosas eccentricas con millionarios Venezolanos y Americanos."

27. O'Hanlon, *In Trouble Again,* pp. 1–40.

28. E.S., "Charles Brewer Carías," p. 65.

29. "For a while, from 1990 till May 1993, they enjoyed a lot of power and let the rest of the world know it should keep off Yanomami territory which for research purposes and applied work was only theirs (Chagnon, Brewer, Matos): The Salesian missionaries were appalled by the abuses of Matos, the illegal extraction of gold from Yanomami territory in military airplanes; perhaps they were also fearful of the power Chagnon and Brewer had acquired via Cecilia Matos. . . . 2) the airforce pilots were also mad and outraged (or envious, I do not know) to be forced to fly airplanes full of gold following orders given by Cecilia Matos. Those pilots, or some of them at least, came to hate Brewer for his association with Matos and the abuse of power. The two aborted military coup-d'etats of 1992 were carried out by junior army people who were fed up by the economic and political abuses committed by Pérez' friends, including his lover Cecilia Matos." Nelly Arvelo Jiménez. "The Repudiation of Brewer Carías and Chagnon Is Due to Their Intimate Association with Goldmining" (Caracas: IVIC, 1994), p. 7.

30. John Quiñones, "A Window on the Past," *Prime Time Live,* July 26, 1991. This segment featured a map of the 18,000-square-mile area that FUNDAFACI would have administered.

31. César Dimanawa, "Carta abierta a Napoleon Chagnon," *La Iglesia en Amazonas,* March 1990, p. 20.

32. James Brooke, "In an Almost Untouched Jungle, Gold Miners Threaten Indian Ways," *NYT,* Sept. 19, 1990.

33. Venevisión's account was fairly typical: the television show claimed that mortality at the Salesian missions was two and a half times higher than that of the "purer" Indians of the Siapa Highlands. Marta Miranda, Venevisión, Caracas, Fundación Cultural Venevisión, July 24, 1991.

34. Charles Brewer Carías, "Una futura zona en reclamación," *El Nacional,* May 10, 1987.

35. Edgar López, in *El Diario de Caracas,* Sept. 2, 1993.

36. Misioneros del Alto Orinoco, "Consideraciones a un documento de Charles Brewer Carías" (Mavaca: Salesian Mission, 1991), p. 12.

37. Napoleon Chagnon has maintained that because President Pérez knew about the flights and tacitly approved them, they were legal. The *Fiscalía,* the attorney general's office, maintained that these flights were simply one more illegal activity that Pérez permitted. The legal argument seems to favor the *Fiscalía.* Federal Law 250 required scientists to receive approval for any expeditions to indigenous reserves. Although it would have been within President Pérez's power to issue a new federal decree empowering Chagnon, Brewer, and Matos, there is no record in the Official Gazette of his ever having done so. These expeditions and their fallout are the subject of chapter 11.

38. They were expelled from the village of Haximu-teri on Sept. 29, but flew out of the reserve on a small plane the next day from a nearby military base. Josefa Camargo, assistant attorney general for indigenous affairs, phone interview, Dec. 23, 1995; Terence Turner, "The Yanomami: Truth and Consequences," *Anthropology Newsletter,* May 1994, p. 48.

39. "Indígenas del Amazonas rechazan presencia de Brewer Carías y Chagnon," *El Nacional,* Sept. 14, 1993.

40. Arvelo Jiménez. "The Repudiation of Brewer Carías and Chagnon," p. 3.

41. María Yolanda García, "Cecilia Matos no iba a proteger indígenas sino a sacar oro del Amazonas," *El Nacional,* Jan. 15, 1993; Leda Martins, "Ciúme na floresta: Chagnon viola a ética médica (1990–1992)," *A Gazeta de Roraima,* March 18–24, 1996, p. 7.

42. "The promise of many presents caused some Indians to collaborate with Mr. Chagnon. This caused a division between communities: those who were on Mr. Chagnon's side and those who were against his visit. It reached the point where they fought a war resulting in three deaths, including Antonio and his brother, collaborators of Chagnon. Of course, the widows of Antonio are still waiting for the outboard motor promised to their dead husband." Juan Finkers, "Aclaraciones al Sr. Chagnon," *La Iglesia en Amazonas,* Dec. 1994, pp. 7–10.

43. "Tuvimos la oportunidad de leer las escandalosas declaraciones que dieron algunos de los pilotos que participaron en la bochornosa intentona golpista . . . para justificarse frente a la opinión pública, cuando señalaron desde su refugio en Perú, que habían participado en el intento de golpe 'indignados'

porque Charles Brewer y Cecilia Matos sacaban oro en unos bidones desde Amazonas." Issam Madi, *Conspiración al sur del Orinoco* (Caracas: self-published, 1998), p. 72.

44. Janer Cristaldo, "Os Bastidores do Ianoblefe," *A Fôlha de São Paulo,* April 24, 1994.

45. Barry Bortnick, "From Amazon Jungle to Ivory Tower," *Santa Barbara News-Press,* April 19, 1999.

46. Napoleon Chagnon, "Killed by Kindness?" *TLS,* Dec. 24, 1993, p. 11.

47. Monaghan, "Bitter Warfare in Anthropology," A19.

48. Jacques Lizot. "N. A. Chagnon, o sea: Un presidente falsificador," *La Iglesia en Amazonas,* March 1994, p. 14.

49. Monaghan, "Bitter Warfare in Anthropology," A10.

50. Terence Turner, interview, Pittsburgh, March 29, 1995. Turner was recalling his statements to colleagues at the American Anthropological Association's annual meeting in Atlanta, Dec. 1994.

51. Lesley Sponsel, Univ. of Hawaii, phone interview, Aug. 22, 1995.

52. Asch, "Bias in Ethnographic Reporting," p. 4.

53. Geoffrey W. Wrangham and Paul G. Peterson, *Demonic Males: Apes and the Origins of Human Violence* (Boston: Houghton Mifflin, 1996), chap. 5.

54. Kenneth Taylor, an anthropologist who received his Ph.D. at the Univ. of Wisconsin-Madison, told me that Chagnon's reputation for bar fighting was still remembered several years afterward. Phone interview, Jan. 27, 1995. See chapter 7 for Chagnon's bar-fighting abilities at Penn State.

55. James V. Neel, "On Being Headman," *Perspectives in Biology and Medicine* 23 (1980): 277–94.

56. Neel, *Physician to the Gene Pool,* p. 134.

57. Walter Raleigh, *The Discoverie of the Large, Rich, and Bewtiful Empire of Guiana,* ed. V. T. Harlow (London: Haklyut Society, 1928), p. 5.

58. Alain Gheerbrant, *The Amazon: Past, Present and Future* (London: Thames and Hudson, 1992), pp. 39–58.

59. Napoleon Chagnon, *Yanomamo: The Fierce People,* 2d ed. (New York: Holt, Rinehart and Winston, 1977), pp. vii–viii.

60. Ibid., pp. 113–24.

61. Johannes Wilbert, *Survivors of El Dorado: Four Indian Cultures of South America* (New York: Praeger, 1972), p. 4.

62. Smole, *The Yanoama Indians,* p. 18.

63. Napoleon Chagnon, "Yanomamo Warfare, Social Organization and Marriage Alliances" (Ph.D. diss., Univ. of Michigan, 1966), p. 137.

64. Wrangham and Peterson, *Demonic Males,* chap. 4.

65. Jodie Dawson, interview at Platanal, with Mahekoto-teri and Patanowa-teri elders, June 11, 1996.

66. Andy Jillings, director, *Warriors of the Amazon, Nova,* WGBH, Boston, 1996.

67. Interviews at Karohiteri, June and Sept. 1996. See chapter 13.

68. *Anthro* now means anyone who seems more dedicated to studying the Yanomami than to helping them—anyone who goes around with a notepad or camera or blood-collecting equipment. See chapter 3.

69. Kenneth Good, *Into the Heart: One Man's Pursuit of Love and Knowledge among the Yanomama* (New York: Simon & Schuster, 1991), pp. 313–14.

70. E.S., "Charles Brewer Carías," p. 66.

71. Mark Ritchie, *Spirit of the Rainforest: A Yanomamo Shaman's Story* (Chicago: Island Lake Press, 1995), pp. 246–50; Mark Ritchie, video interview with Kaobawa, Padamo River, Jan. 1995.

72. Alcida Rita Ramos, "Reflecting on the Yanomami: Ethnographic Images and the Pursuit of the Exotic," *Cultural Anthropology* 2 (1987): 284–304.

73. " 'Turkey's village [Bisaasi-teri] finally got a smart naba to help them and he thinks he can reproduce himself from the back-end of boys?' . . . That's how the naba became known as A.H., meaning Ass Handler." Ritchie, *Spirit of the Rainforest,* p. 144.

74. Gary Dawson, head of the Padamo mission, a translator for Mark Ritchie and also for *National Geographic Explorer,* interview, June 4, 1996.

75. Jesús Cardozo, president of the Venezuelan Anthropological Institute, phone interview, Aug. 31, 1995.

76. Bortnick, "From Amazon Jungle to Ivory Tower."

77. Frank Salamone, *The Yanomami and Their Interpreters: Fierce People or Fierce Interpreters* (Lanham, Md.: (Univ. Press of America, 1997), p. 15.

78. Jesús Cardozo, phone interview, Dec. 20, 1994.

79. Colonel Sergio Milano, IVIC, interview, Dec. 12, 1994.

80. Napoleon Chagnon, "Conversation with Jesús Cardozo," March 23, 1994, p. 6.

81. Bortnick, "From Amazon Jungle to Ivory Tower."

82. Michael Dawson, interview. Padamo mission, June 4, 1996.

83. See chapters 5 and 6 for the deaths of people filmed in *The Feast* and *The Ax Fight;* Napoleon Chagnon and Thomas Melancon, "Epidemics in a Tribal Population," in *The Impact of Contact: Two Yanomamo Case Studies,* ed. K. Kensinger (Cambridge: Cultural Survival, 1983), pp. 53–78.

84. David Thomassen, DOE, Office of Energy Research, personal correspondence in reply to FOIA request No. 9501260003. March 13, 1994, p. 3.

85. James Neel, Timothy Asch, and Napoleon Chagnon, *Yanomama: A Multidisciplinary Study,* 43 min. (DOE, 1971).

86. Anna Mitus, Ann Holoway, Audrey Evans, and John Enders, "Attenuated Measles Vaccine in Children with Acute Leukemia," *AJDC* 103 (1962): 413–17.

87. The Yanomami population in Venezuela was ten thousand at this time, and measles had a mortality rate of 25–30 percent outside the immediate mission stations of Ocamo, Mavaca, and Kosh. As I show in chapter 5, the epidemic swept the whole Orinoco from the Padamo confluence to above the Guaharibo Rapids—more than a hundred miles—and ran into the heart of the Parima Massif on the Padamo, Ocamo, and Manaviche tributaries. See also Smole, *The Yanoama Indians,* p. 50, and Ferguson, *Yanomami Warfare,* p. 309.

88. Kenneth Good, phone interview, Feb. 27, 1998.

89. Mokarita-teri, Sept. 7, 1996.

Chapter 3: The Napoleonic Wars

1. Napoleon Chagnon, letter from the field, in *The Human Condition in Latin America,* ed. Eric Wolf and Edward Hansen (New York: Oxford Univ. Press, 1972), p. 67.

2. "This particular war got started the day I arrived in the field (cause: woman stealing), and it is getting hotter and hotter." Ibid., p. 68.

3. Brian Ferguson, *Yanomami Warfare: A Political History* (Santa Fe: School of American Research Press, 1995), chaps. 13–14.

4. Brian Ferguson, phone interview, Jan. 3, 1995.

5. Jared Diamond, *Guns, Germs, and Steel: The Fates of Human Societies* (New York: W. W. Norton, 1997).

6. Ibid., p. 76.

7. John Hemming, *The Search for El Dorado* (New York, E. P. Dutton, 1978), p. 441.

8. Felipe Salvador Gilij, *Ensayo de historia americano,* vol. 2 (Caracas: Biblioteca de la Academia Nacional de Historia, 1965), p. 289.

9. John Hemming, *Amazon Frontier: The Defeat of the Brazilian Indians* (Cambridge: Harvard Univ. Press, 1987), pp. 36–37.

10. Ibid., p. 29.

11. Jacques Lizot, "Population, Resources, and Warfare among the Yanomami," *Man* 12 (1977): 497–517.

12. "For many miles the Orinoco is confined within a narrow bed of stone. Only short stretches are navigable." Inga Steinvorth Goetz, *Uriji Jami!: Life and Belief of the Forest Waika in the Upper Orinoco,* trans. Peter Furst (Caracas: Asociación Cultural Humboldt, 1969), p. 139.

13. Ibid., pp. 196–97, 194.

14. William J. Smole, *The Yanomama Indians: A Cultural Geography* (Austin: Univ. of Texas Press, 1976), pp. 220–21, n. 36.

15. Hamilton A. Rice, "The Rio Negro, the Casiquiare Canal, and the Upper Orinoco, September 1919–April 1920," *Geographical Journal* 58 (1921): 340–41.

16. Charles Hitchcock, *La región Orinoco-Ventuari: Relato de la expedición Phelps al Cerro Yavi* (Caracas: Ministerio de Educación, Nacional Dirección de Cultura, 1948), p. 34.

17. Ettore Biocca, *Yanoama* (New York: Kodansha International, 1996), pp. 206–28.

18. Luis Cocco, *Iyewei-teri: Quince años entre los Yanomamas* (Caracas: Editorial Salesiana, 1973), p. 60.

19. "Las dificultades comenzaron con Hamilton Rice, quien por miedo, los ametralló. Hamilton les ofreció desde lejos unas baratijas los guaharibos corrieron a tomarlas sin abandonar las flechas entonces aquel creyéndose atacado, hizo funcionar la ametralladora que en su lancha llevaba. Muchos guaharibos murieron en la oportunidad, y desde esa época Rodríguez Franco y yo sufrimos las consecuencias. De tiempo en tiempo pretenden prender fuego a nuestros poblados." Carlos Alamo Ibarra, *Río Negro* (Caracas: Tipografía Vargas, 1950), quoted in Cocco, Iyewei-teri, p. 60.

20. Earl Hanson, "Social Regression in the Orinoco and Amazon Basins: Notes on a Journey in 1931 and 1932," *Geographical Review* 23 (1933): 588.

21. Napoleon Chagnon, *Yanomamo: The Fierce People,* 3d ed. (New York: Holt, Rinehart and Winston, 1983), p. 29.

22. Ibid., pp. 8–9. Attitude of women, p. 114.

23. "Beastly or Manly," *Time,* May 10, 1975.

24. Napoleon Chagnon, *Yanomamo: The Fierce People,* 2d ed. (New York: Holt, Rinehart and Winston, 1977), p. 9.

25. Konrad Lorenz, *On Agression* (New York: Harcourt Brace Jovanovich, 1966), pp. 232–34.

26. Wilson, quoted in Barbara Burke, "Infanticide," *Science 84,* May 1984, p. 31.

27. Michael Harner, *The Jivaro: People of the Sacred Waterfall* (Berkeley: Univ. of California Press, 1984), pp. 134–169.

28. J. Larrick et al., "Patterns of Health and Disease among Waorani Indians of Eastern Ecuador," *Medical Anthropology* 3 (1979): 147–89.

29. Napoleon Chagnon, "Life Histories, Blood Revenge, and Warfare in a Tribal Population," *Science* 239 (1988): 985.

30. Bruce Albert, "Yanomami 'Violence': Inclusive Fitness or Ethnographer's Representation?" *Current Anthropology* 30 (1989): 631.

31. "A major difficulty in characterizing rates of violence in tribal societies with this kind of statistic is the fact that violence waxes and wanes radically over relatively short periods of time in most tribal societies, and grossly different estimates of homicide rates for the same population can be obtained from studies done of the same local group at two different periods of time, or neighboring groups at the same point in time." Chagnon, "Life Histories," p. 991, n. 24.

32. Napoleon Chagnon, "Chronic Problems in Understanding Tribal Violence and Warfare," in *Genetics of Criminal and Antisocial Behavior,* ed. G. R. Bock and J. A. Goode (Chichester, N.Y.: John Wiley, 1996), p. 217; Napoleon Chagnon. "To Save the Fierce People." *Santa Barbara* magazine, Jan.–Feb. 1991, p. 36.

33. Albert, "Yanomami 'Violence,' " p. 631.

34. Frederic Golden, "Scientist a Fierce Advocate for a Fierce People," *Los Angeles Times,* May 15, 1997.

35. "Figure 9: A Yanomamo village on a large river, atypical of their settlement locations." Napoleon Chagnon, *Yanomamo Interactive User's Guide,* p. 10.

36. Smole, *The Yanoama Indians,* p. 52.

37. Ferguson, *Yanomami Warfare,* p. 101.

38. Smole, *The Yanoama Indians,* pp. 51, 76, 72, 50.

39. The five mountain villages had 7.7, 21.4, 15.4, 0, and 12.5 percent, respectively, of *unokai* (killers) among adult men. This averages to 11 percent, though the exact number of *unokai* per village is not available. Chagnon, "Chronic Problems in Understanding Tribal Violence and Warfare," p. 224.

40. Chagnon, "Life Histories," p. 986.

41. Smole, *The Yanoama Indians,* pp. 31–32.

42. Ibid., p. 233, n. 94.

43. Napoleon Chagnon, *Yanomamo,* 3d. ed., p. 175.

44. John D. Early and John F. Peters, *The Population Dynamics of the Mucajaí Yanomama* (San Diego: Academic Press, 1990), p. 23.

45. Ibid., p. 24.

46. Ibid., pp. 74, 67.

47. Ibid., pp. 67–68.

48. Ibid., pp. 79–80.

49. Smole, *The Yanoama Indians,* p. 51.

50. Ferguson, *Yanomami Warfare,* pp. 224–54.

51. "The government malaria post invited the Bisaasi-teri to move to Boca Mavaca, where they were joined in 1959 by New Tribes missionaries down from Platanal." Ibid., p. 265.

52. Helena Valero, *Yo soy Napeyoma: Relato de una mujer raptada por los indígenas Yanomami* (Caracas: Fundación La Salle de Ciencias Naturales, 1984), pp. 24–30.

53. Ibid., pp. 12–13.

54. Ibid., 140–51.

55. Chagnon, *Yanomamo,* 2d ed., p. 79, n. 17; 3d ed., p. 38.

56. Brian Ferguson, phone interview, Jan. 3, 1995.

57. Napoleon Chagnon, *Yanomamo: Last Days of Eden* (San Diego: Harcourt Brace Jovanovich, 1992), p. xv.

58. Chagnon, *Yanomamo,* 2d ed., pp. 96, 14, 13.

59. Kaobawa, video interview by Mark Ritchie, trans. Michael Dawson, Padamo mission, Jan. 1995.

60. Mark Ritchie, phone interview, Feb. 6, 1995.

61. Chagnon, in *The Human Condition in Latin America,* p. 67.

62. James V. Neel, *Physician to the Gene Pool: Genetic Lessons and Other Stories* (New York: John Wiley, 1994), p. 146.

63. John Hemming, *The Conquest of the Incas,* (New York: Harcourt Brace Jovanovich, 1970), p. 26.

64. Terence Turner, unedited tape from an interview with Davi Kopenawa, "I fight because I am alive," Boa Vista, Brazil, 1991. Other portions of this interview were published in *Cultural Survival Quarterly* 15, no. 3 (1991): 59–64, and in the *AAA Newsletter* 32, no. 6 (1991): 52.

65. Brian Ferguson, phone interview, Jan. 3, 1995.

66. Juan Finkers, "Aclaraciones al Sr. Chagnon," *La Iglesia en Amazonas,* Dec. 1994. pp. 7–10.

67. Napoleon Chagnon, "Yanomamo Warfare, Social Organization and Marriage Alliances" (Ph.D. diss., Univ. of Michigan, 1966), p. 17.

68. Napoleon Chagnon, *Studying the Yanomamo* (New York: Holt, Rinehart and Winston, 1974), p. 91.

69. Chagnon, *Yanomamo,* 2d ed., p. 12.

70. Chagnon, *Studying the Yanomamo,* pp. 29, 95.

71. Ibid., p. 23.

72. Chagnon, "Yanomamo Warfare," p. 213.

73. Ferguson, *Yanomami Warfare,* pp. 286–88.

74. Chagnon, "Chronic Problems in Understanding Tribal Violence and Warfare," p. 217.

75. Chagnon, "Yanomamo Warfare," p. 62. Eight deaths occurred in fighting over the Shihota garden, a conflict that coincided with James Barker's arrival at Platanal in 1949, according to Helena Valero, whose chronology I am following here. Chagnon has offered two distinct chronologies, one in his Ph.D. thesis and one in the later editions of his text, *Yanomamo.* I consider both chronologies in more detail in chapter 16. However, in Chagnon's current construction, the violence that I have portrayed here between 1949 and 1951 would be an even sharper spike—all condensed into the year 1950–51. The picture for war deaths also differs substantially in Valero's and Chagnon's accounts. For the purpose of this analysis, I am accepting Chagnon's account only. "The raiders killed one of the Wanidima-tedi men, but in doing so angered both the Hasabowa-tedi and their cognate group, the Ashadowa-tedi. With the aid of the last two villages, the Wanidima-tedi [Valero and Fusiwe's Nomowei splinter] raided the Shihota-tedi [Rashawe and the young Kaobawa's Namowei splinter] and killed a younger brother of Makuwa. The Shihota-tedi reciprocated by raiding the Wanidima-tedi, killing Husiwa (Nabayoma's husband), Hoari, and Siayeikema. This in turn was followed by raids from the Wanidima-tedi in which the Shihota-tedi lost Ushuenawa and the two younger brothers of Paruriwa, and had one of their women, Bhiomi, seized." Ibid., p. 155. "In 1950 these villages [Bisaasi-teri and Monou-teri, western Namowei] were victimized by the Mowaraoba-tedi and the Iwahikoroba-tedi in a treacherous feast. They lost approximately 15 men." Ibid., p. 60. The actual date of the massacre was Feb. 1951, which Chagnon gives elsewhere in the same chapter (p. 21). The important point is that the number of deaths, in Chagnon's Ph.D. thesis history of the Namowei, comes to 23 by 1951. During Chagnon's fieldwork, the Patanowa-teri "suffered about eight deaths." Ferguson, *Yanomami Warfare,* p. 303. In Jan. 1965, the Mounou-tedi killed one Patanowa-teri man, Bosibrei. Chagnon, "Yanomamo Warfare," p. 178. The Patanowa-teri retaliated by killing Damowa, the Monou-teri headman, in March 1965. Ibid., p. 179. The Bisaasi-teri and Monou-teri then united to kill another Patanowa-teri man, unnamed, at the Shihota garden, in late April 1965. Ibid., p. 181. Sometime between Nov. 1965 and Feb. 1966, the Bisaasi-teri and Monou-teri, who now formed one group, killed one additional Patanowa-teri man. Ibid., p. 189. There were about nine war deaths among the Namowei—which included Bisaasi-teri, Monou-teri, and Patanowa-teri. Since one victim was a woman, the total male death count, which is what Chagnon gives, stood at eight. Ferguson, *Yanomami Warfare,* pp. 300–306.

Chapter 4: Atomic Indians

1. Napoleon Chagnon, *Studying the Yanomamo* (New York: Holt, Rinehart and Winston, 1974), p. 191.

2. This trip took place on Aug. 28, 1996, following a great flood on the Upper Orinoco.

3. James Neel, Timothy Asch, and Napoleon Chagnon, *Yanomamo: A Multidisciplinary Study,* 43 min. (DOE, 1971).

4. Napoleon Chagnon, *Yanomamo: The Fierce People,* 3d. ed. (New York: Holt, Rinehart and Winston, 1983), p. 199n.

5. James V. Neel, *Physician to the Gene Pool: Genetic Lessons and Other Stories* (New York: John Wiley, 1994), p. 85.

6. Napoleon Chagnon, *Yanomamo: The Fierce People,* 3d ed. (New York: Holt, Rinehart and Winston, 1983), p. 199.

7. Napoleon Chagnon, personal letter printed in *Brown Gold* 24, no. 7 (Nov. 1989): 10.

8. Pablo Mejía, interview, village of Toki, Aug. 28, 1996.

9. David Weatherall, "The Mutation Man," *New Scientist,* July 9, 1994, p. 42.

10. Neel, *Physician to the Gene Pool,* pp. 1–130.

11. Ibid., p. 392 and p. 394.

12. Ibid., pp. 372–73. "Treatment of children with this disease has the potential for increasing the frequency of mental retardation! The disease also illustrates an additional problem created by medical advances: an individual with a genetic disease who would not previously have reproduced will now do so, transmitting a gene for phenylketonuria to all his or her children."

13. Geoffrey W. Wrangham and Paul G. Peterson, *Demonic Males: Apes and the Origins of Human Violence* (Boston: Houghton Mifflin, 1996), chap. 5.

14. Neel, *Physician to the Gene Pool,* p. 129.

15. Claude Lévi-Strauss, *Saudades do Brasil* (Seattle: Univ. of Washington Press, 1995), p. 15.

16. Terence Turner, interview, Pittsburgh, March 29, 1995.

17. James V. Neel, "On Being Headman," *Perspectives in Biology and Medicine* 23 (1980): 277–94.

18. James Neel, phone interview, March 9, 1997.

19. Neel, *Physician to the Gene Pool,* p. 134.

20. Barry Bortnick, "From Amazon Jungle to Ivory Tower," *Santa Barbara News-Press,* April 19, 1999.

21. Timothy Asch, "Bias in Ethnographic Reporting" (MS), p. 6.

22. "It was 1957, and the Russians had just stunned the world with their Sputnik satellite. Like so many other competitive young Americans, Chagnon wanted to study physics and do his part for the country." Bortnick, "From Amazon Jungle to Ivory Tower."

23. Asch, "Bias in Ethnographic Reporting," p. 6.

24. William Allman, "A Laboratory for Human Conflict," *U.S. News & World Report,* April 11, 1988, pp. 57–58.

25. Neel, *Physician to the Gene Pool,* p. 408, n. 8. and p. 310.

26. Ibid., p. 138.

27. Jesús Cardozo, phone interview, Dec. 20, 1994.

28. Neel, *Physician to the Gene Pool,* p. 122.

29. Frederic Golden, "Scientist a Fierce Advocate for a Fierce People," *Los Angeles Times,* May 15, 1997.

30. Chagnon, *Yanomamo,* 3d ed., pp. 206–10.

31. David Thomassen, DOE, Office of Energy Research, personal correspondence in reply to FOIA request No. 9501260003, March, 13, 1994, pp. 3–4. There was also a control group in Ann Arbor.

32. Alejandro Arenas, M.D., interview, Bisaasi-teri, Upper Orinoco, June 7, 1997.

33. Thomassen, reply to FOIA request, p. 3.

34. John Earle, Ph.D., interview, Pittsburgh, March 7, 1997.

35. Ysbran Poortman, Ph.D., interview, Rio Negro near Manaus, Brazil, Aug. 21, 1996.

36. "Don't Let It Happen Again," *Economist,* May 17, 1997, p. 27.

37. Philip J. Hilts, "Secret Radioactive Experiments to Bring Compensation by U.S.," *NYT,* Nov. 20, 1996.

38. The Salesian Catholic order protested the AEC protocol in Aug. 1998. Until the early 1990s, the Salesian bishop in Puerto Ayacucho was the legal guardian of the Yanomami, because of a treaty between the government of Venezuela and the Catholic Church. A Salesian spokesperson declared that neither the Salesians nor the Venezuelan government was aware of what Neel and Chagnon were doing with the Yanomami blood. Javier Ignacio Mayorca, in *El Nacional,* Aug. 19, 1998.

39. Gary Dawson, head of the Padamo Mission, interview, Padamo River, Aug. 29, 1996.

40. Napoleon Chagnon, "Filming the Ax Fight," *Yanomamo Interactive CD: The Ax Fight* (New York: Harcourt Brace, 1997).

41. Timothy Asch, "Ethnographic Filming and the Yanomamo Indians," *Sightlines,* Jan.–Feb. 1972, pp. 6–12.

42. Chagnon, "Filming the Ax Fight."

43. Ibid.

44. Chagnon, *Studying the Yanomamo,* p. 168.

45. Chagnon, "Filming the Ax Fight."

46. Pablo Mejía, interviews at Toki, Aug. 28, 1996, and at Shakita, Sept. 1, 1996.

47. Kenneth Good, *Into the Heart: One Man's Pursuit of Love and Knowledge among the Yanomama* (New York: Simon & Schuster, 1991), p. 90.

48. Napoleon Chagnon, *Yanomamo,* 5th ed. (Fort Worth: Harcourt Brace, 1997), pp. 53–55.

49. Chagnon, *Yanomamo,* 3d ed., pp. 208–9.

50. Ibid., pp. 209–10.

51. Mark Ritchie, video interview with Kaobawa, trans. Gary Dawson, Padamo River, Jan. 1995.

52. Ibid.

53. Pablo Mejía, addressing the assembly at Toki, Aug. 28, 1996.

54. Neel, Asch, and Chagnon, *Yanomamo: A Multidisciplinary Study.* Neel is the narrator of the film.

55. Neel, *Physician to the Gene Pool,* p. 390.

56. Ibid., p. 310.

57. Neel, "On Being Headman," p. 287 and p. 286.

58. Neel, *Physician to the Gene Pool,* p. 200.

59. Jared Diamond, *Guns, Germs, and Steel: The Fates of Human Societies* (New York: W. W. Norton, 1997), p. 375.

60. Neel, Asch, and Chagnon, *Yanomamo: A Multidisciplinary Study.*

61. Chagnon, *Yanomamo,* 3d ed., p. 205.

62. Ferguson, *Yanomami Warfare,* p. 146.

63. Brian Ferguson, phone interview, Jan. 3, 1995.

64. Chagnon, *Yanomamo,* 3d ed., p. 200.

65. Paul Salopek, "Basically We Are All the Same," *Chicago Tribune,* April 27, 1997. This article won the Pulitzer Prize.

66. Napoleon Chagnon, "Yanomamo Warfare, Social Organization and Marriage Alliances" (Ph.D. diss., Univ. of Michigan, 1966), p. 62.

Chapter 5: Outbreak

1. James Neel, Willard Centerwall, Napoleon Chagnon, and Helen Casey, "Notes on the Effect of Measles and Measles Vaccine in a Virgin Soil Population of South American Indians," *American Journal of Epidemiology* 91 (1970): 425.

2. Vitalino Balthasar, interview, Puerto Ayacucho, Oct. 1, 1996.

3. Ibid.; Padre Jose Berno, "Crónica de la casa de Mavaca," Feb. 15, 1968; Vitalino Balthasar, interview, Puerto Ayacucho, Oct. 1, 1996.

4. Napoleon Chagnon, letter from the field, in *The Human Condition in Latin America,* ed. Eric Wolf and Edward Hansen (New York: Oxford Univ. Press, 1972) p. 68.

5. Neel et al., "Notes on the Effect of Measles," p. 421.

6. Ibid., p. 422.

7. G. S. Wilson, "Measles as a Universal Disease," *AJDC* 53 (1962): 219–23.

8. Neel et al., "Notes on the Effect of Measles," p. 421.

9. Ibid.

10. Napoleon Chagnon, *Yanomamo: The Last Days of Eden* (San Diego: Harcourt Brace Jovanovich, 1992), p. 284.

11. Napoleon Chagnon, *Yanomamo: The Fierce People,* 3d ed. (New York: Holt, Rinehart and Winston, 1983), p. 199.

12. Neel et al., "Notes on the Effect of Measles," p. 422.

13. Chagnon, *Last Days of Eden,* p. 284.

14. Fred McCrumb et al. "Studies with Live Attenuated Measles-Virus Vaccine," *AJDE* 101 (1961): 45.

15. M. Hoekenga, A. Schwartz, H. Carrizo-Palma, and P. Boyer, "Experimental Vaccination against Measles II: Tests of Live Measles and Live Distemper Vaccine in Human Volunteers during a Measles Epidemic in Panama," *JAMA* 173 (1960): 862–68.

16. P. Nagler, A. R. Foley, J. Furesz, and G. Martineau, "Studies on Attenuated Measles-Virus Vaccine in Canada," *BWHO* 32 (1965): 791–801.

17. Saul Krugman, Joan Giles, and Milton Jacobs, "Studies with Live Attenuated Measles-Virus Vaccine," *AJDC* 103 (1962): 353–63.

18. Hilleman et al., "Development and Evaluation of the Moraten Measles Virus Vaccine," *JAMA* 206 (1968): 587–90.

19. Wilson, "Measles as a Universal Disease," p. 220.

20. Richard Hornick et al., "Vaccination with Live Attenuated Measles Virus," *AJDC,* 103 (1962): 344–47.

21. J. Alaistair Dudgeon and William A. M. Cutting, *Immunization: Principles and Practice* (London and New York: Chapman and Hall Medical, 1991), p. 163.

22. Francis Black, J. P. Woodall, and P. Pinheiro, "Measles Vaccine Reactions in a Virgin Population," *American Journal of Epidemiology* 89 (1969): 168–75.

23. Jacob A. Brody et al., "Measles Vaccine Field Trials in Alaska," *JAMA* 189 (1964): 339–42.

24. Anna Mitus, Ann Holoway, Audrey Evans, and John Enders, "Attenuated Measles Vaccine in Children with Acute Leukemia," *AJDC* 103 (1962): 417.

25. Jared Diamond, *Guns, Germs, and Steel: The Fates of Human Societies* (New York: W. W. Norton, 1997), pp. 195–214.

26. Andrew David Cliff, *Measles: An Historical Geography of a Major Human Viral Disease from Global Expansion to Local Retreat, 1840–1990* (Oxford: Blackwell Reference, 1993), p. 7.

27. "The attack rate in measles is higher than for any other infectious diseases. . . . In virgin populations that have not experienced a previous visitation, susceptibility appears to be almost complete." Wilson, "Measles as a Universal Disease," pp. 219–23.

28. "It is doubtful whether the immunity resulting from an attack of any other disease is quite so strong and persistent." Ibid.

29. Stanley A. Plotkin and Edward A. Mortimer, *Vaccines* (Philadelphia and London: W. B. Saunders, 1994), p. 236.

30. Cliff, *Measles.*

31. In 1951, 99.9 percent of the people in a part of Greenland were infected by a single outbreak; in 1952, over 99 percent of all Eskimos were infected in part of the Canadian Arctic. "The difference between 100% attack rate in virgin populations and 85% attack rate in more civilized populations raises the question of genetic immunity." Wilson, "Measles as a Universal Disease," p. 222.

32. Black, Woodall, and Pinheiro, "Measles Vaccine Reactions," p. 169.

33. Charles Cockburn, Joseph Pecenka, and T. Sundaresan, "WHO-Supported Comparative Studies of Attenuated Live Measles Virus Vaccines," *BWHO* 34 (1966): 223–31; Saul Krugman, Joan Giles, Milton Jacobs, and Friedman, "Studies with a Further Attenuated Live Measles-Virus Vaccine," *Pediatrics* 31 (1963): 919–28.

34. Francis Black, Ph.D., Yale Medical School, phone interview, March 17, 1997.

35. "Overall, an average maximum temperature elevation of 1.61 C was observed in the Tiriyo and 0.60 C in 331 vaccines in other populations." Black, Woodall, and Pinheiro, "Measles Vaccine Reactions," p. 174.

36. Francis Black, phone interview, March 17, 1997.

37. Samuel Katz, M.D., phone interview, March 21, 1997.

38. Plotkin and Mortimer, *Vaccines,* pp. 238–39.

39. Hilleman et al., "Development and Evaluation of the Moraten Measles Virus Vaccine," pp. 587–90.

40. Hendrickse, R. G., D. Montefiore, G. Sherman, and G. O. Sofoluwe, "A Further Study on Measles Vaccination in Nigerian Children," *BWHO* 32 (1965): 803–8.

41. Timothy Asch, "Ethnographic Filming and the Yanomamo Indians," *Sightlines,* Jan.–Feb. 1972, p. 8.

42. G. F. Hayden, "Measles Vaccine Failure: A Survey of Causes and Means of Prevention," *Clinical Pediatrics* 18 (1979): 155–67.

43. James V. Neel, *Physician to the Gene Pool: Genetic Lessons and Other Stories* (New York: John Wiley, 1994), p. 163.

44. Diamond, *Guns, Germs, and Steel,* pp. 211–12.

45. Neel, *Physician to the Gene Pool,* p. 150.

46. "Use and Interpretation of Anthropometric Indicators of Nutritional Status," *BWHO* 64 (1986): 929–41.

47. R. Holmes, "Nutritional Status and Cultural Change in Venezuela's Amazon Territory," in *Change in the Amazon Basin,* ed. J. Hemming (Manchester: Univ. of Manchester, 1985), pp. 237–55.

48. Napoleon Chagnon and Thomas Melancon, "Epidemics in a Tribal Population," in *The Impact of Contact: Two Yanomamo Case Studies,* ed. K. Kensinger (Cambridge, Mass.: Cultural Survival, 1983), pp. 53–78.

49. Darna L. Dufour, "Diet and Nutritional Status of Amazonian People," in *Amazonian Indians,* ed. Anna Roosevelt (Tucson: Univ. of Arizona Press, 1994), p. 152.

50. "Among the Tiriyo Indians there is, however, a high incidence of malarial and heminthic infections. . . . As shown in this study, acute coincident infection can affect the reaction to measles; chronic

disease may have a similar effect." Black, Woodall, and Pinheiro, "Measles Vaccine Reactions," p. 174. See also Neel et al., "Notes on the Effect of Measles," p. 421.

51. Neel's data on Amerindians' response to the Edmonston B without gamma globulin was particularly rare. Even the WHO had trouble getting data on the Edmonston B vaccine without gamma globulin, because such a practice was not accepted in various countries by the time field trials were instituted in 1964. Cockburn et al., "WHO-Supported Comparative Studies of Attenuated Live Measles Virus Vaccines," pp. 223–31.

52. Neel et al., "Notes on the Effect of Measles," p. 421.

53. James Neel, phone interview, March 18, 1997.

54. Neel et al., "Notes on the Effect of Measles," p. 423.

55. Ibid., 421.

56. According to Neel's chronology, he had still not seen any measles rash by this date. Ibid., p. 426.

57. "Crónica de Ocamo," Feb. 4, 1968.

58. Black, Woodall, and Pinheiro, "Measles Vaccine Reactions," p. 170.

59. Timothy Asch Collection, 1968, sound tapes 4 and 6, NAA.

60. Marcel Roche, M.D., interview at *Interciencia,* Caracas, June 20, 1996.

61. Neel et al., "Notes on the Effect of Measles," p. 426.

62. Neel et al., "Notes on the Effect of Measles," p. 421.

63. Plotkin and Mortimer, *Vaccines,* p. 238.

64. Samuel Katz, M.D., phone interview, March 21, 1997.

65. Neel et al., "Notes on the Effect of Measles," p. 421.

66. Marcel Roche, M.D., interview at *Interciencia,* June 20, 1996.

67. "Crónica de Ocamo," Jan. 23, 1968: "Guinge ora la Spedizione scientifica acconpagnata dal Dottor Roche et al."

68. Asch, "Ethnographic Filming," pp. 8–9.

69. Neel et al., "Notes on the Effect of Measles," p. 422.

70. Sister Nora Gonzalez, "Crónica de Mavaca," Jan. 24, 1968.

71. Jean Pier Poirier, in "Crónica de la casa de Mavaca," Jan. 31, 1968.

72. Asch, "Ethnographic Filming," pp. 6–12.

73. "Crónica de Ocamo," Feb. 4, 1968: "Il Dottor roche mette la vacuna por il morbillo ma la reazione senbra un po' forte."

74. Padre José Berno, "Crónica de la casa de Mavaca," Feb. 15, 1968.

75. Vitalino Balthasar, interview, Puerto Ayacucho, Oct. 1, 1996.

76. Luis Cocco, *Iyewei-teri: Quince años entre los Yanomamos* (Caracas: Editorial Salesiana, 1972), p. 481.

77. Padre José Berno, "Crónica de la Casa de Mavaca," Feb. 15, 1968.

78. Padre José Berno, interview, Puerto Ayacucho, Oct. 5, 1996.

79. Cliff, *Measles,* p. 22.

80. Saul Krugman, Joan Giles, and Milton Jacobs, "Studies on an Attenuated Measles Vaccine-Virus," *New England Journal of Medicine* 263 (July 28, 1960): 174. S. Katz, phone interview, March 21, 1997.

81. Carlos Botto, interview at CAICET, Oct. 6, 1996.

82. Francis Black, phone interview, March 17, 1997.

83. Constantino and Paul Georgescu Pipera, *Del Orinoco al Río de la Plata* (Barcelona: Ediciones del Serbal, 1987), p. 53.

84. James Neel, phone interview, March 18, 1996.

85. "The ratio of sub-clinical to clinical responses varies greatly from one infectious disease to another. Measles occupies a distinctive position within the table with over 99 percent of those infected showing clinical features. Thus . . . measles is a readily recognizable disease with a low proportion of both misdiagnosed cases and of sub-clinical cases." Cliff, *Measles,* p. 7.

86. "Transmission from exposed immune asymptomatic persons has not been demonstrated. . . ." Plotkin and Mortimer, *Vaccines,* p. 235.

87. "Measles is transmitted primarily from person to person by large respiratory droplets but can also be spread by the airborne route as aerosolized droplet nuclei." Ibid.

88. John Enders, K. McCarthy, Anna Mitus, and W. J. Cheatham, "Isolation of Measles Virus at Autopsy in Cases of Giant-Cell Pneumonia without Rash," *New England Journal of Medicine* 261 (Oct. 29, 1959): 875–81.

89. Neel et al., "Notes on the Effect of Measles," p. 421.

90. Vitalino Balthasar, interview, Puerto Ayacucho, Oct. 1, 1996.

91. Neel et al., "Notes on the Effect of Measles," p. 423.

92. Samuel Katz, phone interview, March 19, 1997. "[A] valid comparison must make allowance for

the difference in method of recording temperature. The difference between axillial and rectal temperature has been variously estimated as about 2 C." Black, Woodall, and Pinheiro, "Measles Vaccine in a Virgin Population," p. 171. The Yanomami's temperature at Ocamo was especially striking since it was recorded on a single day. In any group study, children reach their peak temperature over several days, so the maximum temperature on any given day always understates the overall maximum temperature. For instance, the highest temperature recorded with the Edmonston B vaccine in the United States was 102.9 degrees Fahrenheit. But the highest daily temperature was recorded on the eighth day after vaccination, and it was only 101.1. Krugman, Giles, and Jacobs, "Studies on an Attenuated Measles Virus-Vaccine," p. 174.

93. Inga Steinvorth Goetz, *Uriji Jami!: Life and Belief of the Forest Waika in the Upper Orinoco,* trans. Peter Furst (Caracas: Asociación Cultural Humboldt, 1969), p. 56.

94. Padre José Berno, March 22–April 14, 1968. "Crónica de la Casa de Mavaca," See also May 8: "Dottor Pereira Direttor del Servizzio Cooperativo il Dottor Daboin e un infermiere proseguere al per il Platanal dove l'epidemia continua a for vittime."

95. Carlos Botto, interview at CAICET, Oct. 6, 1996.

96. Although hepatitis has been present for over a generation among the Yanomami, patients have shown "infection without developing antibodies against the surface antigen, which indicates an absence of immune protection." Nahir Martinez, *Health Problems in Isolated Yanomami Communities: Viral Hepatitis in the Upper Orinoquito River"* (Puerto Ayacucho: CAICET, 1996).

97. Arango et al. "Asociación de Antigenemia con Depresión de la Hipersensibilidad Cutánea Retardada en la Onconcercosis," *Proicet Amazonas* (Caracas), no. 2 (1983): 101–8.

98. "The Swiss Professor Louis Agassiz noted in 1865 that the Indians suffered most from malaria. 'It is a curious thing that the natives seem more liable to the maladies of the country than strangers.' " Hemming, *Amazon Frontier,* p. 280.

99. James Neel, Timothy Asch, and Napoleon Chagnon, *Yanomamo: A Multidisciplinary Study,* 43 min. (U.S. DOE, 1971).

100. "Gary Dawson, interview, Kosh, Padamo River, Aug. 29, 1996.

101. Vitalino Balthasar, interview, Puerto Ayacucho, Oct. 1, 1996.

102. Juan González, interview, Padamo mission, June 14, 1996.

103. Maria Wachtler, phone interview, June 20, 1996.

104. Padre José Berno, "Crónica de la casa de Mavaca," Feb. 15, 1968: "Hoy se vacunaron contra el sarampión 71 personas."

105. Sound Roll 3, Measles Epidemic, Feb. 18, 1968, Timothy Asch Collection, NAA.

106. Sound Roll 9, Feb. 23, 1968, Timothy Asch Collection, NAA.

107. Juan González, interview, Kosh, Padamo River, June 14, 1996.

108. All the preceding quotations to the beginning Neel's camera instructions to Asch are from Sound Roll 3, Feb. 18, 1968, Timothy Asch Collection, NAA.

109. Ibid.

110. "Parece que de los sibarioteri se salvaron sólo los que pudieron llegar a Manavice y fueron atendidos por los enfermeros venidos del Tamatama. Que Dios proteja a estos pobres indios Yanomami." Padre José Berno, in "Crónica de la casa de Mavaca," April 14, 1968.

111. Timothy Asch Collection, Sound Roll 3, Feb. 18, 1968, NAA.

112. Ramsay and Emond, quoted in Cliff, *Measles,* p. 23.

113. All of the quotations here were part of the long dialogue already cited. Sound Roll 3, Feb. 18, 1968, Timothy Asch Collection, NAA.

114. Rousseau: "Por los efectos de la vacuna ahora si vienen brotes de sarampión." Chagnon: "Bueno, la vacuna da un efecto casi igual como él." Rousseau: "Es igual, no? Si hay brotes, verdad, veremos. Estando aquí el médico, se lleva el médico. . . ." Ibid.

115. Ibid.

116. This will be laid out in detail in chapter 6. The plane bringing additional medicine and a Venezuelan doctor did not arrive until Monday, Feb. 26. Sound Roll 5, Radio Conference, Patanowa-teri, Feb. 22, 1968, Timothy Asch Collection, NAA.

117. "Marcel and Nap were at Ocamo but somehow came without the gamma globulin that should have been administered with the vaccine." James Neel, phone interview, March 18, 1997.

118. Padre José Berno, "Crónica de la casa de Mavaca," April 14, 1968.

119. Sound Roll 3, Feb. 18, 1968, Timothy Asch Collection, NAA.

120. This sequence was filmed. Neel, Asch, and Chagnon, *A Multidisciplinary Study.*

121. This is a brief conversation, still on Sound Roll 3, which corresponds to Film Roll 5, a part of which appears in the establishing scenes of *A Multidisciplinary Study.* The images of Father Sánchez and his two Brazilian helpers were divorced from the sound track in the final version of the film. In reality,

Sánchez exchanged some pleasantries with Chagnon and Brewer. They asked him when he was going downriver, he said he would be going downriver when "la doctora," meaning Inga Goetz of IVIC, would return from her expedition to the Orinoco headwaters. There is no hint of an ongoing crisis. And, since Sánchez did not have a radio—all radio communication went through Mavaca—he would not have known about the measles epidemic unless they had told him.

122. Neel et al., "Notes on the Effect of Measles," p. 422.

123. Chagnon, *Last Days of Eden,* p. 286.

124. "Two Brazilians employed at the Salesian Mission at Platanal, the mission highest on the Orinoco and closest to the Yanomama heartland, visited the Ocamo Mission at the height of the outbreak there; approximately two weeks later, back in Platanal, both of them developed measles, thereby initiating a new focus." Neel et al., "Notes on the Effect of Measles," p. 422.

125. Brewer says to the bishop, "Tenemos necesidad . . . de que el gobierno tome parte en este programa porque aparentemente la situación se esta poniendo más grave cada vez y en este momento no podemos predecir cuál será el curso en que seguirá esta epidemia, o este ataque de sarampión. . . ." The bishop answers, "El asunto de esta mañana ha sido que ha habido problemas allí que no se daban. . . ." Sound Roll 3, Feb. 18, 1968, Timothy Asch Collection, NAA.

126. Ibid. Translation by *The New Yorker,* Sept. 20, 2000.

127. See chapter 6, and Sound Roll 5, Radio Conference, Patanowa-teri, Feb. 22, 1968, Timothy Asch Collection, NAA.

128. Padre José Berno, "Crónica de la casa de Mavaca," March 13, 1968.

129. Neel et al., "Notes on the Effect of Measles," p. 427, table 4.

130. Mark Papania, M.D., Centers for Disease Control, Atlanta, phone interview, May 22, 1996.

131. Neel, *Physician to the Gene Pool,* p. 162.

132. Neel et al., "Notes on the Effect of Measles," p. 422.

133. Vitalino Balthasar, interview, Puerto Ayacucho, Oct. 1, 1996.

134. Sound Roll 3, Feb. 18, 1968, Timothy Asch Collection, NAA.

135. Wilson, "Measles as a Universal Disease," pp. 219–23.

136. Mark Papania, M.D., phone interview, May 22, 1996.

137. Ibid.

138. Samuel Katz, Letter to Terence Turner, Sept. 28, 2000.

139. Mark Papania, phone interview, May 22, 1996.

140. Joseph E. Jackson, M.D., M.P.H., Dow Chemical, Letter to James Neel, American Philosophical Society, Jan. 6, 1971. The purpose of Neel's proposed study was fourfold: (1) to test the effectiveness of the vaccine; (2) to determine reactions; (3) to estimate the differences in either reactions or immune responses between South American Indians and North American urban populations; (4) to provide anti-measles protection. James V. Neel, "An Evaluation of the Safety and Efficacy of Live, Attenuated Rubeola, Rubella and Mumps Vaccine Administered in Combination," Jan. 18, 1971, James V. Neel Collection, American Philosophical Society.

141. Irving Devore, Letter to *The New Yorker,* Sept. 25, 2000.

142. Neel et al, "Notes on the Effect of Measles," p. 423.

143. The histories of how people died in villages on the Ocamo River is unknown. Even the devastation of Shubariwa-teri, mentioned in the mission chronicles and remembered by Protestant missionaries, is a mystery. Many people fled into the forest and were never seen again.

144. Neel et al, "Notes on the Effect of Measles," pp. 426–427.

145. Sound Roll 3, Feb. 18, 1968, Timothy Asch Collection, NAA.

146. Adelfa Betancourt, Venezuelan Ministry of Health, phone interview, April 6, 2000.

147. Neel et al, "Notes on the Effect of Measles," p. 425. Toototobi suffered twelve deaths and Mucajai one death.

Chapter 6: Filming the Feast

1. Sound Roll 8, Patanowa, Shanishani River, Feb. 23, 1968, Timothy Asch Collection, NAA.

2. This account was given to the explorer Alain Gheerbrant in 1951. Luis Cocco, *Iyeweit-teri: Quince años entre los Yanomamos* (Caracas: Editorial Salesiana, 1973), pp. 82–83.

3. Napoleon Chagnon, *Yanomamo,* 4th ed. (Fort Worth: Harcourt Brace, 1992), pp. 236–37.

4. Jay Ruby, "Out of Sync: The Cinema of Tim Asch," *Visual Anthropology Review* 11, no. 1 (Spring 1995): 22.

5. Ibid. This conference was coordinated by Jean Rouch.

6. Mark Ritchie has recorded Yanomami testimonies about a number of staged films, some comic and

some tragic. In one case, a Yanomami woman who refused to take her clothes off angered "the antros" at Iyewei-teri; in another, a Venezuelan official insisted on filming Yanomami violence at the village of Kosh and helped create a conflict that eventually resulted in two deaths. Mark Ritchie, *Spirit of the Rainforest: A Yanomamo Shaman's Story* (Chicago: Lake Island Press, 1995), pp. 187, 195–212.

7. "Foreigners of all varieties ask and pay the Yanomamo Indians to take off their clothes for the camera in order to look more primitive. This happened almost yearly on the Ocamo and Orinoco Rivers during the decades of the '70's, '80's and '90's." Ibid., p. 252.

8. "The idea of Yanomami using cameras to film others is something that bewilders outsiders to the area and we have constantly to explain the goals of the project. . . . José Seripino, our master cameraman, also told me that while visiting La Esmeralda, a member of the Ministry of the Environment approached him asking why the Yanomami did not allow tourists to film them while, on the other hand, they (the Yanomami) used cameras themselves with which they filmed tourists and other *nape* who visited the area." Jesús Cardozo, report to Timothy Asch, April 9, 1992, pp. 3–4, Timothy Asch Collection, NAA.

9. Napoleon Chagnon, *Studying the Yanomamo* (New York: Holt, Rinehart and Winston, 1974), pp. 113, 111.

10. Waloiwa, interview, village of Guarapana, Upper Orinoco, trans. Jodie Dawson, June 9, 1996.

11. Chagnon, *Yanomamo,* 4th ed., pp. 236–37.

12. César Dimanawa, "Carta abierta a Napoleon Chagnon," *La Iglesia en Amazonas,* March 1990, p. 20.

13. César Dimanawa, interview, Mavakita, Sept. 2, 1996.

14. Patricia Asch, phone interview, May 16, 1997.

15. Karl Heider, *Grand Valley Dani: Peaceful Warriors* (New York: Holt, Rinehart and Winston, 1979), p. 18.

16. Timothy Asch, "Bias in Ethnographic Reporting" (MS), p. 3.

17. "Some, if not most, of the biomedical people held openly contemptuous views of anthropology and anthropologists and regularly made insulting comments to me about anthropological research amounting only to 'collecting anecdotes' as compared to their loftier work, the collection of 'scientific data.' Some of them regarded me as hardly more than a well-educated assistant who could speak both Yanomamo and Spanish, cook meals, wash dishes, repair outboard motors in the dark, keep the aggressive Yanomamo at bay, and do other useful things that advanced their own scientific careers." Napoleon Chagnon, "Filming the Ax Fight," *Yanomamo Interactive CD: The Ax Fight* (New York: Harcourt Brace, 1997).

18. Chagnon, *Studying the Yanomamo,* p. 260.

19. Asiawe, interview, Platanal, Upper Orinoco, Sept. 26, 1999.

20. Napoleon Chagnon, *Yanomamo: The Fierce People,* 2d ed. (New York: Holt, Rinehart and Winston, 1977), p. 127.

21. James Neel, Timothy Asch, and Napoleon Chagnon, *Yanomamo: A Multidisciplinary Study,* 43 min. (DOE, 1971). This feast was also filmed by Chagnon with his 16mm Bolex camera in 1965, and later included in the ethnographic section of *A Multidisciplinary Study.*

22. "Some even put on bright red loincloths which I had traded to them, as the warrior line-up is a spectacle in which the younger men can show off to the girls." Chagnon, *Yanomamo,* 2d ed., p. 130.

23. Joe Dawson, interview, Padamo mission, June 14, 1996.

24. Napoleon Chagnon, "Yanomamo Warfare, Social Organization and Marriage Alliances" (Ph.D. diss., Univ. of Michigan, 1966), pp. 133, 58.

25. Chagnon, *Yanomamo,* 2d ed., p. 1.

26. Chagnon, "Yanomamo Warfare," p. 58.

27. "Thus, I conclude that any factor that increases the sexual activity of females or decreases the female/male ratio reduces the frequency of fighting." Ibid., p. 199.

28. Although Chagnon has stressed the differences between Shamatari Yanomami and Namowei Yanomami, his sex-ratio statistics show great internal divergence between the Patanowa-teri Namowei, who did not have canoes, steel wealth, or shotguns, and the Bisaasi-teri/Monou-teri Namowei, who possessed all of these advantages. The Patanowa-teri sex ratios can be deduced from the overall Namowei figures when the Bisaasi-teri population is subtracted out. In Chagnon's early writing, Villages 8 and 18 are Patanowa-teri. In summary, although Bisaasi-teri and Monou-teri had a combined surplus of marriageable women, the overall Namowei had a deficit, meaning that Patanowa-teri had a severe deficit. Chagnon, *Studying the Yanomamo,* p. 159, table 4.9.

29. Chagnon, "Yanomamo Warfare," pp. 174–91.

30. Although the villages are not identified, the feast sequence in the film was not filmed at the same village as the raid sequence. The first was at Bisaasi-teri's new, large *shabono* on the Orinoco; the second was inland, at Monou-teri. Neel, Asch, and Chagnon, *A Multidisciplinary Study.*

31. Chagnon, *Yanomamo,* 2d ed., p. 135.

32. Brian Ferguson, *Yanomami Warfare: A Political History* (Santa Fe: School of American Research Press, 1995), p. 300.

33. Chagnon, *Yanomamo,* 2d ed., p. 136.

34. Ibid., p. 135.

35. Alfredo Aherowe, interview, Platanal, Sept. 26, 1996.

36. Mahekoto-teri was involved in two filming events with Chagnon. One, described in this chapter, became *The Feast;* the other, an alliance celebration with Bisaasi-teri, became the subject of feasting in *The Fierce People.* In both instances, Alfredo Aherowe's father, Asiawe, played the role of the "visitor's delegate." At Kaobawa's feast, "Asiawa entered the clearing, touching off an explosion of wild cheering that marked the opening of the dance." Chagnon, *Yanomamo,* 2d ed., p. 110.

37. Alfredo Aherowe, interview, Platanal, June 11, 1996.

38. "Shortly after I left the field, Paruriwa and his group separated from Kaobawa's and moved across the Orinoco." Chagnon, "Yanomamo Warfare," p. 185.

39. "The Feast (1970) was based on my then somewhat unusual ethnological argument that the Yanomamo feast was not a 'first-fruits' or 'harvest' ceremony as had been claimed by several anthropologists, but was primarily a *political* event." Chagnon, "Filming the Ax Fight."

40. Their work of art, *The Feast,* eventually won first prize in every film competition at which it was entered, including the American Film Festival and Cine 1970. Ibid.

41. Timothy Asch, "Ethnographic Filming and the Yanomamo Indians," *Sightlines,* Jan.–Feb. 1972, p. 8.

42. "He had, it seeme[d] to me, begun to change in the last few hours. I felt he was taking on attributes of the people he had studied so long, and it seemed I was all the more alone. . . . They looked like a very grim bunch of friends indeed, painted black with charcoal." Ibid., p. 9.

43. Ibid., p. 9.

44. Asch, "Bias in Ethnographic Reporting," p. 14.

45. James Neel, Willard Centerwall, Napoleon Chagnon, and Helen Casey, "Notes on the Effect of Measles and Measles Vaccine in a Virgin Soil Population of South American Indians," *American Journal of Epidemiology* 91 (1970): 422.

46. Here I am accepting the Ocamo mission chronicle's record that Roche, Asch, and Chagnon arrived on Tuesday, Jan. 23, the day following the arrival of a cargo plane at Esmeralda. This account coincides with Asch's article in *Sightlines.* If they had arrived on Tuesday and, according to Asch, left "the next day" for the feast at Reyabobowei-teri, it would have been Jan. 24.

47. "Piu di trenta con febre a 40, fra cui sette cosi gravi con pericolo di morte." Sister Nora González, "Crónica de Mavaca," Jan. 24, 1968.

48. Neel et al., "Notes on the Effect of Measles," p. 421.

49. Napoleon Chagnon, *Yanomamo: The Last Days of Eden* (San Diego: Harcourt Brace Jovanovich, 1992), p. 284.

50. Asch, "Ethnographic Filming," pp. 9, 11.

51. Asch, "Bias in Ethnographic Reporting," p. 14.

52. Chagnon, *Studying the Yanomamo,* p. 266.

53. Asch, "Ethnographic Filming," p. 10.

54. Ibid.

55. Ibid., p. 11.

56. Danny Shaylor, interview, Tama Tama, Upper Orinoco, Sept. 28, 1996.

57. Asch, "Ethnographic Filming," p. 11.

58. Juan de la Cruz y Olmedilla, *Mapa geográfico de América Meridional* (Madrid, 1771, 1775), reproduced in Luis Cocco, *Iyewei-teri:* Quince años entre los Yanomamos (Caracas: Editorial Salesìana, 1972), pp. 35–36.

59. Chagnon, *Studying the Yanomamo,* p. 84.

60. Ferguson, *Yanomami Warfare,* p. 383, n. 1.

61. James Brooke, "In an Almost Untouched Jungle, Gold Miners Threaten Indian Ways," *NYT,* Sept. 19, 1990.

62. Chagnon "Yanomamo Warfare," p. 183; Ferguson, *Yanomami Warfare,* pp. 302–3.

63. "They had managed to kill two Patanowa-teri and abduct two women. The Patanowa-teri only killed one Monou-teri, the headman. Hence, the Monou-teri, at least for the time being, came out ahead. The Patanowa-teri will not cease raiding them until they kill at least one more Monou-teri. . . ." Napoleon Chagnon, *Yanomamo,* 2d ed., p. 137.

64. "If any raids had occurred after early 1966, they did not produce any deaths on either side, and the score remained two or three to one against Patanowa-teri. . . ." Ferguson, *Yanomami Warfare,* p. 318.

65. Chagnon "Yanomamo Warfare," p. 62.
66. Ruby, "Out of Sync," pp. 19–35.
67. Ibid.
68. "The Patanowa-teri had moved into the village of their only ally, the Ashadowa-tedi." Chagnon, "Yanomamo Warfare," p. 181.
69. Asch, "Ethnographic Filming," p. 11.
70. Asiawa, interview, Platanal, trans. Jodie Dawson, June 11, 1996.
71. Timothy Asch and Napoleon Chagnon, *Kaobawa Trades with Reyabobowei-teri,* 8 min. (Somerville, Mass: DER, 1971).
72. Cocco, *Iyewei-teri,* p. 343.
73. Kayopewe, interview, Platanal, trans. Jodie Dawson, June 11, 1996.
74. Jepewe, Karohiteri, Manaviche River, trans. Jodie Dawson, June 7, 1996.
75. Neel, Asch, Chagnon, *A Multidisciplinary Study.*
76. Chagnon, "Filming the Ax Fight."
77. The sound rolls are helpful because each one is labeled, making it possible to match the visual scenes with places and events more specifically than the narrative permits. Sound Roll 2, Platanal, Feb. 19, 1968, Timothy Asch Collection, NAA.
78. Although the expedition went up the Shanishani River early on Monday, Feb. 19, the footage of the expedition's official arrival, with a ceremonial welcome for Chagnon in his feathers, was not until 11:30 A.M., Wednesday, Feb. 21. Sound Roll 4, Feb. 19–Feb. 21, 1968, Timothy Asch Collection, NAA.
79. "Abandoned former shabono of the Pananowueteri. Under the weight of the fast encroaching vegetation many roofs have already caved in, but the plantain gardens continue to bear fruit." From the location, "about a four mile walk from the Orinoco," and from the new *shabono*'s distant relocation "much further inland from the river, close to the distant mountain range in the photo at left," there is no doubt this photo represents Patanowa-teri's old garden site where *The Feast* and *A Multidisciplinary Study* were later filmed. Inga Steinvorth Goetz, *Uriji Jami!: Life and Belief of the Forest Waika in the Upper Orinoco,* trans. Peter Furst (Caracas: Asociación Cultural Humboldt, 1969), pp. 22–23.
80. Timothy Asch Collection, Sound Roll 4, Patanowa-teri, Feb. 21, 1968, NAA.
81. Danny Shaylor, interview, Tama Tama, Upper Orinoco, Sept. 26, 1996.
82. Sound Roll 6, Feb. 21, 1968, Timothy Asch Collection, NAA.
83. Daniel Reff, quoted in *Sanuma Memories: Yanomami Ethnography in Times of Crisis* (Madison: Univ. of Wisconsin Press, 1995), p. 308.
84. Albert L. Hurtado, *Indian Survival on the California Frontier* (New Haven: Yale Univ. Press, 1988), p. 46.
85. "La enfermedad respiratoria aguda es una de las causas más frecuentes de morbilidad y mortalidad tanto en las poblaciones de Parima como en el Alto Orinoco. Esta incluye procesos catarrales—gripe, bronquitis—de las vías respiratorias superiores, asma y neumonía." Carlos Botto, "La situación de salud de la población Yanomami," *La Iglesia en Amazonas* (Puerto Ayacucho, 1991), p. 12.
86. Goetz, *Uriji Jami!,* p. 56.
87. David Atkins and Timothy Asch, *Yanomamo: A Multidisciplinary Study, Field Notes* (Somerville, Mass.: DER, 1975) p. 2.
88. Sound Roll 9, Patanowa-teri, Feb. 23, 1968, Timothy Asch Collection, NAA.
89. Sound Roll 3, Mavaca Measles Epidemic, Feb. 18, 1968, Timothy Asch Collection, NAA.
90. Sound Roll 6, Patanowa-teri, Feb. 21, 1968, Timothy Asch Collection, NAA.
91. Neel, Asch, and Chagnon, *A Multidisciplinary Study.*
92. Chagnon, *Studying the Yanomamo,* p. 20; Chagnon, *The Last Days of Eden,* p. 284.
93. On Feb. 18, Neel mentions that they have given the Protestant missionaries an undisclosed amount of vaccine. Sound Roll 3, Mavaca Measles Epidemic, Feb. 18, 1968, Timothy Asch Collection, NAA. Father Berno mentions that Rerebawa had traveled to the Padamo River—and this had to be with Chagnon—where, according to Berno, Rerebawa came down with measles.
94. Sound Roll 6, Patanowa-teri, Feb. 21, 1968, Timothy Asch Collection, NAA.
95. Sound Roll 5, Patanowa-teri, Feb. 21, 1968, Timothy Asch Collection, NAA.
96. Danny Shaylor, interview, Tama Tama, Upper Orinoco, Sept. 26, 1996.
97. Sound Roll 5, Patanowa-teri, Feb. 21, 1968, Timothy Asch Collection, NAA.
98. "The expedition did similar studies of nearly 40 Yanomami villages, as well as seven Maquiritare villages and three Xavante villages during a two-month period." Atkins and Asch, *A Multidisciplinary Study, Field Notes,* p. 18.
99. Ibid., p. 2.
100. "After they've reacted to the vaccination we've given them, we'll come back and get them [Bisaasi-

teri] and they can come back to the village [Patanowa-teri], and perhaps at that time they can show us where the Ashitowa-teri live. And when they do that we can invite the Ashitowa-teris all to collect their blood samples." Sound Roll 3, Mavaca, Feb. 18, 1968, Timothy Asch Collection, NAA.

101. Sound Roll 3, Mavaca, Feb. 18, 1968, Timothy Asch Collection, NAA.

102. Steinvorth Goetz, *Uriji Jami!,* p. 88.

103. Ferguson's estimate of 25–30 percent mortality is within historical norms. Ferguson, *Yanomami Warfare,* p. 309. But the demographics of Hasupuwe-teri suggest that losses were over one-third of the villages.

104. Atkins and Asch, *A Multidisciplinary Study, Field Notes,* p. 14.

105. Jacob A. Brody et al., "Measles Vaccine Field Trials in Alaska," *JAMA* 189 (1964): 339–42.

106. "Regresaron o trajimos en el bongo a todos los chori, bastante enfermos y se encontraron con la nueva de la muerte por neomonía de la patayoma de los Uitocaueteri." José Berno, "Crónica de la casa de Mavaca," Feb. 29, 1968.

107. Sound Roll 3, Feb. 18, 1968, Timothy Asch Collection, NAA.

108. Neel et al., "Notes on the Effect of Measles," p. 423.

109. Sound Roll 4, Patanowa-teri, Feb. 19, 1968, Timothy Asch Collection, NAA.

110. Ibid.

111. Neel, Asch, and Chagnon, *A Multidisciplinary Study.*

112. Sound Roll 5, Feb. 22, 1968, Timothy Asch Collection, NAA.

113. Sound Roll 6, Patanowa-teri, Feb. 21, 1968, Timothy Asch Collection, NAA.

114. Sound Roll 5, Patanowa-teri, Feb. 21, 1968, Timothy Asch Collection, NAA.

115. Neel, Asch, and Chagnon, *A Multidisciplinary Study.*

116. Asch, "Ethnographic Filming," p. 11.

117. Sound Roll 8, Patanowa-teri, Feb. 23, 1968, Timothy Asch Collection, NAA.

118. Sound Roll 5, Feb. 22, 1968, Timothy Asch Collection, NAA.

119. Sound Roll 8, Feb. 23, 1968, Timothy Asch Collection, NAA.

120. Neel, Asch, and Chagnon, *A Multidisciplinary Study.*

121. Sound Roll 8, Feb. 23, 1968, Timothy Asch Collection, NAA.

122. Sound Roll 6, Patanowa-teri, Feb. 21, 1968, Timothy Asch Collection, NAA.

123. Neel, Asch, and Chagnon, *A Multidisciplinary Study.*

124. This phrase was translated, separately, by José Bórtoli and by Alfredo Aherowe. In Spanish, Aherowe's translation was "Shaki, amas a tu hermano?" Platanal, Sept. 3, 1996.

125. Neel, Asch, and Chagnon, *A Multidisciplinary Study.*

126. Sound Roll 10, Patanowa-teri garden site, Feb. 23, 1968, Timothy Asch Collection, NAA.

127. "While you're over there, could you inquire as to whether or not any of the trade goods for the Indians have come in on the plane that landed last Monday at Esmeralda? We'll pick these up when Dr. Neel comes through Mavaca, and then we'll come back upriver to Patanowa-teri, where I'm going to be staying for several days." Sound Roll 5, Patanowa-teri, Feb. 22, 1968, Timothy Asch Collection, NAA.

128. Asch, "Ethnographic Filming," p. 12.

129. Sound Roll 6, Patanowa-teri, Feb. 21, 1968, Timothy Asch Collection, NAA.

130. Asch, "Bias in Ethnographic Reporting," p. 18.

131. Neel, Asch, and Chagnon, *A Multidisciplinary Study.*

132. Cocco, *Iyewei-teri,* p. 348.

133. Helena Valero, *Yo soy Napeyoma: Relato de una mujer raptada por los indígenas Yanomami* (Caracas: Fundación La Salle de Ciencias Naturales, 1984), pp. 282–283.

134. "El cacique o el dueño del reahu ordena a las mujeres de su grupo que vayan a invitar a las mujeres visitantes para que canten. . . . Una de las mujeres, abrazándose a otras dos, comienza a cantar. . . ." Cocco, *Iyewei-teri,* p. 348.

135. Sound Roll 5, Patanowa-teri, Feb. 21, 1968, Timothy Asch Collection, NAA.

136. Neel, Asch, and Chagnon. *A Multidisciplinary Study.*

137. Ibid.

138. Ibid.

139. Trans. Alfredo Aherowe and Asiawe, Platanal mission, Sept. 3, 1996.

140. Michael Dawson, interview, Padama mission, June 4, 1996.

141. Jacques Lizot, *Tales of the Yanomami: Daily Life in the Venezuelan Forest,* trans. Ernest Simon (Cambridge: Cambridge Univ. Press, 1985), pp. 180–82.

142. "All the able-bodied men are there except the older ones." Ibid., p. 180. See also ibid., pp. 157–85; Cocco, *Iyewei-teri,* pp. 339–64; Valero, *Yo soy Napeyoma,* pp. 147–49.

144. Sound Roll 25, Patanowa-teri, March 4, 1968, Timothy Asch Collection, NAA.

145. Asiawe, interview, Platanal, trans. Jodie Dawson, June 11, 1996.

146. Chagnon, *Last Days of Eden,* p. 286.

147. Atkins and Asch, *A Multidisciplinary Study, Field Notes,* p. 14.

148. David E. Stannard, *American Holocaust: Columbus and the Conquest of the New World* (New York: Oxford Univ. Press, 1992), p. 237.

149. Sound Roll 14, Patanowa-teri, Feb. 27, 1968, Timothy Asch Collection, NAA.

150. In his essay on the *Ax Fight,* Chagnon describes this incident. Chagnon, "Filming the Ax Fight."

151. Sound Roll 14, Patanowa-teri, Feb. 27, 1968, Timothy Asch Collection, NAA.

152. Sound Roll 13, Patanowa-teri, Feb. 27, 1968, Timothy Asch Collection, NAA.

153. Sound Roll 8, Patanowa-teri, Feb. 23, 1968, Timothy Asch Collection, NAA.

154. Neel, Asch, and Chagnon, *A Multidisciplinary Study.*

155. "The Iwahikoroba-teri had also accused me of practicing harmful magic against them, and would kill me on sight for the deaths I had allegedly caused in their village with my *oka* magic." Chagnon, *Studying the Yanomamo,* p. 172.

156. Sister Felicia, interview, Puerto Ayacucho, Oct. 4, 1996.

Chapter 7: A Mythical Village

1. Napoleon Chagnon, *Yanomamo,* 5th ed. (Fort Worth: Harcourt Brace, 1997), p. 226.

2. Napoleon Chagnon, *Yanomamo: The Fierce People,* 3d ed. (New York: Holt, Rinehart and Winston, 1983), p. 200.

3. Napoleon Chagnon, *Studying the Yanomamo* (New York: Holt, Rinehart and Winston, 1974), p. 48.

4. Theodora Kroeber, *Ishi in Two Worlds: A Biography of the Last Wild Indian in North America* (Berkeley: Univ. of California Press, 1961), pp. 116–46, 232, 234.

5. Chagnon, *Studying the Yanomamo,* p. 196.

6. Ibid., p. 125.

7. Ibid., p. 5.

8. Napoleon Chagnon, *Yanomamo: The Fierce People,* 2d ed. (New York: Holt, Rinehart and Winston, 1977), p. 79.

9. Chagnon, *Studying the Yanomamo,* p. 16.

10. "Monou-tedi and Momaribowei-tedi sent a raiding party against them at this location about the time I started my field work and killed one man." Chagnon, "Yanomamo Warfare, Social Organization and Marriage Alliances" (Ph.D. diss., Univ. of Michigan, 1966), p. 173. See also ibid., pp. 174–75.

11. Kaobawa, videotaped interview with Mark Ritchie, Padamo River, Jan. 1995.

12. Chagnon, *Studying the Yanomamo,* p. 18.

13. Ibid., pp. 15, 20, 40–43.

14. Ibid., p. 51.

15. Joseph Grelier, *To the Source of the Orinoco,* trans. H. Schmuckler (London: Herbert Jenkins, 1957), p. 108.

16. Pablo Reyes, head of Malariología, interview, Puerto Ayacucho, June 17, 1996.

17. Helena Valero, *Yo soy Napeyoma: Relato de una mujer raptada por los indígenas Yanomami* (Caracas: Fundación La Salle de Ciencías Naturales, 1984), pp. 81–95.

18. Juan González, interview, Padamo mission, June 13, 1996.

19. Chagnon, *Studying the Yanomamo,* p. 31.

20. Father José Berno at Mavaca refers to "cinco tribus." "En su conjunto se llaman todos Muuereopoteri que a su vez se dividen en: Michimichimapueteri, Yeremaoteri y Taroiteri." Padre José Berno, "Crónica de la casa de Mavaca," June 28, 1972. Elsewhere, he gives the two other "tribes" that made up the five: Nuuereopoue-teri and Taatamupue-teri. Ibid., June 26, 1970.

21. Padre José Berno, "Crónica de la casa de Mavaca," June 28, 1972.

22. "The epidemic on which we now report was the result of an upper respiratory infection which we believe was produced by some form of influenza that afflicted three remote villages in the region of the Upper Mavaca River. No Europeans were present in the villages when the epidemic struck. . . . The villages located in the Upper Mavaca River had been the focus of the senior author's field research between 1968 and 1972. . . . He did not return to the field in 1973, but did so in late 1974, at which point he learned that the largest village in the region (shown in Figure 3 as 'Village 16') had fissioned into three distinct villages (shown on Figure 3 as 'Villages 16, 09 and 49') and had suffered a recent epidemic." Napoleon Chagnon and Thomas Melancon, "Epidemics in a Tribal Population," in *The Impact of Survival: Two Yanomamo Case Studies,* ed. K. Kensinger (Cambridge, Mass.: Cultural Survival, 1983), pp. 58–59.

23. Chagnon, *Yanomamo,* 5th ed., p. 226.

24. "Everyone agreed that the Bisaasi-teri were not the culprits—indeed, the Bisaasi-teri were praised for opposing it and trying to prevent it." Ibid., p. 222. See also Valero, *Yo soy Napeyoma,* pp. 241–45; Chagnon, "Yanomamo Warfare," pp. 153–54, 172.

25. Brian Ferguson, *Yanomami Warfare: A Political History* (Santa Fe: School of American Research Press, 1995), pp. 215–307; Chagnon, "Yanomamo Warfare," p. 158.

26. Chagnon, "Yanomamo Warfare," pp. 173–74.

27. Ibid., p. 173.

28. "These low-lying flat areas also appear to be richer in the kinds of natural resources the Yanomamo traditionally utilize—game animals, plants for food, construction and manufactures, and well-drained, easily cultivated land for gardens." Napoleon Chagnon, *Yanomamo,* 4th ed. (Fort Worth: Harcourt Brace, 1992), p. 83.

29. "Non-Yanomamo were a source of desirable items like machetes, metal cooking pots, axes, etc. . . . [I]t is possible that once these foreigners were 'discovered' by the Yanomamo in the Siapa basin, their settlement patterns and directions of movements were dramatically influenced by them." That is, they left the Siapa Highlands for lowland locations where they could get goods, particularly places in Brazil. Ibid., p. 89.

30. "By contrast, highland villages have fewer abducted women. . . ." Ibid., p. 86.

31. Chagnon, *Studying the Yanomamo,* p. 157.

32. "The Mishimishimabowei-teri. . . . were considering a move to Mavakita or to a site on the lower Mavaca so they could get these items from the Salesian missionaries." Chagnon, *Yanomamo,* 4th ed., p. 223. "Groups that live in the lowlands have to be large and bellicose in order to control the large, desireable, and wide-open ecological niche they live in." Ibid., p. 87. Residence data at Bisaasi-teri show that three Mishimishimabowei-teri adult men and three adult women have moved in. The men have joined their in-laws, a sign of a subordinate ally; the women have been ceded, another sign of subordination. María Eguillor García, *Yopo, shamanismo y hekura* (Caracas: Editorial/Librería Salesiana, 1984), p. 54.

33. In 1942, the Wanitama-teri killed Ruwahiwe and five others. See Valero, *Yo soy Napeyoma,* pp. 240–45. In 1960, the Bisaasi-teri, Monou-teri, and Momaribowei-teri killed three Mishimishimabowei-teri. See Chagnon, "Yanomamo Warfare," p. 173. In 1965, Monou-teri and Momaribowei-teri killed one more man. Ibid.

34. Chagnon, *Yanomamo,* 5th ed., pp. 217, 219, 216, 218–19.

35. Napoleon Chagnon, *Magical Death,* 28 min. (Watertown, Mass.: DER, 1973).

36. It took place "some weeks" before the alliance was consecrated by a raid against Patanowa-teri, on June 26–28, 1970. Chagnon, *Yanomamo,* 5th ed., p. 223.

37. Ibid., p. 216.

38. Napoleon Chagnon, *Yanomamo: Last Days of Eden* (San Diego: Harcourt Brace Jovanovich, 1992), p. 286.

39. "He ultimately seemed to despise *Magical Death.* . . . [H]e used it in classes at USC to persuade inner-city students that it was an example of ethnocentrism in filming, and twice tried to persuade me that we should consider withdrawing it from distribution because 'his black students from the inner city were turned off and abhorred by it." Chagnon, "Filming the Feast," *Yanomamo Interactive CD.*

40. Linda Rabben, *Unnatural Selection: The Yanomami, the Kayapó and the Onslaught of Civilization* (London: Pluto Press, 1998), p. 108.

41. See chapter 6.

42. Chagnon, *Yanomamo,* 4th ed., p. 223.

43. Padre José Berno, "Crónica de la casa de Mavaca," June 28, 1970.

44. "Some serious fights occurred, and one visitor was accused of trying to seduce a local woman and severely beaten with an ax. He died of his injuries after returning home." Chagnon, *Yanomamo,* 5th ed., p. 224.

45. Chagnon, *Yanomamo,* 4th ed., p. 223.

46. Footage from this lineup is also included in *Magical Death.* Ibid.

47. http://www.sscf.ucsb.edu/anth/projects/axfight/index.

48. "At least two Patanowa-teri were killed with shotguns, including Kumaiewa, the headman, the prominent leader shown in the film *The Feast."* Chagnon, *Yanomamo,* 4th ed., p. 204n.

49. Ibid., pp. 190–91.

50. John Peters, "The Effect of Western Material Goods upon the Social Structure of the Family among the Shirishiana" (Ph.D. diss., Western Michigan Univ., 1973), p. 125.

51. Misioneros del Alto Orinoco, "Consideraciones a un documento de Charles Brewer Carías (Puerto Ayacucho, 1991), pp. 28–32.

52. Timothy Asch and Napoleon Chagnon, *The Ax Fight,* 30 min. (Somerville, Mass.: DER, 1975).

53. Ibid.

54. Peter Biella, "Introduction to the Ax Fight," *Yanomamo Interactive CD.*

55. Ibid.

56. Chagnon, *Studying the Yanomamo,* p. 168.

57. Jay Ruby, "Out of Sync: The Cinema of Tim Asch," *Visual Anthropology Review* 11, no. 1 (Spring 1995): 28.

58. "Kama e the hami naprushi peni pe ta sheyo, pemaki noreshi toape, thi pemaki noreshi ha toani pemaka no koamape. Wama the pe re puhiwei ye the pe hipepe. I ha kuuni, tha tehe huya piewe yamaki iyeamayope, tha tehe wamaki iriamorahi, pe tta heyo wamaki ha hushuwamayoni. Pe ta sheyo pe ta hira! Huya pe iriamou te he suwe pe kai hirape, a kuma." Gustavo Konoko, interview, Mishimishimabowei-teri, Sept. 2, 1996, translated in situ by Pablo Mejilla, transcribed by Marco Jiménez.

59. Leda Martins, interview, Pittsburgh, Sept. 15, 1995.

60. Gustavo Konoko, interview, Mavakita, trans. Pablo Mejilla, Sept. 2, 1996.

61. Asch and Chagnon, *The Ax Fight.*

62. The first impression, in the field, was that the fight originated in an incest accusation. The second conclusion, at the Harvard sound lab, was that two factions *within* the same patrilineal descent group had fought over village dominance. This was the explanation that the film actually presented, with lineage charts showing how the antagonistic blood relatives fought each other. Then, six years later, Chagnon and a mathematician worked out an opposite interpretation—namely, that the opposing "teams" in *The Ax Fight* reflected the deep logic of kinship. Gary Seaman, "First Comments" and "Second Comments," *Yanomamo Interactive CD.*

63. "Agnates not only enter into competition with each other because of the nature of their rights in the women of their group, there are no other institutions based on agnatic descent that promote amity and solidarity among them." Chagnon, "Yanomamo Warfare," p. 81.

64. "Yanomamo marriage is peculiar in that the ties between husband and wife are very weak, but that ties between a man and his sister's husband are very strong." Ibid., p. 91.

65. Napoleon Chagnon and Paul Bugos, "Kin Selection and Conflict: An Analysis of a Yanomamo Ax Fight," in *Evolutionary Biology and Human Social Behavior: An Anthropological Perspective,* ed. Napoleon Chagnon and William Irons (North Scituate, Mass.: Duxbury Press, 1979), pp. 213–38.

66. Gary Seaman, "Blow-by-Blow," *Yanomamo Interactive CD.*

67. "It was simulated in a Massachusetts film lab. A watermelon was hit with an ax." Gary Seaman, "First Comments," *Yanomamo Interactive CD.*

68. Ruby, "Out of Sync," p. 28.

69. "Unedited Footage," *The Ax Fight, Yanomamo Interactive CD.*

70. Wilton Martínez, "The Challenges of a Pioneer: Tim Asch, Otherness, and Film Reception," *Visual Anthropology Review* 11, no. 1 (Spring 1995): 53–82.

71. Chagnon, *Studying the Yanomamo,* p. 166.

72. "A principios de enero salieron los bichasiteri a visitar a los mumanipueteri. Empezaron a enfermarse los niños de gripe complicada con el paludismo. Alla se murió un mumanipueteri, dos niños coroteri y un orateri. Regresaron en febrero todos enfermos. Murieron otro orateri—cinco Bichasiteri más el viejo patanoteri. En Dayariteri murieron tres niños y dos niños más y Carojiteri. Total quince. Los médicos se lucieron por su ausencia." José Berno. "Crónica de la casa de Mavaca," 1971.

73. Chagnon, *Studying the Yanomamo,* pp. 172–74.

74. Chagnon, "Filming the Ax Fight."

75. Chagnon, *Studying the Yanomamo,* pp. 172–77.

76. Chagnon, "Filming the Ax Fight."

77. "Malaria had clearly regressed. But it made a brutal and unexpected reappearance in 1971, in a form which might be fatal. This was a case of Falciparum paludism, resistant to the usual treatment. . . ." Jacques Lizot, *The Yanomami in the Face of Ethnocide* (Copenhagen: IWGIA 1976), p. 26.

78. Chagnon, "Filming the Ax Fight."

79. Juan Finkers, phone interview, Jan. 11, 1995.

80. Chagnon, *Last Days of Eden,* p. 286.

81. "He learned that the largest village in the region (shown in Figure Three as 'Village 16') had fissioned into three distinct villages. . . ." Chagnon and Melancon, "Epidemics in a Tribal Population," pp. 58–59.

82. "This appendix gives a summary of marriage and reproductive performance of all those individuals in Village 16, Mishimishimabowei-teri. . . ." Chagnon, *Studying the Yanomamo,* p. 211.

83. Chagnon and Melancon, "Epidemics in a Tribal Population," p. 59; Chagnon, *Yanomamo,* 3d ed., p. 200.

84. Misioneros del Alto Orinoco, "Consideraciones a un documento de Charles Brewer Carías," p. 21.

85. "Contacto de Juan . . . a la altura de Mavakita. Va con ellos a la altura de [indecipherable] más abajo de Mrakapiwei. Regreso de Juan con Sor Nora a la zona (de 28, 24 con falciparo y hepatítis)." José Berno "Crónica de la casa de Mavaca," July–Aug. 1971.

86. The Salesian nuns at Ocamo mention, without any details, the death of more than forty members of Mishimishimabowei-teri in the summer of 1973. I searched the Mavaca chronicle for corroboration but could not find it. "Crónica de Ocamo," Aug. 10, 1973.

87. Chagnon and Melancon, "Epidemics in a Tribal Population," p. 71.

88. Chagnon, *Yanomamo,* 5th ed., pp. 216–26.

89. Ibid., p. 217.

90. Kim Hill and Hillard Kaplan. "Population and Dry-Season Subsistence Strategies of the Recently Contacted Yora of Peru," *National Geographic Research* 5 (1989): 317–34.

91. Chagnon and Melancon, "Epidemics in a Tribal Population," p. 59.

92. Chagnon, *Last Days of Eden,* p. 286.

93. This is only the adult mortality at Patanowa-teri; total mortality was certainly higher but impossible to reconstruct. See the appendix.

Chapter 8: Erotic Indians

1. Jacques Lizot, *Tales of the Yanomami: Daily Life in the Venezuelan Forest,* trans. Ernest Simon (Cambridge: Cambridge Univ. Press, 1985), p. xiv.

2. "El antropólogo francés Claude Bourquelot, confinado junto a su colega Jacques Lizot por un trabajo de campo en la aldea de los adulimawateri de la sierra Parima, repentinamente perdió el juicio. Entre sus desvaríos empuñaba el machete para amenazar a los indígenas y al propio Lizot, quien no dudó en impartir el S.O.S. a la civilización occidental." E.S., "Inventario de supervivencia," *ExcesO,* April 1990, p. 66.

3. Napoleon Chagnon, interview, Univ. of California at Santa Barbara, Oct. 3, 1995.

4. Ibid.

5. According to Mark Ritchie's principal informant, it was the territorial governor Pablo Anduce who had Lizot arrested and removed in handcuffs. Ritchie, *Spirit of the Rainforest: A Yanomamo Shaman's Story* (Chicago: Island Lake Press, 1995), p. 235. According to Good, the arrest took place toward the beginning of his fieldwork, sometime in the late 1970s. Kenneth Good, phone interview, Aug. 5, 1999.

6. Claude Lévi-Strauss, *Tristes Tropiques,* trans. John and Doreen Wightman (New York: Atheneum, 1974), p. 264.

7. Ibid., p. 182.

8. Ibid., p. 189.

9. Ibid., p. 184.

10. Ibid., p. 375.

11. Ibid., p. 384.

12. Linda Rabben, *Unnatural Selection: The Yanomami, the Kayopó and the Onslaught of Civilization* (London: Pluto Press, 1998), p. 34.

13. Jacques Lizot, *The Yanomami in the Face of Ethnocide* (Copenhagen: IWGIA, 1976), author's bio.

14. Timothy Asch, "Ethnographic Filming and the Yanomamo Indians," Sightlines, Jan.–Feb. 1972, p. 9.

15. Marie Dawson, interview, Padamo mission, June 5, 1996.

16. Ritchie, *Spirit of the Rainforest,* p. 150. "After he learned our talk, this second naba began traveling to other villages on the river. The boys of the village went with him to help and were able to earn many valuable things in trade for their help. . . . One night they were in the jungle of the upper Orinoco. Everyone was away from the shelters hunting except the naba and a boy named Lizzard. The new naba came to Lizzard's hammock and sat on it with him. . . . Lizzard would have screamed, but there was no one to hear. . . . Now he understood why he and all his friends had suddenly become so rich." Ibid., p. 141.

17. *Yanomami Homecoming, National Geographic Explorer,* 48 min. (Washington, D.C.: National Geographic Society, 1994).

18. Gary Dawson, interview, Padamo mission, June 4, 1996.

19. Napoleon Chagnon, *Studying the Yanomamo* (New York: Holt, Rinehart and Winston, 1974), p. 18.

20. Ritchie, *Spirit of the Rainforest,* p. 146.

21. Ibid., p. 146.

22. Ibid., p. 147.

23. Ibid., p. 148. Kenneth Good arrived shortly after this fight. He heard, however, that the actual fistfight occurred inside the mission, over dinner, not in the *shabono* plaza. He also heard that González,

after being assaulted by Lizot, gave at least as good as he got. "González was a big tough guy. He wouldn't take any shit off anyone." Kenneth Good, phone interview, Aug. 5, 1999.

24. The anthropologist Frank Salamone explained the history of the Salesian order and gave me some background to Don Juan Bosco's philosophy, and how this was carried on by Padre Luis Cocco. Frank Salamone, phone interview, Dec. 22, 1994.

25. Luis Cocco, *Iyewei-teri: Quince años entre los Yanomamos* (Caracas: Librería Editorial Salesiana, 1972), p. 463.

26. Napoleon Chagnon, interview, Univ. of California at Santa Barbara, Oct. 3, 1995.

27. Napoleon Chagnon, *Yanomamo: The Fierce People,* 3d ed. (New York: Holt, Rinehart and Winston, 1983), pp. 201–2.

28. Fran L. Paver, attorney-adviser, National Science Foundation, FOIA No. 95-004, July 13, 1995.

29. Kenneth Good, phone interview, January 10, 1995.

30. Kenneth Good, *Into the Heart: One Man's Pursuit of Love and Knowledge among the Yanomama* (New York: Simon & Schuster, 1991), p. 33.

31. Ibid., p. 19.

32. Ibid.

33. Chagnon, *Studying the Yanomamo,* p. xiv.

34. Good, *Into the Heart,* p. 22.

35. "As Eric and I were busy working with our hammocks and nets, all of a sudden out of the night two big figures burst into the hut screaming, 'Aaaaaaaaaahhhhhhh!' grabbing us, and shoving us toward our hammocks, ripping the mosquito netting. My heart skipped a beat. I heard Eric gasp. Bracing myself against a table to keep from falling, I twisted around and saw in the glow of the Coleman Chagnon and the French anthropologist, both of them completely drunk." Ibid., p. 23.

36. Kenneth Good, "A Race against Time," *Américas* April 1998, p. 31.

37. Kenneth Good, phone interview, Jan. 10, 1995.

38. In his studies among the Catrimani villages, a Yanomami subgroup, Giovanni Saffirio reported five raids over a fifty-seven-year period. Saffirio, "Ideal and Actual Kinship Terminology among the Yanomama Indians of the Catrimani River Basin (Brazil)" Ph.D. diss., Univ. of Pittsburgh, 1985), pp. 95–100. "[R]elatively little warfare reported in the Ocamo River Basin over the past 100 years." Erik Fredlund, "Shitari Yanomami Incestuous Marriage: A Study of the Use of Structural, Lineal and Biological Criteria When Classifying Marriages" (Ph.D. diss., Pennsylvania State Univ., 1982), p. 37. Hames, Chagnon's closest collaborator, attributes Yanomami migration to intervillage raiding, but his careful history shows that there were few war deaths and long lulls in fighting (lasting up to two decades), and that actual raids figured in only five out of twenty-one migratory moves analyzed. Raymond Hames and W. Vickers, "The Settlement Pattern of a Yanomamo Population Block: A Behavioral Ecological Interpretation." In *Adaptive Responses of Native Amazonians,* ed. Hames and Vickers (New York: Academic Press, 1983), pp. 393–427.

39. Kenneth Good, phone interview, Aug. 8, 1995.

40. Kenneth Good, phone interview, Jan. 10, 1995.

41. Jesús Cardozo, phone interview, Sept. 1, 1995.

42. Kenneth Good, phone interview, Aug. 8, 1995.

43. Good, *Into the Heart,* pp. 116–19.

44. Kenneth Good and Jacques Lizot, letter to *Science* appended to Marvin Harris, "Culture Materialist Theory of Band and Village Warfare," in *Warfare, Culture, and Environment,* ed. R. B. Ferguson (Orlando: Academic Press, 1984), pp. 111–40.

45. Kenneth Good, phone interview, Feb. 27, 1998.

46. Lizot, *Tales of the Yanomami,* p. 31.

47. Ibid., pp. 31–36.

48. "We were also already aware that sex between male and female cross cousins was expected. What we had not realized was that sex between male cross cousins was also common. We were probably slow in recognizing this because most people, most of the time, restrained any overt expressions of sexuality, either heterosexual or homosexual." Clayton Robarchek and Carole Robarchek, *Waorani: The Contexts of Violence and War* (New York: Harcourt Brace, 1998), p. 56.

49. Brian Ferguson, *Yanomami Warfare: A Political History* (Santa Fe: School of American Research Press, 1995), pp. 393–94.

50. "La homosexualidad no me consta que se dé entre los yanomamos como hábito permanente. Existen casos de esta anomalía tanto entre varones como entre hembras, pero que se dan como deslices propios de la juventud y muy esporádicamente en la edad adulta. . . . Helena Valero, respecto a la homosexualidad, dice que era muy esporádica en los grupos entre quienes convivió." Cocco, *Iyewei-teri,* p. 211.

51. "No me ensucien mi ano. No quiero. Déjenme tranquilo. . . . Voy a hacerme un hueco en la pantorrilla, a ver si crío mujer, para que esos animales no me ensucien más y me dejen tranquilo." Ibid., pp. 468–69.

52. Alcida Rita Ramos, "Reflecting on the Yanomami: Ethnographic Images and the Pursuit of the Exotic," *Cultural Anthropology* 2 (1987): 284–304.

53. "Parenthetically, romantic as it may be, this passage would very likely get a chuckle of disbelief from the Yanomamo, for no one in their right mind would remain in bed, or rather in the hammock, by the time the sun is up . . . the loving couple would have been disturbed many times over before the bright disk of the sun emerged from behind the tall trees of the forest surrounding the village." Ibid., p. 290.

54. The exception is Chagnon, who does refer to common homosexual practices among young men in his Ph.D. thesis. "Most of the young men in Bisaasi-teri were having homosexual relationships with each other, but no stigma was attached to this behavior." Chagnon, "Yanomamo Warfare, Social Organization and Marriage Alliances" (Ph.D. diss., Univ. of Michigan, 1966) pp. 61–63. In 1977, he wrote, "Some of the teen-age males have homosexual affairs with each other. . . ." N. Chagnon, *Yanomamo: The Fierce People,* 2d ed. (New York: Holt, Rinehart and Winston, 1977), p. 76. When I spoke to Chagnon personally at the Univ. of California at Santa Barbara, he said, "I have never seen any homosexuality down there." Chagnon, interview, Oct. 3, 1995.

55. Kenneth Good, phone interview, March 25, 1998.

56. Kenneth Good, phone interview, Aug. 8, 1995.

57. Garry Dawson, interview, Padamo mission, June 4, 1996.

58. I am indebted to Kenneth Good for this detailed analysis. Good, phone interview, Aug. 5, 1999.

59. Kenneth Good, phone interview, Feb. 1, 1995.

60. Kenneth Good, phone interview, Aug. 5, 1999.

61. Mark Ritchie, video interviews, Padamo River, Jan. 21, 1995.

62. Ritchie, *Spirit of the Rainforest,* p. 248: "Translator Gary Dawson had an unrecorded conversation with a boy hired by A.H. to do sex acts."

63. Ibid., p. 211.

64. Ibid., p. 142.

65. Ibid., p. 149.

66. Ibid.

67. Kaobawa, interview, Shakita, Upper Orinoco, June 12, 1996.

68. Nelly Arvelo Jiménez, "The Repudiation of Brewer-Carías and Chagnon Is Due to Their Intimate Association with Goldmining, (Caracas: IVIC, Oct. 1994), p. 6.

69. Arvelo Jimenez was present at this talk by Chagnon, and it shaped her opinion, as it had an influence on the other scientists at IVIC. "During a period in which he desperately needed help to get affiliation to a Venezuelan institution, nobody volunteered to help him. He is arrogant, he does not honestly care about the problems or issues or interest of the anthropological community here, thus nobody cares for him." Ibid., p. 3.

70. Kenneth Good, phone interview, Jan. 10, 1995.

71. Leslie Sponsel, Univ. of Hawaii, phone interview, Jan. 12, 1995.

72. "An official policy is approved by the Venezuelan government that terminates anthropological research permits for non-Venezuelans in the Upper Orinoco area 'until further notice.' Announcement of the policy caused one Venezuelan anthropologist to exclaim 'At last we have something to get Chagnon with.' Chagnon attempts to develop a collaboration with Venezuelan anthropologists at IVIC at the suggestion Marcel Roche, a prominent Venezuelan medical researcher. The attempt is sabotaged by UCV [Universidad Central de Venezuela] anthropologists who circulate rumors that Chagnon is 'bribing' IVIC anthropologists with money in order to obtain permits." Napoleon Chagnon, "Notes on Chronology of Recent Attacks on Members of the Venezuelan Presidential Commission by Salesian Missionaries, French, Brazilian and Venezuelan Anthropologists" (MS, May 18, 1994), p. 1.

73. "1985–1988. . . . Chagnon and Hames develop a 3-year project with NSF funding and collaborate legally with Venezuelan archicent Grazziano Gasparini and his American anthropologist wife, Luisa Margolies. . . ." Ibid., p. 2.

74. Chagnon, *Yanomamo,* 2d ed., p. 74.

75. María Eguillor García, *Yopo, shamanes y hekura* (Caracas: Editorial/Librería Salesiana, 1984), p. 56.

76. "Of those who arrived between 1959 and 1968, 66 were male and 78 female. Of those who arrived between 1969 and 1983, 47 were male, and 34 female." Ferguson, *Yanomami Warfare,* pp. 318–19.

77. Napoleon Chagnon, "L'ethnologie du déshonneur: Brief Response to Lizot," *American Ethnologist* 22 (1995): 187–89.

78. On the Mucajaí River, female out-migration virtually ceased, and Mucajaí men were able to ob-

tain wives from other villages after a much shorter bride service than customary. "All 13 new wives came to live in the Mucajaí community without the Mucajaí men spending several years of bride service with the bride's family." John D. Early and John F. Peters, *Population Dynamics of the Mucajaí Yanomama* (San Diego: Academic Press, 1990), p. 67.

79. "El Padre Luis Cocco hizo la experiencia entre los iyeweitheri de Ocamo que cuando éstos comenzaron a tener escopetas que favorecieron la cacería, en la misma medida comenzó a aumentar la poliginia." Eguillor García, *Yopo, shamanes y hekura,* p. 84.

80. The decline in Bisaasi-teri's trade wealth immediately translated into an increase of outside bachelors seeking Bisaasi-teri women—a way of the Bisaasi-teri's holding their alliance together. "There were many more Shamatari men living in the group because they had been given women; that is, the exchanges were more balanced." Chagnon, *Yanomamo,* 2d ed., p. 80. Ferguson observes, "After around 1967, it seems, the Bisaasi-teri were no longer able to demand women from would-be partners." Ferguson, *Yanomami Warfare,* p. 319.

81. Mark Ritchie, video interviews, Padamo River, Jan. 21, 1995. Lizot gives two other examples of young men leaving Karohi-teri to go and live in Tayari-teri. Lizot, *Tales of the Yanomami,* pp. 38, 51.

82. Chagnon, "Yanomamo Warfare," pp. 59–60.

83. Rerebawa, from Karohi-teri, was doing his bride service at Bisaasi-teri when Chagnon arrived in 1964. Chagnon, *Yanomamo,* 2d ed., pp. 11–12.

84. After just one year of Lizot's largesse, Karohi-teri on the Manaviche River, which was Tayari-teri's sister village and Lizot's alternate residence, had more material wealth than the richest Bisaasi-teri group or village, led by Paruriwe, which had moved next to the Salesian mission at Mavaca. Jacques Lizot, "Aspects économiques et sociaux du changement culturel chez les Yanómami," *L'Homme* 11, no. 1 (1971): 45. Eguillor García, *Yopo, shamanes y hekura,* p. 236.

85. Hebewe, Tohowe, and Hishokoiwe were all from Karohi-teri before going to Tayari-teri. But there was a whole faction from Karohi-teri that had moved, en masse, to Tayari-teri. Interestingly, Tohowe, like Fama, had been from a more distant village, Wayabotorewe, before he was able to enter Tayari-teri. Lizot, *Tales of the Yanomami,* pp. 37–50.

86. "A few days ago a boy of about fourteen arrived at Karohi. His home group is situated far away, several days' walking distance, in a mountainous region on the banks of the 'river of honey.' He arrived with a group of visitors who spent only one night. As he expressed a desire to remain a while at Karohi, Kaomawe invited him to stay with his family." Ibid., p. 36.

87. William Smole, *The Yanoama Indians: A Cultural Geography* (Austin: Univ. of Texas Press, 1976), p. 72.

88. "His [Hebewe's] only discomfort arises from the scarcity of women. . . . [I]t is certainly not at Karohi that he can hope to find a wife." Lizot, *Tales of the Yanomami,* pp. 38–39.

89. Chagnon, "Yanomamo Warfare," p. 37.

90. Lizot, *Tales of the Yanomami,* p. 39.

91. Ibid., pp. 42–46.

92. Ibid., p. 51.

93. Ramos, "Reflecting on the Yanomamo," p. 292.

94. Jesús Cardozo, phone interview, Aug. 31, 1995.

95. "Their heart is missing, and one has the impression they have no further taste for life." Lizot, *The Yanomami in the Face of Ethnocide,* p. 23.

96. Kenneth Good, phone interview, Aug. 8, 1995.

97. Napoleon Chagnon, "Reproductive and Somatic Conflicts of Interest in the Genesis of Violence and Warfare among Tribesmen," in *The Anthropology of War,* ed. J. Haas (Cambridge: Cambridge Univ. Press, 1990), p. 99.

98. Eguillor García, *Yopo, shamanes y hekura,* p. 25.

99. Ibid., p. 26.

100. Ibid.

101. Brian Ferguson, phone interview, March 4, 2000; Ritchie, *Spirit of the Rainforest,* p. 149.

102. Jesús Cardozo, phone interview, Sept. 1, 1995.

103. "Tim felt that he should put responsibility and resources into Jesús and Hortensia's Yanomami Video Project." Patricia Asch, phone interview, May 16, 1997.

104. Jesús Cardozo, phone interview, Aug. 31, 1995.

105. Jesús Cardozo, phone interview, Sept. 1, 1995.

106. Jesús Cardozo, phone interview, Aug. 31, 1995.

107. Kenneth Good, phone interview, Aug. 3, 2000.

108. Catherine Alés, interview, Caracas, Aug. 21, 1996.

109. Jesús Cardozo, phone interview, Aug. 31, 1995.

110. Jesús Cardozo, phone interview, Sept. 1, 1995.

111. Davi Kopenawa, interview, office of the Committee for the Creation of the Yanomami Park, Boa Vista, Brazil, Nov. 3, 1990.

112. Dr. John Walden, of Marshall Univ., told me he had videotapes of Yanomami boys near Ocamo describing sexual activities with Lizot. Walden, phone interview, Jan. 3, 1995. Lizot's alleged child molesting is often invoked by Chagnon and his allies as a way of discrediting all of the academic credentials of their opponents. Charles Brewer Carías, phone interview, Jan. 3, 1995. "Lizot threatened to burn down the Salesian mission and kill one priest. He buys really young children from their parents." Kim Hill, phone interview, Jan. 17, 1995. "Lizot did a lot of screwing around, with both young girls and young boys." Terence Turner, phone interview, Sept. 21, 1995.

113. Jesús Cardozo, phone interview, Aug. 31, 1995.

114. Ibid.

115. Giovanni Saffirio, interview, Pittsburgh, June 25, 1994.

116. José Bórtoli, interview, Platanal mission, Sept. 26, 1996.

117. Timothy Asch, foreword, *Tales of the Yanomami,* p. xi.

118. Lizot, *Tales of the Yanomami,* p. xiv.

Chapter 9: That Charlie

1. E.S., "Inventario de supervivencia," *ExcesO,* April 1990, p. 66.

2. James V. Neel, *Physician to the Gene Pool: Genetic Lessons and Other Stories* (New York: John Wiley, 1994), p. 408, n. 8.

3. Charles Brewer Carías, *Roraima: montaña de cristal* (Caracas: Oficina Central de Información, 1975), preface, unpag.

4. Redmond O'Hanlon, *In Trouble Again* (London: Hamish Hamilton, 1988), p. 39.

5. "Charles Brewer Carías, naturalista, explorador, fotógrafo, escritor. Investigador Asociado al New York Botanical Garden y a la Universidad de California, Santa Bárbara." OMEGA, THE SIGN OF EXCELLENCE (Venezuelan magazine advertisement, n.d.).

6. "Para mí el Omega Speedmaster es un robusto instrumento que utilizo constantemente en mis expediciones y me ha ayudado a obtener la precisión que requiero bajo las condiciones más exigentes." Ibid.

7. "Como artesano o mas bien como inventor, ha creado el Cuchillo de Supervivencia Brewer, fabricado por MARTO de Toledo, el cual es utilizado por todos los Cuerpos Comandos del País. . . ." Charles Brewer-Carías, Curriculum en Antropología, Sept. 3, 1993, p. 24.

8. O'Hanlon, *In Trouble Again,* p. 15.

9. John Walden, phone interview, March 4, 1994.

10. Rafael and Teresa Salazar, Pittsburgh, April 30, 1996.

11. O'Hanlon, *In Trouble Again,* p. 39; E.S., "Inventario de supervivencia," p. 71.

12. Ana Ponte, "Charles Brewer Carías: Informe para el Comité del Medio Ambiente del Senado" (Caracas, Jan. 1997), pp. 1–4.

13. "No soy más que un asceta." E.S., "Inventario de supervivencia," p. 68.

14. Ibid., p. 71.

15. Charles Brewer Carías, "Una futura zona en reclamación," *El Nacional,* May 10, 1987.

16. Terence Turner, interview, Cornell Univ., Jan. 27, 1996.

17. E.S., "Inventario de supervivencia," p. 71.

18. Ibid., p. 66.

19. Walter Raleigh, *The Discoverie of the Large, Rich, and Bewtiful Empire of Guiana,* ed. V. T. Harlow (London: Hakluyt Society, 1928), p. 5.

20. Tania Vegas, "Brewer Carías ha devastado zonas protectoras en Guayana," *El Universal,* April 13, 1992.

21. Ibid.

22. Ibid.

23. Permiso Provisional para Transportar y Usar Combustibles y Aceite de Motor en Sitio. Direccion General Sectorial de Hidrocarburos Dirección de Mercado Interno. Sept. 22, 1988.

24. Saidia Alvarez Silvera, Juez, "Subcontrato con Tawinco" (Tumeremo: República de Venezuela, 1992), No. 2010237, Nov. 11, 1992.

25. Marcus Colchester with Fiona Watson, *Venezuela: Violations of Indigenous Rights: Report to the International Labour Office on the Observation of ILO Convention 107* (Oxford: World Rainforest Movement and Survival International, 1995), pp. 19–28.

26. Nelly Arvelo Jiménez. "The Repudiation of Brewer Carías and Chagnon Is Due to Their Intimate Association with Goldmining" (Caracas: IVIC, 1994), p. 3.

27. Orlando Utrera, "Brewer denuncia el 'Plan Gadhafi,' " *El Diario de Caracas,* Aug. 15, 1984.

28. Brewer Carías, "Una futura zona en reclamación."

29. Olgalinda Pimentel. "Denunciaron ante el Fiscal al ex-ministro Brewer Carías," *El Diario de Caracas,* Aug. 4, 1984.

30. Sergio Milano, phone interview, Dec. 12, 1994.

31. Mayor Sergio Rafael Milano (Jefe), Teniente Luis Alberto Godoy y Geraldi Antonio Villaroel (Secretario), Expediente de la Comisión de la Guardia Nacional, Fuerzas Armadas de Cooperación, Comando Regional 6, Destacamento de Frontera No. 61. Puerto Ayacucho, April 18, 1984. The helicopter pilot, Ricardo Antonio Trivisi Muñoz, explained that he falsified the flight receipts at Brewer's request, and in order to hide the fact that they were flying to a place—Kanaripo—outside the scientific expedition's work area. "Me supongo que como él está trabajando con la fundación no quiso que apareciera Kanaripó—por cuanto no es la zona de trabajo."

32. Utrera, "Brewer denuncia el 'Plan Gadhafi.' "

33. David Ayala, "Informe de Comisión de Diputados ratifica denuncias contra Charles Brewer Carías," *Ultimas Noticias* (Caracas), Oct. 5, 1993.

34. O'Hanlon, *In Trouble Again,* pp. 14, 19.

35. Charles Brewer Carías, phone interview, Jan. 3, 1995.

36. Charles Brewer Carías, letter to Andrés Caldera Pietri, Ministerio de la Secretaría de la Presidencia, Nov. 18, 1994.

37. Ibid.

38. V. S. Naipaul, *The Loss of El Dorado* (New York: Penguin Books, 1987), p. 69.

Chapter 10: To Murder and to Multiply

1. Geoffrey W. Wrangham and Paul G. Peterson, *Demonic Males: Apes and the Origins of Human Violence* (Boston: Houghton Mifflin, 1996), p. 68.

2. Napoleon Chagnon, "Life Histories, Blood Revenge, and Warfare in a Tribal Population," *Science* 239 (1988): 985–92.

3. Richard Dawkins, *The Selfish Gene* (Cambridge: Cambridge Univ. Press, 1975).

4. John Horgan, "The Violent Yanomamo," *Scientific American,* March 1988, p. 18.

5. Chagnon is correct when he claims that his *unokai* study "is now very widely cited by scientists all over the world, overwhelmingly with approval." Napoleon Chagnon, "Notes on Chronology of the Recent Attacks" (MS, May 18, 1994), p. 2.

6. William Booth, "Warfare over Yanomamo Indians," *Science* 243 (1989): 1138–40; Maria Manuela Carneiro da Cunha, letter to the editor, *Anthropology Newsletter,* Jan. 1989, p. 3; Brian Ferguson, "Do Yanomamo Killers Have More Kids?" *American Ethnologist* 16 (1989): 564–65; Jacques Lizot, "On Warfare: An Answer to N. A. Chagnon," trans. Sarah Dart, *American Ethnologist* 21 (1994): 845–62; Napoleon Chagnon, "L'ethnologie du déshonneur: Brief Response to Lizot," *American Ethnologist* 22 (1995): 187–89. Jeffrey Rifkin, "Ethnography and Ethnocide," *Dialectical Anthropology* 19 (1994): 295–327; Bruce Albert, "On Yanomami Warfare: Rejoinder," *Current Anthropology* 31 (1990) 558–63; Jacques Lizot, "Sobre la guerra: Una respuesta a N. A. Chagnon (Science, 1988)," *La Iglesia en Amazonas* 44 (1989): 23–34; Chris J. Van Vuner, "To Fight for Women and Lose Your Lands: Violence in Anthropological Writings and the Yanomami of Amazonia," *Unisa Largen* 10 (July 1994): 10–20; Bruce Albert and Alcida Rita Ramos, "O exterminio academico dos Yanomami," *Humanidades* (Brasília), 18 (1988): 85–89.

7. Boyce Rensberger, "Sexual Competition and Violence," *Washington Post,* Feb. 29, 1989; William Allman, "A Laboratory for Human Conflict," *U.S. News & World Report,* April 11, 1988, pp. 57–58.

8. James V. Neel, "On Being Headman," *Perspectives in Biology and Medicine* 23 (1980): 277–94. It was also one of the primary goals of Chagnon's research, since, in the 1968 film, *Yanomamo: A Multidisciplinary Study,* Neel, the narrator, stated that one of their goals was learning the secrets of how male leaders achieve biological success.

9. This entire theory had been forwarded by the philosopher-eugenist Christian von Ehrenfels, an Austrian who corresponded with Freud and helped found Gestalt psychology. Ehrenfels was "part of the anti-Semitic circle around Cosima Wagner in Bayreuth, and had developed some unique ideas about racial engineering. His pet peeve was that some of the best men in the race were its soldiers; hence the finest genetic stock was lost in battle. His solution was that brave soldiers returning from the front should have sexual access to as many women as possible. Largely because of such ideas, he never became a household name. . . ." Tom Reiss, "The Man from the East," *New Yorker,* Oct. 4, 1999, p. 73.

10. "The concern is due simultaneously to the content and to the unfortunate political consequences of the theory expounded, these two aspects being quite obviously intimately intertwined." Lizot, "On Warfare," p. 564.

Ibid.

11. Napoleon Chagnon, "The View from the President's Window," *Human Behavior and Evolution Society Newsletter* 2 no. 3 (Oct. 1993): 2.

12. Chagnon, "Life Histories," 989.

13. Brian Ferguson, phone interview, July 13, 1995.

14. "When the 15,000 Yanomamo are not hunting animals and collecting wild honey, they are often killing each other, says UC Santa Barbara anthropologist Napoleon A. Chagnon. . . . In fact, the Yanomamo are one of the most violent cultures known, he said." Thomas Maugh II, "Homicidal Streak in S. American Tribe Studied by Anthropologist," *Los Angeles Times,* Feb. 26, 1988.

15. Allman, "A Laboratory for Human Conflict," pp. 57–58.

16. "The Yanomamo Indians of the Amazon rain forest have long been known as one of the most violent societies on Earth." Rensberger, "Sexual Competition and Violence."

17. "Antropologo aponta violencia entre indios," *O Globo,* March 1, 1988.

18. "Violência, marca dos Yanomami," *O Estado de São Paulo,* March 1, 1988.

19. "Thus less than a year after the *Time Magazine* piece came out, top-level officials of the Brazilian Indian Service (Fundacao Nacional do Indio-FUNAI) referred to the Yanomami 'violence' as sufficient justification for a plan to cut up their lands into 21 micro-reserves. . . ." Carneiro da Cunha, letter, *Anthropology Newsletter.*

20. Kenneth Taylor, phone interview, Jan. 27, 1995.

21. Booth, "Warfare over Yanomamo Indians," p. 1140.

22. Phone interviews, oct. 1988.

23. Interviews with Pedro of the Opik-teri and Mario of Pacu took place on June 21 and 22, 1989, Catrimani mission. The translator was Father Guillerme Damiolli.

24. "At Karohi, no one is surprised by their return; it is the fate of three raids out of four not to be carried to completion. No one blames them for their caution; the omens were obviously unfavorable, and people are glad to see them return alive." Jacques Lizot, *Tales of the Yanomami: Daily Life in the Venezuelan Forest,* trans. Ernest Simon (Cambridge: Cambridge Univ. Press, 1985), p. 183.

25. Patrick Tierney, *The Highest Altar: The Story of Human Sacrifice* (New York: Viking, 1989), pp. 310–11.

26. Bruce Albert, interview, Toototobi, Dec. 5, 1990.

27. Father Guillerme Damioli, Catrimani mission, June 19, 1989.

28. Giovanni Saffirio and Raymond Hames, "The Forest and the Highway," in *The Impact of Contact: Two Yanomamo Case Studies,* ed. K. Kensinger (Cambridge, Mass.: Cultural Survival, 1983), p. 12.

29. Ibid., pp. 15–16.

30. "The traditional bride service is no longer fashionable among the highway villages because the in-laws prefer goods rather than services. For example, during 1978 and 1980 two men of Opiktheri 135 and another from Opiktheri 132 obtained wives from Yanomama villages at Km 37 and 49 by substituting Brazilian goods. They did less than six months of bride service. This trend weakens the relationship between in-laws." Ibid., p. 27.

31. I spent two weeks near Zeca Diabo's camp on the Rio Branco below Caracarai in June 1992. My outboard motors had both failed. One night, the Opik-teri wanted to stage a chest-pounding duel to satisfy the honor of a young man whose wife was angered that she had received, from Zeca, a pair of panties that did not fit her.

32. Christopher Boehm, phone interview, April 9, 1996.

33. "The killer was thought to carry within him the essence of the vulture spirits feasting on the soul of the victim. His breath would stink with the smell of putrefaction and his skin would become greasy with his victim's body fat." Paul Henley, *Yanomami: Masters of the Spirit World* (San Francisco: Chronicle Books, 1995), p. 51. "When a Yanomamo man kills he must perform a ritual purification called *unokaimou,* one purpose of which is to avert any supernatural harm. . . . Men who have performed the *unokaimou* ceremony are referred to as *unokai.* . . ." Chagnon, "Life Histories," p. 987.

34. Chagnon, "Life Histories," p. 987.

35. Ibid., p. 991, n. 15.

36. Chagnon had 137 *unokais,* or killers, 151 deaths, but 345 instances of individual claims of participation in these 151 killings. Ibid., p. 986, fig. 1, and p. 987.

37. Pressed by Albert on this score, Chagnon offered a second version of the same data for *Current Anthropology.* After removal of five young killers whose ages were unknown, and their four murders, the number of homicides dropped from 151 to 147 and the number of "killers" from 137 to 132. Napoleon Chagnon, "On Yanomamo Violence: Reply to Albert," *Current Anthropology* 31 (1990): 50, fig. 1. "More than half (52%) had a single killer, 76% had either one or two killers, and 82% had three or fewer." Deciphered, this means that a small number of murders (27) produced *most* (209, or 54 percent) of the killing events. Ibid., p. 50.

38. Ibid., p. 51.

39. "On the other hand, from a total of slightly more than 400 marriages . . . only 0.3 percent of the women were taken from an enemy group." Lizot, "On Warfare," p. 854.

40. "Only those who killed a Yanomamo, those who were unokai [killers of men] can not rape

women because they are [ritually] contaminated and must undergo unokaimou [ritual purification] for over 15 days. But everyone else joined in the raping." This is a sentence written by Brewer Carías, with parenthetical comments by Chagnon. It concerned a raid on the Upper Orinoco, in which the attackers were members of the Bisaasi-teri and Upper Mavaca groups whom Chagnon studies. There is also a transcribed statement by one of Chagnon's informants, Alberto Karakawe: "All the others [who were not unokai] did it. . . ." Charles Brewer Carías with Napoleon Chagnon, "The Massacre at Lechoza, September 1992: Brewer's account, 12/92" (MS), p. 13.

41. "No hemos aprovechado para hacer el amor con ellas cuando podíamos. Por lo menos se habrían ido a morir allá por Kamakari [spirits of the dead] y a nosotros nos hubieran curado nuestros brujos." Helena Valero, *Yo soy Napeyoma: Relato de una mujer raptada por los indígenas Yanomami* (Caracas: Fundación La Salle de Ciencias Naturales, 1984), p. 80.

42. "Most of the applied anthropologists working with the Yanomamo believe that only 'politically correct' data—data useful to their cause—should be collected and published, and that any other kind of data is 'bad.' An example of what they consider 'bad' is much of what you have been reading here. . . . I take the peculiar position that *all* facts about the Yanomamo are relevant to their future." Napoleon Chagnon, *Yanomamo: Last Days of Eden* (San Diego: Harcourt Brace Jovanovich, 1992), pp. 244–45.

43. "The biological genitor of a child with several recognized fathers is unknown and socially irrelevant." John D. Early and John F. Peters, *The Population Dynamics of the Mucajaí Yanomama* (San Diego: Academic Press, 1990), p. 41. Nancy Howell, of the Univ. of Toronto, widely respected as the leading demographer in small-population anthropology (and a friend of Chagnon), told me, "I've argued with Napoleon for years that to do a study on comparative reproductive success he would need a hundred-year time frame. I don't think the episodic cross sections he offers give you a good idea of reproductive success, because small populations like the Yanomami's have such big fluctuations. There just isn't enough data. The whole thing is very shaky." Howell, phone interview, Feb. 1, 1995.

44. Lizot. "On Warfare," p. 853.

45. Garden site No. 5 was identified as "Bisaasiteri 2 31 65 10." Biella, Chagnon, and Seaman, *Yanomamo Interactive,* Excel file, Garden Locations, No. 4. In Chagnon's epidemic profile of Yanomami villages, Boca Mavaca was the only village with "maximum contact," and it had a mortality rate of 4.5 percent. That matches No. 5. Chagnon, *Last Days of Eden,* pp. 267–68. Villages 5, 6, and 7 appeared together on an earlier map of the Bisaasi-teri's migratory history. Napoleon Chagnon and Thomas Melancon, "Epidemics in a Tribal Population," in *The Impact of Contact,* p. 58, and Chagnon, *Yanomamo,* 5th ed. (Fort Worth: Harcourt Brace, 1997), p. 82. In these graphs, Chagnon distinguishes between the Namowei villages, Nos. 5,6,7,18, and 19, and the Shamatari villages. Villages 18 and 19 are two Patanowa-teri offshoots, Dorita-teri and Sheroana-teri. "18 Patanowateri 2 18 64 43." Biella, Chagnon, and Seaman, "Garden Locations," No. 11. This division shifted after 1965. Monou-teri joined Upper Bissasi-teri, but almost immediately afterward Upper Bisaasi-teri fissioned and Paruriwa led a faction across the river. A group of Patanowa-teri defectors joined Monou-teri and Upper Bisaasi-teri. Napoleon Chagnon, "Yanomamo Warfare, Social Organization and Marriage Alliances," (Ph.D. diss., Univ. of Michigan, 1966), pp. 182–87. For a graph of the new locations of the three villages, see Chagnon, *Yanomamo,* 5th ed., p. 73.

46. "Kedebabowei 2 02 65 09." Biella, Chagnon, and Seaman, "Garden Locations," No. 19.

47. "Nasikiboweiteri 2 29 65 11." Ibid., No. 21. For Dorita-teri, see chapter 12.

48. "Haoyaboweiteri 2 11 65 08." Ibid., No. 22.

49. Garden site No. 92 identified as "Mishimishi 1 36 65 16." Biella, Chagnon, and Seaman, Garden Locations, No. 23.

50. "Dakowa's Village 2 02 65 04." Ibid., No. 24. For the location of "Dakowa's Village" on the Washewa River, a tributary of the Upper Mavaca, see Charles Brewer Carías, Napoleon Chagnon, and Brian Boom, "Forest and Man" (MS; Caracas: Fundación Explora), p. 12. And see the coincidence between Village 93 and Washewa-teri's 15 percent mortality rate from 1987 to 1991. Chagnon, *Last Days of Eden,* pp. 266–67.

51. By 1983, they had become half a dozen separate *shabonos,* all located around Boca Mavaca. So Lizot was technically correct in saying that Chagnon's "villages" did not exactly match physical locations. The villages were Musiu-teri, Thanahi-teri, Wayawa-teri, Mothoremawe-teri, Hayapeka-teri, Cinctheri and Koro-teri. María Eguillor García, *Yopo, shamanismo y hekura* (Caracas: Editorial/Librería Salesiana, 1984), p. 45. Kaobawa was in Musiutheri, also known as Shakita, "Chagnon's Village." Chagnon's total for Boca Mavaca—414 individuals, when all three groups were pooled—was close to the 422 counted by the Salesian anthropologist María Eguillor García for the Bisaasi-teri splinters. Eguillor García, *Yopo, shamanismo y hekura,* p. 56.

52. "The area of focus includes the drainages of the Mavaca and Bocon Rivers, both tributaries of the upper Orinoco. . . ." Chagnon, "Life Histories," p. 991, n. 21. This is slightly confusing because Venezuelan maps identify Caño Bocon as being west of the Shanishani drainage, and there were no groups living there when I checked. But Chagnon and Brewer identify Caño Bocon with the Shanishani River, and that

is where three Patanowa-teri splinters—Dorita-teri, Sheroana-teri, and Shanishani-teri all live. Brewer Carías, Chagnon, and Boom, "Forest and Man," p. 15. These three villages, which are located in the Shanishani Basin and which Chagnon visited in 1990–91, have a combined population similar to that of Villages 52, 53, and 54. Dorita-teri and Sheroana-teri had 228, according to the FUNDAFACI census (ibid., p. 9), and Shanishani-teri had 130, according to the journalist Marta Miranda for Venevisión, giving a total of 358. Villages 52, 53, and 54 come to 339. All three are included in Chagnon's study of mission impact. Chagnon, *Last Days of Eden,* p. 267.

53. Chagnon, "Life Histories," p. 986.

54. "The most common explanation given for raids (warfare) is revenge *(no yuwo)* for a previous killing, and the most common explanation for the initial cause of the fighting is 'women' *(suwa ta nowa ha)."* Ibid.

55. Ibid., p. 987.

56. Brian Ferguson, *Yanomami Warfare: A Political History* (Santa Fe: School of American Research Press, 1995), pp. 243–343.

57. Bruce Albert, "Yanomami 'Violence': Inclusive Fitness or Ethnographer's Representation?" *Current Anthropology* 30 (1989): 637.

58. "Vengeance motivation persists for many years. In January 1965, for example, the headman of one of the smaller villages (about 75 people) was killed by raiders in retaliation for an earlier killing. . . . In 1975, 10 years after his death, several gourds of his ashes remained, and the villagers were still raiding the group that killed him. . . ." Chagnon, "Life Histories," p. 986.

59. Napoleon Chagnon, *Yanomamo: The Fierce People,* 3d ed. (New York: Holt, Rinehart and Winston, 1983), p. 203. See also Chagnon, "Guns of Mucajaí: The Immorality of Self-deception." (MS, Sept. 1992), p. 3.

60. Chagnon, "Life Histories," p. 991, n. 24.

61. Chagnon, "Guns of Mucajaí," p. 4.

62. Chagnon, "Life Histories," p. 991, n. 25.

63. Chagnon, "Guns of Mucajaí," p. 2.

64. Chagnon, *Yanomamo,* 3d ed., p. 203.

65. Eguillor García, *Yopo, shamanes y hekura,* p. 53.

66. Chagnon, *Yanomamo,* 5th ed., p. 226.

67. Ibid., pp. 245–46.

68. "It seems, indeed, that this intensity varies from one place to another and that it is least where the Yanomami are most acculturated; in the regions where Chagnon and I have worked, I have observed a sudden diminishing of warlike activities during the past five years or so, following an acceleration of the acculturation process." Lizot, "On Warfare," p. 853.

69. Napoleon Chagnon, *Studying the Yanomamo* (New York: Holt, Rinehart and Winston, 1994), pp. 125–32; Chagnon, "Yanomamo Warfare," p. 167. "I showed that the Shamatari Yanomami of the Mavaca-Siapa region . . . studied by Chagnon present a much higher percentage of male mortality in warfare than any other Yanomami subgroup on which such data is available. . . . I stated that this may be related to early historical changes introduced in this population well before 'first contact' with whites . . . based on Chagnon's own attribution of the Shamatari's demographic expansion to their early acquisition of steel tools and their free access to open territories. . . ." Albert, "On Yanomami Warfare," p. 558.

70. Valero, *Yo soy Napeyoma,* pp. 69–81.

71. Ibid., pp. 229–45.

72. Hector Acebes, *Orinoco Adventure* (New York: Doubleday, 1954), p. 242.

73. Chagnon, "Yanomamo Warfare," p. 158.

74. Napoleon Chagnon, *Yanomamo: The Fierce People,* 2d ed. (New York: Holt, Rinehart and Winston, 1977), p. 103.

75. Chagnon, *Yanomamo,* 5th ed., p. 214.

76. Chagnon, "Guns of Mucajaí," p. 4.

77. Eguillor García, *Yopo, shamanes y hekura,* p. 26.

78. Lizot, "On Warfare," p. 852.

79. "Although the map published by Chagnon [*Yanomamo,* 4th ed., p. 167] suggests that Iwahikoroba-teri was a direct outgrowth of Ruwahiwa's people at Konabuma, his other writings [Ph.D. thesis, pp. 172–73; *Studying the Yanomamo,* pp. 9, 86; *Yanomamo,* 2d ed., pp. 102–3] indicate that Ruwahiwe's people formed a different group, led by Sibarariwa, called Mowaraoba-teri (and, more recently, Mishimishimabowei-teri)." Ferguson, *Yanomami Warfare,* p. 240.

80. N. Chagnon, *Yanomamo,* 4th ed., p. 167.

81. Chagnon, *Studying the Yanomamo,* pp. 195–97.

82. "All the *unokais* come from the villages under discussion, but not all of the victims do; some are from villages in adjacent areas beyond the focus of my field studies." Chagnon, "Life Histories," p. 987.

83. Chagnon, *Studying the Yanomamo,* p. 1.

84. "Venía llorando, pero falsamente . . . Allí se puso a preguntar. Quería saber a quién habían flechados los Pishaasi-their. Quería asegurarse de que había sido Husiwe el que había flechado a Wapurawe." Valero, *Yo soy Napeyoma,* p. 347.

85. Kenneth Good, phone interview, Feb. 1, 1995.

86. No. 2130, Yahohoiwa. Chagnon, *Studying the Yanomamo,* appendixes A and B.

87. Chagnon, *Yanomamo,* 5th ed., p. 213.

88. Chagnon, "Life Histories," p. 989.

89. Brian Ferguson, phone interview, June 14, 1997.

90. Napoleon Chagnon, "Response to Ferguson," *American Ethnologist* 16 (1989): p. 566.

91. Chagnon, "Life Histories," table 3.

92. Ibid., p. 988.

93. N. Chagnon, "About the Yanomamo," user's guide to accompany *Yanomamo Interactive,* p. 22.

94. For eight highland villages, there were 9 polygamous men, total: 1.1 per village. Since the total population was 479, that would work out to somewhat less than 10 percent. Chagnon, *Last Days of Eden,* p. 107.

95. "In all my time, I never saw anybody with more than three wives—and that was one guy." Kenneth Good, phone interview, April 17, 1997. Lizot found that 10 percent of all marriages were polygamous for his area in the adjacent Parima Mountain foothills. Lizot, "Sobre la guerra," p. 31.

96. Eguillor García, *Yopo, shamanes y hekura,* p. 84.

97. "I reconstruct Paruriwa's changing interests this way: Before being faced down in July [when Chagnon intervened to take visitors Paruriwa wanted to kill back to their home village in his boat], he hoped that by putting himself at the head of the Monou-teri, in addition to his own Bisaasi-teri faction, he could replace Kaobawa as leader of upper Bisaasi-teri and be treated accordingly. . . . After backing down from Kaobawa and in the process both losing the Monou-teri's allegiance and seeing Chagnon act in support of Kaobawa, he knew his plan would fail. At this point he began work on his house by the mission, but he held off moving while Chagnon remained in upper Bisaasi-teri, in the meantime allowing the priest to make him generous gifts." Ferguson, *Yanomami Warfare,* p. 304. "I saw Paruriwa in 1967 when I stopped to greet the priest; he was proudly bearing a presumptuous Spanish title, beating his chest with his fist, urging me to pay attention: 'Me capitán! Me capitán! Me capitán!' He swaggered off after I acknowledged that I understood that he was a leader now, carelessly shouldering a rusty 16-gauge shotgun. . . ." Chagnon, *Yanomamo,* 2d ed., p. 151.

98. Asiawe, interview, Platanal mission, Upper Orinoco, June 11, 1996.

99. Valero recounted how her husband, Fusiwe, would go and beg machetes from Kasiawe. "Bring me, bring me things from the *nape,* you who are friend of the *nape.*" Valero, transcribed statements, in Luis Cocco, *Iyewei-teri: Quince años entre los Yanomamos* (Caracas: Editorial Salesiana, 1972), pp. 205, 385.

100. Pablo Anduze, *Shailili Ko: Relato de un naturalista que también llegó a las fuentes del río Orinoco* (Caracas: Talleres Gráficos Ilustrados, 1960), p. 246.

101. "El Padre Luis Cocco hizo la experiencia entre los iyeweitheri de Ocamo que cuando éstos comenzaron a tener escopetas que favorecieron la cacería, en la misma medida comenzó a aumentar la poliginia." Eguillor García, *Yopo, shamanes y hekura,* p. 84.

102. Biella, Chagnon, and Seaman, "Genealogies," *Yanomamo Interactive.*

103. Chagnon, *Studying the Yanomamo,* p. 147.

104. "While polygyny and occasionally polyandry do occur, multiple marriages are much more likely to be consecutive than concurrent." Clayton Robarchek and Carole Robarchek, *Waorani: The Contexts of Violence and War* (New York: Harcourt Brace, 1998), p. 132.

105. Chagnon, *Studying the Yanomamo,* pp. 198–218. The village includes three women not actually present, Nanokawa's extra wives.

106. I am counting Village 84, just above the confluence of the Orinoco-Mavaca, along with Villages 5, 6, and 7. Chagnon apparently does the same for his study of mission mortality. Chagnon, *Yanomamo,* 5th ed., p. 246.

107. Chagnon, *Studying the Yanomamo,* appendix B, No. 340.

108. Ibid., No. 777.

109. Ibid., No. 651.

110. Ibid., No. 1929.

111. Ibid., No. 178.

112. Ibid., No. 1240.

113. Ibid., No. 1335.

114. No. 340, Dedeheiwa, 60 years old, sixteen children; No. 777, Ishiweiwa, 70 years old, thirteen; No. 1929, Wadoshewa, 46 years old, nine; No. 256, Borosoteri, 42 years old, eight; No. 2248, Yoinakuwa, 42 years old, nine; No. 1335, Nanokawa, 35 years old, nine.

115. "To qualify for a wife, a young man must have proved himself a capable hunter, and he ought

to have had some experience in gardening. . . . A marriage agreement generally includes a suitable bride price. . . . Inexperienced young men are also at a disadvantage in competition with older men for wives, since the older men have had time to make the suitable arrangements with the girl's parents long before she reaches marriageable age." William Smole, *The Yanoama Indians: A Cultural Geography* (Austin: Univ. of Texas Press, 1976), p. 75.

116. "Some of the apparent correlation of *unokai* status with higher numbers of wives and children may be a result of covariation with age *within* the four age categories. A 40-year-old man, for example, is more likely to be *unokai* than a 31-year-old man." Ferguson, *Yanomami Warfare,* p. 360.

117. Photo plate 4: "Moawa and his father-in-law, Dedeheiwa—the secular and spiritual authorities of Mishimishimabowei-teri." Chagnon, *Studying the Yanomamo,* opposite p. 1.

118. The Hasupuweteri had two factions, each led by great shamans, Orawe and Yarimowe. Kenneth Good, *Into the Heart: One Man's Pursuit of Love and Knowledge among the Yanomama* (New York: Simon & Schuster, 1991), p. 66 and 79. See also Lizot, "Sobre la guerra," p. 31. Among the Iyewei-teri, there was a shamanic lineage from grandfather (Badaxiwe) to son (Hioduwa) to grandson (Renato), which Redmond has analyzed as an incipient chieftainship. Elsa Redmond, "In War and Peace," in *Chiefdoms and Chieftaincy in the Americas,* ed. Redmond (Gainesville: Univ. Press of Florida, 1998), pp. 96–97.

119. Saffirio and Hames, "The Forest and the Highway," p. 25.

120. "In the hammock next to me lay a young, muscular, handsome man of obvious importance. . . . A year later I learned that he was *the* headman, and his renown had eclipsed that of Sibariwa. . . ." Ibid. See also Chagnon, *Studying the Yanomamo,* p. 10, fig. 1.8: "Sibarariwa, the fabled headman of Mishimishimabowei-teri."

121. Chagnon, *Studying the Yanomamo,* p. 193.

122. "Nanokawa, Moawa's classificatory brother, was himself a man of renown and had fought a great deal with Moawa." Ibid., p. 192, fig. 5.11. "Many readers will be tempted to reduce my conflict with Moawa to a simple explanation that it represents only an ethnographer, in possession of valuable goods, giving them away in such a fashion that it was perceived by the headman as a threat to his authority and an undermining of it." Moawa saw it that way. Ibid., p. 197.

123. Ibid. Nos. 876, 1188, and 2115 are listed in appendix B as Nanokawa's wives, but they are not listed in appendix A.

124. James Neel, Timothy Asch, and Napoleon Chagnon, *Yanomamo: A Multidisciplinary Study,* 43 min. (DOE, 1971).

125. Napoleon Chagnon, M. Flinn and Thomas Melancon, "Sex-Ratio Variation among the Yanomamo Indians," in *Evolutionary Biology and Human Social Behavior: An Anthropological Perspective,* ed. Napoleon Chagnon and William Irons, (North Scituate, Mass.: Duxbury Press, 1979), p. 308.

126. All the headmen Chagnon has identified were over 30. But Chagnon tends to focus on young men competing for his goods, men who are in their mid-30s like Nanakowa and Moawa, whose ages he estimated at 35. See Chagnon, *Studying the Yanomamo,* appendix A, Nos. 1335 and 1240. The established headmen of true renown were all over 40. Reromawa of Lower Bisaasiteri was 53. See Chagnon, "Yanomamo Warfare," p. 212. Sibarariwa of Mishimishimabowei-teri was in his 50s. See Chagnon, *Studying the Yanomamo,* pp. 25–27. Kasiawe of Maheko-teri and Kumaiwe of Patanowa-teri were both well over 50, because Helena Valero had known them both as adults for decades. See Cocco, *Iyewei-teri,* pp. 205, 134.

127. "The greatest respect and authority is given to older men who have the authority to make such decisions as choosing mates for their daughters, selecting the location of the next village, giving advice in regards to raiding other villages, or the scheduling of a feast and related activities. An older age means years of experience, which makes a person more reliable." Saffirio and Haymes, "The Forest and the Highway," p. 25. Cocco summarizes the observations of explorers from the first half of the twentieth century—including Spencer Dickey, Marqués de Wavrin, and others—who argue that the Yanomami have always had an older man directing a younger war leader who acts as a sort of field commander: "entre los jovenes guerreros descollaba siempre un individuo más ardoroso y gallardo que parecía dirigirlos pero que, por encima de él, ejercía el mando un anciano que parecía ser el verdadero jefe." Cocco, *Iyewei-teri,* p. 430.

128. Chagnon, "Life Histories," p. 989.

129. Ferguson, "Do Yanomamo Killers Have More Kids?" p. 564.

130. The killers in *Science* aged 20–24 have .17 wives and .23 children each. The men in Mishimishimabowei-teri in 1971 had .67 wives and .67 children. Chagnon, "Life Histories," p. 986.

131. Eguillor García, *Yopo, shamanes y hekura,* p. 84.

132. "Es rarísima la perspectiva de soltería entre los adultos sin que quieran casarse. Ninguno tiene vocación de solterón . . . en nuestra investigación encontramos un caso: Horiwe, en Patanowatheri, residiendo ahora en Mavaca, ya mayor, que, oficialmente, nunca tuvo esposa." Ibid., p. 87.

133. No. 950: Kodedeari, born 1949, three children, one shared wife; No. 517: Hemoshabuma, born 1945, one wife. Chagnon, *Studying the Yanomamo,* appendixes A and B.

134. Gary Seaman, "Blow by Blow," accessed under "Ruwamowa: People Screen," *Yanomamo Interactive.*

135. Mohesiwa: No. 1246. Chagnon, *Studying the Yanomamo,* p. 204, appendix A. Ruamowa: No. 1568. Ibid., p. 205.

136. Chagnon, "Life Histories," p. 989, table 2.

137. Chagnon, "On Yanomamo Violence," p. 50, n. 5.

138. William Allman, *Stone Age Present* (New York: Simon & Schuster, 1994), p. 148.

139. Martin Daly and Margot Wilson, *Homicide* (Hawthorne, N.Y.: Aldine de Gruyter, 1988), p. 170.

140. Margot Wilson, phone interview, April 10, 1996.

141. "Of 15 recent killings, four of the victims were females. . . . Nine of the males were under 30 years of age, of whom four were under an estimated 25 years of age." Chagnon, "Life Histories," p. 990.

142. Ibid.

143. Ibid., p. 989, fine print above table 2.

144. Ferguson, *Yanomami Warfare,* p. 367.

145. Chagnon, "Response to Ferguson," p. 566.

146. Chagnon, "Life Histories," p. 986, fig. 1.

147. Valero, *Yo soy Napeyoma,* p. 237.

148. John H. Moore, "The Reproductive Success of the Cheyenne War Chiefs: A Contrary Case to Chagnon's Yanomamo," *Current Anthropology* 31 (1990): 322–30.

149. Robarchek and Robarchek, *Waorani,* pp. 131–37.

150. Elsa Redmond, *Tribal and Chiefly Warfare in South America* (Ann Arbor: Univ. of Michigan, Museum of Anthropology, 1994).

151. Kenneth Good, phone interview, April 17, 1997.

152. Frans De Waal, *Good Natured: The Origins of Right and Wrong in Humans and Other Animals* (Cambridge: Harvard Univ. Press, 1996), pp. 154–62.

153. Chagnon, "Reply to Albert."

154. Chagnon, "Response to Ferguson," p. 567.

155. Ibid., p. 569.

156. Ferguson, *Yanomami Warfare,* p. 407, n. 12.

157. Clark McCauley, "Conference Overview," in *The Anthropology of War,* ed. J. Haas (Cambridge: Cambridge Univ. Press, 1990), pp. 2–6.

Chapter 11: A Kingdom of Their Own

1. María Yolanda García, "Cecilia Matos no iba a proteger indígenas sino a sacar oro del Amazonas," *El Nacional,* Jan. 15, 1993.

2. Nelly Arvelo Jiménez and Andrew L. Cousins, "False Promises: Venezuela Appears to Have Protected the Yanomami, But Appearances Can Be Deceiving," *Cultural Survival Quarterly,* Winter 1992, pp. 10–14. This article includes a map of the Chagnon-Brewer Siapa biosphere as apparently presented in Caracas in 1991, at a meeting called "A Future for the Orinoquia-Amazonas." James Brooke, however, mentions the area as about "8000 square miles" ("In an Almost Untouched Jungle, Gold Miners Threaten Indian Ways," *NYT,* Sept. 19, 1990), while ABC *Prime Time Live* showed a map of the Chagnon-Brewer reserve calculated at 18,000 square miles. In every case, however, a large majority of the Yanomami would have been left without protection under the Chagnon-Brewer plan.

3. Edgar López, in *El Diario de Caracas,* Sept. 2, 1993.

4. Napoleon Chagnon, *Yanomamo:* 4th ed. (Fort Worth: Harcourt Brace, 1992), p. xv.

5. James Neel, phone interview, March 18, 1997.

6. "The logo emblazoned on the side of the helicopter bodes ill for Yanomamo culture: Conquest of the South." Napoleon Chagnon, *Yanomamo: The Fierce People.* 3d ed. (New York: Holt, Rinehart and Winston, 1983), p. 202, fig. 7.1.

7. Chagnon's last feature article in a refereed anthropological journal was "Genealogy, Solidarity and Relatedness: Limits to Local Group Size and Patterns of Fissioning in an Expanding Population," *Yearbook of Physical Anthropology* 19 (1975): 95–110; Napoleon Chagnon, *Yanomamo,* 5th ed. (Fort Worth: Harcourt Brace, 1997), pp. 264–66.

8. Napoleon Chagnon, personal correspondence, Jan. 24, 1990.

9. Chagnon, *Yanomamo,* 4th ed., pp. 218–19.

10. "It seems to me that the work of the Salesians is very important for the Yanomami because the Salesians are very practical. . . . The Salesian men and women missionaries have a mixture of theology and

love for the Indians, a way of thinking that envisions the future of the Yanomami in a practical way. I admire that." N. Chagnon, quoted and translated by Father E. J. Cappelletti, director, Salesian missions, New Rochelle, N.Y., letter to the editor, *NYT,* Jan. 18, 1994.

11. Napoleon Chagnon, letter to Padre José Bórtoli, July 19, 1988.

12. I had five phone interviews with Jesús Cardozo between Dec. 1994 and Sept. 1995. The remarks in this first paragraph were made on Dec. 20, 1994. Jesús made some minor changes, particularly in the spellings and names of the Yanomami groups, on Aug. 8, 1995.

13. Jesús Cardozo, phone interview, June 21, 1995.

14. Jesús Cardozo, phone interview, Aug. 8, 1995.

15. Ibid., Dec. 20, 1994.

16. Napoleon Chagnon, *Studying the Yanomamo* (New York: Holt, Rinehart and Winston, 1974), p. 30.

17. Sound Roll 9, Patanowa-teri, Feb. 23, 1968, Timothy Asch Collection, NAA.

18. Juan Finkers, phone interview, Jan. 24, 1995.

19. Ibid.

20. César Dimanawa, "Carta abierta a Napoleon Chagnon," *La Iglesia en Amazonas,* March 1990, p. 20.

21. *Shereka siwaka,* literally, "fire arrow." César Dimanawa, interview, Mavakita, Sept. 2, 1996. These were members of the Washewa-teri, formerly Iwahikoroba-teri. Chagnon had a particularly difficult relationship with this village. Its people not only shot an effigy of him full of arrows; they also tried to kill him while he rested in his hammock, and desisted only at the last moment when he turned on his flashlight. Similar stories were told by Yanomami at several other villages, but always from *shabonos* some distance from the missions and the main course of the Orinoco. Etilio, who came from Momaribowei-teri, the first Shamatari village Chagnon visited in the interior, had this recollection: "El decia, 'Yo tengo bomba.' Llevaba aqui chiquitico, asi. [Showed like a small container on his belt.] 'Con ese, si Ud. se porta mal conmigo, yo voy a tirar este en frente del shabono y voy a quemar todo. Yo acabo con todo un shabono.' Por eso nadie hacia mal con el. El decia, 'Yo soy hombre guerrillero. Yo se matar gente. Yo soy peligroso. Ustedes salvages no me comen nada porque con ese yo puedo matar BOOOOM. Se mueren rapido.' Yo lo ví.. Con su pistola siempre disparaba contra palo: BOOOM BOOOM BOOOM. Y hasta con su escopeta: BOOOM BOOOM. La gente no puede creer eso. Pero nuestra comunidad Yanomami trabajo mucho con Shaki. Conocemos mucho a Shaki. Los que cargaban, si hacían mal, el disparaba por un lado, por el otro, para asustar. No paso nada pero asustaba." Etilio, Guarapana, Boca Mavaca, June 9, 1996.

22. Ramon Bokoramo, Kedebabowe-teri, Mrakapiwei, June 8, 1996.

23. Napoleon Chagnon, letter to Padre José Bórtoli, July 19, 1988.

24. Napoleon Chagnon, letter to Padre José Bórtoli, April 16, 1990.

25. Charles Brewer Carías, Curriculum en antropología, Sept. 3, 1993, p. 3. The UCSB public relations department confirmed the appointment when I called in June 1995, but I was told it had been only for a year, 1991–92, so that Brewer's continued claim to be a UCSB research associate appears doubtful.

26. Brooke, "In an Almost Untouched Jungle."

27. Brian Ferguson, phone interview, Feb. 3, 1995.

28. John Quiñones, "A Window on the Past," *Prime Time Live,* July 26, 1991; James Brooke, "Venezuela Befriends Tribe, But What's Venezuela?" *NYT,* Sept. 11, 1991.

29. Napoleon Chagnon, José Bórtoli, and María Eguillor García, "Una aplicación antropológica práctica entre los yanomami: Colaboración entre misioneros y antropólogos," *La Iglesia en Amazonas,* 1988, pp. 75–83.

30. "La alianza entre los teólogos de la liberación y los marxistas al Sur del Orinoco esta conduciendo a la creación de un peligroso vacío del poder que está sentando lás bases para la disolución territorial de Venezuela." Issam Madi, *Conspiración al sur del Orinoco* (Caracas: self-published, 1998), jacket.

31. When Brewer Carías attacked "the Yanomami Reserve" and "the biosphere," he was referring to two different but related proposals. One was put forward in 1983 by the Catholic La Salle Foundation—Los yanomami venezolanos: Propuesta para la creación de la Reserva Indígena Yanomami (Caracas, Fundación La Salle, 1983). This proposal would have set aside 15,000 square miles for the Yanomami. The following year, a plan was introduced that defined a core area and outlying buffer zones around a new concept called "a biosphere"—Nelly Arvelo Jimenez, "La Reserva de Biósfera Yanomami: Una auténtica estrategia para el ecodesarrollo nacional" (Caracas: IVIC, 1984). Brewer's attacks on the concept of indigenous land rights came at a crucial time. In June 1984, Piaroa Indians were violently treated by cattle ranchers who were illegally encroaching on their lands. Brewer not only attributed the Yanomami Reserve to a Communist plot but also took the opportunity to defend the cattle ranchers in question. Brewer's motives have been questioned, and not only because of his proven gold-mining activities in the area. In 1982, the Venezuelan government granted 173 square miles of tin concessions along the Upper

Orinoco in Yanomami territory. Whether or not Brewer was involved with these tin concessions, as Salesian missionaries charged, his jingoistic denunciation of Indian lands was part of "a resurgence of anti-Indian rhetoric in 1984" that "buried both proposals." Marcus Colchester, *Sustainability and Decision-making in the Venezuelan Amazon: The Yanomami in the Upper Orinoco-Casiquiare Biosphere Reserve* (Oxford: World Rainforest Movement, 1995), pp. 16–17.

32. Napoleon Chagnon, letter to Timothy Asch, Feb. 10, 1994, Timothy Asch Collection, NAA.

33. Napoleon Chagnon, letter to Timothy Asch, March 22, 1991, Timothy Asch Collection, NAA.

34. Timothy Asch, letter to Napoleon Chagnon, June 18, 1991, Timothy Asch Collection, NAA.

35. Chagnon, *Yanomamo,* 4th ed., p. 243.

36. Timothy Asch, letter to Jesús Cardozo and Hortensia Caballero, June 20, 1991, Timothy Asch Collection, NAA.

37. "I did want to say, Jesús, that I was most impressed with your abilities to mediate. I really think that this is a professional talent that you should develop. . . . I think you are really very good because you are able to get the information that is needed in a calm way and work sensibly with it." Ibid.

38. Jesús Cardozo, phone interview, Dec. 20, 1994.

39. Ibid.

40. Kim Hill, phone interview, Jan. 17, 1995.

41. Charles Brewer Carías, Napoleon Chagnon, and Brian Boom, "Forest and Man" (MS; Caracas: Fundación Explora, 1993), pp. 10–20.

42. Kim Hill and Hillard Kaplan. "Population and Dry-Season Subsistence Strategies of the Recently Contacted Yora of Peru." *National Geographic Research* 5 (1989): 332.

43. Bruce Albert, interview, Toototobi, Demini River, Brazil, Dec. 5, 1990.

44. Josefa Camargo, phone interview, Dec. 19, 1994.

45. Raymond Hames, phone interview, Dec. 29, 1994.

46. Kim Hill, phone interview, Jan. 17, 1995.

47. "Braorewa-teri y Doshamosha-teri: 8 de enero al 12 de enero. . . . Participantes: Charles Brewer-Carías, Napoleon A. Chagnon, Darius Chagnon y el medico Maxilimiano Ravard." Brewer, Chagnon, and Boom, "Forest and Man," p. 11.

48. Ibid., pp. 12–13.

49. García, "Cecilia Matos no iba a proteger indigenas sino a sacar oro del Amazonas."

50. Brewer, Chagnon, and Boom, "Forest and Man," p. 12.

51. Carlos Botto, interview, CAISET, Puerto Ayacucho, Oct. 6, 1996.

52. García, "Cecilia Matos no iba a proteger índigenas sino a sacar oro del Amazonas."

53. Leslie Illiman, "Intrigues Hinder Yanomami Massacre Probe," *Daily Journal* (Caracas), Sept. 20, 1993.

54. Jota Rodriguez Flores, "Yo acuso a Charles Brewer Carías y a Cecilia Matos," *El Mundo,* Sept. 14, 1993.

55. Edgar Lopéz, in *El Diario de Caracas,* Sept. 2, 1993.

56. Misioneros del Atto Orinoco, *Consideraciones a un documento de Charles Brewer Carías* (Mavaca: Salesian Mission, 1991), p. 12.

57. Patrick J. O'Donoghue, "Cecilia Matos' Lawyer Accuses Venezuelan Foreign Minister of Harassing His Client," Vheadline.com, March 25, 1998.

58. See Introduction, n. 21. Patrick J. O'Donoghue, "Disgraced Ex-President Carlos Andres Pérez Angered over CSJ Ruling Ratifying His House Arrest," Vheadline.com, Aug. 12, 1998.

Chapter 12: The Massacre at Haximu

1. Napoleon Chagnon, "Covering Up the Yanomamo Massacre," op-ed, *NYT,* Oct. 23, 1993.

2. Bruce Albert, "The Massacre of the Yanomami at Hashimu" (MS, based on an article in *Fôlha de São Paulo,* Oct. 10, 1993), pp. 5–6.

3. "Amazon Murder Mystery," *Los Angeles Times,* Aug. 30, 1993.

4. James Brooke, "Attack on Brazilian Indians Is Worst since 1910," *NYT,* Aug. 29, 1993.

5. James Brooke, "Raids on Miners Follow Killings in Amazon," *NYT,* Sept. 9, 1993; "Un grito do fundo da selva," *Veja,* Aug. 25, 1993, p. 24.

6. Leda Martins, interview, Univ. of Pittsburgh, March 7, 1995.

7. Decreto No. 1635, *Gaceta Oficial de la Republica de Venezuela,* Aug. 1, 1991.

8. Napoleon Chagnon, *Last Days of Eden* (New York: Harcourt Brace, 1992), p. 252.

9. Decreto No. 3127, *Gazeta Oficial de la República de Venezuela,* No. 35.292, Sept. 8, 1993.

10. "Indígenas del Amazonas rechazan presencia de Brewer Carías y Chagnon," *El Nacional* (Caracas), Sept. 14, 1993.

11. Ibid,

12. Ibid.

13. Juan Ignacio Cortinas, "Las nuevas tribus sí han hecho daño al yanomami," *El Diario de Caracas,* Oct. 23, 1993.

14. "Además no recibiremos indicaciones de gente que, como Chagnon, ha sido vetada por el Colegio de Antropólogos y el Conicit." José Visconti, "Los salesianos vetan a Brewer Carías y Chagnon," *El Diario de Caracas,* Sept. 18, 1993.

15. "Most members of that sub-committee were protested by different sectors of Venezuelan society, the Catholic Church, the indian organizations, the Colegio de Antropólogos de Venezuela, the main Department of Anthropology, UCV Department of Anthropology, ect., ect., there were many voices voiced against that Sub-Committee and most especially against Charles Brewer and Nap. Chagnon. People knew about the latter two connection with the hated Carlos Andrés Pérez (under trial) and his lover Cecilia Matos. For the first time in its history, this Department, IVIC's Anthropology Department also made a move to protest Brewer Carías and Chagnon's appointment. . . ." Nelly Arvelo Jiménez, letter to Dr. Gale Goodwin-Gómez, Sept. 29, 1994.

16. " 'La comisión presidencial designada por el Presidente Velásquez y presidida por Charles Brewer Carías, constituye un irrespeto para el gremio de profesionales de antropólogos y sociólogos.' Así lo destacaron varios especialistas en la materia de la Escuela de Sociología y antropología de la UCV, quienes calificaron a Brewer Carías como un odontólogo-aventurero ególatra, egocéntrico y con intereses muy particulares en la región Orinoco-Amazónica." Anabel Flores, "Sociólogos y antropólogos objetan presencia de Brewer Carías en comisión presidencial," *Ultimas Noticias,* Oct. 5, 1993, p. 41.

17. Napoleon Chagnon, "Killed by Kindness?: The Dubious Influence of the Salesian Missions in Amazonas," *TLS,* Dec. 24, 1993, p. 11.

18. "El diputado causaerredista Carlos Azpurua, uno de los comisionados de la Cámara de Diputados, coadyuvó a la elaboración de un pormenorizado informe relacionado con el problema fronterizo, que se refiere, en su parte esencial, a algunas personalidades políticas venezolanas, que están vinculadas a la explotación del oro y otras riquezas, en el territorio amazónico de Venezuela. . . . El informe explica expresamente cómo el doctor Charles Brewer Carías le impone un velo de legalidad a sus actividades que constituyen una explotación inmisericorde a los indígenas, contrariando principios esenciales de respeto a los derechos humanos." David Ayala, "Informe de Comisión de Diputados ratifica denuncias contra Charles Brewer Carías," *Ultimas Noticias,* Oct. 5, 1993, p. 22.

19. "El último toque contra Brewer Carías lo proporcionó el gobernador del Amazonas, quien ayer le presentó un informe sobre la masacre de los 73 yanomamis y las protestas de las fuerzas del estado contra Brewer Carías." Exequíades Chirinos Q., "El presidente Velásquez revocó designación de Charles Brewer Carías," *El Universal,* Sept. 14, 1993.

20. "Se pudo conocer que el gobernador Edgar Sayago, de amazonas, en su intervención en la reunión de ayer, rechazó categóricamente—con golpe de mano en la mesa—'por indeseable' al explorador Brewer Carías, señalando que 'en Amazonas no lo aceptamos.' " Adela Leal, "Deja Comisión Yanomami Charles Brewer Carías: Reestructurán el decreto presidencial," *El Nacional,* Sept. 14, 1993.

21. Cardenal José A. Lebrún, monseñores Ovidio Pérez Morales, Baltasar Porras y Mario Moronta, "Documento Oficial de la Conferencia Episcopal Venezolana" (Universidad Católica Andrés Bello, Sept. 11, 1993).

22. "No es confiable comisión que investigará caso de yanomamis: Monseñor Ovidio Pérez Morales," *Ultimas Noticias,* Oct. 5, 1993, p. 41.

23. Orlando Utrera, "Brewer denuncia el 'Plan Gadhafi,' " *El Diario de Caracas,* Aug. 15, 1984.

24. "Napoleon Chagnon dijo que toda persona que realiza cualquier actividad se beneficia de ella, pero él ha realizado labor en beneficio de los indígenas y citó el caso de dos yanomami que se han manifestado publicamente contra él, a quienes vacunó cuando eran niños y dijo que posiblemente vivían gracias a esa vacuna." Victor Manuel Reinoso, " 'Me rechazan por envidia' asegura Charles Brewer Carías," *El Nacional,* Sept. 16, 1993.

25. "El candidato presidencial Oswaldo Alvarez Paz (COPEI) solicitó al Presidente de la República, Ramón J. Velásquez, revisar el nombramiento de la comisión especial. . . ." Comisión Investigadora Venezolana, "Supuesto asesinato de ciudadanos venezolanos de la etnia yanomami por ciudadanos brasileños" (Caracas, 1993), p. 12.

26. "During a period of some 6 or 7 weeks, approximately 600 newspaper articles appeared in the major Venezuelan newspapers discussing the massacre and the Presidential Commission appointed to investigate it. Many of the articles were critical of Chagnon and Brewer, attempting to implicate them in gold mining activities in Amazonas and to corruption in the former Venezuelan government. . . ." Napoleon Chagnon, "Notes on the Chronology of the Recent Attacks on Members of the Venezuelan Presidential Commission" (MS, May 18, 1994), p. 4.

27. Napoleon Chagnon. "The View from the President's Window," *Human Behavior and Evolution Society Newsletter* 2, no. 3 (Oct. 1993): 1.

28. Reinoso, " 'Me rechazan por envidia' asegura Charles Brewer Carías."

29. Ibid.

30. Minas Guariche C.A., "Modificación de estatutos," Registrador Mercantíl de la Circunscripción Judicial del Distrito Federal y Estado Miranda, May 12, 1993.

31. "Public Lands, Private Profits." *Frontline,* WGBH, Boston, 1994.

32. Kenneth Gooding, "Race to Move Mountain of Waste in the Rockies," *Financial Times* (London), Nov. 8, 1993, p. 8.

33. Friedland initially decided to acquire the controlling interest of the Golden Star company after learning that its gold deposits at Omai were suitable for cyanide concentration. "Golden Star's stock was beat when I came on the scene because of Placer Dome's withdrawal from Omai," Friedland told the *Northern Miner* in a March 29, 1993, interview. "It was a bird with a broken wing, and I helped it mend." It was Friedland who also arranged for a partnership with the Cambior to develop the Omai deposits. Vivian Danielson, "Friedland Strong Supporter of Guiana Shield Gold Rush," *Northern Miner,* March 29, 1993. See also "Cyanide Spill Poisons River," *Latin American Press,* July 13, 1995; "Gold Mine Loses Its Luster," ibid., Aug. 31, 1995.

34. Diana Jean Schemo, "Legally Now, Venezuelans to Mine Fragile Lands," *NYT,* Dec. 8, 1995.

35. Elizabeth Kline, *Mining Abuses Tarnish Venezuela's Environmental Image* (Caracas: Sociedad Conservacionista Audubon de Venezuela, 1994), p. 4.

36. Ana Ponte, "Charles Brewer Carías: Informe para el Comité del Medio Ambiente" (MS, Jan. 1997), p. 4.

37. Napoleon Chagnon, letter to Ramades Muñoz León, ministro de defensa, Oct. 2, 1993.

38. "El 27 de septiembre de 1993, el Comando Aéreo de esta fuerza trasladó en horas de la tarde en una de sus aeronaves hasta el sector de Parima B del estado Amazonas a un grupo de personas entre las que destacaba el Dr. CHARLES BREWER CARIAS, quien en forma displicente adujo ser el 'Presidente de la Comisión Presidencial Yanomami, y que requería de los medios aéreos necesarios para cumplir su labor.' El Coronel le explicó lo limitado que estaba a este respecto, ya que se estaba ejecutando una operación para evacuar el personal que había aparecido ese día en la mañana, que también estaba apoyando una Comisión del Ministerio Público que estaba en Ocamo y tenía escasez de combustible. No obstante y pese a las limitaciones, a tempranas horas del 28 de septiembre de 1993 se trasladó el Dr. BREWER CARIAS y su Comitiva hasta Haximu, indicándosele que serían buscados al día siguiente." Pedro José Romero Farias, General de División Comandante de la Guardia Nacional, nota informativa al vicealmirante ministro de la Defensa, Oct. 7, 1993, pp. 1–2.

39. Ibid., p. 2.

40. Napoleon Chagnon, letter to Ramades Muñoz León, Oct. 2, 1993.

41. Josefa Camargo, phone interview, Dec. 19, 1994.

42. Ibid.

43. Chagnon, "Covering Up the Yanomamo Massacre."

44. "Holy War in the Amazon," *Newsweek* (international ed.), Oct. 11, 1993, p. 3.

45. "While en route back from Venezuela via Miami, Chagnon [referring to himself in the third person] was asked by *Newsweek* to give them an account of the investigation of the Hashimo-teri massacre. A brief article appeared shortly after in *Newsweek*'s International edition. The allusion to the Salesian control over Amazonas as a 'theocracy' appeared here for the first time." Chagnon, "Notes on the Chronology of the Recent Attacks," p. 5.

46. Spencer Reiss, "The Last Days of Eden," *Newsweek,* Dec. 3, 1990, pp. 40–42.

47. Chagnon, "Killed by Kindness?" pp. 11–12.

48. Ibid., p. 11.

49. Napoleon Chagnon, letter to Robin Hanbury-Tennison, Oct. 29, 1993.

50. Terence Turner, "The Yanomami: Truth and Consequences," *Anthropology Newsletter,* May 1994, p. 48.

51. *Gaceta Oficial de la Republica de Venezuela,* No. 36.123, Jan. 10, 1997.

52. E. J. Cappelletti, "Venezuela Mine Scheme Targets Salesians," letter to *NYT,* Jan. 18, 1994.

53. Bruce Albert, personal correspondence, Paris, Dec. 15, 1994.

54. "To denounce the mission in a slanderous way could make their work very difficult, and this could easily have a disastrous effect on the Yanomami." Irenaus Eibl-Eibesfeldt and Gabriele Herzog-Schroder, "In Defense of the Mission" (MS), pp. 4–5.

55. Jacques Lizot. "N. A. Chagnon, o sea: Un presidente falsificador" (letter, Caracas, Dec. 13, 1993), p. 4.

56. Kim Hill, "Response to Cardozo and Lizot" (MS, March 1994), p. 4.

57. Eric R. Wolf, "Demonization of Anthropologists in the Amazon," *Anthropology Newsletter,* March 1994, p. 2.

58. Chagnon, "View from the President's Window," p. 3.

59. " 'As observações e a ciencia de Chagnon são basicamente correctas,' afirma Edward Wilson, renomado biológo da Universidade de Harvard. 'Ele esta na linha de frente da moderna sociobiologia. Por isso, talvez, a polémica o persiga.' " Cited in Eurípedes Alcántara, "Indio também é gente," *Veja,* Dec. 6, 1995, p. 7. "American anthropologists, both individually and through their organization, should rally to the support of Chagnon and the absolute value of his courageous and brilliant field studies of Yanomamo culture as well as his practical efforts to save it." Robin Fox, "Evil Wrought in the Name of Good," *Anthropology Newsletter,* March 1994, p. 2.

60. Matt Ridley, fax to Napoleon Chagnon, Aug. 16, 1994.

61. Napoleon Chagnon, "Notes on the Chronology of the Recent Attacks on Members of the Venezuelan Presidential Commission" (MS), p. 11.

62. Lizot. "N. A. Chagnon, o sea," p. 4.

63. Konrad Lorenz, *On Aggression,* 11th ed. (New York: Bantam Books, 1971), pp. 123, 152, 156.

64. "When we read Chagnon's letter of the President in the Newsletter of the Human Behaviour and Evolution Society, we found it very strange and annoying. A newsletter serving scientific communication is not the place to vent highly personal frustrations. We are happy that Lizot replied to Chagnon. Ever since I, Eibl-Eibesfeldt, started my documentary work amongst the Yanomami in 1969, I learned with utmost respect about the pioneer effort of the Salesian Mission to prepare the Yanomami for the inevitable impact of modern civilization and to brace them to survive as an ethnic group. Catholic missions amongst the Yanomami in Brazil had to suffer quite a lot under government administration for this very reason, which is less sympathetic to the fate of the Yanomami and which resented missionaries for siding with the interests of the Yanomami. Fortunately, the situation is different in Venezuela. . . . The mission nowadays are the only present force who protect the highly vulnerable traditional societies against exploitation and domination by ruthless wordly powers. . . . The work of the Salesian Mission on the Upper Orinoco has in particular gained my highest appreciation and support." Irenaus Eibl-Eibesfeldt and Gabriele Herzog-Schroder, letter to Bishop Ignacio Velasco, Feb. 28, 1994.

65. Eibl-Eibesfeldt and Herzog-Schroder, "In Defense of the Mission," p. 3.

66. "While the Salesians claim they no longer attract converts by offering shotguns, that was their policy until 1991. Over the past five years there has been a rash of shotgun killings." Chagnon, "Covering Up the Yanomamo Massacre."

67. "Their kindness and good example have, in a number of cases with which I am familiar, prevented bloodshed among the Indians." Napoleon Chagnon, "Yanomamo Warfare, Social Organization and Marriage Alliances" (Ph.D. diss., Univ. of Michigan, 1966), p. 198. "The missionaries are very cautious about loaning the Yanomamo firearms, knowing that these would be used in the wars." Napoleon Chagnon, *Yanomamo: The Fierce People,* 2d ed. (New York: Holt, Rinehart and Winston, 1977), p. 122n. "Warfare has recently diminished in most regions due to the increasing influence of missionaries and government agents and is almost nonexistent in some villages." Napoleon Chagnon, "Life Histories, Blood Revenge, and Warfare in a Tribal Population," *Science* 239 (1988): 986.

68. Napoleon Chagnon, *Yanomamo:* 4th ed. (Fort Worth: Harcourt Brace, 1992), p. 220.

69. Napoleon Chagnon, *Studying the Yanomamo* (New York: Holt, Rinehart and Winston, 1974), p. xiv.

70. "The evening before the men returned from the trip, one of the Salesian missionaries, Padre Luis Cocco, visited me, having traveled up the Orinoco River by dark—a dangerous undertaking at that time of the year. Padre Cocco had just received word by shortwave radio from the mission at Mahekodo-teri that a large party of men had left for Bisaasi-teri intent on capturing women. They had learned of the poorly guarded women from the six visitors and were determined to take advantage of the situation." Napoleon Chagnon, *Yanomamo: The Fierce People,* 3d ed. (New York: Holt, Rinehart and Winston, 1983), p. 154.

71. Kim Hill, phone interview, Feb. 3, 1995.

72. This was Father José Bórtoli's purpose in going to the Atlanta convention of the AAA in Dec. 1994. José Bórtoli, interview, Mavaca mission, June 11, 1996.

73. Frank Salamone, *The Yanomami and Their Interpreters: Fierce People or Fierce Interpreters* (Lanham, Md.: Univ. Press of America, 1997), pp. 1–8.

74. Frank Salamone, phone interview, Dec. 22, 1994.

75. "It was Chagnon's charge that the Salesians were making guns available to the Yanomami that heated up their dispute. It is not illegal for any Venezuelan citizen to have a gun. The Yanomami, as citizens, can use guns for hunting. However, behind the charge was the implication that the Salesians aided

the Yanomami who lived near their mission stations in their wars against other Yanomami. The Salesians have turned this argument around. They claim that it is Chagnon who brought weapons to the Yanomami." Salamone, *The Yanomami and Their Interpreters,* p. 84.

76. Ibid., endnote 8.

77. Chagnon, "Covering Up the Yanomamo Massacre."

78. Michael D'Antonio, "Napoleon Chagnon's War of Discovery," *Los Angeles Times Sunday Magazine,* Jan. 30, 2000.

79. The anthropologist Ferguson refers to "the sphere of mission beneficence and protection" Brian Ferguson, *Yanomami Warfare: A Political History* (Santa Fe: School of American Research Press, 1995) p. 146. Lizot wrote, "It is near the missions where the population conserves a certain dynamism and increases." Jacques Lizot, "N.A. Chagnon o sea: Un presidente falsificador" (letter, Caracas, Dec. 13, 1993), p. 4.

80. Napoleon Chagnon, *Yanomamo,* 5th ed. (Fort Worth: Harcourt Brace, 1987), p. 246.

81. "The Dorita-teri had suffered a 25% mortality since my last census of them four years earlier, from an epidemic that killed mostly children and old people. This is the highest mortality rate I documented for the 17 villages I censused in 1987 and again in 1991." Chagnon, *Yanomamo,* 4th ed., p. 239.

82. FUNDAFACI's helicopter descent upon Dorita-teri was the subject of chapter 1. See also, Chagnon, *Yanomamo,* 5th ed., pp. 235–39.

83. "In January 1992, when I returned for a brief visit to this area, the alarming news reached me that a major epidemic had struck the Dedabobowei-teri; 21 people or so had died within a week or so just before I arrived. . . ." Chagnon, *Yanomamo,* 4th ed., p. 224. According to Juan Finkers, who actually went to the assistance of the Kedebabowei-teri, these deaths occurred in Dec. 1991. Juan Finkers, phone interview, Jan. 24, 1995.

84. Raymond Hames, Kim Hill, and Ana Magdalena Hurtado, "Defamation Campaign against Napoleon A. Chagnon," rhames@unlinfo.unl.edu, May 1994.

85. "Ray Hames just informed me of your possible interest in doing a story about some of the Byzantine events that are going on regarding my 30-odd-years of research. . . ." Napoleon Chagnon, e-mail to Liz McMillen, *Chronicle of Higher Education,* Aug. 18, 1994.

86. Ibid., Aug. 23, 1994.

87. Ibid., Aug. 18, 1994.

88. Peter Monaghan, "Bitter Warfare in Anthropology," *Chronicle of Higher Education,* Oct. 26, 1994, A19.

89. "Parties in Bitter Dispute over Amazonian Indians Reach a Fragile Truce," *Chronicle of Higher Education,* Dec. 14, 1994, A18.

90. Chagnon, "Covering Up the Yanomamo Massacre."

91. Chagnon, "Killed by Kindness?" p. 11.

92. "36 died of malaria 87–89. . . ." Bruce Albert, personal correspondence, Dec. 14, 1994.

93. Chagnon, "Covering Up the Yanomamo Massacre."

94. "On the second day of our visit seven Yanomamo from a nearby village arrived, and I questioned four of them." Chagnon, "View from the President's Window," p. 1.

95. "At the same moment a group of Yanomamo—seven men and four women—walked toward us from beyond our camp." Chagnon, "Killed by Kindness?" p. 11.

96. Issam Madi, *Conspiración al sur del Orinoco* (Caracas: self-published, 1998), p. 63.

97. I have this recollection from Leda Martins, who organized Bezerra's press conference. Martins, interview, Pittsburgh, March 7, 1995.

98. Bruce Albert, "La fumée du métal: Histoire et représentations du contact chez les Yanomami (Brésil)," *L'Homme* 28, nos. 2–3 (1988): 87–119.

99. Janer Cristaldo, "Uma teocracia na Amazonia," *A Fôlha de São Paulo,* Feb. 12, 1995.

100. "Antes de atribui-las a garimpeiros a Polícia Federal e a Funai fariam melhor ter lido *Yanomamo* do antropologo americano Napoleon Chagnon. . . ." Janer Cristaldo, "Os bastidores do ianoblefe," *A Fôlha de São Paulo,* April 24, 1994.

101. Bruce Albert, personal correspondence, Dec. 15, 1994, p. 3.

102. Chagnon, *Yanomamo,* 4th ed., p. 220.

103. *The Population Dynamics of the Mucajai Yanomama,* by John Early and John Peters, has been praised as "the most comprehensive demographic study done for any Yanomami group." See Ferguson, *Yanomami Warfare,* p. 352. It has also been called "the most complete and detailed ethno-demographic account of a lowland South American Indian group." See Warren M. Hern, in *Population Studies* 45 (1991): 359–71. Nancy Howell, of the Univ. of Toronto, widely respected as the leading demographer in anthropology, has described the Early and Peters book as "a jem." See Howell, in *Canadian Review of Sociology and Anthropology* 28 (1991): 151–52.

104. John D. Early and John F. Peters, *The Population Dynamics of the Mucajaí Yanomama* (San Diego: Academic Press, 1990), pp. 64–68.
105. Aurora Anderson, Interview, Mucajai mission, May 1990.
106. John Peters, phone interview, Jan. 3, 1995.
107. Bob Cable, phone interview, Jan. 3, 1995.
108. Gay Cable, phone interview, Jan. 3, 1995.
109. Kenneth Taylor, phone interview, Jan. 27, 1995.
110. Milton Camargo, phone interview, Feb. 14, 1996.

Chapter 13: Warriors of the Amazon

1. *Warriors of the Amazon, Nova,* WGBH, Boston, 1996.
2. "Among the several dozen films and videos on the Yanomami, in my opinion by far the most balanced and humanistic is *Warriors of the Amazon,* which Lizot made in collaboration with the television science series *Nova."* Leslie E. Sponsel, "Yanomami: An Arena of Conflict and Aggression in the Amazon," *Aggressive Behavior* 24 (1998): 99. However, Sponsel had many reservations about the film.
3. *Warriors of the Amazon,* narration.
4. Brian Ferguson, phone interview, April 19, 1996.
5. Wilma Dawson, interview, Puerto Ayacucho, June 3, 1996.
6. Michael Dawson, interview, Padamo Mission, June 4, 1996.
7. "I would like my book to help revise the exaggerated representation that has been given of Yanomami violence. The Yanomami are warriors; they can be brutal and cruel, but they can also be delicate, sensitive, and loving. Violence is only sporadic; it never dominates social life for any length of time, and long peaceful moments can separate two explosions." Jacques Lizot, *Tales of the Yanomami: Daily Life in the Venezuelan Forest,* trans. Ernest Simon (Cambridge: Cambridge Univ. Press, 1985), p. xiv.
8. Lizot's accounts of events at one small Yanomamo village (approximately 70 people) between 1968 and 1976 indicates that mortality due to violence was very low to almost nonexistent that time period." Napoleon Chagnon, "Life Histories, Blood Revenge, and Warfare in a Tribal Population," *Science* 239 (1988): 991, n. 24.
9. María Eguillor García, *Yopo, Hękura y Shamanes* (Caracas: Editorial/Librería Salesiana, 1984), p. 25.
10. Jacques Lizot, "On Warfare: An Answer to N. A. Chagnon," trans. Sarah Dart, *American Ethnologist* 21 (1994): 853.
11. *Warriors of the Amazon,* translation by the Dawsons.
12. Michael Dawson, interview, Padamo mission, June 4, 1996.
13. Paul Griffiths, interview, Puerto Ayacucho, June 3, 1996.
14. Michael Dawson, interview, Padamo mission, June 4, 1996.
15. Pablo Mejía, interview, Padamo mission, June 4, 1996.
16. Renaldo, interview, Manaviche River, June 7, 1996.
17. Kenneth Good, phone interview, Jan. 31, 1997.
18. José Bórtoli, interview, Mavaca mission, June 6, 1996.
19. *Survivors of the Amazon,* BBC 4, 1996.
20. Andy Jillings, phone interview, Feb. 18, 1997.
21. Ibid.
22. Andy Jillings, letter to Patrick Tierney, Feb. 20, 1997.
23. Brian Ferguson, phone interview, May 26, 1996.
24. Timoteo, interview, Padamo mission, June 4, 1996.
25. Marinho de Souza, interview, Karohiteri, Sept. 1, 1996.
26. Renaldo, interview, Manaviche River, June 7, 1996.
27. Brian Ferguson, phone interview, April 19, 1996.

Chapter 14: Into the Vortex

1. Juan Finkers, "Aclaraciones al Sr. Chagnon," *La Iglesia en Amazonas,* Dec. 1994, pp. 7–10.
2. Brooke cites Charles Brewer as saying that twenty-one "wild Yanomami" had been killed by "mission Yanomami" over the preceding twelve months. James Brooke, "Venezuela Befriends Tribe, But What's Venezuela?" *NYT,* Sept. 11. However, Chagnon's account makes it clear that the wars actually began in early July 1990. Napoleon Chagnon, "The Guns of Mucajaí: The Immorality of Self-deception" (MS, Sept. 1992), p. 5.

3. Charles Brewer Carías with Napoleon Chagnon, "The Massacre at Lechoza, September 1992: Brewer's Account, 12/92."

4. "The number of (living) *unokais* in the current population is 137, 132 of whom are estimated to be 25 or older, and represent 44% of the men age 25 or older. A retrospective perusal of the data indicates that this has generally been the case in those villages whose *unokais* have not killed someone during the past 5 years." In other words, there were so few young men involved in fighting in the twelve villages in his study area because there had been very few killings for five years (1982–87). Napoleon Chagnon, "Life Histories, Blood Revenge, and Warfare in a Tribal Population," *Science* 239 (1988): 987. Elsewhere, Chagnon noted that the Mishimishimabowei-teri, by far the largest and most violent of the groups he studied, experienced peace with all their former enemies from 1977 onward. Napoleon Chagnon, *Yanomamo,* 5th ed. (Fort Worth: Harcourt Brace, 1997), p. 226. According to Lizot, there was a dramatic decline in violence among the groups he and Chagnon studied from 1984 to 1989. Jacques Lizot, "Sobre la guerra: Una respuesta a N. A. Chagnon (Science, 1988)," *La Iglesia en Amazonas,* April 1989, pp. 23–34. There were only three war deaths from 1968 to 1983 among all the seven groups near the Mavaca mission. Two of these three were killed by Tayari-teri archers in early 1981; the other, by Tayari-teri archers in 1979. María Eguillor García, *Yopo, Shamanismo y Hekura* (Caracas: Editorial/Librería Salesiana, 1984), pp. 24–26, 53. Moreover, Chagnon's unpublished reconstruction of violence originating at missions, indicates no instances of killings from 1979 until 1990. Chagnon, "The Guns of Mucajaí," pp. 3–4.

5. James Brooke, "In an Almost Untouched Jungle, Gold Miners Threaten Indian Ways," *NYT,* Sept. 19, 1990.

6. Brian Ferguson, phone interview, July 13, 1995.

7. Napoleon Chagnon, *Yanomamo: The Last Days of Eden* (San Diego: Harcourt Brace Jovanovich, 1992), p. xv.

8. " 'People' at Platanal did not like me because I brought trade goods—machetes, axes, fishhooks, etc.—directly into them without going through the Yanomamo (and priests) at the mission and gave these things away freely to them." Napoleon Chagnon, *Yanomamo,* 4th ed. (Fort Worth: Harcourt Brace, 1992), p. 238.

9. "I was unthreatened and was prepared to defend my claims in a court of law. I pointed out that this has gotten so large that perhaps the court should be something like the United Nations." Napoleon Chagnon, "Conversation with Jesús Cardozo;" (part of his press kit, March 24, 1994), p. 6.

10. "Before I went to Santa Barbara I spoke to Bórtoli. I asked him if he'd be willing to have Chagnon file a formal complaint to the General Attorney's Office—Fiscal General de la República—on everything he's accused you of. Would you be willing to have a neutral party designated by the Fiscalía to investigate this. And in case it comes out *en contra*—if it comes out that, yes, you have been giving out shotguns and shells and, yes, you have been urging the Yanomami to kill Chagnon and, yes, you have been helping some Yanomami raid others or at least turning your backs on things you don't want to deal with—if that is in fact determined, would you accept the judge's determinations and findings even if it means being expelled from the Upper Orinoco? So he said, 'Yes, we are willing to do that.' So I said I would invite him to come down, which is exactly what I did. So in the midst of my conversation with Chagnon, I told him, 'Chagnon, this controversy has been going on for a long time. It's very high profile and it's doing nobody any good. Why don't we put an end to it. I am authorized by the Salesians. . . . The Salesians are willing; in fact they are inviting you to go to Venezuela to file a formal complaint against them. To have this matter all cleared up, they are willing to accept a neutral committee set up by whoever, whether it be local institutions. And in fact Bórtoli later said at the AAA they were willing to accept an international fact-finding team. And they are willing to accept the consequences of what the verdict might be. But they would also like for you to accept the consequences. They would like you to accept the fact that if the verdict finds that they are not engaging in the kinds of things you accuse them of, then you may be subject to legal charges, criminal charges, on the basis of defamatory slander and that kind of thing.' And he said, 'Well.' He got really upset, he asked me if I was threatening him. I said, 'Why don't we just clear this up once and for all. . . . But you do have to assume responsibility for what you are saying.' " Jesús Cardozo, phone interview, Aug. 8, 1995.

11. "The important thing is not what I say or what Chagnon says but that the American Anthropological Association sends a commission to investigate the facts on site. . . . I've asked the AAA to send a commission to investigate, but they say it would be impossible to find an impartial group." José Bórtoli, phone interview, Dec. 6, 1994.

12. Jodie Dawson, interview, Padamo mission, June 6, 1996.

13. Juan Finkers, *Los Yanomami y su sistema alimenticio: Yanomami ni i pe* (Puerto Ayacucho: Vicariato Apostólico, 1986).

14. Brewer Carías with Chagnon, "The Massacre at Lechoza."

15. "Timanawe no es un santo, todo lo contrario: es un bestia violenta. . . ." Jacques Lizot, personal correspondence, Jan. 20, 1995.

16. "As customary among the Waika, it is the shaman who has the real authority." Inga Steinvorth von Goetz, *Uriji Jami!: Life and Belief of the Forest Waika in the Upper Orinoco,* trans. Peter Furst (Caracas: Asociación Cultural Humboldt, 1969), p. 145.

17. César Dimanawa, interview, Mavakita, June 8, 1996.

18. Bokoramo, interview, Mrakapiwei, Upper Mavaca, June 8, 1996.

19. Chagnon, *Last Days of Eden,* p. 274.

20. César Dimanawa, interview, Mavakita, June 8, 1996.

21. Ibid., Sept. 2, 1996.

22. Juan Finkers, interview, Mavaca Mission, June 12, 1996.

23. Bokoramo, interview, Mrakapiwei, June 8, 1996.

24. Chagnon, "Life Histories," p. 987.

25. Bokoramo, interview, Mrakapiwei, June 8, 1996.

26. Jesús Cardozo, phone interview, Dec. 20, 1994.

27. Chagnon, "The Guns of Mucajaí," p. 5.

28. Bokoramo, interview, Mrakapiwei, June 8, 1996.

29. Chagnon, "The Guns of Mucajaí," p. 5.

30. Chagnon, *Yanomamo,* 4th ed., p. 225.

31. Frank Salamone, phone interview, Dec. 22, 1994.

32. César Dimanawa, interview, Mavakita, Sept. 2, 1996.

33. Raymond Hames, phone interview, Dec. 29, 1994.

34. Jesús Cardozo, phone interview, Dec. 20, 1994. See chapter 12.

35. César Dimanawa, interview, Mavakita, June 8, 1996.

36. Brooke, "Venezuela Befriends Tribe."

37. Kenneth Good, phone interview, Feb. 22, 1995.

38. Chagnon, *Last Days of Eden,* p. 262.

39. Ibid.

40. Decreto No. 3127, *Gazeta Oficial de la República de Venezuela,* No. 35,292, Sept. 8, 1993.

41. Marta Miranda, Venevisión, Caracas, Fundación Cultural Venevisión, July 24, 1991.

42. Napoleon Chagnon, "Killed by Kindness?: The Dubious Influence of the Salesian Missions in Amazonas," *TLS,* Dec. 24, 1993, p. 12.

43. Brewer Carías with Chagnon, "The Massacre at Lechoza."

44. Ibid, p. 11 and n. 19.

45. Alberto Karakawe, interview, Ocamo mission, Aug. 31, 1996. Nelson, Mavaca, June 6, 1996.

46. Chagnon, "Killed by Kindness?" p. 12.

47. Ibid.

48. César Timanaxie, "Carta enviada por un yanomami a N. A. Chagnon," *La Iglesia en Amazonas,* Feb. 1994, p. 19.

49. This is reportedly a quotation that appeared in a Caracas daily, *El Nacional,* on Nov. 18, 1991. Brewer Carías with Chagnon, "The Massacre at Lochoza," p. 18.

50. Ibid., p. 15.

51. Juan Finkers, phone interview, Jan. 24, 1995.

52. Bokoramo, interview, Mrakapiwei, June 8, 1996.

53. Chagnon, *Last Days of Eden,* p. 274.

54. Juan Finkers, interview, Mavaca mission, June 12, 1996.

55. Ibid.

56. Bokoramo, interview, Mrakapiwei, June 8, 1996.

57. Bokoramo, interview, Mavakita, Sept. 2, 1996.

58. "Chagnon, kamijeri motoro ya puhi. Ohote ipe heroye ibe motor wama e hipeape, wama re ohotemotyonowei the no wa he ya puhi shatio shoaa. Kamiye suwe ya re kui ya puhi shatio shoaa." Translation into Spanish by the anthropologist Javier Cabrera: "Chagnon, quiero mi motor. Por el trabajo de mi esposo usted nos dar un motor, porque con usted trabajamos mucho. Yo estoy pendiente todavía. Quiero mi motor, por eso estoy pendiente con mi motor todavía. Yo quería que mi marido consiguiera el motor. Sin embargo me quedé triste y sin nada." Isabela, interview, Mavakita, Sept. 2, 1996.

59. Pablo Mejía, interview, Mavakita, June 8, 1996.

60. Brian Ferguson, phone interview, July 13, 1995.

Chapter 15: In Helena's Footsteps

1. "Nosotros nos quedamos pensando en tigres todo el tiempo." Helena Valero, *Yo soy Napeyoma: Relato de una mujer raptada por los indígenas Yanomami* (Caracas: Fundación La Salle de Ciencias Naturales, 1984) p. 300.

2. Luis Cocco, *Iyewei-teri: Quince años entre los Yanomamos* (Caracas: Editorial Salesians, 1973), p. 105.

3. Helena Valero, interview, Upper Orinoco, Aug. 31, 1996.

4. Ibid.

5. "At first the apparently total recall of this illiterate woman for these 20 years, accurate wherever it could be checked against Chagnon's taperecorded material, stretched my imagination. Slowly I realized that the events she recounts had been discussed so many times in the constricted world of the villages in which she lived that they had become much more ingrained in her memory than the events of the varied, somewhat helter-skelter lives we live are for us." James V. Neel, *Physician to the Gene Pool: Genetic Lessons and Other Stories* (New York: John Wiley, 1994), p. 407, n.1.

6. Ettore Biocca, *Yanoamo* (New York: Kodansha International, 1996), p. xii.

7. Helena Valero, interview, Upper Orinoco, Aug. 31, 1996.

8. Ibid.

9. Ibid.

10. Napoleon Chagnon, "Filming the Ax Fight," *Yanomamo Interactive CD* (New York: Harcourt Brace, 1997).

11. Napoleon Chagnon, *Yanomamo: The Fierce People,* 3d ed. (New York: Holt, Rinehart and Winston, 1983), pp. 18–19.

12. Napoleon Chagnon, *Yanomamo,* 5th ed. (Fort Worth: Harcourt Brace, 1997), p. 90.

13. Helena spent "about a month" by the tributary where she was captured with the Kohoroshiwetari; then she traveled for eleven days and was captured anew by the Karawe-tari. Biocca, *Yanoama,* pp. 23–37.

14. Valero, *Yo soy Napeyoma,* pp. 31–69.

15. Biocca, *Yanoama,* pp. 52–66.

16. Chagnon, *Yanomamo,* 5th ed., pp. 2–3.

17. "I never asked her any questions about her 'life' among the Yanomamo, and simply told her what I knew of the many Yanomamo with whom she had lived and who asked me to give her messages." Chagnon, "Filming the Ax Fight."

18. Napoleon Chagnon, *Studying the Yanomamo* (New York: Holt, Rinehart and Winston, 1974), p. 95.

19. Napoleon Chagnon, "Yanomamo Warfare, Social Organization and Marriage Alliances" (Ph.D. diss., Univ. of Michigan, 1966), p. 22.

20. Valero, *Yo soy Napeyoma,* pp. 21–30.

21. Chagnon, "Yanomamo Warfare," p. 152.

22. Napoleon Chagnon, *Yanomamo,* 4th ed. (Fort Worth: Harcourt Brace, 1992), p. 3.

23. For a detailed comparison of Chagnon's and Valero's accounts, see Brian Ferguson, *Yanomami Warfare: A Political History* (Santa Fe: School of American Research Press, 1995), pp. 393–95.

24. Chagnon, "Yanomamo Warfare," pp. 24–25.

25. Napoleon Chagnon, "Life Histories, Blood Revenge, and Warfare in a Tribal Population," *Science* 239 (1988): 991, n. 15.

26. Chagnon, *Yanomamo,* 4th ed., p. 3.

27. Valero, *Yo soy Napeyoma,* pp. 234–36.

28. Biocca, *Yanoama,* p. 197.

29. Valero, *Yo soy Napeyoma,* pp. 352–55.

30. Chagnon, "Yanomamo Warfare," p. 155.

31. Valero, *Yo soy Napeyoma,* p. 354.

32. Jacques Lizot, "El río de los periquitos," *Antropológica* 37 (1974) 3–23.

33. Napoleon Chagnon, "Yanomamo," in *Primitive Worlds* (Washington, D.C.: National Geographic Society, 1974), pp. 141–83.

34. Chagnon, "Yanomamo: The True People," *National Geographic,* Aug. 1976, pp. 210–21.

35. Pablo Mejía, interview, Patahama-teri, Sept. 5, 1996.

36. Yarima, interview, Irokai-teri, Sept. 7, 1996.

37. Ibid.

38. Kenneth Good, phone interview, Jan. 30, 1997.

39. Ibid., Jan. 10, 1995.

40. Kenneth Good, *Into the Heart: One Man's Pursuit of Love and Knowledge among the Yanomama* (New York: Simon & Schuster, 1991), pp. 102–5.

41. Ibid., p. 202.

42. Yarima, quoted ibid., pp. 308–9.

43. Kenneth Good, phone interview, Feb. 4, 1997.

44. Ibid., May 14, 1997.

45. "I know that you and Jesús and the Yanomami could make a video which they would be proud of and which they would like the modern western world to see. . . . Now are there any hidden agendas on my part in doing this project? Yes. There is the fact that 37 Yanomami films exist and that people use 10 of them a lot and that these 10 tend to reinforce every basic prejudice we're trying to enlighten in introductory anthropology courses. . . . The best one could do would be to take all the films off the market and hide them for a hundred years perhaps, but realistically that's not going to happen either. So why not make a film that's exactly the film the Yanomami would want shown which could then put all the other films in context?" Timothy Asch, letter to José Bórtoli, Jan. 17, 1991, Timothy Asch Collection, NAA.

46. Jesús Cardozo, letter to Timothy Asch, April 9, 1992, (Timothy Asch Collection), NAA.

47. Amy Wray, phone interview, May 10, 1996.

48. Kim Hill and Hillard Kaplan, "Population and Dry-Season Subsistence Strategies of the Recently Contacted Yora of Peru." *National Geographic Research* 5 (1989): 332.

49. *Yanomami Homecoming, National Geographic Explorer,* 48 min. (Washington, D.C.: National Geographic Society, 1994).

50. Yarima, interview, Irokai-teri, Sept. 7, 1996.

51. "The Orinoco," *National Geographic,* April 1998.

52. Ulisses Capozoli, "Yarima, cinderela rebelde," *O Estado de São Paulo,* March 3, 1997.

53. "An American anthropologist intends to go to the Amazonian jungle to search for his Stone Age tribeswoman wife and persuade her to rejoin him in the West. . . . Their unprecedented union was hailed as one of the greatest love stories of all time." Gabriela Gamini and Quentin Letts, "American Plans Jungle Trip to Win Back Wife," *Times* (London), Jan. 31, 1997. An accompanying piece, reported by the *Times*'s New York stringer on the basis of interviews in Yarima's New Jersey neighborhood, stated, "Modern devices such as washing-machines, television and the telephone were as foreign to her as they would have been to a Neanderthal man and her arrival in well-to-do New Jersey caused a worldwide sensation." Yarima's former English teacher was quoted as saying that Yarima was four feet tall and had no conception of time ("She did not know if it was morning or afternoon, or when she would next see her husband"). And, although Yarima had made progress in learning English, "One thing you noticed about her was that she could not co-ordinate colors." Maritza Nelson, in Quentin Letts, "Spurning the Good Life for Call of the Wild," ibid. A third article in the *Times* included responses from leading anthropologists attacking the imaginary expedition. Gabriela Gamini, "Search for Jungle Wife Condemned by Amazon Experts," ibid., Feb. 1, 1997. Finally, a short, less prominently featured piece acknowledged that the expedition had never been planned and that Good had been in New Jersey, not the Amazon, all along. Quentin Letts, "Tribal Wife Is Home for Good," ibid., Feb. 17, 1997.

Chapter 16: Gardens of Hunger, Dogs of War

1. Helena Valero, *Yo soy Napeyoma: Relato de una mujer raptada por los indígenas Yanomami* (Caracas: Fundación La Salle de Ciencias Naturales, 1984), p. 395.

2. In the Siapa region, the Yanomami spend over 40 percent of their time on foraging treks; but they often undertake treks when garden production falters. "Thus, among the 'traditional' Yanoama, collecting and trekking form alternative means of acquiring *calories* when agriculture fails them. . . ." Marcus Colchester, "Rethinking Stone Age Economics: Some Speculations regarding the Pre-Columbian Yanoama Economy," *Human Ecology* 12 (1984): 301. During his 1973 visit, Lizot reported just such a generalized crop failure among the Yanomami of the Siapa valley, followed by a long, "semi-hungry" period of trekking. The Indians were skinny, but healthy, according to a medical examiner. "Las planiaclones que escaparon de la catástrofe fueron muy escasas. A estos siguió una penuria de productos alimenticios cultivados, aunque no hubo escasez completa, la selva ofrecia rofiosamente recursos casi suficientes. Entonces, dejando las plantaciones, la mayor parte de los grupes indios abandonaron la vivienda semi-permanente para dedicarse a una economía de nomadismo, explotando sucezivamente zonas de la solva. Trabajando más que de costumbre, los indios seguían en un estiado semi-hinnbriento, pero susistían." Jacques Lizot, "El río de los periquitos," *Antropológica* 37 (1974): 7.

3. These estimates by Marinho De Souza were in line with the conclusions of CAICET. "La desnutri-

ción es altamente prevalente en las comunidedes indígenas. . . . Aunque posiblemente tiene una gran importancia en la mayor susceptibilidad que presenta la población indígona a diferentes enfermedades endémicas, no ha sido adecuadamente documentada y existe un subregistro importante." Carlos Botto, "Impactos ambientales en Salud: La experiencia de CAICET" (Belem, Brazil, June 6, 1996), p. 10. Malnutrition was listed as the fifth-leading cause of death among indigenous communities of Amazonas state. G. Rodríguez Ochoa, "Situación de salud en el Territorio Federal Amazonas, Venezuela," *Enfoque Integral de la Salud Humana en la Amazonia,* vol. 10 (Montevideo, Uruguay: Editorial Trilce, 1992), pp. 407–26.

4. Valero, *Yo soy Napeyoma,* pp. 282–83.

5. Although FUNDAFACI never came to Mokarita-teri—a descent into this valley would have been risky—that did not stop Mokarita-teri from going to FUNDAFACI at two other villages, Shanishani-teri and Ashtitowa-teri. "Shaki paid us for our spirits [pictures] and to draw our blood. He gave us machetes, knives and fish line. We were not as sick at that time." Interview, Mokarita-teri, transl. Marco Jimenez, Sept. 9, 1996.

6. Darna L. Dufour, "Diet and Nutritional Status of Amazonian People," in *Amazonian Indians,* ed. Anna Roosevelt (Tucson: Univ. of Arizona Press, 1994), p. 157.

7. "Of the total caloric yield from just the musaceous plants in this garden region at least half is provided by the *cowata* plantains, while as much as 98 percent may come from all plantains." William J. Smole, *The Yanoama Indians: A Cultural Geography* (Austin: Univ. of Texas Press, 1976), p. 151.

8. Edward O. Wilson, preface to Napoleon Chagnon, *Yanomamo: The Last Days of Eden* (San Diego: Harcourt Brace Jovanovich, 1992), p. x.

9. L. Keeley, *War before Civilization* (Oxford: Oxford Univ. Press, 1995).

10. Napoleon Chagnon, "Chronic Problems in Understanding Tribal Violence and Warfare," in *Genetics of Criminal and Antisocial Behavior* (Chichester, N.Y.: John Wiley, 1996), p. 213.

11. Napoleon Chagnon, *Yanomamo,* 5th ed. (Fort Worth: Harcourt Brace, 1997), p. 93.

12. James V. Neel, "Lessons from a Primitive People," *Science* 170 (1970): 815–22.

13. "The males, in general, present a picture of exuberant vitality, an impression confirmed by their dancing and chanting frequently extending through most of the night. . . . When I write of a picture of exuberant vitality, it must be remembered we are seeing only a snapshot in time. Furthermore, some of those who were ill might not present themselves for physical examinations. The true picture of health and disease can only be derived from a longitudinal study." James V. Neel, *Physician to the Gene Pool: Genetic Lessons and Other Stories* (New York: John Wiley, 1994), p. 150.

14. "Anthropologists often describe Amazonian Indians as healthy and well-nourished, but the health status of Indian groups is commonly inferred from visual assessments of adult men's health, which is usually comparatively favorable because of their preferential access to resources and greater tolerance for nutritional inadequacies than is the case with young children and pregnant or lactating women." Anna C. Roosevelt, "Strategy for a New Synthesis," in *Amazonian Indians,* p. 14.

15. Sound Roll 6, Patanowa-teri, Feb. 21, 1968, Timothy Asch Collection, NAA.

16. "There is no body weight data available for the Yanomami." Dufour, "Diet and Nutritional Status of Amazonian People," p. 168.

17. "Los Guicas que yo he medido tenfan una talla mediana de 4 pies y 7 pulgadas a 4 pios y 8 pulgadas." Alexander von Humboldt, cited in Luis Cocco, *Iyewei-teri: Quince años entre los Yanomamos* (Caracas: Editorial Salesiana, 1973), pp. 47–48.

18. R. Holmes, "Nutritional Status and Cultural Change in Venezuela's Amazon Territory," in *Change in the Amazon Basin,* ed. John Hemming (Manchester: Univ. of Manchester, 1985), p. 251.

19. "Use and Interpretation of Anthropometric Indicators of Nutritional Status," *BWHO* 64 (1986): 929–41.

20. James Neel mentions data nearly identical for the Parima Mountains. "In some villages in the northern aspects of their distribution [i.e., Parima Mountains], the Yanomama are very small, warranting the term pygmoid. For instance, in one village the men averaged only 147.7 cm in height and the women only 137.5 cm." Neel, *Physician to the Gene Pool,* p. 205.

21. Dufour, "Diet and Nutritional Status of Amazonian People," p. 156.

22. Alain Gheerbrant, *Journey to the Far Amazon* (New York: Simon & Schuster, 1954). Gheerbrant was one of the first Europeans to cross the Sierra Parima's central massif (1948–50), where he recorded horrible scenes of hunger and sickness at a large Yanomami village named Okomatidi. "Las mujeres, los ancianos y los niños con el vientre hinchado de parásitos intestinales (esos terribles niños de ojos abiertos demasiado fijamente y llenos de moscos) nos observaban, inmóviles, a través de los tabiques de sus refugios. . . . Detrás de tanta suciedad, de tanta enfermedad, detrás de la apariencia de esos cuerpos esqueléticos roídos por treinta y seis lepras . . . [había] una realidad absolutamente humana. . . ." Alain Gheerbrant, *La expedición Orinoco-Amazonas (1948–1950)* (Caracas: Banco de la República/Áncora Editores, 1997), pp. 305–6.

23. Colchester, "Rethinking Stone Age Economics," pp. 291–314.

24. Hamilton Rice, "The Rio Branco, Uraricoera, and Parlma, Part 3," *Geographical Journal* 71 (1928): 354.

25. Pablo Anduze, *Shilili-Ko: Relato de un naturalista que también llegó a las fuentes del río Orinoco* (Caracas: Talleres Gráficos Ilustrados, 1960), p. 203.

26. Dufour, "Diet and Nutritional Status of Amazonian People," p. 156.

27. Napoleon Chagnon, *Studying the Yanomamo* (New York: Holt, Rinehart and Winston, 1974), appendix E.

28. Kim Hill, phone interview, Jan. 3, 1995.

29. The smallest Amerindians recorded by T. D. Stewart, the longtime physical anthropologist at the Smithsonian Institution, were the Brazilian Guarani. The Guarani males measured 153 cm. Among North American tribes, the smallest were the Otomi of Central Mexico, whose males measured 158.5 cm. Thomas Dale Stewart, *The People of America* (London: Weidenfeld and Nicolson, 1973), pp. 123, 111. A more focused study on South American Indians reported that the smallest group was the Yupa of the Perijá Mountains of Venezuela, whose men measured 151.1 cm; the women, 139.1 cm. Francisco M. Salzano and Sidia M. Callegari-Jacques, *South American Indians: A Case Study in Evolution* (Oxford: Clarendon Press, 1988), p. 116.

30. The most detailed analysis of "pygmy" populations in America was undertaken by the anthropologist Juan Comas of Mexico's Universidad Nacional Autónoma. He reviewed reports and studies going back over a century and described the three smallest groups for which there were acceptable data: (1) the Shiriana Yanomami of the Upper Ventuari River, whose males measured 150.9 cm and females 138.7 cm; (2) the Yupa of Venezuela's Perijá Mountains, whose males measured 151.1 cm and females 139.1; and (3) the Ayamanes of the Tocuyo River in Venezuela, whose males measured 152.4 cm and females 138.8 cm. Comas wrote his book before Holmes had studied the Parima Yanomami. The Shiriana Yanomami are thus the second-smallest group in the New World, after the Parima Yanomami. And the four smallest groups in the Americas all come from the rain forest highlands of Venezuela, though the Ayamanes and Yupa are much closer to the Caribbean than to the Amazon. Juan Comas, *¿Pigmeos en América?* (Mexico City: Universidad Nacional Autónoma de México, 1960), pp. 6–34.

31. Among the Western Pygmies, males measured 152.7 cm and females 145 cm. Among the Eastern Pygmies, or Ituri, males were 144.2 and females 137.4, according to Turnbull. According to Sforza, males were 144.4 and females, 136. The Ituri are, on average, about the same size as the Parima Yanomami: males 145.3 and females 136.2. Luigi Luca Cavalli-Sforza, "Anthropometric Data," in *African Pygmies,* ed. Luigi Luca Cavalli-Sforza (Orlando: Academic Press, 1986), pp. 83–85.

32. Stewart placed the smallest peoples of the Americas in the northwest Amazon and the contiguous rain forests of Central America. He showed that some animals, such as the puma, also vary in size accordingly, and he related smaller size in the tropics, of both humans and animals, to soils poor in nutrients. Stewart, *The People of America,* pp. 48–51. Salzano and Callegari-Jacques also showed that tribes in the northwestern part of South America tend to be smaller than the rest of the continent, even though they omitted the Parima Yanomami, the Shiriana Yanomami, and the Ayamanes from their map, which would have made the study more striking. They also described an attempt to measure anthropometric and genetic traits in assigning village and cluster membership "of 520 Yanomam Indians from 19 villages in nine clusters. . . . On the basis of anthropometrics alone, 36% of the individuals were allocated to the right village and 60% to the right cluster. The monogenic traits did not discriminate as well (the values being 16% and 26%, respectively)." Salzano and Callegari-Jacques, *South American Indians,* p. 120 fig. 6.1, and p. 123.

33. Cocco, *Iyewei-teri,* p. 123.

34. Dufour, "Diet and Nutritional Status of Amazonian People," p. 155, fig. 7.2.

35. Roosevelt, "Strategy for a New Synthesis," pp. 4–15.

36. Napoleon Chagnon, *Yanomamo,* 5th ed., pp. 91–97.

37. Napoleon Chagnon, "Yanomamo Warfare, Social Organization and Marriage Alliances" (Ph.D. diss., Univ. of Michigan, 1966), pp. 53–79.

38. Ibid., p. 167.

39. Marvin Harris, review of *Yanomami Warfare: A Political History,* by Brian Ferguson, *Human Ecology* 24 (1996): 413.

40. "First, young couples do not relish the prospect of a long period of celibacy that pregnancy and lactation taboos impose on them." Chagnon, "Yanomamo Warfare," p. 53.

41. Daly and Wilson, *Homicide* (Hawthorne, N.Y.: Aldine de Gruyter, 1988), p. 58.

42. Ibid., p. 56. ". . . Divale and Harris were surely right to ask why some societies have fallen into the trap and not others. Moreover, they were probably right to seek an ecological answer, even though their own proposal—protein shortage—was a failure." Ibid.

43. "Barbarians find it difficult to support themselves and their children, and it is a simple plan to kill their infants." Charles Darwin, *The Descent of Man and Selection in Relation to Sex* (Detroit: Gale Research, 1974), p. 586.

44. Napoleon Chagnon, M. Flinn and Thomas Melancon, "Sex-ratio Variation among the Yanomamo Indians," in *Evolutionary Biology and Human Social Behavior: An Anthropological Perspective,* ed. Napoleon Chagnon and William Irons (North Scituate, Mass.: Duxbury Press, 1979), pp. 290–320.

45. Neel, *Physician to the Gene Pool,* p. 176.

46. Of 484 live births around the Mavaca mission, 39 were known cases of infanticide, 24 of them female. María Eguillor García, *Yopo, shamanismo y hekura* (Caracas: Editorial/Librería Salesiana, 1984), pp. 50–51.

47. The most detailed demographic study, among the Mucajaí Ninam, found that "preferential female infanticide is practiced precisely to hasten another pregnancy and birth, hopefully a male." For a twenty-nine-year period of precontact, there were 77.4 percent males to 22.6 percent females. John D. Early and John F. Peters, *The Population Dynamics of the Mucajai Yanomama* (San Diego: Academic Press, 1990), pp. 136–37, 19–23.

48. Napoleon Chagnon, *Yanomamo,* 4th ed. (Fort Worth: Harcourt Brace, 1992), p. 93. For a discussion, see Brian Ferguson, *Yanomami Warfare: A Political History* (Santa Fe: School of American Research Press, 1995), p. 352.

49. In his recent Pulitzer Prize–winning book, *Consilience: The Unity of Knowledge* (New York: Alfred A. Knopf, 1998), p. 170, Edward O. Wilson summarized Darwinian theory regarding resources, food, and territory as follows: "The territorial instinct arises during evolution when some vital resource serves as a 'density-dependent factor.' That is, the growth of population density is slowed incrementally by an increasing shortage of food, water, nest sites, or the entire local terrain available to individuals searching for these resources."

50. Randy Bird and Garland Allen "Charles Darwin," *Encarta Encyclopedia* (Microsoft, 1999).

51. Matt Ridley, *The Red Queen* (New York: Penguin, 1993), pp. 203–4.

52. Chagnon, *Studying the Yanomamo,* p. 123.

53. Johannes Wilbert, *Survivors of El Dorado: Four Indian Cultures of South America* (New York: Praeger, 1972), pp. 15–16.

54. Eugene Hammel, "Demographic Constraints on Population Growth in Early humans," *Human Nature* 7 (1996): 217–55.

55. Joseph Birdsell, "Some Predictions for the Pleistocene Based on Equilibrium Systems among Recent Hunter-gatherers," in *Man the Hunter,* ed. Richard B. Lee and Irv DeVore (Chicago: Aldine, 1968), pp. 229–49.

56. "At fourteen this !Kung girl is just reaching puberty. Under the harsh conditions of the Kalabari, she is still several years away from menstruating for the first time. . . . Adolescent sub-fertility, not chastity, ensures that she will be nearly twenty before she gives birth." Susan Blaffer Hrdy, *Mother Nature: A History of Mothers, Infants and Natural Selection* (New York: Pantheon, 1999), p. 187, fig. 8.2.

57. Herbert Spencer, *Principles of Biology,* vol. 2 (London: William and Norgate, 1867), p. 486.

58. Hrdy, *Mother Nature,* p. 9.

59. An interesting question is whether Yanomami girls, like those of the Kalahari !Kung, reach sexual maturity later than girls in wealthy countries. There is no data on the age of menarche in the highland communities; it is 12.4 years among the Mucajaí Yanomami—almost identical to that of girls in the United States and in southern Europe. See Early and Peters, *The Population Dynamics of the Mucajaí Yanomama,* p. 138. But among the Mucajaí Yanomami life expectancy (varying between 32.9 and 55.1 years, depending on the contact period) and population growth (3.5 percent per year for the postcontact period when data on first menstruation was gathered) differ dramatically from those of the Yanomami as a whole, and remote, highland Yanomami in particular. Girls in some Scandanavian countries experienced menarche at about the same age as the Kalahari !Kung even as population increased.

60. Wilbert, *Survivors of El Dorado.*

61. John Hemming, *The Search for El Dorado* (New York: E. P. Dutton, 1978), p. 441.

62. " 'Good,' he said. 'And since I am very poor, when you come back, be sure to bring me another machete and a pot.' " Kenneth Good, *Into the Heart: One Man's Pursuit of Love and Knowledge among the Yanomama* (New York: Simon & Schuster, 1991), p. 98. " 'Yes! We are poor in machetes," he said cynically, implying that I ought to know better than to ask stupid questions when I could plainly see that they were poor." Chagnon, *Studying the Yanomamo,* p. 29.

63. In the Surucucu area, the word the miners used was *Beriu,* which was the word the local Yanomami repeated when they wanted something. Michelle Rodríguez Costa, interview, Surucucus army post, Nov. 1990.

64. Cocco, *Iyewei-teri,* p. 34.

65. "The Yanomama have traditionally relied on the Makiritare for steel tools. . . ." Napoleon Chagnon, James Neel, Lowel Weitkamp, Henry Gershowitz, and Manuel Ayres, "The Influence of Cultural Factors on the Demography and Pattern of Gene Flow from the Makiritare to the Yanomama Indians," *American Journal of Physical Anthropology* 32 (1970): 343.

66. Alfonso Vinci, *Red Cloth and Green Forest,* trans. James Cadell (London: Hutchinson, 1959), pp. 123–24.

67. Daniel de Barandarian and Aushi Walalam, *Hijos de la luna: Monografla antropológica sobre los indios sanemá yanomama* (Caracas: Editorial Arte, 1983), p. 103.

68. " 'Give me a white woman,' said one, though what he meant was a machete." Good, *Into the Heart,* p. 97.

69. N. Chagnon, *Yanomamo: The Fierce People,* 3d ed. (New York: Holt, Rinehart and Winston, 1983), p. 85.

70. Kenneth Good, phone interview, Feb. 10, 1997.

71. Good, *Into the Heart,* p. 115.

72. Napoleon Chagnon and Raymond Hames, "Protein Deficiency and Tribal Warfare in Amazonia: New Data," *Science* 203 (1979): 910–13.

73. Chagnon, *Yanomamo,* 3d ed., pp. 84, 86.

74. When the explorer Bobadilla went up this stretch of the Padamo, he found no Yanomami in 1756; nor did the German explorer Robert Schomburgk in the nineteenth century. Cocco, *Iyewei-teri,* pp. 38–50.

75. Raymond Hames, "Behavioral Account of the Division of Labor among the Yekwana Indians" Ph.D. diss., Univ. of California at Santa Barbara, 1978), pp. 19–23.

76. Chagnon and Hames, "Protein Deficiency and Tribal Warfare in Amazonia," pp. 910–13.

77. Kenneth Good, phone interview, Jan. 10, 1995.

78. Valero, *Yo soy Napeyoma,* pp. 319–21.

79. Napoleon Chagnon, "Reproductive and Somatic Conflicts of Interest in the Genesis of Violence and Warfare among Tribesmen," in *The Anthropology of War,* ed. J. Haas (Cambridge: Cambridge Univ. Press, 1990), p. 99.

80. Timothy Asch, "Ethnographic Filming and the Yanomamo Indians," *Sightlines,* Jan.–Feb. 1972, p. 11.

81. James Neel, Timothy Asch, and Napoleon Chagnon, *Yanomama: A Multidisciplinary Study,* 43 min. (DOE, 1971).

82. Timothy Asch and Napoleon Chagnon, *Kaobawa Trades with Reyabobowei-teri,* 8 min. (Somerville, Mass.: DER, 1971).

83. Sound Roll 8, Faianowa-teri, Feb. 23, 1968, Timothy Asch Collection, NAA.

84. Kenneth Good, "Yanomami Hunting Patterns: Trekking and Garden Relocation as an Adaptation to Game Availability in Amazonia (Ph.D. diss., Univ. of Florida, 1989), p. 95.

85. Chagnon, *Yanomamo,* 5th ed., p. 96.

86. "That Hames' research village does not trek is the strongest indication that this group has abandoned its traditional shapono residence, married into the Ye'kwana community and adopted Ye'kwana gardening practices." Good, "Yanomami Hunting Patterns," p. 162.

87. "The very low fat content (<9% of calories) of Amerindian diets . . . is also a potential constraint for children. It is generally assumed that diets in which fat contributes <15% of the calories are not concentrated enough for a child." Dufour, "Diet and Nutritional Status of Amazonian People," p. 167.

88. Chagnon, "Yanomamo Warfare," pp. 71–72.

89. "The Yanomamo frequently return to their old, abandoned sites in order to harvest the peach palm crop. This is usually a dangerous undertaking, as their old enemies are also aware of the fruit, and exploit it themselves." Ibid, p. 72. "Ownership rules stipulate that peach palm trees belong to the individual who plants them. A man from one of the factions took the palm fruit belonging to a man from the other faction one year; the latter reciprocated by stealing plantains from the garden of the former, precipitating a club fight between the two villages." Ibid., p. 154.

90. "In these highland areas, crossing he rugged terrain, making gardens, and finding game animals and other resources are all more difficult. Just collecting firewood and hauling water each day takes much more energy than in the lowlands." Napoleon Chagnon, *Yamomamo: The Last Days of Eden* (San Diego: Harcourt Brace Jovanovich, 1992), p. 105.

91. Frank A. Salamone, "Chagnon's Response to His Critics," in *The Yanomami and Their Interpreters: Fierce People or Fierce Interpreters?* (Lanham, Md.: Univ. Press of America, 1977), p. 111.

92. Ibid., endnote 6.

93. Redmond O'Hanlon, *In Trouble Again* (London: Hamish Hamilton, 1988), p. 18.

94. "At higher elevations . . . resources are presumably less abundant and more energetically costly to

produce or acquire. For example, many species of useful plants and animals are not even found there, such as caiman or tapir. Gardening is extremely difficult on the steep slopes and energetically very costly in comparison with lowland cultivation." Chagnon, "Chronic Problems in Understanding Tribal Violence and Warfare," p. 224.

95. "Female infanticide primarily affects the sex ratio at birth, which is distinct from the sex ratio of the total population. The latter is affected by the additional factors entering into differential mortality and out-migration. Therefore even if a Yanomama population has a dominant male sex ratio, it cannot be concluded that there is high preferential female infanticide. . . . The sex ratio for the combined 17 villages with quantitative data is 109, which is not an unusual ratio. These data show that not only has the incidence of female infanticide been exaggerated, but so has the extent of male-dominated sex ratios of the population at the village level." Nevertheless, the authors recognized that preferential female infanticide has taken place among the Mucajaí, and was certainly a factor in the 36.4 percent of females at first contact. What they cannot say is exactly what role female infanticide played, or how the practice has varied in the precontact period, which is the focus of all general discussions. They say Chagnon dropped the argument because he knew the weakness of his data. Early and Peters, *The Population Dynamics of the Mucajaí Yanomama,* pp. 134–35 (quotation) and 98–100 (sex imbalance at first contact).

96. "My data for the Parima highland core of Yanoama territory show a general balance in the sex ratio, with a tendency toward a slight surplus of females. This fact is noteworthy, since females are not nearly as visible to the foreign visitor as are males. Females tend to be shy and reticent when strangers arrive, and sometimes men order attractive girls to hide from foreigners." Smole, *The Yanomama Indians,* p. 72.

97. Chagnon, *Yanomamo,* 4th ed., p. 85.

98. Chagnon, "Chronic Problems in Understanding Tribal Violence and Warfare," p. 224, table 1.

99. These are simple averages because the number of actual cases is not given. Ibid., p. 86.

100. Smole, *The Yanoama Indians,* p. 32, n. 36, and 233, n. 94.

101. Ettore Biocca, *Yanoama* (New York: Kodansha International, 1996), p. 218.

102. "Had I known in 1972 what I now know about the role of steel in Yanomami life, I strongly doubt that I would have pushed for an ecological, population-pressure model to explain Yanomami warfare." Harris, review of *Yanomami Warfare,* p. 416.

103. Jacques Lizot, "El río de los periquitos," *Antropología* 37 (1974): 5.

104. Marco, interview, path to Hokomapiwe-teri, Sept. 9, 1996.

105. "Presently, the most common articles of exchange are Western goods by communities residing near the Orinoco River and other points of foreign settlement and hunting dogs by the communities living in more remote areas. . . . Likewise, the communities living nearer to the source of manufactured goods will trade the dog it acquired to communities living at the source of goods and replace the items it gave up." Kenneth Good, "Demography and Land Use among the Yanomamo of the Orinoco-Siapa Block in Amazon Territory, Venezuela" (MS, 1984), pp. 6–7.

106. Napoleon Chagnon, "The Guns of Mucajaí: The Immorality of Self-deception" (MS, Sept. 1992), p. 5.

107. Chagnon landed near Doshamosha-teri, on Aug. 8, 1990, according to Charles Brewer's expedition log and UCSB report. ("8 de agosto al 28 de agost de 1990. Exploración del río Siapa medio utilizando como campamento base el poblado de Doshamosha-teri al lado del raudal Shukúmenn-ka-bora." Charles Brewer Carías, Napoleon Chagnon, and Brian Boom, "Forest and Man" (MS; Caracas: Fundación Explora, 1993), p. 11. According to my interviews with the headman of Narimobowei-teri, Chagnon generously distributed trade goods and received a delegation from the Narimobowei-teri. He gave the Narimobowei-teri some machetes and promised them many more presents if they would clear a helicopter landing site for him near their own *shabono.* They obliged. Chagnon flew to Narimabowei-teri three weeks later, on Aug. 26, 1990, as is attested to by Chagnon's article in *Santa Barbara* magazine. See Napoleon Chagnon, "To Save the Fierce People," *Santa Barbara,* Jan.–Feb. 1991, p. 37, 40. However, the Doshamosha-teri were miffed at FUNDAFACI's departure, so miffed that they sent a raiding party against the Narimobowei-teri. Wauparuwe, interview, Narimobowei-teri, trans. Marco Jiménez, Sept. 11–12, 1996.

108. Helena Valero, interview, Upper Orinoco, Aug. 31, 1996.

109. Lizot, "El río de los Periquitos," p. 5.

110. Good's first visit occurred around the end of 1976 or in Jan. 1977. Good, *Into the Heart,* pp. 91–98. "In reference to the village of 'Narimabowe-teri' it should be noted that I have made three expeditions to the village and have spent more than two months with them." Kenneth Good, letter to the editor of *NYT,* Sept. 29, 1990.

111. Marinho de Souza, interview, path to Hokomapiwe-teri, Sept. 9, 1996.

112. Narimabowei-teri, Sept. 11, 1996.

Chapter 17: Machines That Make Black Magic

1. Quoted in Jeffrey Rifkin, "Ethnography and Ethnocide," *Dialectical Anthropology*, 19 (1994): 295.

2. It was the Comisión de Límites that realized its overflights between 1941 and 1944. Luis Cocco, *Iyewei-teri: Quince años entre los Yanomamos* (Caracas: Editorial Salesiana, 1973), p. 78.

3. Helena Valero, *Yo soy Napeyoma: Relato de una mujer raptada por los indígenas Yanomami* (Caracas: Fundación La Salle de Ciencias Naturales, 1984), (New York: Oxford Univ. Press, 1992), p. 53.

4. David E. Stannard, *American Holocaust: Columbus and the Conquest of the New World* (New York: Oxford Univ. Press, 1992), p. 53.

5. Ettore Biocca, *Yanoama* (New York: Kodansha International, 1996), p. 213.

6. Bruce Albert, "La Fumée du métal: Histoire et représentations du contact chez les Yanomami (Brésil)," *L'Homme* 28, nos. 2–3 (1988): 87–119.

7. Bruce Albert, interview, Toototobi, Brazil, Dec. 5, 1990.

8. Narimobowei-teri, Ashidowa-teri, Shanishani-teri, and Hiomita-teri.

9. Napoleon Chagnon, "To Save the Fierce People." *Santa Barbara*, Jan.–Feb. 1991, p. 36.

10. Waupuruwe. "Shaki hizo xawara para maternos. No queremos que regresa." In-situ translation by Marco Jimenez, Narimobowei-teri, Sept. 12, 1996.

11. Hetoyaw, interview, temporary shelter between Narimobowei-teri and Ashidowa-teri, Sept. 16, 1996.

12. Yanowe, interview, Ashidowa-teri, Sept. 16, 1996.

13. "Campamento basado en el Shapone de Waborawa-teri para estudio de los pobladores del área y del Shapono de Narimobowei-teri." Charles Brewer Carías, Napoleon Chagnon, and Brian Boom, "Forest and Man" (MS; Caracas: Fundación Explora, 1993), p. 15.

14. Javier Carrera, phone interview, Feb. 1998.

15. Waupuruwe, Narimobowei-teri, Sept. 13, 1996.

16. Marco, interview, Toobatotoi-teri, first *shabono*, Sept. 13, 1996.

17. Brewer, Chagnon, and Boom, "Forest and Man," p. 6.

18. Cargo, or *Kago*, is a term laden with more millennial expectations than the Yanomami *madohe*. "The concept Cargo (in pidgin *Kago*) implies a totality of material, organizational and spiritual welfare, collectively desired as a replacement for current inadequacy, and projected into the imminent future as a coming 'salvation.' . . . [W]hat the 'whitemen' or foreigners possess in contrast to the lowly villager is so extraordinary that it already implies for him some miracle of transcendence." G. W. Trompf, introduction, *Cargo Cults and Millenarian Movements: Transoceanic Comparisons of New Religious Movements*, ed. G. W. Trompf (New York: Mouton de Gruyter, 1990), pp. 10–11.

19. Yanowe, interview, Dorita-teri, Sept. 21, 1996.

20. Napoleon Chagnon, "The Guns of Mucajai: The Immorality of Self-deception" (MS, Sept. 1992), p. 5.

21. Ibid., pp. 4–5.

22. Toobatotoi-teri, second *shabono*, Sept. 14, 1996.

23. Brewer, Chagnon, and Boom, "Forest and Man," p. 14.

24. Frank A. Salamone, *The Yanomami and Their Interpreters: Fierce People or Fierce Interpreters?* (Lanham, Md.: Univ. Press of America, 1997), p. 78.

25. Ibid., pp. 12–14.

26. Toobatotoi-teri, second *Shabono*, Sept. 15, 1996.

27. Brewer, Chagnon, and Boom, "Forest and Man," p. 11.

28. "Ashidowa-teri: 1 al 7 de Septiembre de 1991. Cabecerns del rio Shani-shani (callo Bocón). . . . N 02 04′17″—W 064 35′ 36″." Brewer, Chagnon, and Boom, "Forest and Man," p. 16.

29. Marco, interview, Ashidowa-teri, Sept. 17, 1997.

30. "Entonces Shaki aterrizo aqui, se fue, surgio la shawaray comenzamos a enfermarnos todos de verdad." Translation by Javier Carrera.

31. James Brooke, "Venezuela Befriends Tribe, But What's Venezuela?" *NYT*, Sept. 11, 1991.

32. Irota-teri was the mother village of both Hasupuwe-teri and Ashidowa-teri. Irota-teri became Fusiwe's great ally. He was on the way to Irota-teri when he was assassinated, and it was at their village at the foot of a mountain in the Rahuawe valley, a few miles east of the Shanishani, that he was cremated. Valero, *Yo soy Napeyoma*, pp. 284, 322–23, 353–54.

33. Kenneth Good, letter to the editor of *NYT*, Sept. 13, 1991.

34. "I was quite surprised, however, to read that Mr. Chagnon had 'discovered 10 Yanomami villages . . . and visited three of them in the Siapa River Valley. . . .' " Kenneth Good, letter to the editor of *NYT*, Sept. 29, 1990.

35. Kenneth Good, phone interview, Jan. 3, 1995.
36. Carlos Botto, M.D., interview, CAICET, Puerto Ayacucho, Oct. 6, 1996.
37. Ibid.
38. Mirapewe, interview, Ashidowa-teri, Sept. 17, 1996.
39. "En el viaje de entrada a Ashidowa-teri el dia I[de Septiembre, 1991] también fueron en el helicóptero . . . James Brooke: New York Times, Kelvin Noblet: Associated Press, Diego Gidice: Associated Press." Brewer, Chagnon, and Boom, "Forest and Man," p. 18.
40. America Perdamo, interview, CAISET, Puerto Ayacucho, June 17, 1996.
41. Brewer, Chagnon, and Boom, "Forest and Man," p. 17.
42. Brooke, "Venezuela Befriends Tribe."
43. Theodora Kroeber, *Ishi in Two Worlds: A Biography of the Last Wild Indian in North America* (Berkeley: Univ. of California Press, 1961).
44. Carlos Botto, interview, M.D., CAISET, Puerto Ayacucho, Oct. 6, 1996.
45. Napoleon Chagnon, *Studying the Yanomamo* (New York: Holt, Rinehart and Winston, 1974), p. 125.
46. Napoleon Chagnon, *Yanomamo,* 4th ed. (Fort Worth: Harcourt Brace, 1992), p. 236.
47. Isaam Madi, *Conspiración al Sur del Orinoco,* p. 71. (Caracas: self-published, 1998), p.
48. Charles Brewer Carías, Napoleon Chagnon, and Brian Bloom, "Forest and Man" (MS; Caracas: Fundación Explora, 1993), pp. 10–20.
49. Chagnon, *Studying the Yanomamo,* p. 181.
50. Napoleon Chagnon, "The View from the President's Window," *Human Behavior and Evolution Society Newsletter* 2, no. 3 (Oct. 1993): 4.
51. Irenaus Eibl-Eibesfeldt and Gabriele Herzog-Schroder, "In Defense of the Mission" (MS, Feb. 28, 1994), p. 6.
52. Alfredo Aherowe, interview, Mahekoto-teri, Sept. 24, 1996.
53. Kayopewe, interview, Mahekoto-teri, June 11, 1996.
54. Enrique Lucho, interview, Kosh, Padamo River, June 14, 1996.
55. Napoleon Chagnon, personal letter printed in *Brown Gold* 24, no. 7 (Nov. 1989): 10.
56. César Dimanawa, interview, Mavakita, June 8, 1996.
57. Rifkin, "Ethnography and Ethnocide," p. 325.

Chapter 18: Human Products and the Isotope Men

1. Roger Bacon, *The Mirror of Alchemy.*
2. John Hemming, *The Search for El Dorado* (New York: E. P. Dutton, 1978), pp. 1–20.
3. David Weatherall, "The Mutation Man," *New Scientist,* July 9, 1994, p. 42.
4. James Neel, *Physician to the Gene Pool: Genetic Lessons and Other Stories* (New York: John Wiley, 1994), p. 227.
5. *To Each His Farthest Star: University of Rochester Medical Center* (Rochester: Univ. of Rochester Medical Center, 1975), p. 507.
6. Stephane Groueff, *Manhattan Project: The Untold Story of the Making of the Atomic Bomb* (London: Andre Deutsch, 1963), p. 152.
7. "The Rochester Story" (n.d., no author or date appears on document DOE: No. 707326); See also the general overview of Rochester in DOE's Roadmap to the Human Radiation Experiments. http://hrex.dis.anl.gov.
8. "Actually, we have got the results of an enormous experiment. We have the experiment involving over 200,000 people in the Nagasaki and Hiroshima areas, and I think that those results are real." Quoted in Eileen Welsome, *The Plutonium Files: America's Secret Medical Experiments in the Cold War* (New York: Dial Press, 1999), p. 324.
9. Wright Langham, "Revised Plan of 'Product' of Rochester Experiment" (DOE, 0719208, HREX), p. 6.
10. Welsome, *The Plutonium Files,* pp. 126, 131.
11. Langham, "Revised Plan of 'Product,' " p. 2.
12. Bassett, quoted in Welsome, *The Plutonium Files,* p. 129.
13. "Comments on Meeting with Dr. Hempelmann on April 17, 1974," FOIA, p. 1, cited ibid., pp. 127, 510.
14. Samuel Bassett, "Excretion of Plutonium Administered Intravenously to Man" (Advisory Committee on Human Radiation Experiments, DOE 121294-D-19), p. 2.
15. Langham, "Revised Plan of 'Product,' " p. 6.

16. Neel, *Physician to the Gene Pool,* p. 39; James Neel and William Valentine, "Hematologic and Genetic Study of the Transmission of Thalassemia (Cooley's Anemia, Mediterranean Anemia)," *Archives of Internal Medicine* 74 (1944): 185–96; "The Frequency of Thalassemia," *American Journal of Medical Science* 209 (1945): 568–72.

17. "Elected to NAS: 1977. Scientific Field: Medical Genetics, Hematology, and Oncology." National Academy of Sciences—Members.

18. Corydon Ireland, "Radiation Records at UR Missing," *Rochester Democrat & Chronicle,* Jan. 1, 1995.

19. Corydon Ireland, phone interview, Feb. 4, 2000.

20. Corydon Ireland, " 'No Bad Guys' in Study: Doctor Defends Good Intentions," Rochester *Democrat & Chronicle,* Dec. 16, 1994.

21. "At the request of the Manhattan Engineer District, Bale had 'activated' a metabolic ward at Rochester's Strong Memorial Hospital to carry out 'certain tracer studies' with long-lived radioisotopes. The ward, at least in the early years, appears to have been used exclusively for the radioisotope studies." Welsome, *The Plutonium Files,* p. 125.

22. Ireland, " 'No Bad Guys' in Study."

23. Welsome, *The Plutonium Files,* pp. 127, 445.

24. Ibid., p. 131.

25. Neel, *Physician to the Gene Pool,* p. 22.

26. Ibid., p. 27.

27. "One day I spoke to Joe: if Col. Warren knew of any long-range planning within the military regarding follow-up studies of Hiroshima and Nagasaki, might he throw in the hopper the name of Lieutenant-to-be-Neel, as someone interested in the genetic aspects. . . . Incredibly, my talk with Joe Howland had resulted in an assignment to the Manhattan Engineering District. . . ." Ibid., p. 27.

28. Welsome, *The Plutonium Files,* p. 85.

29. Thomas Powers, "Die Hard," *Los Angeles Times,* latimes.com, Jan. 2, 2000, p. 4.

30. Welsome, *The Plutonium Files,* p. 165.

31. Ibid., p. 389.

32. Eileen Welsome, phone interview, Feb. 3, 2000.

33. John Dower, *Embracing Defeat: Japan in the Wake of World War II* (New York: W. W. Norton, 1999), p. 205.

34. Neel, *Physician to the Gene Pool,* p. 88.

35. "In connection with the discussion on genetics the Committee agreed unanimously that the analysis of the genetic data should be made in the United States and the University of Michigan, under the direction of Dr. James Neel." Minutes of the Division of Biology and Medicine, ABCC, 1954 (DOE No. 1073291), p. 5.

36. Neel, *Physician to the Gene Pool,* pp. 317–39.

37. A glimpse of funding levels is offered by a DOE archive press release describing a $4.6 million contract from 1976 to 1981. "ERDA to Fund Mutation Monitoring Program" (Washington, D.C.: Energy Research and Development Foundation, Feb. 16, 1976).

38. "Concerned that its very existence was threatened if the public believed that there was an increased risk of cancer at these low levels of exposure, the nuclear-industrial complex determined that it would vigorously respond to all challengers." Karl Z. Morgan and Ken M. Peterson, *The Angry Genie: One Man's Walk through the Nuclear Age* (Norman: Univ. of Oklahoma Press, 1999), pp. 112–13.

39. Dower, *Embracing Defeat,* p. 493.

40. Powers, "Die Hard."

41. Eileen Welsome, phone interview, Feb. 3, 2000.

42. "Research Projects Approved during December 1949," Biology and Medicine Division, AEC, (DOE Archives, Unclassified, 4005484), p. 5.

43. Welsome, *The Plutonium Files,* p. 248.

44. Van Middlesworth, quoted ibid., p. 303.

45. Libby, quoted ibid., pp. 302, 489.

46. Merril Eisenbud, Jan. 26, 1995 (DOE/EH-0456).

47. The Organization of American States provided the names of some useful contacts. "Inasmuch as I am sure you will be acquainted with specialized scientists of the different countries which have invited you for lectures I guess you would like to be acquainted with some other who would introduce you to some more having that way a chance of meeting the outstanding scientists of South America." Cortés Fla, chief, Science and Technology Section, Organization of American States, "Proposed Itinerary for P. C. Aebersold's Trip to South America. June 2, 1954" (DOE, No. 0716665, HREX.dis.anl.gov.), p. 1.

48. *Operation Tumbler-Snapper,* 47 min. (Las Vegas: Coordination and Information Center, DOE, 1952).

49. "Ninth Annual Report of the Oak Ridge Institute of Nuclear Studies, June 30, 1955" (DOE, No. 0712485, HREX.dis.anl.gov), pp. 22–61.

50. Welsome, *The Plutonium Files,* p. 217.

51. Untitled in Human Radiation Index. The page heading reads, "Section 9C: Radioactive Isotopes and Nuclear Radiations in Medicine: Diagnosis and Studies of Disease, by Marshall Brucer" (DOE, No. 0720486).

52. R. Riviére, D. Comar, M. Colonomos, J. Desenne, and M. Roche, "Iodine Deficiency without Goiter in Isolated Yanomama Indians, Preliminary Note," *Biomedical Challenges Presented by the American Indian* (Washington: Pan American Health Organization, 1968). "In repeat studies in 1962 and 1968, Roche and colleagues found that the uptake [of radioactive iodine] had dropped to 60% in two different villages in contact with missions, but in a third, still very isolated village, the uptake was as high as before." Neel, *Physician to the Gene Pool,* p. 156.

53. Jay Stannard, phone interview, Feb. 4, 2000.

54. Terence Collins, phone interview, Feb. 13, 2000.

"For a study by the contractor of the metabolism of the human bone marrow. . . . During this period studies will be continued on the metabolism of mature and immature stages of human leukocytes obtained from normal and diseased patients." The salary for Dr. T. Arends was $825.00 from Jan. to March. "Contract No. AT-(40-11-1081) Modification No. 2, January 30, 1953" (DOE, No. 0717188, http://hrex.dis.anl.gov).

55. "Biology and Medicine Semiannual Report for October 1959 through March, 1960" (DOE, No. 0724963, HREX), pp. 111–14. Later, the U.S. Public Health Service sponsored a study of how healthy males from a number of countries, including Venezuelan natives, responded to the injection of radioactive iron. "A collaborative study was undertaken in an attempt to document obligatory iron losses in adult male subjects, using a variety of isotopic and chemical methods. Total body excretion was measured in four groups of subjects by injecting Fe55 intravenously and following the decline in red cell activity over several years. . . . Healthy male subjects belonging to different ethnic groups were studied. They lived in the United States of America, Venezuela or South Africa, and were employed in either manual or sedentary occupations." W. Green et al., "Body Iron Excretion in Man: A Collaborative Study," *American Journal of Medicine* 45 (1968): 336–37.

56. "Your invitation to participate in the U.S. Atomic Energy Commission's exhibit program in Latin America is one which we are happy to accept. We have selected two applications of radioisotopes in medicine and biological research which we would like to cover in fairly comprehensive scope with photographs and written material. . . . Another current study involves the anemia secondary to hookworm infection, and is being conducted in collaboration with Dr. M. Layrisse, Banco de Sangres del distrito Federal, Caracas, Venezuela." Letter of John H. Lawrence to Edward R. Gardner, Sept. 29, 1960 (DOE, No. 0724853, HREX).

57. "During 1960–61 a major U.S. scientific exhibit will be presented in Argentina, Brazil, Venezuela and Peru. The exhibit is to include technical facilities and equipment necessary to illustrate radioisotope applications in industry, agriculture, and medicine. In addition, a low power operating reactor and a 2000 curie, Cobalt-60 pure facility will provide the local scientific community with an opportunity to train and perform actual research. An estimated four million people will see this exhibition of which many will actually work and participate in technical symposia and meetings." Edward R. Gardner, AEC, Office of Special Projects, to Dr. John H. Lawrence, Director, Donner Laboratory, Univ. of California at Berkeley, Sept. 12, 1960 (DOE, No. 0715342, HREX).

58. "When the nuclear powered aircraft thing left Convair, he [Gallimore] went to Caracas, Venezuela—one of those one-year reactor things there." "Interview with Robert G. Thomas," Sept. 22, 1981 (DOE, No. 0702930, HREX), p. 22. See also "Oral History of Cell Biologist Don Francis Petersen, Ph.D.," Nov. 29, 1994 (DOE, No. 0727845, HREX).

59. "Quarterly Progress Report to the Joint Committee on Atomic Energy, July–September, 1957, Part VII" (DOE, No. 0719025, HREX), pp. 33–34.

60. "Project Sunshine: Annual Report, March 31, 1955–April 1, 1956" (DOE, No. 0710270, HREX), p. 50.

61. Quarterly Progress Report to the Joint Committee on Atomic Energy, July–September, 1957, Part VII (DOE, No. 0719025, HREX), p. 33.

62. Kulp, quoted in Welsome, *The Plutonium Files,* p. 302.

63. "Radioisotopes in Science and Industry: Shipment of Radioisotopes to Foreign Countries," (DOE, 0716934), p. 122.

64. Eduardo Galeano, *Open Veins of Latin America: Five Centuries of the Pillage of a Continent* (New York: Monthly Review Press, 1973).

65. Stafford Warren, "Radioactivity, Health and Safety," April 1, 1947, (Yale Univ. Library, George Darling Collection, MSS 770), p. 29, cited in Welsome, *The Plutonium Files,* pp. 209, 520.

66. Matthew L. Wald, "U.S. Acknowledges Radiation Killed Weapons Workers," *NYT,* Jan. 29, 2000.

67. Eileen Welsome, phone interview, Feb. 3, 2000.

68. Welsome, *The Plutonium Files,* p. 5.

69. William Valentine denied that he ever gave plutonium injections. See p. 299.

70. Walter Burkert, *Homo Necans: The Anthropology of Ancient Greek Sacrificial Ritual and Myth* (Berkeley: Univ. of California Press, 1983).

71. Joseph Howland, "An Experience in Nuclear Medicine" (Univ. of Rochester, Edward C. Miner Library, Joseph Howland Collection), p. 3, cited in Welsome, *The Plutonium Files,* p. 85.

72. "Se dan abundantes casos de anemia, algunos de sapillo, menos de hernia y de peritonitis." Luis Cocco, *Iyewei-teri: Quince años entre los yanomami* (Caracas: Editorial Salesiana, 1973), p. 418.

73. J. Enders, K. McCarthy, A. Mitus, and W. J. Cheatham, "Isolation of Measles Virus at Autopsy in Cases of Giant-Cell Pneumonia without Rash," *New England Journal of Medicine* 261 (Oct. 29, 1959): 875–81.

74. Marcel Roche, M.D., interview at *Interciencia,* Caracas, June 20, 1996.

75. Roche told me he did not have time to check the reaction to measles in the Ocamo and Mavaca Yanomami, because he was doing an iodine study. Marcel Roche, M.D., interview at *Interciencia,* Caracas, June 20, 1996. James Neel confirmed part of Roche's account. Roche administered radioactive iodine to both mission and isolated Yanomami in 1968. Neel, *Physician to the Gene Pool,* pp. 155–56.

76. Marcel Roche, M.D., interview at *Interciencia,* Caracas, June 20, 1966.

77. All the preceding quotations, to the beginning Neel's camera instructions to Asch, are from the same sound roll. Sound Roll 3, Feb. 18, 1968, Timothy Asch Non-film Collection, NAA.

78. Sound Roll 9, Patanowa-teri, Feb. 23, 1968, Timothy Asch Collection, NAA.

79. Juan Gonzalez, interview, Kosharowa-teri, Padamo River, June 14, 1996.

80. George and Louise Spindler, foreword to *Yanomamo: The Fierce People,* 2d ed. (New York: Holt, Rinehart and Winston, 1977), pp. vii–viii.

81. Neel, *Physician to the Gene Pool,* p. 346.

82. "Report of the Visiting Committee: Brookhaven National Laboratory, May 3 and 4, 1977" (DOE, No. 718287, HREX), pp. 5–6.

Appendix: Mortality at Yanomami Villages

1. Napoleon Chagnon, "Covering Up the Yanomamo Massacre," op-ed., *New York Times,* Oct. 23, 1993.

2. Irenaus Eibl-Eibesfeldt and Gabriele Herzog-Schroder, "In Defense of the Mission" (MS, Feb. 28, 1994), p. 4.

3. "It is not uncommon to find 85 to 90 percent of any given Indian group destroyed by a rapid series of epidemics." Darrel Addison Posey, "Environmental and Social Implications of Pre- and Postcontact Situations on Brazilian Indians," in *Amazonian Indians* (Tucson: Univ. of Arizona Press, 1994), p. 273.

4. "One suspects his neighbors of sorcery, witchcraft and chicane, and treats them accordingly." "With distance there is security and sovereignty." "[N]eighboring villages constitute a barrier as effective as a chain of mountains or a desert." Napoleon Chagnon, *Studying the Yanomamo* (New York: Holt, Rinehart and Winston, 1974), pp. 77, 78, 76.

5. Francis Black, "Infecção, mortalidade e populações indígenas: Homogeneidade biológica como possível razão para tantas mortes," in *Saúde e Povos Indígenas,* ed. Ricardo V. Santos and Carlos E. A. Coimbra Jr. (Rio de Janeiro: Editora Fiocruz, 1994), pp. 63–87.

6. James Neel originally estimated growth between 0 and 2 percent. James Neel, "Progress Report" (MS, Washington, D.C.: DOE, March 17, 1973), p. 5. In the end, he picked a figure between 0.5 and 1 percent. James Neel, *Physician to the Gene Pool: Genetic Lessons and Other Stories* (New York: John Wiley, 1994), p. 182. Neel's final estimate was too low because his overall population estimate for the Yanomami was off by about 10,000. He thought there were, at most, 15,000 Yanomami. Johannes Wilbert was much closer to the right figure when he guessed 25,000–30,000, though Wilbert believed there were more Yanomami in Brazil than in Venezuela, which turned out to be wrong. Johannes Wilbert, *Survivors of El Dorado: Four Indian Cultures of South America* (New York: Praeger, 1972), p. 16. Although Chagnon did not hazard a specific figure, his data suggests that two large villages, of unknown size but probably not exceeding 250 people each, gave rise to the Namowei and Shamatari villages, 700 and

2,000 individuals, respectively, in a hundred-year period. That would be growth on the order of 2 percent yearly for the combined populations. Chagnon, *Studying the Yanomamo,* pp. 129–37.

7. See chapter 5 for an account of the collapse of the Ocamo River population.

8. Hasupuwe-teri's village history illustrated this strange disparity. Its inhabitants numbered over 300 in 1966. There were two Hasupuwe-teri subdivisions, and the smaller one had up to 150 individuals. Inga Steinvorth Goetz, *Uriji Jami!: Life and Belief of the Forest Waika in the Upper Orinonco,* trans. Peter Furst (Caracas: Asociación Cultural Humboldt, 1969), p. 88. A year later, 30 percent of them—about 100—died in the measles epidemic. Brian Ferguson, *Yanomami Warfare: A Political History* (Santa Fe: School of American Research Press, 1995), p. 309. Luis Cocco, *Iyewei-teri: Quince años entre los Yanomamos* (Caracas: Editorial Salesiana, 1973), p. 417. Good joined the Hasupuwe-teri in 1975, when they had not fully recovered; they still numbered only 208 individuals. Kenneth Good, "Yanomami Hunting Patterns: Trekking and Garden Relocation as an Adaptation to Game Availability in Amazonia" (Ph.D. diss., Univ. of Florida, 1989), p. 6. Irregular visits from the Platanal mission and the Malaria Department started about this time. That help, plus the relative isolation conferred by the Guaharibo Rapids, enabled them to survive, but not to prosper. By 1990, Hasupuwe-teri, including its recently fissioned sister village, Patahama-teri, had only 225 people. Charles Brewer Carías, Napoleon Chagnon, and Brian Boom, "Forest and Man" (MS; Caracas: Fundación Explora, 1993), p. 9. The Hasupuwe-teri's erratic demographic pattern—explosion followed by collapse followed by stagnation—revealed the sharp inequalities of uncontrolled contact and uneven access to medical care. If they had continued their historical growth, by the most conservative estimate there would be at least 400 Hasupuwe-teri today.

9. A survey of villages affected by the gold rush in the Brazilian Parima region of Surucucu found that mortality was as high as 60 percent. Overall death rates for the estimated 4,000 Yanomami in the highlands was 1,100 individuals, over 25 percent. Oneron de Abreu Pithan, M.D., "A situação de Saúde dos Yanomami de Roraima" (MS; Boa Vista: Fundação Nacional de Saúde, Oct. 1990).

10. Half the population of five villages in the mountainous headwaters was wiped out by measles. Giovanni Saffirio and Raymond Hames, "The Forest and the Highway," in *The Impact of Contact: Two Yanomami Case Studies,* ed. K. Kensinger (Cambridge, Mass.: Cultural Survival, 1983), p. 12.

11. The Mucajaí Ninam grew at 3.5 percent from 1958 to 1983. John D. Early and John F. Peters, *Population Dynamics of the Mucajai Yanomama* (San Diego: Academic Press, 1990), p. 25. In 1968, there were about 100 Yanomami at the Padamo mission; today, there are over 350. Discounting immigration, it still amounted to a population increase of 3 percent a year (about normal for the twelve mission stations in Yanomamiland). At the New Tribes Parima B, population increased from 250 in 1968 to over 700 in 1995, even though out-migration exceeded in-migration. Greg Sanford, "Who Speaks for the Yanomami?: A New Tribes Perspective," in Frank A. Salamone, *The Yanomami and Their Interpreters: Fierce People or Fierce Interpreters?* (Lanham, Md.: Univ. Press of America, 1997), p. 59. Among the Iyewei-teri, population increased from 58 to 135 in fifteen years of mission work. Cocco, *Iyewei-teri,* pp. 480–81. Since the founding of the Catholic mission at Mavaca, in 1965, to 1984, population increased from 202 (see Chagnon, "Yanomamo Warfare, Social Organization and Marriage Alliances" [Ph.D. diss., Univ. of Michigan, 1966], p. 209) to 422, with 80 percent of the surge the result of natural increase. María Eguillor García, *Yopo, shamanismo y hekura* (Caracas: Editorial/Librería Salesiana, 1984), p. 56. Among the Lower Catrimani villages, which are well cared for by the Consolata missionaries, population boomed from about 130 in 1968 to 465 in 1995. Even though it is in a mosquito-ridden, lowland area with some gold miners, there were no cases of malaria at Catrimani in 1993.

12. Carlos Botto, M.D., interview, CAICET, Puerto Ayacucho, Oct. 6, 1996.

13. "[Y]la población del alto Orinoco ubicada cerca de las medicaturas de Ocamo, Mavaca y Platanal o accesible durante las penetraciones alcanza aproximadamente a 3000 Yanomami. . . ." "100% de la población estudiada en el Alto Orinoco tiene evidencias de infección malárica activa o pasada, presentando más de una tercera parte de ellos una complicación grave conocida con el nombre de esplenomegalia malárica hiperreactiva." Carlos Botto, "La situación de salud de la población Yanomami," *La Iglesia en Amazonas* (Puerto Ayacucho, 1991), pp. 14, 12.

14. Napoleon Chagnon, "Killed by Kindness?: The Dubious Influence of the Salesian Missions in Amazonas," *TLS,* Dec. 24, 1993, p. 11.

15. Napoleon Chagnon, *Yanomamo: The Fierce People,* 3rd ed. (New York: Holt, Rinehart and Winston, 1983) p. 199.

16. Marta Rodríguez Miranda, Venevisión, Caracas, Fundación Cultural Venevisión, July 24, 1991.

17. Chagnon, "Covering Up the Yanomamo Massacre."

18. Chagnon, "Killed by Kindness?," p. 11.

19. Ferguson, *Yanomami Warfare,* p. 315.

20. This is why Lizot complained about Chagnon's not identifying his villages. Jacques Lizot. "On Warfare: An answer to N. A. Chagnon," trans. Sarah Dart, *American Ethnologist* 21 (1994): 853.

21. Napoleon Chagnon, "Life Histories, Blood Revenge, and Warfare in a Tribal Population," *Science* 239 (1988): 986.

22. Napoleon Chagnon, *The Last Days of Eden* (San Diego: Harcourt Brace Jovanovich, 1992), pp. 267–68.

23. The mortality average for the 480 individuals at Boca Mavaca is therefore not derived on a village-by-village basis. These are simple averages.

24. The villages were Musiu-teri (56), Thanahi-teri (105), Wayawa-teri (65), Mothoremawe-teri (44), Hayapeka-teri (55), Cinc-teri (36), and Koro-teri (61). This total comes to 422, compared with Chagnon's 414 for Villages 5, 6, and 7. But family groups often come and go, so minute changes like this are normal. This census was taken in 1983. The eighth village may be Monou-teri (which Eguillor García sometimes considers a separate entity, just as Chagnon did earlier). By 1987, one of these villages, Thanahi-teri, had fissioned, and one of its fissions had moved upstream with Kaobawa as its head. This group is called Shakita today. Actually, Kaobawa's group has been called Shakita, at least informally, for many years. Eguillor García, *Yopo, shamanismo y hekura,* p. 45.

25. Napoleon Chagnon, *Yanomamo,* 5th ed. (Fort Worth: Harcourt Brace, 1997), p. 246.

26. Timothy Asch, "Ethnographic Filming and the Yanomamo Indians," *Sightlines,* Jan.–Feb. 1972, pp. 6–12.

27. "We spent the first three days in Shanishani-teri filming what a 'remote' village was like. We then tried to move to one of two other, less known groups, but found their *shabonos* empty. We therefore decided to spend the next few days in Dorita-teri to document an example of a Yanomamo community that had limited contact but was still in a remote area." Napoleon Chagnon, *Yanomamo,* 4th ed. (Fort Worth: Harcourt Brace, 1992), p. 235.

28. Padre Nelson, camcorder interview with Marta Rodríguez Miranda, Ocamo, May 13, 1991.

29. Chagnon, *Yanomamo,* 4th ed., p. 224.

30. Ibid., p. 226.

31. Ibid., p. 224.

32. Juan Finkers, phone interview, Jan. 24, 1995.

33. "If the residents of Ironasi-teri (09) and Mishimishimabowei-teri (16) were pooled together to reflect what their composition was like before they fissioned, . . . the village would include over 400 people." Chagnon, *Studying the Yanomamo,* p. 139.

34. "When I visited them in 1972, they numbered 179 people. I did not return to them for two years; in 1975 when I went inland to update my census on them, I found that they had suffered a 40% mortality the previous year and there were very few children below the age of 10 years." Chagnon, *Yanomamo,* 3rd ed., p. 200.

35. Of the fifty adults in *The Ax Fight,* ten died in 1986. They were living at several different sites, including Village 92 (Mishimishimabowei-teri) and Village 51 (Kedebabowei-teri). Chagnon, "People," *Yanomamo Interactive.*

36. Chagnon, *Yanomamo,* 4th ed., pp. 224–27.

37. María Yolanda García. "Cecilia Matos no iba a proteger indígenas sino a sacar oro del Amazonas." *El Nacional,* Jan. 15, 1993. Brewer, Chagnon, and Boom, "Forest and Man," pp. 12–14.

38. Brewer, Chagnon, and Boom, "Forest and Man," pp. 12–13.

39. Carlos Botto, M.D., interview, CAISET, Puerto Ayacucho, Oct. 6, 1996.

40. America Perdamo, interview, CAISET, Puerto Ayacucho, June 17, 1996.

41. James Brooke, "Venezuela Befriends Tribe, But What's Venezuela?," *NYT,* Sept. 11, 1991.

42. Gare Smith, the State Department's deputy undersecretary for human rights, recently sounded the Venezuelan government about Yanomami health care. He was told that the government did not want to disturb the Yanomami's traditional lifestyle. "We do not completely agree with this," Smith wrote to me. "We believe the Venezuelan government could do more for the Yanomami." Gare A. Smith, personal correspondence, Nov. 16, 1998.

43. Carlos Botto, M.D., interview, CAISET, Puerto Ayacucho, Oct. 6, 1996.

44. John Walden, M.D., phone interview, March 4, 1997.

45. Ibid.

46. Marinho de Souza, interview, Ashidowa-teri, Malaria Census, CAISET, Sept. 1996.

47. Chagnon, *The Last Days of Eden,* p. 271.

48. Mirapewe, interview, Ashidowa-teri, Sept. 18, 1996.

Bibliography

Books, Dissertations, and Magazine Articles

Acebes, Hector. *Orinoco Adventure.* New York: Doubleday, 1954.

Albert, Bruce. "La fumée du métal: Histoire et représentations du contact chez les Yanomami (Brésil)." *L'Homme* 28 nos. 2–3 (1988): 87–119.

———. "Yanomami 'Violence': Inclusive Fitness or Ethnographer's Representation?" *Current Anthropology* 30 (1989): 637–40.

———. "On Yanomami Warfare: Rejoinder." *Current Anthropology* 31 (1990): 558–63.

———. "The Massacre of the Yanomami at Hashimu." Translation based on an article in *Fôlha de São Paulo,* Oct. 10, 1993.

Albert, Bruce, and Alcida Rita Ramos. "O exterminio academico dos Yanomami." *Humanidades* (Brasília) 18 (1988): 85–89.

Alcántara, Eurípides. "Indio também é gente." *Veja,* Dec. 6, 1995, pp. 7–10.

Allman, William. "A Laboratory for Human Conflict." *U.S. News & World Report,* April 11, 1988, pp. 57–58.

———. *Stone Age Present.* New York: Simon & Schuster, 1994.

Anduze, Pablo. *Shailili-Ko: Relato de un naturalista que también llegó a las fuentes del río Orinoco.* Caracas: Talleres Gráficos Ilustrados, 1960.

Arango, E. Lugo, A. Ouaissi, I. des Moutis, A. Capron, and L. Yarzabal. "Asociación de antigenemia con depresión de la hipersensibilidad cutanea retardada en la onconcercosis." *Proicet Amazonas,* no. 2 (1983): 101–8.

Arvelo Jiménez, Nelly, and Andrew L. Cousins. "False Promises: Venezuela Appears to Have Protected the Yanomami, But Appearances Can Be Deceiving." *Cultural Survival Quarterly,* Winter 1992, pp. 10–14.

Asch, Timothy. "Ethnographic Filming and the Yanomamo Indians." *Sightlines,* Jan.–Feb. 1972, pp. 6–12.

———. Foreword. *Tales of the Yanomami,* by Jacques Lizot. New York: Cambridge Univ. Press, 1985.

Asch, Timothy, and David Atkins. *Yanomamo: A Multidisciplinary Study, Field Notes.* Somerville, Mass.: DER, 1975.

Ayala, David. "Informe de comisión de diputados ratifica denuncias contra Charles Brewer Carías." *Ultimas Noticias* (Caracas), Oct. 5, 1993.

Barandarian, Daniel de, and Aushi Walalam. *Hijos de la Luna: Monografia antropológica sobre los indios sanemá yanomama.* Caracas: Editorial Arte, 1983.

Biocca, Ettore. *Yanoama.* New York: Kodansha International, 1996.

Bird, Randy, and Garland Allen. "Charles Darwin." *Encarta Encyclopedia.* Microsoft, 1999.

Birdsell, Joseph. "Some Predictions for the Pleistocene Based on Equilibrium Systems among Recent Hunter-Gatherers." In *Man the Hunter,* ed. Richard B. Lee and Irv DeVore, pp. 229–49. Chicago: Aldine, 1968.

Black, Francis. "Infeção, mortalidade e populações indígenas: Homogeneidade biológica como possível razão para tantas mortes." *Saúde e Povos Indígenas,* ed. Ricardo V. Santos and Carlos E. A. Coimbra Jr., pp. 63–87. Rio de Janeiro: Editora Fiocruz, 1994.

Black, Francis, J. P. Woodall, and P. Pinheiro. "Measles Vaccine Reactions in a Virgin Population." *American Journal of Epidemiology* 89 (1969): 168–75.

Booth, William, "Warfare over Yanomamo Indians." *Science* 243 (1989): 1138–40.

Bortnick, Barry. "From Amazon Jungle to Ivory Tower." *Santa Barbara News-Press,* April 19, 1999.

Botto, Carlos. "La situación de salud de la población yanomami." *La Iglesia en Amazonas.* Puerto Ayacucho: 1991.

———. "Impactos ambientales en salud: La experiencia de CAICET." Belem, Brazil: June 6, 1996.

Brewer Carías, Charles. *Roraima: montaña de cristal.* Caracas: Oficina Central de Información, 1975.

———. "Una futura zona en reclamación." *El Nacional,* May 10, 1987.

———. "Teocracia y soberanía de Amazonas." *El Nacional,* Nov. 25, 1995.

Brody, Jacob A., R. McAlister, I. Emanuel, and E. R. Alexander. "Measles Vaccine Field Trials in Alaska." *Journal of the American Medical Association* 189 (1964): 339–42.

Brooke, James. "In an Almost Untouched Jungle, Gold Miners Threaten Indian Ways." *NYT,* Sept. 19, 1990.

———. "Reserve for Primitive Tribe Promised in 6 Months." *NYT,* Sept. 25, 1990.

———. "Stone Age Village Found: Venezuela to Protect Yanomami Indians," *Gazette* (Canada), Sept. 27, 1990.

———. "Venezuela Befriends Tribe, But What's Venezuela?" *NYT,* Sept. 11, 1991.

———. "Attack on Brazilian Indians Is Worst since 1910." *NYT,* Aug. 29, 1993.

———. "Raids on Miners Follow Killings in Amazon." *NYT,* Sept. 9, 1993.

Burke, Barbara. "Infanticide," *Science 84,* May 1984, p. 31.

Capozoli, Ulisses. "Yarima, cinderela rebelde." *O Estado de São Paulo,* March 3, 1997.

Cappelletti, E. J. "Venezuela Mine Scheme Targets Salesians." Letter to the editor, *NYT,* Jan. 18, 1994.

Carneiro da Cunha, Maria Manuela. Letter to the editor. *Anthropology Newsletter,* Jan. 1989, p. 3.

Cavalli-Sforza, Luigi Luca. "Anthropometric Data." In *African Pygmies,* ed. Luigi Luca Cavalli-Sforza. Orlando: Academic Press, 1986.

Chagnon, Napoleon. "Yanomamo Warfare, Social Organization and Marriage Alliances." Ph.D. diss., Univ. of Michigan, 1966.

———. *Yanomamo: The Fierce People.* 1st ed. New York: Holt, Rinehart and Winston, 1968.

———. Letter from the field. In *The Human Condition in Latin America,* ed. E. Wolf and E. Hansen, pp. 65–69. New York: Oxford Univ. Press, 1972.

———. *Studying the Yanomamo.* New York: Holt, Rinehart and Winston, 1974.

———. "Yanomamo." In *Primitive Worlds,* pp. 141–83. Washington, D.C.: National Geographic Society, 1974.

———. "Genealogy, Solidarity and Relatedness: Limits to Local Group Size and Patterns of Fissioning in an Expanding Population." *Yearbook of Physical Anthropology* 19 (1975): 95–110.

———. *Yanomamo: The Fierce People.* 2d ed. New York: Holt, Rinehart and Winston, 1977.

———. *Yanomamo: The Fierce People.* 3d ed. New York: Holt, Rinehart and Winston, 1983.

———. "Life Histories, Blood Revenge, and Warfare in a Tribal Population." *Science* 239 (1988): 985–92.

———. Letter to the editor. *Anthropology Newsletter,* Jan. 1989, pp. 3, 24.
———. "Response to Ferguson." *American Ethnologist* 16 (1989): 565–70.
———. "Reproductive and Somatic Conflicts of Interest in the Genesis of Violence and Warfare among Tribesmen." In *The Anthropology of War,* ed. J. Haas, pp. 77–104. Cambridge: Cambridge Univ. Press, 1990.
———. "On Yanomamo Violence: Reply to Albert." *Current Anthropology* 31 (1990): 49–53.
———. "To Save the Fierce People." *Santa Barbara,* Jan.–Feb. 1991, pp. 36–43, 70–71.
———. *Yanomamo.* 4th ed. Fort Worth: Harcourt Brace, 1992.
———. *Yanomamo: The Last Days of Eden.* San Diego: Harcourt Brace Jovanovich, 1992.
———. "The View from the President's Window." *Human Behavior and Evolution Society Newsletter* 2, no. 3 (Oct. 1993).
———. "Covering Up the Yanomamo Massacre." *NYT,* Oct. 23, 1993.
———. "Killed by Kindness?" *Times Literary Supplement,* Dec. 24, 1993, pp. 11–12.
———. "L'ethnologie du déshonneur: Brief Response to Lizot." *American Ethnologist* 22 (1995): 187–89.
———. "Chronic Problems in Understanding Tribal Violence and Warfare." In *Genetics of Criminal and Antisocial Behavior,* ed. G. R. Bock and J. A. Goode. pp. 202–36. New York: John Wiley, 1996.
———. *Yanomamo.* 5th ed. Fort Worth: Harcourt Brace, 1997.
———. "Filming the Ax Fight." *Yanomamo Interactive CD: The Ax Fight,* by Peter Biella, Napoleon Chagnon, and Gary Seaman. New York: Harcourt Brace, 1997.
Chagnon, Napoleon, James Neel, Lowel Weitkamp, Henry Gershowitz, and Manuel Ayres. "The Influence of Cultural Factors on the Demography and Pattern of Gene Flow from the Makiritare to the Yanomama Indians." *American Journal of Physical Anthropology* 32 (1970): 339–49.
Chagnon, Napoleon, and Raymond Hames, "Protein Deficiency and Tribal Warfare in Amazonia: New Data." *Science* 203 (1979): 910–13.
Chagnon, Napoleon, and Paul Bugos. "Kin Selection and Conflict: An Analysis of a Yanomamo Ax Fight." *Evolutionary Biology and Human Social Behavior: An Anthropological Perspective,* ed. Napoleon Chagnon and William Irons, pp. 213–38. North Scituate, Mass.: Duxbury Press, 1979.
Chagnon, Napoleon, and Thomas Melancon. "Epidemics in a Tribal Population." In *The Impact of Contact: Two Yanomamo Case Studies,* ed. K. Kensinger, pp. 53–78. Cambridge, Mass.: Cultural Survival, 1983.
Cleary, David, *The Anatomy of the Amazon Gold Rush.* Oxford: Macmillan and St. Antony, 1990.
Cliff, Andrew David. *Measles: An Historical Geography of a Major Human Viral Disease from Global Expansion to Local Retreat, 1840–1990.* Oxford: Blackwell Reference, 1993.
Cocco, Luis, *Iyeweit-teri: Quince años entre los Yanomamos.* Caracas: Editorial Salesiana, 1973.
Cockburn, Charles, Joseph Pecenka, and T. Sundaresan. "WHO-Supported Comparative Studies of Attenuated Live Measles Virus Vaccines." *Bulletin of the World Health Organization* 34 (1966): 223–31.
Colchester, Marcus. "Rethinking Stone Age Economics: Some Speculations regarding the Pre-Columban Yanoama Economy." *Human Ecology* 12 (1984): 291–314.
Colchester, Marcus, with Fiona Watson. *Venezuela: Violations of Indigenous Rights: Report to the International Labour Office on the Observation of ILO Convention 107.* Oxford: World Rainforest Movement and Survival International, 1995.
Comas, Juan. *¿Pigmeos en América?* Mexico City: Universidad Nacional Autónoma de México, 1960.
Cortinas, Juan Ignacio. "Las nuevas tribus sí han hecho daño al yanomami." *El Diario de Caracas,* Oct. 23, 1993.
Cristaldo, Janer. "Os bastidores do ianoblefe." *A Fôlha de São Paulo,* April 24, 1994.
———. "Uma teocracia na Amazonia." *A Fôlha de São Paulo,* Feb. 12, 1995.
Daly, Martin, and Margot Wilson. *Homicide.* Hawthorne, N.Y.: Aldine de Gruyter, 1988.
Danielson, Vivian. "Friedland Strong Supporter of Guiana Shield Gold Rush." *Northern Miner,* March 29, 1993.
D'Antonio, Michael. "Napoleon Chagnon's War of Discovery." *Los Angeles Times Sunday Magazine,* Jan. 30, 2000, pp. 16–19, 36–37.
Darwin, Charles. *The Descent of Man and Selection in Relation to Sex.* Detroit: Gale Research, 1974.
Dawkins, Richard. *The Selfish Gene.* Cambridge: Cambridge Univ. Press, 1975.
de Abreu, Oneron Pithan. "A situacao de Saúde dos Yanomami de Roraima." Boa Vista: Fundaçao Nacional de Saúde, Oct. 1990.
Dennett, Daniel C. *Darwin's Dangerous Idea: Evolution and the Meaning of Life.* New York: Touchstone, 1995.
De Waal, Frans. *Good Natured: The Origins of Right and Wrong in Humans and Other Animals.* Cambridge: Harvard Univ. Press, 1996.
Diamond, Jared. *Guns, Germs, and Steel: The Fates of Human Societies.* New York: W. W. Norton, 1997.

Dimanawa, César. "Carta abierta a Napoleon Chagnon." *La Iglesia en Amazonas,* March 1990, p. 20.

———. "Carta enviada por un Yanomami a N. A. Chagnon." *La Iglesia en Amazonas,* Feb. 1994, p. 19.

Dower, John. *Embracing Defeat: Japan in the Wake of World War II.* New York: W. W. Norton, 1999.

Dudgeon, J. Alaistair, and William A. M. Cutting. *Immunization: Principles and Practice.* London and New York: Chapman and Hall Medical, 1991.

Dufour, Darna L. "Diet and Nutritional Status of Amazonian People." In *Amazonian Indians,* ed. Anna Roosevelt, pp. 151–76. Tucson: Univ. of Arizona Press, 1994.

E.S. "Charles Brewer Carías: Inventario de supervivencia." *ExcesO,* April 1990, pp. 65–71.

Early, John D., and John F. Peters. *The Population Dynamics of the Mucajai Yanomama.* San Diego: Academic Press, 1990.

Eguillor García, María. *Yopo: shamanes y hekura.* Caracas: Editorial/Librería Salesiana, 1984.

Eibl-Eibesfeldt, Irenaus, and Gabriele Herzog-Schroder. "In Defense of the Mission." Paper, Forschungsstelle für Humanetheologie in der Max-Planck Gesellschaft, Feb. 28, 1994.

Enders, J., K. McCarthy, A. Mitus, and W. J. Cheatham. "Isolation of Measles Virus at Autopsy in Cases of Giant-Cell Pneunomia without Rash." *New England Journal of Medicine* 261 (Oct. 29, 1959): 875–81.

Esposito, Rubens, *Yanomami: Um povo ameacado de extinção.* Rio de Janeiro: Qualitymark Editora, 1998.

Ferguson, Brian. "Do Yanomamo Killers Have More Kids?" *American Ethnologist* 16 (1989): 564–65.

———. *Yanomami Warfare: A Political History.* Santa Fe: School of American Research Press, 1995.

Finkers, Juan. *Los Yanomami y su sistema alimenticio: Yanomami ni i pe.* Puerto Ayacucho: Vicariato Apostólico, 1986.

———. "Aclaraciones al Sr. Chagnon." *La Iglesia en Amazonas,* Dec. 1994, pp. 7–10.

Flores, Anabel. "Sociólogos y antrópologos objetan presencia de Brewer Carías en comisión presidencial." *Ultimas Noticias* (Caracas), Oct. 5, 1993.

Fox, Robin. "Evil Wrought in the Name of Good." *Anthropology Newsletter,* March 1994, p. 2.

Fredlund, Erik. "Shitari Yanomami Incestuous Marriage: A Study of the Use of Structural, Lineal and Biological Criteria When Classifying Marriages." Ph.D. diss., Pennsylvania State Univ., 1982.

Galeano, Eduardo. *Open Veins of Latin America: Five Centuries of the Pillage of a Continent.* New York: Monthly Review Press, 1973.

Gamini, Gabriela. "Search for Jungle Wife Condemned by Amazon Experts." *Times* (London), Feb. 1, 1997.

Gamini, Gabriela, and Quentin Letts. "American Plans Jungle Trip to Win Back Wife." *Times* (London), Jan. 31, 1997.

García, María Yolanda. "Cecilia Matos no iba a proteger indígenas sino a sacar oro del Amazonas." *El Nacional,* Jan. 15, 1993.

Gheerbrant, Alain. *Journey to the Far Amazon.* New York: Simon and Schuster, 1954.

———. *The Amazon: Past, Present and Future.* London: Thames and Hudson, 1992.

———. La expedición Orinoco-Amazonas (1948–1950). Caracas: Banco de la República/Áncora Editores, 1997.

Gilij, Felipe Salvador. *Ensayo de historia americano.* Vol. 2. Caracas: Biblioteca de la Academica Nacional de Historia, 1965.

Golden, Frederic. "Scientist a Fierce Advocate for a Fierce People." *Los Angeles Times,* May 15, 1997.

Good, Kenneth. "Yanomami Hunting Patterns: Trekking and Garden Relocation as an Adaption to Game Availability in Amazonia, Venezuela." Ph.D. diss., Univ. of Florida, 1989.

———. *Into the Heart: One Man's Pursuit of Love and Knowledge among the Yanomama.* New York: Simon & Schuster, 1991.

———. "A Race against Time." *Américas,* Oct. 1998, pp. 28–37.

Good, Kenneth, and Jacques Lizot. Letter to *Science.* Appended to Marvin Harris. "Culture Materialist Theory of Band and Village Warfare." In *Warfare, Culture, and Environment,* ed. R. B. Ferguson, pp. 111–40. Orlando: Academic Press, 1984.

Gooding, Kenneth. "Race to Move Mountain of Waste in the Rockies." *Financial Times* (London), Nov. 8, 1993, p. 8.

Green, W., and R. Charlton, H. Beftel, T. Bothwell, P. Mayer, R. Adams, C. Finch, and M. Layrisse. "Body Iron Excretion in Man." *American Journal of Medicine* 45 (1968): 336–53.

Grelier, Joseph. *To the Source of the Orinoco.* Translated by H. Schmuckler. London: Herbert Jenkins, 1957.

Groueff, Stephane. *Manhattan Project: The Untold Story of the Making of the Atomic Bomb* (London: Andre Deutsch, 1963).

Hames, Raymond. "Behavioral Account of the Division of Labor among the Yekwana Indians." Ph.D. diss., Univ. of California at Santa Barbara, 1978.

Hames, Raymond, and W. Vickers. "The Settlement Pattern of a Yanomamo Population Block: A Behavioral Ecological Interpretation." In *Adaptive Responses of Native Amazonians,* ed. Hames and Vickers, pp. 393–427. New York: Academic Press, 1983.

Hammel, Eugene. "Demographic Constraints on Population Growth in Early Humans." *Human Nature* 7 (1996): 217–55.

Hanson, Earl. "Social Regression in the Orinoco and Amazon Basins: Notes on a Journey in 1931 and 1932." *Geographical Review* 23 (1933): 578–98.

Harner, Michael. *The Jivaro: People of the Sacred Waterfall.* Berkeley: Univ. of California Press, 1984.

Harris, Marvin. Review of *Yanomami Warfare: A Political History,* by Brian Ferguson. *Human Ecology* 24 (1996): 413.

Hayden, G. F. "Measles Vaccine Failure: A Survey of Causes and Means of Prevention." *Clinical Pediatrics* 18 (1979): 155–67.

Heider, Karl. *Grand Valley Dani: Peaceful Warriors.* New York: Holt, Rinehart and Winston, 1979.

Hemming, John. *The Conquest of the Incas.* New York: Harcourt Brace Jovanovich: 1970.

———. *The Search for El Dorado.* New York: E. P. Dutton, 1978.

———. *Amazon Frontier: The Defeat of the Brazilian Indians.* Cambridge: Harvard Univ. Press, 1987.

Hendrickse, R. G., D. Montefiore, G. Sherman, and G. O. Sofoluwe, "A Further Study on Measles Vaccination in Nigerian Children." *Bulletin of the World Health Organization* 32 (1965): 803–8.

Henley, Paul. *Yanomami: Masters of the Spirit World.* San Francisco: Chronicle Books, 1995.

Hill, Kim, and Hillard Kaplan. "Population and Dry-Season Subsistence Strategies of the Recently Contacted Yora of Peru." *National Geographic Research* 5 (1989): 317–34.

Hilleman, Maurice, Eugene Buynak, Robert Weibel, Joseph Stokes, James Whitman, and M. Bernice Leagus. "Development and Evaluation of the Moraten Measles Virus Vaccine." *Journal of the American Medical Association* 206 (1968): 587–90.

Hilts, Philip J. "Secret Radioactive Experiments to Bring Compensation by U.S.," *NYT,* Nov. 20, 1996.

Hitchcock, Charles. *La región Orinoco-Ventuari: Relato de la expedición Phelps al Cerro Yavi.* Caracas: Ministerio de Educación Nacional, Dirección de Cultura, 1948.

Hoekenga, M., A. Schwartz, H. Carrizo-Palma, and P. Boyer. "Experimental Vaccination against Measles II: Tests of Live Measles and Live Distemper Vaccine in Human Volunteers during a Measles Epidemic in Panama." *Journal of the American Medical Association* 173 (1960): 862–68.

Holmes, R. "Nutritional Status and Cultural Change in Venezuela's Amazon Territory." In *Change in the Amazon Basin,* ed. J. Hemming, pp. 237–55. Manchester: Univ. of Manchester, 1985.

Horgan, John. "The Violent Yanomamo." *Scientific American,* March 1988, p. 18.

Hornick, Richard, Ann Schluederberg, and Fred McCrumb. "Vaccination with Live Attenuated Measles Virus." *American Journal of Diseases of Children* 103 (1962): 344–47.

Howell, Nancy. Review of *The Population Dynamics of the Mucajai Yanomama,* by John D. Early and John F. Peters. *Canadian Review of Sociology and Anthropology* 28 (1991): 151–52.

Hrdy, Susan Blaffer. *Mother Nature: A History of Mothers, Infants and Natural Selection.* New York: Pantheon, 1999.

Hurtado, Albert L. *Indian Survival on the California Frontier.* New Haven: Yale Univ. Press, 1988.

Ibarra, Carlos Alamo. *Río Negro.* Caracas: Tipografía Vargas, 1950.

Ireland, Corydon. " 'No Bad Guys' in Study: Doctor Defends Good Intentions." *Rochester Democrat & Chronicle,* Dec. 16, 1994.

———. "Radiation Records at UR Missing." *Rochester Democrat & Chronicle,* Jan. 1, 1995.

Keeley, L. *War before Civilization.* Oxford: Oxford Univ. Press, 1995.

Kline, Elizabeth. *Mining Abuses Tarnish Venezuela's Environmental Image.* Caracas: Sociedad Conservacionista Audubon de Venezuela, 1994.

Kroeber, Theodora. *Ishi in Two Worlds: A Biography of the Last Wild Indian in North America.* Berkeley: Univ. of California Press, 1961.

Krugman, Saul, Joan Giles, and Milton Jacobs. "Studies on an Attenuated Measles Vaccine-Virus," *New England Journal of Medicine* 263 (July 28, 1960): 174.

———. "Studied with Live Attenuated Measles-Virus Vaccine." *American Journal of Diseases of Children* 103 (1962): 353–63.

Krugman, Saul, Joan Giles, A. Milton Jacobs, and Harriet Friedman. "Studies with a Further Attenuated Live Measles-Virus Vaccine." *Pediatrics* 31 (1963): 919–28.

Larrick, J., J. A. Yost, J. Kaplan, G. King, and J. Mayhall. "Patterns of Health and Disease among Waorani Indians of Eastern Ecuador." *Medical Anthropology* 3 (19): 147–89.

Leal, Adela. "Deja Comisión Yanomami Charles Brewer Carías: Reestructurán el decreto presidencial." *El Nacional,* Sept. 14, 1993.

Letts, Quentin. "Spurning the Good Life for Call of the Wild." *Times* (London), Jan. 31, 1997.
Lévi-Strauss, Claude. *Tristes Tropiques.* Translated by John and Doreen Wightman. New York: Atheneum, 1974.
———. *Saudades do Brasil.* Seattle: Univ. of Washington Press, 1995.
Lizot, Jacques. "Aspects économiques et sociaux du changement culturel chez les Yanómami." *L'Homme* 11, no. 1 (1971): 32–51.
———. "El rio de los Periquitos: Breve relato de un viaje entre los yanomami del alto Siapa." *Antropológica* 37 (1974): 3–23.
———. *The Yanomami in the Face of Ethnocide.* Copenhagen: IWGIA, 1976.
———. "Population, Resources, and Warfare among the Yanomami." *Man* 12 (1977): 497–517.
———. *Tales of the Yanomami: Daily Life in the Venezuelan Forest.* New York: Cambridge Univ. Press, 1985.
———. "Sobre la guerra: Una respuesta a N. A. Chagnon (Science, 1988)." *La Iglesia en Amazonas,* April 1989, pp. 23–34.
———. "On Warfare: An Answer to N. A. Chagnon." Translated by Sarah Dart. *American Ethnologist.* 21 (1994): 845–62.
———. "N. A. Chagnon, o sea: Un presidente falsificador." *La Iglesia en Amazonas,* March 1994, p. 14.
López, Edgar. In *El Diario de Caracas,* Sept. 2, 1993.
Lorenz, Konrad. *On Aggression.* New York: Harcourt Brace Jovanovich, 1966.
Madi, Isaam. *Conspiración al sur del Orinoco.* Caracas: self-published, 1998.
Martínez, Nahir. "Health Problems in Isolated Yanomami Communities: Viral Hepatitis in the Upper Orinoquito River." CAICET: Puerto Ayacucho, 1996.
Martínez, Wilton. "The Challenges of a Pioneer: Tim Asch, Otherness, and Film Reception." *Visual Anthropology Review* 11, no. 1 (Spring 1995): 53–82.
Martins, Leda. "Ciúme na floresta:" *A Gazeta de Roraima,* March 18–24, 1996, pp. 5–8.
Martin, Leda, and Patrick Tierney, "El Dorado: Lost Again?" *NYT,* April 7, 1995.
Maugh II, Thomas. "Homicidal Streak in S. American Tribe Studied by Anthropologist." *Los Angeles Times,* Feb. 26, 1988.
Mayorca, Javier Ignacio. In *El Nacional,* Aug. 19, 1998.
McCauley, Clark. "Conference Overview." In *The Anthropology of War,* ed. J. Haas, pp. 2–6. Cambridge: Cambridge Univ. Press, 1990.
McCrumb, Fred, Sheldon Kress, Elijah Saunders, Meril Snyder, and Ann Schluederberg, "Studies with Live Attenuated Measles-Virus Vaccine," *American Journal of Diseases of Children* 101 (1961): 45.
Misioneros del Alto Orinoco. *Consideraciones a un documento de Charles Brewer Carías.* Mavaca: Salesian Mission, 1991.
Mitus, Anna, Ann Holoway, Audrey Evans, and John Enders. "Attenuated Measles Vaccine in Children with Acute Leukemia." *American Journal of Diseases of Children* 103 (1962): 413–18.
Monaghan, Peter. "Bitter Warfare in Anthropology." *Chronicle of Higher Education,* Oct. 26, 1994, p. A10.
Moore, John H. "The Reproductive Success of the Cheyenne War Chiefs: A Contrary Case to Chagnon's Yanomamo." *Current Anthropology* 31 (1990): 322–30.
Morgan, Karl Z., and Ken M. Peterson. *The Angry Genie: One Man's Walk through the Nuclear Age.* Norman: Univ. of Oklahoma Press, 1999.
Nagler, P., A. R. Foley, J. Furesz, and G. Martineau. "Studies on Attenuated Measles-Virus Vaccine in Canada." *Bulletin of the World Health Organization* 32 (1965): 791–801.
Naipaul, V. S. *The Loss of El Dorado.* New York: Penguin, 1987.
Neel, James V. "Lessons from a Primitive People." *Science* 170 (1970): 815–22.
———. "On Being Headman." *Perspectives in Biology and Medicine* 23 (1980): 277–94.
———. *Physician to the Gene Pool: Genetic Lessons and Other Stories.* New York: John Wiley, 1994.
Neel, James V., and William Valentine. "Hematologic and Genetic Study of the Transmission of Thalassemia (Cooley's Anemia, Mediterranean Anemia). *Archives of Internal Medicine* 74 (1944): 185–96.
———. "The Frequency of Thalassemia." *American Journal of Medical Science* 209 (1945): 568–72.
Neel, James V., and W. J. Schull. *The Effect of Exposure to the Atomic Bombs on Pregnancy Termination in Hiroshima and Nagasaki.* Washington, D.C.: National Academy of Sciences-National Research Council, 1956.
Neel, James V., Willard Centerwall, Napoleon Chagnon, and Helen Casey. "Notes on the Effect of Measles and Measles Vaccine in a Virgin Soil Population of South American Indians." *American Journal of Epidemiology* 91 (1970): 418–29.
O'Hanlon, Redmond. *In Trouble Again.* London: Hamish Hamilton, 1988.
Ochoa, G. Rodríguez. "Situación de salud en el Territorio Federal Amazonas, Venezuela." *Enfoque inte-*

gral de la salud humana en la Amazonia. Vol. 10, pp. 407–26. Montevideo, Uruguay: Editorial Trilce, 1992.

Peters, John. "The Effect of Western Material Goods upon the Social Structure of the Family among the Shirishiana." Ph.D. diss., Western Michigan Univ., 1973.

Pimentel, Olgalinda. "Denunciaron ante el Fiscal al ex-Ministro Brewer Carías." *El Diario de Caracas,* Aug. 4, 1984.

Plotkin, Stanley A., and Edward A. Mortimer. *Vaccines.* Philadelphia and London: W. B. Saunders, 1994.

Posey, Darrel Addison. "Environmental and Social Implications of Pre- and Postcontact Situations on Brazilian Indians." *Amazonian Indians,* ed. Anna Roosevelt, pp. 271–86. Tucson: Univ. of Arizona Press, 1994.

Powers, Thomas. "Die Hard." *Los Angeles Times,* http://latimes.com, Jan. 2, 2000.

Rabben, Linda. *Unnatural Selection: The Yanomami, the Kayapó and the Onslaught of Civilization.* London: Pluto Press, 1998.

Ramos, Alcida. "Reflecting on the Yanomami: Ethnographic Images and the Pursuit of the Exotic." *Cultural Anthropology* 2 (1987): 284–304.

Redmond, Elsa. *Tribal and Chiefly Warfare in South America.* Ann Arbor: Univ. of Michigan, Museum of Anthropology, 1994.

———. "In War and Peace." In *Chiefdoms and Chieftaincy in the Americas,* ed. Elsa Redmond, pp. 68–103. Gainesville: Univ. Press of Florida, 1998.

Reinoso, Víctor Manuel. " 'Me rechazan por envidia' asegura Charles Brewer Carías." *El Nacional* (Caracas), Sept. 16, 1993.

Reiss, Spencer. "The Last Days of Eden: The Yanomamo Indians Will Have to Adapt to the 20th Century—or Die." *Newsweek,* Dec. 3, 1990, pp. 40–42.

Reiss, Tom. "The Man from the East." *New Yorker,* Oct. 4, 1999, pp. 68–79.

Rensberger, Boyce. "Sexual Competition and Violence." *Washington Post,* Feb. 29, 1989.

Rice, Hamilton A. "The Rio Negro, the Casiquiare Canal, and the Upper Orinoco, September 1919–April 1920," *Geographical Journal* 58 (1921): 321–44.

———. "The Rio Branco, Uraricoera, and Parima, Part 3." *Geographical Journal* 71 (1928): 345–56.

Ridley, Matt. *The Red Queen.* New York: Penguin, 1993.

Rifkin, Jeffrey. "Ethnography and Ethnocide." *Dialectical Anthropology* 19 (1994): 295–327.

Rita Ramos, Alcida. "Reflecting on the Yanomami: Ethnographic Images and the Pursuit of the Exotic." *Cultural Anthropology* 2 (1987): 284–304.

Ritchie, Mark. *Spirit of the Rainforest: A Yanomamo Shaman's Story.* Chicago: Island Lake Press, 1995.

Riviére, R., D. Comar, M. Colonomos, J. Desenne, and M. Roche. "Iodine Deficiency without Goiter in Isolated Yanomama Indians, Preliminary Note." *Biomedical Challenges Presented by the American Indian.* Washington, D.C.: Pan American Health Organization, 1968.

Robarchek, Clayton, and Carole Robarchek. *Waorani: The Contexts of Violence and War.* New York: Harcourt Brace, 1998.

Roosevelt, Anna C. "Strategy for a New Synthesis." In *Amazonian Indians,* ed. Anna C. Roosevelt, pp. 1–29. Tucson: Univ. of Arizona Press, 1994.

Ruby, Jay. "Out of Sync: The Cinema of Tim Asch." *Visual Anthropology Review* 11, no. 1 (Spring 1995): 19–35.

Saffirio, Giovanni. "Ideal and Actual Kinship Terminology among the Yanomama Indians of the Catrimani River Basin (Brazil)." Ph.D. diss., Univ. of Pittsburgh, 1985.

Salamone, Frank. *The Yanomami and Their Interpreters: Fierce People or Fierce Interpreters?* Lanham, Md.: Univ. Press of America, 1997.

Salopek, Paul. "Basically We Are All the Same." *Chicago Tribune,* April 27, 1997.

Salzano, Francisco M., and Sidia M. Callegari-Jacques. *South American Indians: A Case Study in Evolution.* Oxford: Clarendon Press, 1988.

Sanford, Greg. "A New Tribes Perspective." In *The Yanomami and Their Interpreters,* ed. Frank Salamone, pp. 57–66. Lanham, Md.: Univ. Press of America, 1997.

Schemo, Diana Jean. "In Brazil, Indians Call on Spirits to Save Land," *NYT,* July 21, 1996.

———. "Legally Now, Venezuelans to Mine Fragile Lands." *NYT,* Dec. 8, 1995.

Seaman, Gary. "First Comments" and "Second Comments." *Yanomamo Interactive CD.* New York: Harcourt Brace, 1997.

———. "Blow-by-Blow." *Yanomamo Interactive CD.*

Smole, William J. *The Yanoama Indians: A Cultural Geography.* Austin: Univ. of Texas Press, 1976.

Spencer, Herbert. *Principles of Biology.* Vol. 2. London: William and Norgate, 1867.

Sponsel, Leslie E. "Yanomami: An Arena of Conflict and Aggression in the Amazon," *Aggressive Behavior* 24 (1998): 97–122.

Stannard, David E. *American Holocaust: Columbus and the Conquest of the New World.* New York: Oxford Univ. Press, 1992.
Steinvorth Goetz, Inga. *Uriji Jami!: Life and Belief of the Forest Waika in the Upper Orinoco.* Translated by Peter Furst. Caracas: Asociación Cultural Humboldt, 1969.
Stewart, Thomas Dale. *The People of America.* London: Weidenfeld and Nicolson, 1973.
Tierney, Patrick. *The Highest Altar: The Story of Human Sacrifice.* New York: Viking, 1989.
Trompf, G. W. Introduction. *Cargo Cults and Millenarian Movements: Transoceanic Comparisons of New Religious Movements,* ed. G. W. Trompf. New York: Mouton de Gruyter, 1990.
Turner, Terence. Interview with Davi Kopenawa, Boa Vista, Brazil, 1991. Portions published in *Cultural Survival Quarterly* 15 59–64, and *Anthropology Newsletter,* p. 52.
———. "The Yanomami: Truth and Consequences." *Anthropology Newsletter,* May 1994, p. 48.
Utrera, Orlando. "Brewer denuncia el 'Plan Gadhafi.' " *El Diario de Caracas,* Aug. 15, 1984.
Valero, Helena. *Yo soy Napeyoma: Relato de una mujer raptada por los indígenas yanomami.* Caracas: Fundación La Salle de Ciencias Naturales, 1984.
Van Vuner, Chris J. "To Fight for Women and Lose Your Lands: Violence in Anthropological Writings and the Yanomami of Amazonia," *Unisa Largen* 10, no. 2 (July 1994): 2.
Vegas, Tania. "Brewer Carías ha devastado zonas protectoras en Guayana." *El Universal,* April 13, 1992.
Vinci, Alfonso. *Red Cloth and Green Forest.* Translated by James Cadell. London: Hutchinson, 1959.
Visconti, José. "Los salesianos vetan a Brewer Carías y Chagnon." *El Diario de Caracas,* Sept. 18, 1993.
Wald, Matthew L. "U.S. Acknowledges Radiation Killed Weapons Workers." *NYT,* Jan. 29, 2000.
Walter Raleigh. *The Discoverie of the Large, Rich, and Bewtiful Empire of Guiana,* ed. V. T. Harlow. London: Haklyut Society, 1928.
Weatherall, David. "The Mutation Man." *New Scientist,* July 9, 1994, p. 42.
Webster, Donovan. "The Orinoco." *National Geographic,* April 1998. pp.
Welsome, Eileen. *The Plutonium Files: America's Secret Medical Experiments in the Cold War.* New York: Dial Press, 1999.
Wilbert, Johannes. *Survivors of El Dorado: Four Indian Cultures of South America.* New York: Praeger, 1972.
Wilson, Edward O. *Consilience: The Unity of Knowledge.* New York: Alfred A. Knopf, 1998.
Wilson, G. S. "Measles as a Universal Disease." *American Journal of Diseases of Children* 53 (1962): 219–23.
Wolf, Eric R. "Demonization of Anthropologists in the Amazon." *Anthropology Newsletter,* March 1994, p. 2.
Wrangham, Richard W., and Paul G. Peterson. *Demonic Males: Apes and the Origins of Human Violence.* Boston: Houghton Mifflin, 1996.
Wyden, P. *Day One: Before Hiroshima and After.* New York: Simon & Schuster, 1984.

Government Documents

Comisión Investigadora Venezolana. "Supuesto asesinato de ciudadanos venezolanos de la etnia yanomami por ciudadanos brasileños." Caracas, 1993.
Gaceta Oficial de la Republica de Venezuela. 1991, 1993, 1997.
Milano (Jefe), Mayor Sergio Rafael, Teniente Luis Alberto Godoy y Geraldi Antonio Villaroel (Secretario). *Expediente de la Comisión de la Guardia Nacional.* Fuerzas Armadas de Cooperación, Comando Regional 6, Destacamento de Frontera No. 61. Puerto Ayacucho, April 18, 1984.
Ministerio da Justica de Brasil, Fundacao Nacional do Indio, Autorizacao para ingreso em área indígena No. 059/CGED/95. September 1, 1995. Memo No. 239/240/CGEP/95. Otília María C. E. Nogueira, Coordenadora Geral de Estudos e Pesquisas.
Paver, Fran L., Attorney-Advisor, National Science Foundation. FOIA No. 95-004, July 13, 1995.
Ponte, Ana. "Charles Brewer Carías: Informe para el Comité del Medio Ambiente del Senado." Caracas: Jan. 1997.
Thomassen, David. DOE, Office of Energy Research. Personal correspondence in reply to FOIA request No. 9501260003, March 13, 1994.
U.S. Department of Energy. Human Radiation Experiments. http://hrex.dis.anl.gov.

Films and Documentaries

Asch, Timothy, and Napoleon Chagnon. *The Feast.* 29 min. Washington, D.C.: DOE, 1970.
———. *Kaobawa Trades with Reyabobowei-teri.* 8 min. Somerville, Mass.: DER, 1971.
Briceño, Nelson. Camcorder interview with Marta Miranda Rodrigues. Ocamo Mission, May 13, 1991.
Chagnon, Napoleon. *Magical Death,* 28 min. Watertown, Mass.: DER, 1973.

Miranda, Marta Rodríguez. Venevisión. Caracas. Fundación Cultural Venevisión, July 24, 1991.
Neel, James, Timothy Asch, and Napoleon Chagnon. *Yanomama: A Multidisciplinary Study.* 43 min. Washington, D.C.: DOE, 1971.
Public Lands, Private Profits. 45 min. *Frontline.* WGBH, Boston, 1994.
Ritchie, Mark. Video of Kaobawa, Padamo River, Jan. 1995.
———. Video Interviews, Padamo River, Jan. 21, 1995.
Survivors of the Amazon. 50 min. Director, Andy Jillings. *BBC 4.* London, 1996.
Warriors of the Amazon. 45 min. Director, Andy Jillings. *Nova.* WGBH, Boston, 1996.
Window on the Past. 12 min. ABC *Prime Time Live.* July 26, 1991.
Yanomami Homecoming. 48 min. *National Geographic Explorer.* Washington, D.C.: National Geographic Society, 1994.

Unpublished Sources

Albert, Bruce. Letter to Patrick Tierney, Dec. 15, 1994.
Arvelo Jiménez, Nelly. "The Repudiation of Brewer Carías and Chagnon Is Due to Their Intimate Association with Goldmining." Caracas: IVIC, 1994.
———. Letter to Dr. Gale Goodwin-Gómez, Sept. 29, 1994.
Asch, Timothy. "Bias in Ethnographic Reporting."
Asch, Timothy. Collection. Washington, D.C.: Smithsonian National Anthropological Archives.
Berno, José. "Crónica de la casa de Mavaca," 1968–72.
Bórtoli, José. "Sumario de la Crónica de la casa de Mavaca para julio/agosto 1971."
Brewer Carías, Charles. Curriculum en Antropología, Sept. 3, 1993.
———. Correspondence to the Office of the Presidency, Nov. 18, 1994.
Brewer Carías, Charles, Napoleon Chagnon, and Brian Boom. "Forest and Man." MS; Caracas: Fundación Explora, 1993.
Chagnon, Napoleon. Letter to Padre José Bórtoli, July 19, 1988.
———. "The Guns of Mucajaí: The Immorality of Self-deception," Sept. 1992.
———. Letter to Ramades Muñoz León, Oct. 2, 1993.
———. Letter to Robin Hanbury-Tennison, Oct. 29, 1993.
———. "Conversation with Jesús Cardozo" (No. 9A, press package), March 23, 1994.
———. "Notes on Chronology of Recent Attacks on Members of the Venezuelan Presidential Commission by Salesian Missionaries, French, Brazilian and Venezuelan Anthropologists" (Overview of press package), May 18, 1994.
———. E-mails to Liz McMillen, Aug. 18 and 23, 1994.
De Souza, Marinho. Malaria Census, CAISET, Sept. 1996.
Eibl-Eibesfeldt, Irenaus, and Gabriele Herzog-Schroder. Letter to Bishop Ignacio Velasco, Feb. 28, 1994.
Galé, Nelson. Letter to president of FUNAI, Márcio Santilli, Sept. 27, 1995.
González, Nora. "Crónica de Mavaca." 1968.
González, Maria Wachtler. "Crónica de Mavaca." 1972.
Good, Kenneth. "Demography and Land Use among the Yanomamo of the Orinoco-Siapa Block in Amazon Territory, Venezuela." 1984.
———. Letter to the editor of *NYT,* Sept. 29, 1990.
———. Letter to the editor of *NYT,* Sept. 13, 1991.
Hames, Raymond, Kim Hill, and Ana Magdalena Hurtado. "Defamation Campaign against Napoleon A. Chagnon." rhames@unlinfo.unl.edu. May 1994.
Hill, Kim. "Response to Cardozo and Lizot." March 1994.
Lebrún, Cardenal José A., monseñores Ovidio Pérez Morales, Baltasar Porras y Mario Moronta. "Documento Oficial de la Conferencia Episcopal Venezolana." Universidad Católica Andrés Bello, Caracas, Sept. 11, 1993.
Lizot, Jacques. Letter to Patrick Tierney. Jan. 20, 1995.
———. "N.A. Chagnon, O Sea: Un presidente falsificador." Letter Dec. 13, 1993.
Mari, Antonio. Letters to Patrick Tierney, May 10 and May 12, 2000.
Misioneros del Alto Orinoco. "Consideraciones a un documento de Charles Brewer-Carías." Puerto Ayacucho, 1991.
Ridley, Matt. Fax to Napoleon Chagnon (No. 18 in Chagnon's press kit), Aug. 16, 1994.
Smith, Gare. Letter to Patrick Tierney, Nov. 16, 1998.
Tierney, Patrick. Letter to Gare Smith, Principal Deputy, Assistant Secretary, Bureau of Democracy Human Rights, and Labor. Aug. 24, 1998.
University of California at Santa Barbara. http://www.sscf.uscb.edu/anth/projects/axfight/index.

Interviews

Aherowe, Alfredo. Platanal, Upper Orinoco, June 11, Sept. 3, 24, and 26, 1996.
Albert, Bruce. Toototobi, Brazil. Dec. 5, 1990.
Alés, Catherine. Caracas, Aug. 21, 1996.
Anderson, Aurora. Mucajaí mission, Brazil, May 1990.
Arenas, Alejandro, M.D. Bisaasi-teri, Upper Orinoco, June 7, 1997.
Asch, Patricia. Phone, May 16, 1997.
Balthasar, Vitalino. Puerto Ayacucho, Oct. 1, 1996.
Berno, José. Puerto Ayacucho, Oct. 5, 1996.
Betancourt, Adelfa. Phone, April 6, 2000.
Black, Francis. Phone, March 17, 1997.
Boehm, Christopher. Phone, April 9, 1996.
Bokoramo, Ramon. Kedebabowei-teri, Mrakapiwei, June 8, 1996.
Bórtoli, José. Mavaca mission, June 6 and 11, 1996; Platanal mission, Sept. 26, 1996.
Botto, Carlos. CAICET, Puerto Ayacucho, Oct. 6, 1996.
Brewer-Carías, Charles. Phone, Jan. 3, 1995.
Cable, Bob and Gay. Phone, Jan. 3, 1995.
Camargo, Josefa. Phone, Dec. 19 and 23, 1995.
Camargo, Milton. Phone, Feb. 14, 1996.
Cardozo, Jesús. Phone, Dec. 20, 1994, June 21, Aug. 8 and 31, and Sept. 1, 1995.
Carrera, Javier. Phone, Feb. 12, 1997.
Chagnon, Napoleon. Univ. of California at Santa Barbara, Oct. 2 and 3, 1995.
Collins, Terence. Phone, Feb. 13, 2000.
Costa, Michelle Rodríguez. Surucucu army post, Brazil, Nov. 1990.
Damiolli, Guillerme. Catrimani mission, June 19, 1989.
Dawson, Gary. Padamo mission, June 4 and Aug. 29, 1996.
Dawson, Jodie. Platanal, with Mahekototeri and Patanowateri elders, June 11, 1996.
Dawson, Joe. Padamo mission, June 14, 1996.
Dawson, Marie. Padamo mission, June 5, 1996.
Dawson, Michael. Padamo mission, June 4, 1996.
Dawson, Wilma. Puerto Ayacucho, June 3, 1996.
De Souza, Marinho. Karohi-teri, Sept. 1, 1996; Hokomapiwe-teri, Sept. 9, 1996.
Dimanawa, César. Mavakita, June 8 and Sept. 2, 1996.
Earle, John. Pittsburgh, March 7, 1997.
Etilio, Guarapana, Boca Mavaca, June 9, 1996.
Felicia, Sister. Puerto Ayacucho, Oct. 4, 1996.
Ferguson, Brian. Phone, Jan. 3 and July 13, 1995, April 19 and May 26, 1996, June 14, 1997. March 4, 2000.
Finkers, Juan. Phone, Jan. 11 24, 1995; Mavaca mission, June 12, 1996.
Gonzalez, Juan. Padamo mission, June 14, 1996.
Good, Kenneth. Phone, Jan. 10, Feb. 1 and 22, and Aug. 8, 1995, Jan. 30–31, Feb. 4, April 17, and May 14, 1997, Feb. 27 and March 25, 1998, Aug. 5, 1999.
Griffiths, Paul. Puerto Ayacucho, June 3, 1996.
Hames, Raymond. Phone, Dec. 29, 1994.
Hetoyaw. Between Narimobowei-teri and Ashidowa-teri, Sept. 16, 1996.
Hill, Kim. Phone, Jan. 17, 1995.
Howell, Nancy. Phone, Feb. 1, 1995.
Ireland, Corydon. Phone, Feb. 4, 2000.
Isabela. Mavakita, Sept. 2, 1996.
Jepewe. Karohi-teri, Manaviche River, June 7, 1996. Translated by Jodie Dawson.
Jillings, Andy. Phone, Feb. 18, 1997.
Jiménez, Marco. Near Hokomapiwe-teri, Sept. 9, 1996; Toobatotoi-teri, Sept. 13, 1996; Ashidowa-teri, Sept. 17, 1996.
Kaobawa. Shakita, Upper Orinoco. June 12, 1996.
Karakawe, Alberto. Ocamo Mission, Aug. 31, 1996.
Katz, Samuel. Phone, March 19, 1997.
Kayopewe. Platanal, June 11, 1996. Translated by Jodie Dawson.
Konoko, Gustavo. Mavakita, Sept. 2, 1996.
Kopenawa, Davi. Boa Vista, Brazil, Nov. 3, 1990.
Lucho, Enrique. Kosh, Padamo River, June 14, 1996.

Maier, John. Macuxi village of Cajú. July 1, 1996.
Martins, Leda. Pittsburgh, March 7 and Sept. 15, 1995.
Mejía, Pablo. Padamo mission, June 4, 1996; Mavakita, June 8, 1996; Toki, Aug. 28, 1996; Shakita, Sept. 1, 1996; Patahama-teri, Sept. 5, 1996.
Milano, Colonel Sergio. Phone, Dec. 12, 1994.
Mirapewe. Ashidowa-teri, Sept. 17, 1996.
Neel, James. Phone, March 18, 1996.
Papania, Mark. Phone, May 22, 1996.
Pedro of the Opiktheri and Mario of Pacu. Catrimani mission, Brazil, June 21 and 22, 1989. Translated by Guillerme Damiolli.
Perdamo, America. Puerto Ayacucho, June 16–17, 1996.
Peters, John. Phone, Jan. 3, 1995.
Poortman, Ysbran. Rio Negro near Manaus, Brazil, Aug. 17, 1996.
Renaldo. Karohi-teri splinter group, Manaviche River, June 7, 1996.
Reyes, Pablo. Puerto Ayacucho, June 17, 1996.
Ritchie, Mark. Phone, Feb. 6, 1995.
Roche, Marcel. Caracas, June 20, 1996.
Saffirio, Giovanni. Pittsburgh, June 25, 1994; Riverside, California, Jan. 1, 2000.
Salamone, Frank. Phone, Dec. 22, 1994.
Salazar, Rafael and Teresa. Pittsburgh, April 30, 1996.
Shaylor, Danny. Tama Tama, Upper Orinoco, Sept. 26, 1996.
Sponsel, Leslie. Phone, Jan. 12 and Sept. 15, 1995.
Stannard, Jay. Phone, Feb. 4, 2000.
Taylor, Kenneth. Phone, Jan. 27, 1995.
Timoteo. Padamo mission, June 4, 1996.
Turner, Terence. Pittsburgh, March 29, 1995; Univ. of Chicago, Sept. 21, 1995.
Valero, Helena. Upper Orinoco, Aug. 31, 1996.
Wachtler, Maria. Phone, June 20, 1996.
Walden, John. Phone, Jan. 3, 1995; March 4, 1997.
Waloiwa, Shakita, Upper Orinoco, June 9, 1996. Translated by Jodie Dawson.
Waupuruwe. Narimobowei-teri, Sept. 11–12, 1996.
Welsome, Eileen. Phone, Feb. 3, 2000.
Wilson, Margot. Phone, April 10, 1996.
Yanowe. Ashidowa-teri, Sept. 16, 1996; Dorita-teri, Sept. 21, 1996.
Yarima. Irokai-teri, Sept. 7, 1996.

Index

ABA (Brazilian Anthropological Association), 160
ABC network, 4–5, 187–88
"Academic Extermination of the Yanomami, The" (Albert and Ramos), 8–9
Achuar people, 23
Adulimawa-teri, 149
Aebersold, Paul, 305–6, 310–13
AEC, *see* Atomic Energy Commission, U.S.
Africa, 19, 63, 264, 297
African Queen, 131
AFVI (American Friends of the Venezuelan Indians), 189
age, 51–53, 159–60, 162–63, 173–77
  estimation of, 176
  in marriage statistics, 163, 173–76, 360*n*–61*n*
  in violence statistics, 162, 175–77, 267, 360*n*
Aguilera, Nilda, 10, 200, 202, 209
Aherowe, Alfredo, 87, 250, 259–60, 293
Aherowe, Marco, 250, 260, 276, 283–85, 289
AIDS testing, 145
Air Force, Venezuelan, 192, 324–25
airplanes, 27, 88, 125, 127, 130–31, 182, 211, 212, 229–30, 244, 280–82
  nuclear, 307, 310, 381*n*
airstrips, xxiv, 8, 208, 237, 270, 285–86, 323–25, 377*n*
Ajuricaba, 19, 269
Albania, 163
Albert, Bruce, 162, 164, 165–66, 176, 179, 191, 202, 210, 281–82, 328*n*–29*n,* 356*n*
Albright, Madeleine, xxv
Alès, Catherine, 144–45
Alexander, Richard, 207
Allen, Elmer, 302
alliances, military, 14, 29–30, 86–88, 92–93, 101–4, 105, 111–14, 121, 142, 216, 220–22, 293
  in pan-Yanomami politics, 235–37, 293, 323
Allman, William, 176

Almada, Lobo de, 19, 91
Altino Machado, Ze, 154
Amazonas State, Venezuela, 155–56, 197–99, 290–91, 325
Amazon gold rush, xxiii–xxv, 8–11, 45, 152–58, 186, 191, 199–201, 208–11
  disease linked to, xxiv, 8, 191, 209, 210, 281–84, 288, 383*n*
  see also Haximu-teri, massacre of; mining
Amazonian civilizations, prehistoric, 156–57, 265
Amazon Indians, Congress of (1993), 197–98
Amazon-Orinoco canal project, 21
America, conquest of, 5, 12, 32, 48, 57, 107–8, 150, 265
American Anthropological Association, 11, 32, 161, 202, 205, 229, 236, 249, 274, 369*n*
American Ethnologist, 164, 179
American Film Festival, 113–14
American Film Quarterly, 102
American Friends of the Venezuelan Indians (AFVI), 189
American Journal of Epidemiology, 61–62, 79
American Journal of Physical Anthropology, 270
American Museum of Natural History, 130, 155–56
Amerindians, 8, 23–25, 28, 126–27, 152, 281
  artistry of, 126–27, 265
  body size of, 262–64, 374*n*
  captives of, 255–56
  decimation of, xxiii, 19–20, 39, 50, 59, 67–69, 72, 94–95, 104–5, 108, 120–22, 190, 291, 317–18
  immune response of, 56–60, 66–69
  murder among, 13, 23, 29, 41, 126–27, 178
  polygamy of, 174–75
  prehistory of, 265
  spiritual movements of, 282
  see also Yanomami people
Amnesty International, 113, 127
amoamou (chanting), 102
Anderson, Aurora, 212
Andrés Pérez, Carlos, 9–10, 181–82, 189, 194, 196–98, 329*n,* 332*n,* 364*n*
Andujar, Claudia, 160
anemia, 311
Angel Falls, Venezuela, 152, 200
animal behavior, 13, 14, 16, 22, 40, 179, 204, 260
animal sacrifices, 163
animal smuggling, 155, 193–94
animal spirits, 161, 285
animal testing, 298, 300–301, 314
anopheles mosquitoes, 153, 318
anthro (exploitive outsider), 14, 48
anthropological films, *see* filmmaking, ethnographic
anthropologists, behavior of, 14–17, 89–90, 105–6, 125–48, 202–8, 248–49, 313–15
  aggressive, 16, 32, 47–48, 88, 142, 144, 147, 203–4
  sexual, 16, 129, 132–37, 143–48, 222–23, 341*n*–43*n*
"Anthropologist Underscores Violence among Indians," 160–61
anthropology:
  advocacy in, xxiii–xxv, 19, 127, 138, 147, 161, 164, 184–85, 252, 325–26
  American vs. French, 15, 84, 126–27, 147–48
  American vs. South American, 160–61
  as black magic, 191–94, 313–15
  controversies in, 8–9, 11–12, 42, 91–92, 113, 127, 157–61, 176–80, 182–86, 201–14, 261–74
  first contact craze in, 5, 107–8, 187, 248–49, 289–90
  literary models for, 147–48
  nostalgic basis of, 108–9, 314
  policy implications of, 8, 14, 42, 127, 159–61, 313–14, 325
  research protocols in, xxii, 32, 121–22, 187, 190–91
  structuralist, 126–27, 147
Anthropology Newsletter, 202
anthropometry, 59–60, 262–64
antibiotics, 55, 63, 68–69, 73, 96, 190, 312, 319
antibody levels, 55–56, 69, 79, 325
Antonio (*motorista*), 250
Apocalypse Now, 143, 219
Arata, José, 199–200
archaeology, 23–24, 155, 261–62, 265
Arenas, Alejandro, 43
Argonne Laboratory, 299, 302
Arimawu-teri, 220, 222–23
armed forces, Venezuelan, 11, 145–46, 181, 190–94, 200, 237, 250
Army, U.S., 21, 28, 91, 111, 281
Art of Speaking Well, The, see Warriors of the Amazon
Arvelo Jiménez, Nelly, 10, 154, 364*n*
Asch, Patricia, 71, 85
Asch, Timothy, 37, 41, 62, 85, 88–106, 110, 112–18, 127, 129, 142, 246, 322, 342*n*
  Chagnon criticized by, 88, 102, 112–13, 116, 118, 188–89, 348*n*
  events choreographed by, 14, 88, 93–94, 119, 293
  hunger endured by, 90, 272
  measles epidemic recorded by, 70–80, 95–96, 104–6, 312
  reputation of, 80, 119
  Yanomami filmmakers trained by, 253–54
Ashidowa-teri, 92, 97, 226, 288–91, 325–26, 336*n,* 378*n*
Asiawe, 104, 172, 293, 344*n*
Associated Press, 4, 325
Atomic Bomb Casualty Commission, 37, 302–4
Atomic Energy Commission, U.S. (AEC), 16–17, 48, 50, 59, 77–78, 301–10
  atomic survivor studies of, 16, 37–39, 42–45, 302–4

bones collected by, 307–9
in Cold War politics, 42, 44–45, 304–6, 310, 314
culture of, 306, 309–15
denial policy of, 303–5, 309
films bankrolled by, 37, 42, 70–71, 93–94, 102–3, 165, 292–93
genetic research sponsored by, 16–17, 36–37, 42–44, 59, 165, 297, 300, 304–10
lawsuits feared by, 309–10
radioisotope studies of, 305–10
results skewed by, 300, 304–5, 309
steel goods supplied by, 16, 51–52, 89, 114, 118, 172, 246, 293
Yanomami studies of, 16–17, 36–37, 42–52, 59, 70–71, 77–78, 93–95, 165, 172, 246, 262, 292–93, 297, 300, 304–10
Atoms for Peace, 305–6
Attorney General's Office, Venezuelan, 197, 200–201
Augustín, 36
Auyán-tepuí, 152
Avila Mountains, 144
axes, 5–6, 157, 168, 291, 292–94
fight with, 113–18, 175–76, 272
society altered by, 228–29, 266
as trade goods, 21, 30, 85, 92, 105–6, 111, 117, 130, 229
Ax Fight, The, 114–21, 168, 233, 272, 324, 349
Ayamanes Indians, 264
Aztec empire, 7

bachelors, permanent, 171–72, 175–76
Bacon, Roger, 296
badao (without cause), 166
Bakotawa, 33
Balthasar, Roberto, 53–54, 63–65, 80
Balthasar, Vitalino, 53–54, 64–65, 66, 69, 74
bananas, 162, 257–58, 261, 275
Barahiwa, 173
Bassett, Sam, 299–300
Batesian mimicry, 179
BBC television, xxi, 14–15, 183, 215, 219–23
"Beastly or Manly?," 160
beetles, collecting of, 45, 100, 106
Bering Strait, 265
Berkeley, University of California at, 108, 291, 297, 307, 309
Berno, José, 64–65, 78, 111, 113, 345
Betancourt, Adelfa, 82
"Bias in Ethnographic Reporting" (Asch), 9, 102
Biella, Peter, 115
Billy the Kid, 171
Biocca, Ettore, 244
biodiversity, 152, 155, 261
Biological Basis of Morality, The (Ridley), 203
biological determinism, *see* sociobiology
biosphere plan, 9–11, 154, 181–82, 187–94, 196–98
power struggle precipitated by, 233–40, 242, 254
birds, collecting of, 151, 182, 193
Bisaasi-teri, 23–24, 28–34, 46–48, 127–29, 137–39, 246, 267, 343*n*
disease in, 28, 51, 72, 74–75, 97, 119, 121, 318, 321–22
filmmaking in, 84–86
polygamy in, 172
wars of, 29, 33–34, 90, 100, 111–14, 141–42, 165–68, 217, 275, 293
"Bitter Warfare in Anthropology" (Monaghan), 11, 208
Black, Francis, 57–58, 318
Black Plague, 52
Blaffer Hrdy, Susan, 268–69
blood, 13, 325
collecting of, xxii–xxiii, 30, 43–52, 97, 99, 104, 106, 119, 193–94, 197, 291–92, 294–95, 299, 313–14, 325, 337*n*
Boas, Franz, 11–12
Boa Vista, Brazil, xxiv, 145, 182–83, 191, 209, 278
Boca Mavaca, 165, 167, 173, 270, 321–22, 357*n*
see also Mavaca mission; Mavaca River
Bocarohi-teri, 223
body temperature, 67–68, 89, 340*n*–41*n*
Boehm, Christopher, 163
Bogotá, Brewer's mission to, 151
Bokoramo, 231, 233–35, 241–42
Bolívar, Simón, 157
Bolívar Goldfields mining company, 200
Bolívar State, Venezuela, 153–54, 199–200
Bolivia, xxii, 229, 265
Boom, Brian, 291
bore (apparitions), 17, 218
Bórtoli, José, 143, 145–47, 205, 219, 254, 369*n*
Bosco, John, 129, 350*n*
Botto, Carlos, 65, 66, 68–69, 192, 290–91, 325
Bourquelot, Claude, 125–26, 147
Braorewa-teri, 286
Brasília, University of, 213
Brazil, xxi–xxv, 19, 39–41, 126–27, 151, 157–58, 325
corruption in, xxiv–xxv
in Haximu massacre affair, 195–96, 202, 208–11
human rights groups in, xxi–xxii, xxiv, 160
media in, xxii–xxiii, 160–61, 196
military in, xxii, 208–9
mining in, xxii–xxv, 181, 195–201, 208–11, 281–84
right-wing interests in, xxii-xxiii, xxv, 210
Brazilian Anthropological Association (ABA), xxi, 160
Brazilian Indian Agency, *see* FUNAI
Brewer Carías, Charles, 17, 83, 99–100, 149–57, 201, 237–40, 273
academic training of, 9, 41, 149, 182

Brewer Carías, Charles (*continued)*
biological specimens collected by, 155, 193–94, 274, 291
books authored by, 149–50, 153, 290
celebrities escorted by, 9, 290
El Dorado sought by, 156–57
expelled from Yanomamiland, 10, 200–201, 209
in FUNDAFACI biosphere plan, 9–10, 154, 181–82, 187–94, 196–98, 237–38, 283, 287, 290–91, 324–25, 332*n*
government positions held by, 150–51
in Haximu massacre affair, 196–202, 365*n*
Indian autonomy denounced by, 154–56, 187–88, 198, 362*n*–63*n*
Indian resistance to, 3–6, 292
in measles vaccine experiment, 75–78, 80, 96
mining interests of, xxv*n,* xxvi, 5, 152–57, 187, 191, 193–94, 197–201, 294, 310, 332*n,* 363*n*–65*n*
missionaries' conflict with, 186–88, 237, 290
public opposition to, 10, 197
species named after, 156
violent methods of, 9, 41, 150–51, 182, 190, 194, 205, 290, 293
Yanomami Reserve entrusted to, 196–98, 237
Brewer Explorer Survival Knife, 150
bride price, 174
bride service, 26–27, 232, 353*n,* 356*n*
bronchopneumonia, 62–65, 68, 89, 99, 108
Brooke, James, 187, 191, 289–91
Brues, Austin, 302, 310
brujos (sorcerers), 46–47
Bruno, 240
Burkert, Walter, 311

C-34 transport plane, 88, 127
Caballeros, Hortensia, 189
Cable, Bob, 212–14
Cable, Gay, 212–13
Cade, Ebb, 311
Caduveo people, 126–28
CAICET (Amazon Center for the Investigation and Control of Tropical Disease), 65, 69, 192–93, 290, 325, 373*n*
California malaria epidemic (1883), 95
Camargo, Josefa, 191, 199–201
Camargo, Milton, 214
cameras, 5, 157, 218, 221–22, 291–93
Canada, measles vaccine in, 55, 339*n*
Canadian mining concerns, 153
Canaima National Park, Venezuela, 200
cancer, 301, 309
Candia, Pedro de, 32
cannibalism, 21, 43, 311
spiritual, 47–48, 112–13, 134
Caracas, Venezuela, xxv, 144–45, 157, 181–83, 189, 191–94, 197–98, 255
coup attempt in, 11, 181, 193–94
Caraca-teri, 284–86, 289
Cardozo, Jesús, 15–16, 41–42, 142–46, 183–85, 188–90, 219, 234, 253–54, 324, 331*n*
cargo cults, 286, 378*n*
cargo planes, 88, 127, 130–31, 211, 212
Carib peoples, 19, 24
Carneiro, Robert, 130
Carneiro da Cunha, María Manuela, 160
Carrera, Javier, 284
Case Studies in Anthropology, 249
Casiquiare Canal, 65, 146, 152
cassiterite (tin ore) deposits, 154–55, 188, 213–14
Catrimani River, tribes of, 124, 161–63, 318, 327*n*–28*n,* 351*n*
CAV (Venezuelan anthropological association), 197
census taking, 11, 66–66, 145–46, 172, 175, 181, 190–94, 200, 237, 250, 295, 382*n*–84*n*
Centers for Disease Control, U.S., 79–80, 82
Centerwall, Willard, 95, 98–99, 102
Central University of Venezuela (UCV), 183, 197
ceramic fragments, 156–57, 265
Cerro Neblina, 12, 155–57
Cessna aircraft, 211, 229–30
Chagnon, Darius, 192, 325
Chagnon, Napoleon:
academic influence of, xxiii-xxiv, 7–9, 21–22, 34, 80, 130, 159, 203–4, 355*n*
academic opposition to, xxi, 8–9, 16, 85, 133, 137–38, 159–66, 175, 178–83, 186, 202–8, 265–67, 271–74, 289–90, 320
alliances brokered by, 14, 30, 86–88, 92–93, 101–4, 111–14, 121, 293
anthropologists denounced by, xxii-xxiii, xxv, 11, 16, 30, 164, 166–67, 179–80, 200–208, 265, 271, 357*n*
anti-Communism of, 11, 40–42, 179–80, 271, 313
attack dogs of, 130, 182
blood collected by, xxii-xxiii, 29–30, 37, 43–48, 97, 106, 119, 291–92, 294–95, 313, 337*n*
blood collected by, xxii-xxiii, 30, 37, 42–48, 97, 106, 119, 291–92, 294–95, 313, 337*n*
bride sought by, 31, 34
child abuse allegations against, 186, 231
in Conquista del Sur plan, 157, 182
data withheld by, 28, 35, 164, 179–80, 194, 267
death threats against, 5–6, 22, 32, 46, 85, 106, 115, 174, 184, 190, 231, 236, 292–94, 362*n*
doctoral dissertation of, 34–35, 117–18, 168, 171, 247, 265–66, 336*n*
education of, 9, 12, 40–41, 42, 204
empirical claims of, 28–29, 35, 120–21, 158–60, 162–69, 172–80, 205–7, 239, 247–48, 271–75, 320–26, 336*n,* 343*n,* 351*n,* 356*n*–60*n*

events choreographed by, 14, 85–88, 93–94, 101–6, 115–16, 166, 291
expelled from Yanomamiland, xxi, 10, 115, 138, 182–83, 200–203, 209, 231–32, 239
fighting prowess of, 12, 130, 182, 351*n*
films made by, 5, 14, 37, 42–43, 46, 49, 70–71, 85–106, 112–22, 129, 182, 186, 220, 272–73, 292–93, 310, 342*n*
first encounters described by, 7–8, 10, 21–22, 110–12, 282, 289–90
first encounters sought by, 107–10, 246, 248–49
fratricidal theory of, 117–18
in FUNDAFACI biosphere plan, 9–11, 181–82, 187–94, 196–97, 282–88, 291, 293–94, 324–25
genealogies collected by, 30–34, 46, 165–66, 185, 231–32, 240–41, 246–47, 291–92, 313
gifts dispensed by, 6, 30–35, 45–46, 48, 85–93, 101–2, 110–11, 115–16, 119, 138, 170–71, 174, 227, 229, 291–92
grants awarded to, 130, 137–38
hallucinogens used by, 15, 43, 46–48, 113, 294
in Haximu massacre affair, 195–214, 238, 365*n*
health of Yanomami asserted by, 262–63, 269, 270–74, 313
human rights groups attacked by, xxii-xxiii, xxvi, 11, 201–2, 207
incest theory of, 115, 117, 349*n*
Indian bodies marked by, 185, 324
Indian opposition to, 5–6, 9–10, 17, 46, 84–85, 87, 106, 109, 115, 183–86, 197–98, 231–42, 254, 283, 286, 289–94, 313, 362*n*
"innate killer" theory of, xxi, 8, 12–14, 39–42, 119
lactation taboos theory of, 266–67, 374*n*
language difficulties of, 105, 115, 202
in measles vaccine experiment, 54, 60–64, 66–67, 70, 72–81, 89, 98–99, 106, 198, 293, 310–13
media tactics of, xxii-xxiii, xxv, 9–10, 159–60, 183, 186–88, 201–14, 319
mining interests supported by, xxii–xxiii, xxv, 8–11, 204, 210, 294, 365*n*
missionaries opposed by, xxii–xxiii, xxvi, 10, 50, 54–55, 114, 122, 130–31, 167, 186–87, 195, 200–208, 210–14, 228–29, 236–37, 293, 317, 319–25, 366*n*–67*n*
objectivity claimed by, 138, 179, 184–85, 207, 238–39
postmodernist sensibility of, 118, 170
promises made by, 45–46, 115, 227, 231, 233, 241, 332*n*
public protests against, xxi, 10, 183, 197, 327*n*–28*n*
rivals of, 15–16, 126, 142–44, 147–48, 159, 164–67, 202–4
shotgun diplomacy of, 32–34, 46, 48, 89, 92, 112–14, 185–86, 205, 235–36, 283, 362*n*
spirits identified with, 15, 43, 46–48, 286–87, 289, 291–95, 313
students of, xxvi, 17, 131–32, 143, 161, 182–84, 236, 271–72
topographies invented by, 8, 110, 211, 213–14
as tribal leader, 15–16, 18, 31–35, 46–47, 88, 93, 111, 138, 142, 166, 294–95
Valero disregarded by, 246–50, 289–90
Vietnam War and, 42
village names obscured by, 104, 120–22, 164–65, 167, 320–22, 357*n*
villages identified with, 15, 111, 137–38, 142, 166–67, 217, 228, 292–93
violence incited by, 10–11, 15–16, 18, 29–35, 54, 87–88, 101–6, 112–19, 166–67, 174, 227–29, 231–41, 255, 293
violence theories of, 9–10, 12–14, 22–23, 26–27, 115–18, 158–64, 166, 169–80, 194, 239, 247, 261–62, 265–76
women's experience disregarded by, 102, 246–50
Yanomami leaders opposed by, xxiv, xxvi, 10–11, 201, 205, 207, 238–39, 293–94, 328*n*–29*n*
Yanomami Reserve entrusted to, 196–98, 294
chanting, 47, 59, 102, 112–13, 245, 250, 258–60, 284
healing by, 250, 258, 286, 289–90
Cheyenne Indians, 178
Chiappino, Jean, 144–45, 147
Chicago, University of, 11, 152, 202, 208
Chico, 161–63, 172, 174
children, xxiv, 3, 33, 47–48, 159, 186, 231, 243–46
death of, 25, 47–48, 51–54, 67–69, 126–27, 183–84, 217, 220–23, 238–39, 312, 324, 326
filming of, 84, 216–17
malnutrition among, 59–60, 67, 258, 262–63, 276, 288–89
sexual exploitation of, 129, 132–37, 143–47, 222–23
spiritual murder of, 47–48, 112–13
see also infanticide
chivakoa (skin disease), 286
Christianity, Indian understanding of, 212
Chronicle of Higher Education, 11, 207–8, 212
cigarettes, 21, 95, 106, 132, 139, 151, 219
clothing, 13, 28, 31, 46–47, 87–88, 117, 130, 219, 231, 252, 270
ceremonial, 47–48, 247–48, 286
as trade goods, 134, 139, 141, 289
coalition theory, 269–70
Cocco, Luis, 129, 133–34, 172, 205, 264, 270, 350*n*, 366*n*
colds, *see* respiratory infections
Cold War, 14, 40–42, 44, 155, 304–6, 310

Collins, Terrence, xviii, 307
Colombia, 151
Coming of Age in Samoa (Mead), 11–12
Communism, 11, 42, 155, 179–80, 304, 310
Conan Doyle, Arthur, 12, 152
Congress, Venezuelan, 154, 156, 193–94, 197, 200
Congress of Amazon Indians (1993), 197–98
CONICIT (Venezuelan National Council of Scientific and Technological Investigation), 197
CONIVE (Indian rights organization), 186
Connell, Mary Jeanne, 300
Conquista del Sur, La (Conquest of the South), 150–51, 155–57, 187
conquistadores, 31–32, 48, 57, 107–8, 280, 296
Conrad, Joseph, 13
Conservation Society of Guyana, 153
Corporación Venezolana de Guayana (CVG), 199–200
Corps of Engineers, U.S. Army, 21, 28, 91, 111, 281
Cortés, Hernán, 48
coups (trophy heads), 163
Couto de Magalhães River, 27, 208–9
"Covering Up the Yanomamo Massacre" (Chagnon), 201–8, 319–20
Cree people, 281
crowd diseases, 57
curare poison, 171, 245, 247–49
curassows, 260
Current Anthropology, 176, 179
Cuyuni River, 153
Cuzco, looting of, 108, 157
CVG (Corporación Venezolana de Guayana), 199–200
cyanide, gold mining and, 199, 365*n*

DAI, see Indian Agency, Venezuelan
Daly, Martin, 266
Damioli, Guillerme, 162
dancing, 31–32, 47, 59–60, 86, 101–2, 113, 219, 286
Dani people, New Guinea, 85
Darwin, Charles, 266, 267
Darwinism, vii, 12, 22–23, 26, 49, 95–96, 179, 182, 203–4, 222, 266–68
  see also sociobiology
Darwin's Dangerous Idea (Dennett), vii
Dawkins, Richard, 203, 207
Dawson, Gary, 44, 47–48, 69, 128–29, 135–37, 147, 255
Dawson, Jodie, 230, 242
Dawson, Joe, 86, 128–29
Dawson, Marie, 127–28
Dawson, Michael, 16, 103, 217–18
Dawson, Wilma, 217
Dawson family, 216–17, 242
DC-3 cargo plane, 130–31

dead, names of, 32–34, 46–48, 170–71, 230–32, 240–41, 291–92, 313
Dead Birds, 85
Dedeheiwa, 173–74
Defense Department, U.S., 152
democracy, Darwinist critique of, 38–40, 49, 313–14
Demonic Males (Wrangham), 14
Dennett, Daniel C., vii
Denys, Bayna, 160
deodorant, 132, 219
depósito (handicrafts distribution point), 230, 234–35
Detroit, homicide rate in, 23
Diamond, Jared, 19, 59
Diario de Caracas, El, 155
Diego factor, 13
"Die Hard," 297
Dimanawa, César, 85, 87, 185–86, 190, 230–33, 235–42, 293–94, 323
Dimorama (Yanomamo woman), 33–34
diphtheria, 57
Dirección de Asuntos Indígenas, *see* Indian Agency, Venezuelan
disease, diseases, 8, 16–17, 19, 48–82, 104–6, 300–301
  death rates from, 10, 17, 20, 50–52, 67–69, 77, 104–5, 205–8, 255, 317–26, 334*n,* 345*n*–46*n,* 382*n*
  eugenic view of, 38–39, 48–49, 95–96
  Old World, 50–52, 56–60, 104–5
  prevention of, xxvi, 63, 74–75, 89, 98–99, 110, 119, 190–91, 221–22, 244, 250–51
  respiratory, 21, 27–28, 51, 63–65, 89–91, 95, 106, 108, 119–21, 211, 250, 347*n*
  spread of, *see* epidemics
  susceptibility to, 56–62, 66–69, 89, 95, 104–8, 311, 317–18, 339*n,* 341*n*
  Yanomami conception of, 5, 21, 24, 27–29, 48, 52, 78, 90–91, 105–6, 120, 183–85, 211, 218, 232, 280–88
  see also epidemics; measles vaccine experiment; *specific diseases*
Doctors of the World, 209–10
Dorita-teri, 3–6, 165, 206, 291, 292–94, 321–22, 325, 384*n*
Doshamosha-teri, 201, 243–44, 277, 279, 283–85, 377*n*
dos Santos, Cícero Hipólito, 25–26
Dos Santos, Suami Percíllio, xxii
Dower, John, 302
drinks, *see* food and drink
Dutch colonialism, 19

Eagle Mountain (Toki), 36–37
Earle, John, 44
Early, John, 26
ebene (hallucinogen), 47–48
Economist, 203–4
Ecuador, 265, 296

Edmonston B measles vaccine, 55–61, 63–69, 73–81, 96–98, 106, 306, 311–13, 340*n*
alternatives to, 58–59, 79–81, 99, 110
febrile response to, 55–56, 58, 67–68
gamma globulin coverage for, 56, 58–61, 68, 75–77, 98–99
see also measles vaccine experiment
effigies, attacks on, 5, 106, 170, 177, 293, 362*n*
Eguillor García, María, 141, 169, 172–73, 175, 240, 331*n*, 384*n*
Eibl-Eibesfeldt, Irenaus, 133, 202, 204, 293, 317, 320, 366*n*
Eisenbud, Merril, 305
Eisenhower administration, 305–6
El Dorado (mining camp), 152
El Dorado legend, 10, 13, 19, 106–7, 156–57, 296
Embracing Defeat: Japan in the Wake of World War II (Dower), 302
Emerald Rain Forest, The, 14
Enders, John, 66
Energy Department, U.S., 16, 43–44, 304, 307
English colonists, 12, 104–5, 157
environmental catastrophes, 5, 153, 199
environmentalists, xxii, xxiv–xxv, 153–54, 197–98
Environmental Protection Agency, U.S., 199
environmental stress, theory of, 263–69
see also hunger
Environment Law, Venezuelan, 153
Environment Ministry, Venezuelan, 197
epidemics, 49–82, 119–22, 167, 317–26
exploration linked to, 16–17, 21, 24, 27–29, 49–53, 75, 88–89, 190–93, 281, 282–83
film projects linked to, 5, 89, 95, 99, 104–7, 119–22, 219–22
gold rush linked to, xxiv, 8, 191, 209, 210, 281–82, 288
immune depression and, 56–62, 66–69, 95
killing spawned by, 21, 24, 27–29, 52, 111
mythology spawned by, 281–83, 291–95
social structure altered by, 28–29, 168
trade routes in, 317–18
see also disease, diseases; measles vaccine experiment
Erotic People stereotype, 15, 84, 133–34
Eskimos, 56, 108, 339*n*
Essay on the Principle of Population, An (Malthus), 267
"Es sou prisoneira," 259
Estado de São Paulo, O, 160, 256
ethnographic films, *see* filmmaking, ethnographic
"Ethnography and Ethnocide" (Rifkin), 8, 295
eugenics movement, 38–40, 355*n*
Eugenics Records Office, 38
European Alliance of Genetic Support Groups, 44
evangelical missionaries, *see* Protestant missionaries
Evangelical Mission of Amazonia (MEVA), 212–14
evolutionary theory, *see* Darwinism; sociobiology
ExcesO, 149, 152
explorers, European, 5, 12, 18–21, 48, 91, 94–95, 104–5, 107–8, 152, 157

falciparum malaria, 94, 119–21, 220–22, 256, 288–89, 318, 325
fallout, radioactive, 304–5
Fama, 139
Feast, The, 5, 91–93, 99–107, 110, 112–13, 120, 151, 166, 172, 182, 223, 253, 272, 293, 312, 344*n*–45*n*
feasts, 88–89, 99–103, 110, 164, 206, 216, 220
singing at, 259–60
violence associated with, 88, 102–3, 168–69, 223, 235, 250, 293, 336*n*
Felicita, Sister, 106
female infanticide, 265–67, 274, 313, 377*n*
Ferguson, Brian, 45, 50–51, 133–34, 159, 171, 175, 177, 179, 187, 203, 331*n*
on epidemics, 50–51, 221
social disruption observed by, 18–19, 30, 32, 142, 223, 228–29, 242, 276
fertility, 88, 161–62, 173–75, 252, 268–69
Fierce People, The, see Yanomamo: The Fierce People
Fierce People stereotype, xxi, 8, 12–14, 20–27, 84, 107, 119, 131–33, 158–61, 164, 215–17, 219, 232, 261–63, 271, 275, 309–15
Fiji, 19, 286
filmmaking, ethnographic, 14–15, 59, 83–106, 112–22, 121, 182, 215–23, 253–56
acclaim for, 5, 81–82, 102, 112–14, 122, 215–16
AEC support of, 37, 42, 70–71, 93–94, 102–3, 118, 165, 292–93, 312
anthropologists' critique of, 81–82, 85, 101–2, 112–13, 116–19, 216, 221
Darwinism implicit in, 222
economic impact of, 84–86, 91–92, 114–17, 217–20, 222, 253–54
epidemics associated with, 5, 88–89, 95, 99, 104–7, 119–22, 219–22, 254–55, 312–13
Indian critique of, 37, 42–43, 46, 48, 85–87, 116, 218, 221, 342*n*–43*n*
in measles experiment, 5, 17, 70–71, 77–78, 80–82, 95–99, 104–7, 293, 312
political use of, 42, 151, 220, 222
postmodernist, 118–19, 170
shabonos created for, 14, 85–88, 216–17, 221–22
stereotypes reproduced in, 84, 102, 113, 116–19, 170, 216–17, 219, 222
violence in, 5–6, 85–88, 101–6, 112–19, 121, 166, 168, 216–17, 219, 223, 342*n*
wars precipitated by, 14, 85, 87, 103–4, 113–14, 222–23, 342*n*
by Yanomami, 84, 253–54, 343*n*, 372*n*
Yanomami fear of, 5, 83–85, 218

Finkers, Juan, 32, 120–21, 145, 167, 185, 227, 230–31, 241, 323
First Congress of Amazon Indians (1993), 197–98
first contact, 7–8, 18–22, 49–53, 91, 104–5, 107–22, 185, 211
  disruption caused by, 5, 18–19, 21, 24, 27–35, 45–46, 114, 119–22, 127, 185, 228–29
  false claims of, 246–49, 289–90
  guidelines for, 190–91
  mystique of, 3–6, 10, 50, 107–10, 187, 248–49
Fiscalía, Venezuelan, 194, 198–201, 332*n*
FLASA (Fundación La Salle), 154
flying boxcar (transport plane), 88, 127
FNS (Brazilian National Health Foundation), xxii, 196, 244
Fôlha de São Paulo, A, 196, 210
food and drink, 87, 89–90, 92, 100, 146, 174, 257–58, 261, 266, 278
  arguments over, 115–16, 128–29, 141–42, 176, 272–74
  see also hunger
Forestry, Soils, and Water Law, Venezuelan, 153
Foundation for Anthropological Investigation, Venezuelan (FUNVENA), 15, 183
Foundation for the Development of Physical and Mathematical Sciences, Venezuelan, 155–56
Four Horsemen of the Apocalypse (gang), 210
Fourth Air Force Group of Amazonia, 192, 324–25
Fox, Robin, 203
France, anthropology in, 15, 84, 126–27
Franco-German expedition, to Yanomamiland (1951), 250, 263
Franklin, Benjamin, 255–56
Freedom of Information Act, 43, 297
Freeman, Derek, 14
French Embassy, Venezuela, 129, 137, 144
French Guyana, xxii, xxiv
Friedland, Robert M., 199–200, 365*n*
Fuentes, Emilio, 132
FUNAI (Brazilian Indian Agency), xxii, 190, 195, 209–10
Fundaçao Nacional de Saúde, Brazilian (FNS), xxii, 196, 244
Fundación La Salle (FLASA), 154
FUNDAFACI, 9–10, 182, 186–94, 206–7, 260, 282–88, 323–25, 373*n*
  warfare unleashed by, 227–28, 234–35, 242, 254, 277, 283
funeral rituals, 25–26, 69–70, 84, 102–4, 107, 218, 312–13, 325
Fusiwe, 246, 248, 250, 259, 272, 275–76, 281, 285, 359*n,* 378*n*

Gadhafi, Muammar Muhammad, 154–55, 198
Galactic mining company, 199
Galeano, Eduardo, 309
Galton, Francis, 38–39
gamma globulin, 53, 56, 58–61, 68, 75–77
García Márquez, Gabriel, 15
Gardner, Robert, 85
garimpeiros (gold miners), 11, 153–54, 201, 205, 208–12, 238, 249, 281–82, 284, 289, 294, 319
  see also Haximu-teri, massacre of; mining
Garimpeiro Union, Brazilian, 154
genealogies, 30–34, 43, 166, 185, 231–32, 244–45, 313
  spurious, 32, 170
genetic mutation, 16, 37, 43
genetic research, xxii–xxiii, 12–14, 304–15
  AEC sponsorship of, 16–17, 36–37, 42–44, 59, 61, 165, 297, 300, 304–10, 380*n*
  ethics of, 43–44, 297
  immunological questions in, 54, 57–60
Geo, 9
Ghost Dance movement, 282
ghosts, 17, 25, 90–91, 161, 218, 245, 280
"Give me" (airstrips), 270
Globo, O, 160, 196
Goetz, Inga Steinvorth, 63, 68, 78, 345*n*
Goffman, John, 309
gold, extraction of, 153–54, 156, 199, 365*n*
  see also Amazon gold rush; *garimpeiros*; mining
Golden Star mining company, 199–200, 365*n*
González, José, 129–30, 132, 147
González, Juan, 70, 72–73, 104, 110, 129, 147, 312–13
Good, Kenneth, 17, 47, 170, 178–79, 222, 237, 277, 289–90
  cross-cultural marriage of, 250–56, 372*n*
  Lizot's village observed by, 130–36, 138, 143–44, 219
  protein intake measured by, 271–74, 331*n*
GPS (Global Positioning Satellite) equipment, 165, 187
Gran Manoa, 12, 19, 269
  see also El Dorado legend
Great Protein Debate, 265–67, 270–74
Greece, ancient, 14, 16, 311
Green Book (Gadhafi), 154
Green Forest mining company, 200
Greenland, 68, 339*n*
Griffiths, Paul, 217
Groves, Leslie, 303–4
Gruber, Alfredo, 199–200
Guaharibo Rapids, 20–21, 134, 249, 318
  see also Orinoco River
Guaharibos, White, *see* Yanomami people
Guarapana, 84–85
Guardia Nacional, Venezuelan, 155–56, 200, 237
Guiana Shield, 12
guides, killing of, 227, 231
guns, 25, 99–100, 102, 129–30, 150, 177, 204–5, 211–14, 247, 252, 260
  Indians intimidated by, 31–34, 46, 48, 89, 92, 185–86, 232

murders committed with, 27, 113–14, 165–68, 195, 201, 204–5, 211–12, 223, 234, 238–41
sex bartered for, 136–37, 139, 141, 143, 223
societies altered by, 8, 19, 27, 112–14, 138, 162, 166–67, 213–14
as status symbol, 143, 162, 172, 216
Guns, Germs, and Steel (Diamond), 19
"Guns of Mucajaí, The" (Chagnon), 212–14
Gus (attack dog), 130, 182
Guyana, Conservation Society of, 153
Guyana, French, xxii, xxiv
Guyanas, 12, 150, 152, 199
Guzmán, Antonio, 197

hakimou (dance), 102
hallucinogens, 7, 15, 43, 46–48, 103, 113, 119, 129, 140–41, 250, 286
Hames, Raymond, 162, 191, 207, 236, 271–73
Haoyabowei-teri, 165, 167, 321
hariri (disease), 105, 283, 288
see also disease, diseases; epidemics
Harokoiwa, 5, 292
Harris, Marvin, 265–67, 270–74, 276
Harvard University, 39, 85, 117
Hasabowa-teri, *see* Hasupuwe-teri
Hasupuwe-teri, 47, 77, 97, 134–35, 147, 168–69, 252–54, 258, 260, 278–79, 293, 336*n*, 360*n*, 378*n*, 383*n*
Haximu-teri, massacre of (1993), 195–214
anthropologists' responses to, 201–14, 238
investigation of, 196–202, 210
press coverage of, 196, 201–8, 364*n*–65*n*
sociobiological interest in, 203–4, 207–8
Yanomami suspected in, 208–11
head-hunting, 163, 178
headmen, 3, 5, 24, 33, 46–47, 95, 103, 124, 137–38, 163, 168–71, 180, 216–17, 232
Chagnon's appointment of, 33, 100–101, 174, 369*n*
children fathered by, 161–62, 173–75
death of, 114, 161, 166, 168, 177–78, 234
as shamans, 93, 100, 174, 282–84
sociobiological theory of, 39–40, 49, 159, 171–74
spirits of, 161, 285
health, measurement of, 59–60, 99, 373*n*
Health Ministry, Venezuelan, 63, 65, 70, 81, 98–99, 312
Heawe, 114
Hebewe, 139–41
hehohi (revolver), 283
Heider, Karl, 85
Heisenberg's War (Powers), 301
hekura (nature sprites), 285
helicopters, xxvi, 9–10, 91, 150, 182, 191–94, 196, 200, 206, 233, 237
illegal use of, 10, 155–56, 187, 192–94, 227, 254, 332*n*
shabonos destroyed by, 3–5, 282, 290–91, 294
superstitions about, 284–87, 289, 291–92
Hemingway, Margot, 9
Hemming, John, 19–20
Henshaw, Paul, 302, 310
hepatitis, 8, 69, 120–21, 325, 341*n*
Hepewe, 146
Herrera, José Antonio, 198–99
Herzog-Schroder, Gabriele, 317
Highest Altar, The (Tierney), xxiii-xxiv, 162
Hill, Kim, 121, 190–91, 203, 205, 263–64
Hiomita-teri, 235, 287
Hiroshima, 16, 37, 43, 298, 302–3, 314, 379*n*–80*n*
Hisiwe, 217
Hokohenawa, 247
Hokomapiwe-teri, 260, 276–79
Homer, 244
Homicide (Daly and Wilson), 177, 266
Homo Necans (Burkert), 311
homosexuality, 128, 133–37, 143–47, 350*n*–52*n*
horemu (fake), 105–6, 116
Howashiwa, 173
Howland, Joseph, 301, 303–4, 310–12
HP (human product) numbers, 298, 300
Huarani people, 23
Hudson Bay Company, 95
Hukoshikuwa, 87
Human Behavior and Evolution Society, 203–4, 366*n*
Human Genetics Department, at University of Michigan, 9, 12, 38–42, 297, 302–3
Human Genome Diversity Project, 51
human nature, theories of, *see* Darwinism; sociobiology
human radiation experiments, 44, 297–314
casualties of, 298–302, 305, 309–10
compensation for, 309–10
Human Radiation Experiments (web site), 304
human rights groups, xxi–xxii, 11, 113, 127, 154–56, 160, 201–2
indigenous, xxiv–xxv, 183, 185–86, 197, 201
Humboldt, Alexander von, 9, 12, 65, 79, 152, 250, 262
hunger, 99–100, 245, 257–79, 288–89
conflict caused by, 115–16, 128–29, 141–42, 176, 272–74
growth stunted by, 60, 262–65
infections caused by, 60, 67
hunting, 13, 90, 92, 100, 141, 163, 172, 232, 258, 260–61, 266, 272–77
Hurtado, Albert, 95
huyas (boys), 139, 278–79

Iaduce, 210
I Am the White Woman (Valero), 244
Iceland, 54, 57
Iglesia en Amazonas, La, 185–86
immune deficiency, 56–62, 66–69
Inca empire, 12, 32, 57, 108, 156–57
incest, 115, 117, 349*n*

Index of Innate Ability (IIA), 12, 40, 45, 49, 314
see also leadership gene, search for
Indian Agency, Brazilian, *see* FUNAI
Indian Agency, Venezuelan (DAI), 9, 144, 147, 186, 191, 193, 232
indigenous rights activists, *see* human rights groups
infanticide, 265–67, 274, 313, 377*n*
infant mortality, 51–52, 138, 183, 324, 326
influenza, 8, 57
informed consent doctrine, 43–45, 307
initiation rites, 285–86
Institute for Advanced Studies, 161
Institute of Higher Studies, Paris, 144, 147
Interciencia, 62
Internet, 203, 304
intestinal parasites, 273, 288
Into the Heart (Good), 131–32, 251–52, 271
In Trouble Again (O'Hanlon), 9, 201
iodine, radioactive, 304, 306–7, 311, 314, 381*n*–82*n*
Ireland, Corydon, 299–300
Irokai-teri, 250–52, 256
iron, radioactive, 307
Ironasi-teri, 115, 117, 119–20, 324, 384*n*
Isabelita, 241
Ishi, 108
Ishiweiwa, 173–74
Ituri pygmies, 264–65
IVIC (Venezuelan Institute of Scientific Investigation), 44, 63, 73, 78, 137–38, 154, 197, 311–12, 325
Iwahikoroba-teri, 106, 165, 168–69, 183–85
Iyewei-teri, 74–75, 172, 306, 311

jaguars, 22, 161, 243, 245, 260, 285
Jaime, 135–36
Japan, atomic bombings of, 16, 37, 43, 298, 314, 379*n*–80*n*
Jatar, Luis Manuel, 193
Jepewe, 93
Jesuits, 49
jewelry, 132, 143, 145, 231, 248, 286
Jillings, Andy, 218–22
Jivaro people, 178
Johnson, Craig, 117–18
journalists, xxii, xxv, xxvi, 17, 107–8, 153, 290–91, 319, 322, 325
behavior of, 196, 249, 256
exclusive rights offered to, 4–5, 10, 156, 187, 201
see also media
Juan, 95, 106

Kanaripó, 155–56
Kaobawa, 47–48, 87–89, 108–9, 112–13, 169, 384*n*
Chagnon's patronage of, 30–31, 33, 87–88, 137–38, 141–42, 359*n*
Kaobawa Trades with the Reyabobowei-teri, 89, 273
Kaplan, Hillard, 121
Karakawe, Alberto, 237–39, 287
Karina, 109–10, 128–29
Karohi-teri, 72, 78, 93, 137, 139–45, 217–23
Kasiawe, 172
Katz, Samuel, 61, 65, 67–68, 340*n*–41*n*
Kayapo people, 39–40, 41
Kayopewe, 92–93, 294
Kedebabowei-teri, 165, 167–68, 170–71, 184–85, 206–7, 233–34, 241, 320, 322–24
Keti-U River, 234–35
"Killed by Kindness?" (Chagnon), 201, 227
"Killing of Ruwahiwa, The" (Chagnon), 247
Kilometer 88, Venezuela, 153–54
Klawaoitheri, 141–42
Kohoroshi-teri, 245–46
Konoko, Gustavo, 116
Kopenawa, Davi, xxiv, 7, 11, 32, 327*n*–29*n*
Kosh, 69, 221
Koshirowa-teri, 135
Kreibowei-teri, 168–69
see also Bisaasi-teri
Kulp, Laurence, 308
Kumadowa, 234
Kumaiewa, 100, 114
Kunata-teri, 275
!Kung people, 268, 375*n*

"Laboratory for Human Conflict, A" (Allman), 160
landing strips, xxiv, 8, 208, 237, 270, 285–86, 323–25, 377*n*
Las Casas, Bartolomé, 280
Last Days of Eden (Chagnon), 201, 231–32, 323, 357*n*
laws, Venezuelan, 153, 156, 196–200, 205, 229
leadership gene, search for, 12, 39–40, 45, 49, 159, 294
Lechoza, massacre at, 238–40
Lederle pharmaceuticals, 59
Lévi-Strauss, Claude, 15, 39, 126–28
Libby, Willard, 305, 310
Libya, 154–55, 198
"Life Histories, Blood Revenge, and Warfare in a Tribal Population" (Chagnon), 158–62, 171, 173–76, 180, 203, 233, 247, 355*n,* 369*n*
literacy, xxii, 46, 135, 185, 232, 239–41, 293
Lizot, Jacques, 15–17, 125–48, 159, 164–67, 169, 228, 248, 272, 276–77, 320, 322, 331*n,* 368*n*
as A. H., 15, 128, 134
filming overseen by, 15, 217–20, 222–23
gifts dispensed by, 127–28, 132–37, 139, 141, 223
in Haximu massacre affair, 202–4, 366*n*
missionaries opposed by, 129–30, 132, 141
names for, 15, 128, 134
pedophilia of, 129, 132–37, 143–48, 222–23, 353*n*–54*n*

political power of, 126, 136, 144–47
as tribal leader, 15–16, 141–43, 166
Yanomami sexuality as described by, 15, 133–45
"Lizot's village," *see* Tayari-teri
Lorenz, Konrad, 22, 204
Los Alamos, N.Mex., 297
Los Angeles, University of California at (UCLA), 269, 300
Los Angeles Times, 160, 206, 297, 301
Lost World, The (Conan Doyle), 152
LSD, 199
Lucho, Enrique, 294

MacArthur, Douglas, 302
McCann, William, 299
McCarthy, Joseph, 40, 180
machetes, 26–30, 110–11, 114–17, 138, 161, 195, 241, 245, 270, 272
as trade goods, xxii, 21, 51, 75, 85, 92–93, 117, 119, 139, 146, 170–72, 174, 217, 229, 277
Macondo, 15
Macuxi people, xxiv–xv, 19
Madera, 151
Madi, Isaam, 292
madohe (stuff), 29, 101, 129, 259–60
see also trade goods
MAF (Missionary Aviation Fellowship), 229–30
Magical Death, 112–14, 116, 120, 348*n*
Magris, Magda, 192
Mahekoto-teri, 68–69, 76–77, 87, 112–13, 120, 122, 141, 172, 250, 259, 293, 318
filming of, 77, 92–93, 102–6, 344*n*
Maipure Rapids, 229
Makiritama, 184
malaria, xxii, xxiv, 8, 51, 60, 63, 89, 94–95, 147, 153, 191, 212, 255, 287–90, 318–19, 325–26
censuses of, 244, 250–51, 326
control of, 28, 87, 221–22, 244, 250–51, 260, 319
falciparum, 94, 119–21, 220–22, 256, 288–89, 318, 325
genetics of, 38, 297
Malaria Department, Venezuelan, 44, 63, 65, 76, 80, 110, 290–91, 325
Malariologia, *see* Malaria Department, Venezuelan
male dominance, theories of, 12, 39–40, 49, 117–18, 314
malnutrition, *see* hunger
Malthus, Thomas, 267–69
Mamikininiwa, 247
Manau people, 19
Manaus, Brazil, 110
Manaviche River, 93, 143, 215
Manhattan Project, 297–301, 303, 309–10
Manu National Park, Peru, 190–91
maps, viii–ix, 109–10, 285
maquinari a ne wakeshibi (sorcery machines), 281–82, 291
Maquiritare people, 36–37, 133, 155, 233, 270–73, 306–9, 311
Yanomami indentured to, 270, 271
Marahuaca, Mount, 36, 229
Marashi-teri, 208–10
Margaret Mead in Samoa (Freeman), 14
Mari, Antonio, xxii
Mariano, 241
marijuana, 132–33
Mario, 124
Márquez, Oscar, 200
marriage, 26–27, 29, 129, 139, 146–47, 162–64, 171–76, 359*n*–61*n*
age factor in, 163, 173–76, 360*n*–61*n*
cross-cultural, 251–56, 372*n*
economics of, 30–31, 162, 172, 174, 360*n*
Martínez, Rafael, 154
Martins, Leda, xxiii–xxv, 116, 196
Matos, Cecilia, xxv*n,* 9–10, 181–82, 187, 190, 192–94, 198, 232, 254, 294, 329*n,* 332*n,* 364*n*
Mavaca Indians, 317
"Mavaca Measles Epidemic," 71–78
Mavaca mission, 90, 93–99, 111, 113, 119–22, 143, 147, 221, 230, 238–39, 242, 267, 320
census taken at, 166–67, 172, 175
measles epidemic at, 62, 64–65, 70–74, 76–78, 89, 95–99, 106, 312
Mavaca River, 23–24, 84–87, 91, 107, 109–10, 120–21, 165–68, 185, 205, 246, 323–24
warfare on, 227–29, 233–35, 254, 276, 287
Mavakita, Venezuela, 165, 183–85, 230–35, 240, 321
Max Planck Institute for Behavioral Physiology, 133, 204
Mead, Margaret, 8, 11–12, 14
measles-smallpox vaccine, 58
measles vaccine experiment (1968), 17, 53–82, 89, 104–7, 198, 293, 306
audio record of, 70–78, 81–82, 95–97, 105–6
Brazilian workers blamed in, 61–62, 65–67, 77–79
dislocation of Yanomami caused by, 68–70, 107, 183
film treatment of, 5, 17, 70–71, 77–78, 81–82, 95–99, 311–12, 314, 341*n*–42*n*
gamma globulin withheld in, 60–61, 68, 75–77, 98–99, 311, 340*n*
inconsistent accounts of, 54–55, 60–61, 65–67, 78–80
motives for, 54, 57–60, 72, 79–82, 95–96, 311–15
protocols ignored in, 60–61, 66–67, 76–77, 80–81, 98–99, 311–14, 340*n*
vaccine choice in, *see* Edmonston B measles vaccine
media, xxii–xxv, 3–6, 16, 108, 159–61, 178, 196, 290–91

Chagnon's appeal to, xxii-xxiii, xxv, 9–10, 159–60, 183, 186–88, 201–8
Dimanawa's use of, 231
first contact sought by, 4–5, 10, 187, 201, 248–49
scandals covered by, 151, 155, 193–94, 198–99, 201–8
stereotypes transmitted by, 3–4, 14, 22, 160–61
medical care, 25, 27–28, 50, 61, 63–65, 68–69, 104, 110, 120–21
withholding of, 30, 76–81, 95–98, 106, 121, 183–85, 190–93, 217–18, 221–22, 311–12, 324–26
medical ethics, 43–44, 55, 301, 306–7
medicine, 55, 63, 68–69, 73, 96, 162, 190, 256, 270, 318–19
mefloquine, 318
Mejía, Pablo, 37, 46–48, 135–36, 139, 218, 230, 242, 250
Mejía, Samuel, 46
MEVA (Evangelical Mission of Amazonia), 212–14
Michigan, University of, 9, 12, 38–42, 85, 149, 182, 302–3
Michigan College of Mining and Technology, 40
Micronesia, 54
Midas Gold mining company, 200
Milano, Sergio, 155–56
military alliances, *see* alliances, military
Minas Guariche Limitado, 152–54, 199–200
Minera Cuyuni, 199–200
Mines Law, Venezuelan, 153
Mines Ministry, Venezuelan, 199
mining, xxii, xxiii–xv, 9–11, 45, 152–57, 195–201, 208–14
environmental impact of, 153–54, 199
of gold, 153–54, 156, 199, 365*n*; *see also* Amazon gold rush; *garimpeiros*
mythology spawned by, 281–82
open-pit, 5, 153
opposition to, xxiv, 5, 154–56
strip-, 5, 153
of tin, 154–55, 188, 213
violence over, *see* Haximu-teri, massacre of
"Mining Invasion of Eastern Bolívar State" (map), 154
mining machinery, 281–82, 291
mining mafias, 209
Ministry of Health, Venezuelan, 63, 65, 70, 82, 98–99, 312
Ministry of Mines, Venezuelan, 199
Ministry of the Environment, Venezuelan, 197
Ministry of Youth, Venezuelan, 150–51
Miranda, Marta Rodríguez, 4, 319, 322, 383*n*
Mirapewe, 326
Mishimishimabowei-teri, 30, 46–47, 110–22, 128, 165–76, 185–86, 231, 240–41, 246, 293, 358*n*, 369*n*, 384*n*
in ax fight, 113–18, 175–76, 272
Fiercer People stereotype of, 107, 113, 118–19, 167
filming of, 85, 112–19, 170, 263
health of, 119–22, 167, 263, 318, 320, 324
polygamy in, 172–75
revenge declined by, 111–12, 168–70
missionaries, 8, 14, 31, 44–45, 49, 95, 110, 128–30, 161–62
Brewer's relations with, 186–87, 200, 237
Chagnon's conflicts with, xxii-xxiii, xxvi, 9–10, 50, 54–55, 114, 120–22, 130, 166–67, 186–88, 200–201, 228–29
education provided by, 167, 187, 230
in Haximu massacre affair, 195, 201–8, 210–14
Lizot's relations with, 129–32, 141
in measles vaccine experiment, 54, 60–67, 69–70, 318
medical care provided by, 27, 29, 50, 61, 63–65, 98, 110, 120–21, 184, 221, 318–19, 323, 325–26
proselytizing de-emphasized by, 50, 187, 212, 230
Protestant, 28–29, 44, 50, 83, 86, 96, 111, 128, 135, 187, 210–14, 271, 326
shotgun distribution blamed on, 114, 204–5, 211–14, 236–37, 366*n*–67*n*, 369*n*
stability brought by, 45, 50, 165, 166–67, 204–5
in Yanomami wars, 45, 165–67, 228–29
Missionary Aviation Fellowship (MAF), 229–30
missions, 28–29, 50, 111, 172, 271–72
death rates at, 10, 121, 205–7, 317, 319–26
Salesian, *see* Mavaca mission; Salesian order
Moawa, 115, 117, 171, 173–74, 178, 239, 360*n*
Mohesiwa, 116–19, 175–76
Mokarita-teri, 17, 257–60, 276–77, 373*n*
Momanipue, 85
Momaribowei-teri, 33–34, 89, 139, 230
Monaghan, Peter, 11, 208
Mongiano, Aldo, 146–47
monkeys, 215, 260
Monou-teri, 86–87, 103, 112, 165–66, 336*n*
Monsoyi, Esteban, 154
Montenegrins, 163
Moraten measles vaccine, 58
morbilliform rash, 61–62, 65–68, 71, 79, 81, 98
"Mortality and Divorce in Yanomamo" (Chagnon), 132
mortality rates, 17, 20, 50–52, 61–70, 74, 76–77, 104–5, 112, 120–22, 334*n*, 382*n*
age factor in, 51–52, 159–60, 162, 176–79, 326, 334*n*, 345*n*–46*n*
at missions, 10, 121, 205–7, 317–26
from violence, 13, 22–24, 29, 35, 41, 51, 85, 102, 113–14, 158–64, 169, 176–79, 227–29, 274–75, 335*n*

mosquitoes, 28, 94–95, 153, 215, 245, 263, 318, 325
Mountain People, The (Turnbull), 184–85
Mowaraoba-teri, 110, 168–69
Mozart, Wolfgang Amadeus, 143
Mrakapiwei, 165, 167, 233–35, 320
Mucajaí Borabuk, 26–28
Mucajaí mission, 210–12
Mucajaí River, 8, 26–28, 212–14
Mucajaí Yanomami, 26–28, 208–16, 249, 267, 352*n*
murder, murders, xxiii, 113–15, 158–80, 233, 245–48, 295
  age factor in, 162, 175–79, 360*n*
  conspiracies for, 22, 137, 140–42, 147, 170
  of guides, 227, 231, 240–41
  with guns, 27, 113–14, 165–68, 195, 201, 204–5, 211–12, 234, 238–41
  prevalence of, 13, 22–24, 29, 41, 102, 113–14, 158–64, 335*n*–36*n*
  spiritual, 47–48, 112–13, 170
  symbolic, 311
  see also Haximu-teri, massacre of; sexual competition; violence; warfare
mythology, 17, 25, 83–85, 112–13, 161, 168, 280–95
  cameras in, 5, 218, 291–93
  Chagnon in, 15, 43, 46–48, 109, 286–87, 289, 291–95, 313
  disease in, 5, 21, 24, 27–29, 48, 52, 78–79, 90–91, 105–6, 120, 183–85, 211, 218, 232, 280–83, 289
  helicopters in, 284–87, 289
  nabah in, 46–48, 101, 131, 137, 146, 259–60
  names in, 32–34, 46–48, 170–71, 230–32, 240–41, 291–92, 313

nabah (outsiders), 46–48, 101, 131, 137, 146, 255, 259–60
  mythological interpretation of, 282–86, 291–94
  prestige conferred by, 138–39, 141–42
Nacional, El, 154–55
Nagasaki, 16, 37, 43, 298, 313, 379*n*–80*n*
names, naming, 32–34, 46–48, 170–71, 230–32, 240–41, 246, 291–92, 313
Namowei Yanomami, 28, 29, 34, 112, 168, 246
Nanokawa, 173–74, 360*n*
Nape, 288–89
"Napoleon Chagnon: O Dossier" (Martins), xxiii, xxv
Narimobowei-teri, 277–79, 283–84, 289, 377*n*
Nasikibowei-teri, 165, 167, 183–84, 321
National Academy of Sciences, U.S., 299, 302
National Archives, U.S., 71, 106, 262
National Conference of Catholic Bishops, Venezuelan, 198
National Council of Scientific and Technological Investigation, Venezuelan (CONICIT), 197
National Council on Aging, U.S., 38, 41
National Film Archives, U.S., 17
National Geographic, 230, 237, 249
National Geographic Research, 121, 128, 190–91, 254–55
National Geographic Society, 249, 255
National Geographic specials, 190, 222, 253–56
National Guard, Venezuelan, 155–56, 200, 237
National Health Foundation, Brazilian (FNS), xxii, 196, 244
National Institute of Mental Health, U.S. (NIMH), 119
National Institutes of Health, U.S. (NIH), 56, 57
National Science Foundation, U.S., 130
native rights activists, *see* human rights groups
Neel, James, 9, 13, 31, 37–45, 53–82, 119, 137, 149–43, 194, 213, 270, 296–97, 310–15
  in atomic survivor studies, 38, 43, 301–3, 380*n*
  eugenic views of, 38–40, 48–49, 79, 95–96, 182, 313–14, 337*n*
  in film productions, 70, 81–82, 93, 95–99, 101–4, 106, 312–13
  genealogies scrutinized by, 244, 246
  in Haximu massacre affair, 203, 207
  health of Yanomami asserted by, 59–60, 262–63, 313, 373*n*
  in human radiation experiments, 299–304, 310
  infanticide reported by, 267, 274, 313
  leadership gene sought by, 12, 39–40, 45, 49, 159, 314
  in measles vaccine experiment, 17, 54–55, 58–62, 65–69, 71–82, 95–99, 106, 310–13
  medical breakthroughs of, 37–38, 297
  modern society denounced by, 12, 39, 41–42, 49, 99, 313–15
  radiation risk discounted by, 302–3
Nelson, Father, 238–39
neutron bombardment experiments, 305, 310
New Guinea, 19, 85
New Jersey, as viewed by Yarima, 251–53, 372*n*
New Mexico, University of, 190, 203, 263
Newsweek, 4, 201, 208, 365*n*
New Tribes missionaries, 44, 328*n*
New York Academy of Sciences, 203
New York Botanical Garden, 4, 150, 291
New York Times, xxii, xxv, 4, 187, 191, 201–2, 204–8, 228–29, 237, 289–90, 309, 319–20, 325
New York University, 55–56
Ninam Yanomami, 216
Nixon, Richard, 198
Noble Savage myth, xxiii, 11–12, 21, 26, 42
nomohori (dastardly trick), 26, 168–69, 235
noreshi (spiritual essence), 84, 218, 285
"Notes on the Effect of Measles" (Neel et al.), 61–62, 79
Nova, 14–15, 215–16, 219–23, 368*n*
nuclear airplane project, 307, 310, 381*n*

Nuñez Montano, Manuel, 153
Nuremberg Code (1947), 44, 309

objectivity, xxiv, 19, 127, 138, 179, 184–85, 207, 238–39
  postmodernism vs., 118–19, 170, 207
Ocamo airstrip, 237
Ocamo Is My Town, 127
Ocamo mission, 106, 130, 172, 223, 232, 237–39, 244
  measles outbreak at, 54, 61–68, 71–72, 74, 77–78, 97–98, 306
Ocamo River, 69–70, 74–75, 242, 264, 318
O'Hanlon, Redmond, 9, 150, 156, 201, 274
O'Leary, Hazel, 44
Omawe (the Creator), 280
On Aggression (Lorenz), 22
On Photography (Sontag), 83
Open Veins of Latin America (Galeano), 309
Opik-teri, 161, 356*n*
Origin of Species (Darwin), 267
Orinoco River, 12, 19–21, 65, 91, 145–46, 167, 243, 249–50, 285, 288
  in biosphere plan, 154, 188, 237–38
  diseases along, 28, 60, 62–63, 89, 94–95, 106, 119–22, 220–22, 318–26
  filmmaking on, 84–88, 92, 119–20, 151, 215–23, 253–56
  flooding of, 229–30, 242, 249
  warfare along, 23–24, 28, 167–68, 223, 237–41, 260, 276
ornaments, body, 132, 143, 145, 231, 248, 286
outboard motors, 36, 65, 131, 141, 215–16, 231, 248–50, 325
  as trade good, 31, 143, 233, 241
  in Yanomami warfare, 30, 33, 113, 166, 238, 323

Paapiu, Brazil, mining camp at, 208–10, 281–82
Padamo missions, 96–97, 128, 218, 242, 271–72, 383*n*
Padamo River, 36–37, 78, 96–97, 229–30, 242, 271
Panama, measles vaccine in, 55–56
Panama Canal, 21
Papania, Mark, 79–80
parasites, 69, 192, 273, 278, 288, 307
  see also malaria
Parima Massif, 263
Parima Mountains, 15, 69, 91, 125–26, 139, 147, 215, 275, 278, 318
  hunger in, 261–65, 274
  massacre in, *see* Haximu-teri, massacre of
  mining operations in, 154–55, 208–11
  Yanomami origin in, 91, 231, 269
Paris, University of, 126, 162, 202, 281
Parke Davis Laboratories, 59
Parma (attack dog), 130, 182
Paruriwa, 137, 142, 172, 359*n*
Patahama-teri, 250, 254, 256
Patanowa-teri, 15, 29–30, 90–106, 122, 151, 175, 245–46, 259, 272–73, 290, 343*n*–45*n*
  in measles vaccine experiment, 77, 95–99, 106, 294, 312–13, 318, 322
  warfare of, 30, 86–87, 90–92, 102–4, 165–67, 293
Peabody Museum, Harvard University, 85
Pedro, 161
Pemon Indians, 152–54, 199–200
penicillin, 68–69, 73, 312
Penn State University, 51, 130
Perdamo, America, 291, 325
Pérez Morales, Ovidio, 198
Perry, Admiral, 108
Peru, 154–55, 190–91
Peters, John, 26, 30, 212
Peterson, Dale, 158
Philips Roxane, 59
photography, xxii, 83–84, 108, 160, 231, 291, 293–95
Physician to the Gene Pool (Neel), 38, 79, 301, 313–14
Pizarro, Francisco, 32, 48, 157
Plains Indians, 178
plantains, 87, 100–101, 103, 115, 176, 257–58, 261, 266, 278, 285, 373*n*
Platanal mission, 76–77, 92, 104, 129, 132, 141, 145–47, 172, 255, 290, 293, 312
"Platanal Yanomami", 104, 120
  see also Mahekoto-teri
Plutonium Files, The (Welsome), 297, 300–302, 309
plutonium injections, 44, 297–301, 310–11
pneumonia, 62–65, 68, 89, 99, 108
Poirier, Jean Pier, 63–65
Poland, 23
polonium injections, 298
polyandry, 27, 171, 175
polygamy, 129, 139, 159, 162–63, 171–75, 232, 359*n*–60*n*
Poortman, Ysbran, 44
Portuguese colonialism, 19, 91, 269
Portuguese language, 209, 244–45
postmodernism, 118–19, 170, 207
Powers, Thomas, 301
primates, 13, 14, 22, 179, 215
Prime Time Live, 4, 187
Príncipe, El, *see* Hebewe
Procuradoria Geral de República, Brazilian, xxiv
Project Sunshine, AEC, 305, 308–10
prostitution, 8, 133–37, 139–41, 239
Protestant missionaries, 28–29, 44, 50, 83, 86, 96, 111, 128, 135, 187, 210–14, 271, 326
protest marches, xxi, 10, 183, 197, 327*n*–28*n*
Public Health Laboratory Service, England, 56
Puerto Ayacucho, 192–93, 197, 218, 229
  bishop of, 63, 73, 78, 337*n*
Puig, Carlos, 110
purification rituals, 163–64

Quetzacoatl, 48
Quiñones, John, 5, 187
Quito, Ecuador, 296

Rabben, Linda, 113, 127
radiation experiments, 44
radioisotope experiments, 305–10, 381*n*–82*n*
see also human radiation experiments
Rahakanariwa (Vulture Spirit), 43, 47–48, 294, 356*n*
Rahras (mythical serpent), 109
raiding, *see* warfare
Raimundo Nenem airstrip, 208
rainforest, exploitation of, 150–51, 156–57, 193–94, 197–200
Rain Forest Action Network, xxiv-xxv
Raleigh, Walter, 12, 152, 157
Ramos, Alcida, 134, 140–41
rape, 11–12, 164, 201, 223, 238, 356*n*–57*n*
Reagan administration, 38, 41
Red Queen (Ridley), 203–4, 268
Reff, Daniel, 94–95
Reich, Wilhelm, 267
Renaldo, 218–19
Rerebawa, 72–74, 78, 114, 137, 345*n*
respiratory infections, 21, 27–28, 51, 63–65, 89–91, 95, 106, 108, 119–21, 211, 250, 347*n*
Reyabobowei-teri, 88–89, 93, 100, 272
Riakowa, 168–69
Rice, Hamilton, 20–21, 250, 263
Ridley, Matt, 203–4, 268
Rifkin, Jeffrey, 9, 295
Rio Negro tribe, 29, 66, 146, 244, 278
depopulation of, 19–20
Ritchie, Mark, 15, 31, 47–48, 128, 135, 142, 342*n*
rituals, 13, 88–89, 99–104, 110, 140–41, 161, 240
funeral, 25–26, 69–70, 84, 102–4, 107, 218
initiation, 284–86, 311
purification, 163–64, 356*n*
soul-eating, 47–48, 112–13
women in, 102–3, 216–17
Roanoke colony, 105
Roche, Marcel, 62–63, 66–67, 71, 78–79, 89, 137–38, 306–7, 310–12, 382*n*
Rochester, University of, 297–98
Rochester Democrat & Chronicle, 299–300
Rochester radiation experiments, 297–301, 307, 309
Rockefeller, David, 9
Rockefeller Foundation, 297
Rodrigues Ferreira, Alexandre, 20
Rodríguez, Marta, *see* Miranda, Marta Rodríguez
Rodríguez Costa, Michelle, 208
Roman Catholic church, xxiv, 197, 201, 208, 230, 364*n*
Roosevelt, Anna, 265–66, 373*n*
Roraima, Brazil, xxi-xxiv, 152
Roraima, Mount, 12
Roraima: montaña de cristal (Brewer Carías), 153
Rousseau (radioman), 73, 75, 78, 80, 312
Rousseau, Jean-Jacques, 12
Rowahiwa, 234
Royal Geographical Society, 20–21, 155–56
Ruby, Jay, 92
Ruwahiwa, 168–69, 245–48, 358*n*
Ruwamowa, 175–76

Saffirio, Giovanni, xxiv, 146–47, 161–62
Salamone, Frank, 205, 235–36, 274, 287, 350*n*
Salazar, Rafael, 150–51
Salazar, Teresa, 150–51
Salesian order, 44, 63, 111, 120–21, 122, 167, 187, 217, 229–31, 319–26, 350*n*
anthropologists' relations with, 129–32, 141, 145–47, 154–55, 166, 183–88, 200–201, 229, 293, 319–20, 362*n*, 366*n*–67*n*
baptism not practiced by, 187
in Haximu massacre affair, 195, 201–8
medical care provided by, 27, 29, 50, 61, 63–65, 98, 110, 120–21, 184, 221, 318–19, 323, 325–26
shotgun distribution blamed on, 114, 204–5, 236–37, 366*n*–67*n*, 369*n*
see also Mavaca mission; Ocamo mission
San Carlos del Río Negro, 65–66
sand fleas, 278
Sanema Yanomami, 269, 270
San Francisco, University of, 305
Sanidad, see Ministry of Health, Venezuelan
Santa Barbara, University of California at, xxv-xxvii, 183, 186–87, 190, 193, 282, 291
Santa Barbara Magazine, 282
Santa Fe, N.Mex., anthropology conference at (1985), 142
Savage, Charlie, 19
"Savage Encounters" (Ferguson), 228
Sayago, Edgar, 198
schools, mission, 167, 187, 230
Schwarz measles vaccine, 55, 58, 82, 99, 110
Science, 9, 133, 158–61, 165–66, 171, 173–76, 180, 203, 233, 247, 262, 369*n*
protein debate in, 271–72, 331*n*
Scientific American, 159
selfish gene theory, 158–59
see also sociobiology
Serbia, 163
serial killers, 178, 233
Seripino, José, 253–54, 343*n*
sexual competition, 22, 24–27, 86, 112, 115–18, 142, 158–80, 265–70
age factor in, 162–63, 175–79, 360*n*–61*n*
alternatives to, 26–27, 29, 164
headmen in, 39–40, 49, 159, 171–74
infanticide in, 265–67, 274, 313, 377*n*
permanent bachelors in, 171–72, 175–76

sexual competition (*continued*)
sociobiological theory of, 13–14, 39–42, 49, 118, 158–59, 164, 169–80, 233, 247–48, 266–68, 313, 343*n*
sexuality, 11–12, 15–16, 102, 164, 171–72, 252, 351*n*–52*n*, 356*n*–57*n*
of anthropologists, 15, 127–29, 132–37, 143–47, 350*n*–54*n*
shabonos (communal houses), 13–14, 24, 28–29, 97, 134, 139–41, 170–71, 184, 230, 240, 250–52, 257, 278, 282–84
abandonment of, 5, 69–79, 104, 218, 223, 233–34, 241, 288, 293, 354*n*
acculturation of, 85, 99, 108–9, 162–63, 216, 219–20, 236, 358*n*
anthropologists identified with, 111, 137–38, 142, 217, 219, 292–93
cooperative ethic in, 251–52
demographic analysis of, 165–67, 172–73, 175, 180, 206–7, 211
destruction of, 3–5, 140–42, 282, 290–91, 294
in film productions, 4, 14, 84–88, 93–94, 114–19, 216–23
measles epidemic in, 63, 69, 72–74, 311–13
in mining operations, 154–55
raids on, *see* warfare
shamanic rites in, 46–48, 284–86
shaburi (initiate), 285
Shaki, *see* Chagnon, Napoleon
Shakita, 137–38, 141, 384*n*
shamans, 7, 15, 39, 46–48, 93, 100, 105, 124, 210, 250–51, 280–86
chants of, 47, 112–13, 250, 258, 285–86, 289–90
prowess of, 163, 172, 174–75
Shamatari Yanomami, 107–11, 139, 230, 358*n*
see also Mishimishimabowei-teri
Shamatha-teri, 233, 240
Shamawe, 100
Shanishani River, 92–94, 165, 288, 320–21, 358*n*
Shanishani-teri, 4, 165, 320, 373*n*, 384*n*
Shashanawa-teri, 104
Shaylor, Danny, 76, 90–91, 94, 96–97, 106
Shaylor, Robert, 96
Sheroana-teri, 93–94, 165, 321
Shiborowa, 33–34
Shining Path movement, Peruvian, 154–55
Shiriana Yanomami, 263–64
Shiri-teri, 27–28
Shubariwa-teri, 74–75
Shukumena ka bora (waterfall on the river of the Parakeets), 243–44
Siapa Highlands, Venezuela, 3–4, 10, 17, 91–93, 170, 172, 201, 318–19, 322
food supply in, 257–58, 272*n*, 273–77
FUNDAFACI expeditions to, 190–93, 227–28, 282–88, 290–91, 323–26, 373*n*
new shamanism in, 284–85
political struggle in, 227–28, 234–35
Siapa River, 91, 109–12, 240, 243–45, 282, 285
Siapa valley, 150, 168, 187, 246, 260
flights into, 190–93, 228
Sibariwa, 168, 358*n*, 360*n*
sickle-cell anemia, 38
Sierra Parima, *see* Parima Mountains
Sinabimi, 115, 176
Sixth Penal Court, Venezuelan, 194
slave trade, 19–20, 24
smallpox, 57
Smithsonian Institution, 155–56
"Smoke of Metals, The" (Albert), 281–82
Smole, William, 24–25, 275, 360*n*
snuff, hallucinogenic, 7, 43, 46–48, 103, 113, 119, 140, 250, 286
Social Darwinism, 12, 38–40, 268–69
sociobiology, 17, 158–80, 265–70
animal studies in, 13–14, 22, 40, 179, 204
contrary cases for, 164, 169–71, 177–78, 247–48, 265
Darwinist basis of, 12, 22–23, 38–42, 179–80, 203–4, 267–68
in Haximu massacre affair, 203–4, 207–8
structuralism vs., 127, 272
violence theories of, 9–14, 22–23, 39–42, 49, 86, 115–18, 158–65, 169–72, 174–80, 233, 239, 248, 266–69, 313
see also Darwinism; sexual competition
Sociobiology (Wilson), 118
Sontag, Susan, 83
soul-eating ritual, 47–48, 112–13
Sound Roll 3, 71–78
Southern California, University of (USC), 113, 115, 118–19, 163
Souza, Marinho De, 221–22, 244, 249–51, 257–58, 260, 277–78, 284, 288, 318, 326, 373*n*–74*n*
Soviet Union, 23, 41, 304
Spanish language, 231–32, 244, 293
Special Commission to Investigate the Situation of the Brazilian Yanomami, 202
Species and Speciation (Darwin), 267
Spencer, Herbert, 268–69
Spindler, George, 249
Spirit of the Rainforest (Ritchie), 15, 31, 128, 342*n*
spirits, 17, 25, 27–28, 43, 46–48, 109, 161, 294
epidemics blamed on, 5, 90, 183–85, 232, 280–83, 313
healing by, 250, 285
stealing of, 83–84, 112–13, 218, 292–93, 313–15
Sponsel, Lesley, 11, 138
Sputnik, 41
Stadt, Janet, 300–301
Stannard, Jay, 307
State Department, U.S., xxv
steel goods, 24, 27–35, 108–11, 130, 261
blood purchased with, 16, 45–46, 51, 75, 119, 291–92
epidemics linked to, 16, 21, 27–29, 50–52, 75, 78, 91, 119

in filmmaking expeditions, 85–86, 88–93, 101, 105–6, 114–19, 217–20, 293
information exchanged for, 32–33, 170–71, 291–92
sex exchanged for, 138–39, 143–44, 146
social disruption caused by, 18, 27–30, 34, 103, 142, 168, 223, 228–29, 266, 270–71, 276
wives bought with, 30–31, 162, 172, 232
Stone Age Present (Allman), 176
Stone Age stereotype, 7–8, 12–14, 22–23, 39, 108, 112, 160, 184, 256, 268, 314
Strong Memorial Hospital, radiation experiments at, 297–301
strontium 90, 305, 307–8, 314
structuralism, 126–27, 147
Studying the Yanomamo (Chagnon), 110, 120, 170–71, 173
subclinical measles, 66, 79
Summitville Mine, Colo., 199
Supamo mountains, 153
Superior Court of Salvaguarda, Caracas, 194
Supreme Court, U.S., 43–44
Supreme Court, Venezuelan, 194
Suriname, 19, 57–58
Surucucu, Brazil, 211–14, 270, 375*n*–76*n*, 383*n*
Survival International, xxiv-xxvi, 154, 160, 197, 201–2
Survivors of El Dorado (Wilbert), 269
Survivors of the Amazon, see Warriors of the Amazon
SUYAO (economic cooperative), 85, 183, 186, 189–90, 197, 205, 250, 253, 259
political violence and, 227, 229–36, 287, 293, 323
swampland, 12, 20, 23, 28, 261

tabletop mountains, 9, 12, 15, 36, 152–53, 229
taboos, 43, 89–90, 133–34, 164, 266, 285, 311, 374*n*
of naming, 32–34, 46–48, 170–71, 185, 230–32, 240–41, 291–92, 313
Tales of the Yanomami (Lizot), 139–41, 147–48, 368*n*
Tapirapecó mountains, 168, 243, 246, 281
tapiris (temporary shelters), 273, 276
Tayari-teri, 129, 132–43, 166, 169, 217
destruction of, 141–43, 272
Taylor, Kenneth, 160, 213
Tennessee radiation experiments, 44, 309
terrorism, 154–55, 188
testicular atrophy, 303–4
texemomou (invitation), 92, 97
thalassemia, 38, 299
"Theocracy in Amazonia, A," 210
Time, 22, 160
Timenes, 256
Times (London), 9, 208, 256
Times Literary Supplement, xxii, 150, 201–2, 208–9, 227, 229, 238–39, 320
Timoteo, 135–36, 221
tin ore (cassiterite) deposits, 154–55, 188, 213
Tiriyo people, 57–58
"To Fight over Women and to Lose Your Lands" (Van Vuner), 9
Tohowe, 140
Toki (Eagle Mountain), 36–37, 42–43, 46, 48
Toki (Maquiritare village), 271–73, 331*n*
Toobatotoi-teri, 234–35, 284–88
Toototobi, Brazil, 191
Torawa, 118
tourism, 50, 162, 190–93, 197, 323
toxic metals, disposal of, 199
tracers, radioactive, 304, 306–7, 311, 314
trade cooperative, *see* SUYAO
trade goods, 6, 18–19, 24, 26, 30–35, 45–46, 127, 138, 286
epidemics linked to, 27–29, 50–52, 75, 78, 222, 281–83, 289, 318
in film production, 84–86, 88–93, 101, 105–6, 110–11, 114–19, 168, 217–20, 253–54
genealogies bought with, 30–34, 46, 246
sexual favors bought with, 129, 132–37, 139–41, 143–47, 341*n*
wars fought over, 18, 27–30, 34, 103, 142, 168, 223, 254, 277, 283, 287
wives bought with, 30–31, 162–63, 174, 232
see also specific items
trading rights, conflict over, 230–31, 234–40
Tristes Tropiques (Lévi-Strauss), 126–27, 272
Triunfo II and III, 153
Tropical Medical Institute, Venezuelan, *see* CAICET
"Truth and Consequences" (Turner), 202
tuberculosis, 69, 108, 288
Tumbes, conquest of, 32
Turnbull, Colin, 184–85, 264
Turner, Terence, 11, 32, 39–40, 152, 202, 208

UCLA (University of California at Los Angeles), 269, 300
United Nations, 229, 303
United Shabonos of the Upper Orinoco, *see* SUYAO
United States:
body size in, 265
measles in, 57–58
mining in, 199
murder in, 23, 176–77
Yanomami awareness in, 160–61
Yanomami perception of, 251–53
Universidad Central, Venezuela (UCV), 183, 197
unokai (men who have killed), 159–65, 170, 177–80, 239, 241, 267, 275, 314, 356*n*–57*n,* 360*n,* 369*n*
unokaimou (purification ritual), 163, 356*n*

Unturán Mountains, 206, 243, 275, 278
Upper Orinoco, *see* Yanomamiland
uranium injections, 44, 297–98, 300
urine specimens, 97, 99, 102, 106, 299, 314
"U.S. Acknowledges Radiation Killed Weapons Workers" (Wald), 309
USC (University of Southern California), 113, 115, 118–19, 163
Ushubiriwa, 231–35, 240–42, 255, 323
U.S. News & World Report, 41, 160, 176
Uuwa, 116–17

Valentine, William, 299–300, 311
Valero, Helena, 91, 110, 133–34, 168, 244–50, 255–56, 272, 277, 280–81, 318, 359*n,* 371*n*
  first contacts made by, 246–49, 290
  hunger endured by, 245, 275–76
  warfare witnessed by, 29, 111, 247–48, 336*n*
Valero, José, 237
Vanderbilt University radiation experiments, 44, 309
Van Middlesworth, Lester, 304
Vegas, Tania, 153
Veja, xxii–xxiii
Velásquez, Ramón, 197–98, 202, 364*n*
Vemeru I-VI mines, 154
Venevisión network, 6, 292–93, 322
Venezuela, xxi, xxiv, 10–11, 37, 99, 126, 150–57, 181–83, 280
  armed forces of, 11, 145–46, 181, 190–94, 200, 237, 250, 324–25, 332*n*
  attempted coup in, 11, 181, 193–94
  corruption scandals in, xxv*n,* 194, 196–98, 329*n*
  foreign embassies in, xxv, 130, 137, 144
  gold mining in, xxiii-xxiv, 9, 152–57, 186, 197–201
  Indian health program in, 325–26, 384*n*
  Orinoco colonization plan of, 129–30
  political parties in, 197–98
  radioisotope experiments in, 305–10
  rainforest exploitation in, 150–51, 156–57, 193–94, 197–200, 291
Venezuelan Institute of Scientific Investigation (IVIC), 44, 63, 73, 78, 137–38, 154, 197, 263, 311–12, 325
Venezuela: Violations of Indigenous Rights (Survival International), 154
Vengold mining company, 200
Ventuari River, 155
Vietnam War, 42, 44, 88
Village 16, 120–21
  see also Mishimishimabowei-teri
Village 51, 165, 167–68, 170–71, 322–24
  see also Kedebabowei-teri
Vinci, Alfonso, 270
violence:
  age factor in, 162, 175–77, 267, 360*n*
  choreographed, 14, 85–88, 101–4, 114–19, 216–17, 342*n*
  copulation theory of, 266–67, 313
  in ethnographic films, 5–6, 85–88, 101–6, 112–19, 166, 168, 216–17, 219, 223, 342*n*–43*n,* 349*n*
  feasts associated with, 88, 102–3, 168–69, 223, 235, 250, 293, 336*n*
  food as cause of, 115–16, 128–29, 176, 272–74
  in industrialized societies, 23, 176–77
  initiation into, 311
  in pan-Yanomami politics, 234–40, 323
  public policy implications of, 8, 14, 41–42, 159–60, 313–14
  scientists and, 309–15
  social disruption theory of, 18–19, 30, 32, 142, 223, 228–29, 242, 276
  sociobiological theories of, 9–10, 12–14, 22–23, 39–42, 49, 86, 115–18, 158–80, 239, 247, 267, 313, 349*n*
  spirit possession as, 47–48, 112–13, 163–64, 313
  against women, 11–12, 115, 164, 176, 201, 223, 238, 252, 361*n*
  women as cause of, *see* sexual competition
  see also warfare
Visual Anthropology Review, 92
Vulture Spirit, 43, 47–48, 294, 356*n*

Waborawa-teri, 282–83
Wachtler, Maria, 70
Wadoshewa, 173
Wagner, Richard, 143, 219
waikas (spirit enemies), 284–87
Walden, John, 150, 325–26
Wallace, Alfred Russell, 267
Waloiwa, 84–85
Wanitama-teri, 111–12
Waorani people, 178
warfare, 7–8, 18–35, 85–86, 111–19, 196, 219, 227–42, 250
  Chagnon's involvement in, 10–11, 15–16, 18, 29–35, 54, 87–88, 101–6, 112–19, 166–67, 174, 227–29, 231–41
  death toll from, 23–24, 29, 34, 35, 51, 85, 102, 111–12, 158–64, 169, 227–28, 358*n,* 382*n*
  film projects followed by, 14, 85, 87, 103–4, 113–14, 222–23
  hunger exacerbated by, 257, 274–76
  inconsistent accounts of, 117–18, 168–71, 238–39, 246–48, 273–74
  Lizot's involvement in, 141–43, 166–67, 228
  missionaries' role in, 45, 165–67, 186, 236–37, 293
  population pressure in, 262, 268–70
  protein theory of, 261–74, 276, 374*n*–75*n*
  ritual purpose of, 161
  spiritual, 47–48, 112–13, 163–64
  technological advances in, 27, 30, 33, 113–14, 165–68

territorial expansion as cause of, 261–62, 266, 268–70, 375*n*
trade goods as cause of, 18, 27–30, 34, 103, 142, 168, 223, 270–71, 276
trading rights as cause of, 234–40
village fissioning and, 29, 234, 266
women as cause of, *see* sexual competition
Yanomami avoidance of, 13–14, 29, 90, 103, 111–12, 131–32, 167, 211, 216, 251–52, 274–76, 356*n*
see also violence
"Warfare over Yanomamo Indians" (Booth), 9, 160–61
Warren, Shields, 304, 310
Warren, Stafford, 298–301, 303–4, 309–11, 314, 380*n*
Warriors of the Amazon, 14, 215–23, 368*n*
Washewa River, 165, 184, 234–35, 321
Washewa-teri, 165, 185, 192, 232, 235–36, 241, 287, 321, 324–25, 362*n*
Washington Post, 160
Watemosiwa, 234
Watson, Fiona, 197, 202
Waupuruwe, 282–84
wayamou (trade chant), 259
wayumi (foraging trek), 185
Welsome, Eileen, 297, 300–302, 304, 309
White Guaharibos, *see* Yanomami people
White Indians, *see* Yanomami people
"white woman" (machete), 270
WHO (World Health Organization), 58–60, 262–63
Wilbert, Johannes, 269
Wild West Vancouver exchange, 153
Wilson, Edward O., xxi, 23, 118, 203, 207, 261, 375*n*
Wilson, G. S., 56, 57, 80
Wilson, Margot, 177, 266
Witokay-teri, 72
women, 3, 6, 49, 84, 86, 89, 100–102, 115–16, 127–28, 146, 250–56, 377*n*
abduction of, 24–27, 29, 33–34, 142, 164, 180, 201, 223, 232, 238, 241–42, 275
conflict over, *see* sexual competition
exchange of, 27, 29, 30–31, 138–39, 162, 232
grief of, 166, 255
health of, 262–63, 375*n*
in rituals, 102–3, 216–17
sharing of, 27, 171, 175, 252, 270, 348*n*
World Health Organization (WHO), 58–60, 262–63
World War I, 23
World War II, 21, 23, 286, 297–98, 303
WQED Pittsburgh, 190
Wrangham, Richard, 14, 158
Wray, Amy, 254

Xavante people, 39, 264
xawara (evil vapors), 5, 183–85, 232, 280–83, 289, 292–94

Yabitawa-teri, 103
Yahi Indians, 108
Yahohoiwa, 170–71
Yanomama: A Multidisciplinary Study, 5, 37, 43, 46, 49, 93–95, 172
fight scene in, 85–86
measles epidemic documented in, 70, 77, 82, 95–96, 312
Yanomami and Their Food System, The (Finkers), 120
"Yanomami Fraud," 210
Yanomami Health District, Venezuelan, 43
Yanomami Homecoming, 253–56
Yanomamiland, xxiii-xxiv, 15–16
administration of, 9–10, 63, 154–55, 196–201
anthropologists banished from, xxi, 10, 115, 138, 182–83, 200–203, 209
highland vs. lowland, 23–24, 28, 51, 91, 172–74, 274–76
topography of, 8, 12, 19–21, 28, 36, 91, 152–53, 206, 211, 213–14, 229–30, 260–61, 272, 276–78, 287–88, 357*n*–58*n*
tourists in, 50, 162, 190–93, 197
Yanomami language, 14–15, 48, 101, 105, 183, 217–18, 244, 260, 284, 290
musicality of, 258–59, 289
regional dialects of, 23, 202, 209
translations of, 115, 128, 163, 217, 231, 283, 289
verb tenses in, 242
see also specific words
Yanomami people:
acculturation of, 85, 99, 108–9, 162–63, 187, 216, 219–20, 236, 239, 358*n*
AEC studies of, 16–17, 36–37, 42–52, 59, 70–71, 77–78, 93–95, 165, 172, 246, 262, 292–93, 297, 300, 304–10
begging strategy of, 162, 269–70
in biosphere plan, 9–11, 154, 181–82, 187–94, 237–40, 254
blood collected from, xxiii, 16–17, 30, 43–52, 83, 97, 99, 104, 106, 119, 193–94, 197, 292, 294–95, 313–14, 325, 337*n*
body painting by, 26, 46, 85, 87–88, 93, 113, 126, 172, 196, 231, 248, 278–79, 286, 289
bones collected from, 307–10
celibacy as practiced by, 163–64, 266–67, 374*n*
club fighting ritual of, 240
counting unknown to, 163, 248, 255
cremation practiced by, 69–70, 84, 102–4, 195, 218
demographic analysis of, 164–65, 172–77, 181, 187, 190–94, 200, 211, 237, 250, 269, 295, 382*n*–84*n*
diet of, 87, 99–101, 103, 115, 176, 257–58, 261–62, 264, 376*n*–77*n*
diseases of, *see* disease, diseases; epidemics; measles vaccine experiment; *specific diseases*

Yanomami people (*continued*)
diversity among, 23–25, 172–73, 202, 209, 230
divorce among, 173, 175
education of, 167, 187, 230, 232
energy as viewed by, 285, 289
Erotic People stereotype of, 15, 84, 133–34
fecal samples from, 97, 99, 299
fetal development theory of, 252
fidelity as viewed by, 252
Fierce People stereotype of, xxi, 8, 12–14, 20–27, 84, 107, 119, 131–32, 158–61, 164, 216–19, 232, 261, 263, 309–10
films made by, 84, 253–54, 343*n*, 372*n*
films viewed by, 37, 43, 218, 221, 312, 342*n*
first contacts with, 3–6, 10, 20–21, 25–26, 45–46, 114, 119–22, 127, 185, 187, 190–93, 211, 228, 248–49
funeral rituals of, 25–26, 69–70, 84, 102–4, 107, 219, 312–13
gardens of, 13, 99–101, 134, 171, 174, 232, 257–58, 261, 263, 272–76, 376*n*–77*n*
generosity of, 25, 90, 92
genetic inheritance of, 4, 13
goods desired by, 27–36, 45–46, 50, 85–93, 101–2, 108–11, 115–19, 132–47, 162–63, 170–71, 174, 211–14, 217–20, 222, 231, 240, 259–60, 270, 278, 287
grieving as viewed by, 166, 255
hallucinogens used by, 7, 46–48, 103, 113, 119, 129, 250
handicrafts sold by, 230–31, 234–35
historical accounts of, 19–21, 91
hunger among, 60, 67, 99–100, 245, 257, 373*n*–74*n*
immunological weakness of, 59–62, 66–69, 89, 95, 311, 317–18, 339*n*, 341*n*
indentured servitude of, 270
infanticide attributed to, 265–67, 274, 313, 377*n*
influenza and, 8, 57
innate killer stereotype of, xxi, 12–14, 39–42, 119
intestinal parasites in, 273, 288
jokes of, 116–17
leadership of, xxiii, xxiv–xxvi, 5, 9, 85, 87, 135, 159, 183, 185–86, 197, 230–42, 293–94, 328*n*–29*n*
life expectancy of, 173
literacy among, xxii, 46, 135, 185, 232, 239–41, 293
marriage among, 26–27, 29, 129, 139, 146–47, 162–63, 171–75, 245, 359*n*–60*n*
massacre of, 21, 195–214, 250
media access to, 4, 10, 84–85, 201
medical care for, 25, 27–28, 50, 61, 63–65, 68–69, 95–98, 104, 110, 119–21, 184–85, 190–93, 221, 230, 288–89
misfits among, 33, 109, 128, 170–71, 240
mortality rates of, *see* mortality rates
murder rates among, 23–24, 29, 41, 51, 85, 102, 113–14, 158–64, 169, 176–77, 233, 274–75, 335*n*–36*n*
mythology of, *see* mythology
nomadic existence of, 69, 155, 273
nostalgia of, 250, 253
origin of, 13, 24, 91, 231, 250, 262, 265, 269
penis strings worn by, 13, 90, 139
physical characteristics of, 13, 59–60, 261–64
political support for, xxi-xxiii, 154, 183, 185–86, 189–90, 197–201, 207–8
population increase of, 261, 268–70
protein intake of, 261–62, 265, 270–74
rituals of, *see* rituals
robbery by, 21, 25–26, 28–29, 83, 128, 209–10, 237, 270–71
sexual mores of, 15, 84, 102, 132–37, 139, 143–47, 164, 252, 313, 351*n*–52*n*, 356*n*–57*n*
spontaneity of, 88–89, 116
Stone Age stereotype of, 7–8, 22–23, 39, 108, 112, 160, 184, 256, 268
superstitions of, 5, 17, 25, 27–28, 46–48, 83–85, 90, 109, 112–13, 161, 168, 218, 280–95
taboos of, 32–34, 43, 46, 89–90, 133–34, 164, 170–71, 185, 218, 240–41, 266
territorial expansion of, 261, 266–70
time sense of, 242, 372*n*
urine collected from, 97, 99, 102, 106, 314
warfare among, *see* murder, murders; sexual competition; violence; warfare
warfare avoided by, 29, 90, 103, 111–12, 131–32, 163, 168–69, 211, 216, 269–70, 356*n*
weapons of, 3–6, 7, 27, 33–34, 47, 83, 113–15, 140, 165–67, 196, 204–5, 211, 234, 239–41
Western haircuts among, 216
witchcraft accusations by, 21, 24, 27–29, 52, 106, 111, 211
Yanomami reserve, Brazilian, xxi-xxiii, 27, 158, 162, 260
epidemics in, 191
mining in, xxiii-xxv, 191, 195–201, 208–14, 249
missionary activity in, 50
partition of, 160
Yanomami reserve, Venezuelan, 3–4, 9–10, 63, 187–94
anthropologists expelled from, xxi, 10, 115, 138, 182–83, 200–201
colonization plan for, 129–30
mining in, 154–57, 186
missionary activity in, 50, 128–32, 167, 185–87, 201–14, 229–31
Yanomami Warfare (Ferguson), 18–19, 32, 45, 177, 359*n*
Yanomamo: The Fierce People (Chagnon), xxiii-xxiv, xxvi, 7–9, 21–22, 31, 37, 47–48, 137, 160, 246–47

Darwinian assumptions in, 11–12, 21–22, 26–27, 169–70
disease in, 50, 54–55, 206, 319–20
fundraising solicitation in, 189
warfare described in, 26, 86–87, 113–14, 169, 171
Yanomamo Interactive, 117–18, 172–73, 175–76, 357n
Yanomamo Survival Fund, 183, 188–89
Yanowe, 283, 286
Yarima, 251–56, 372n
yawaremou (misbehavior), 115
yopo (hallucinogen), 46–48
Yo soy Napeyoma (Valero), 244
Youngbird, see Karina
Youth Ministry, Venezuelan, 150–51
Yupa Indians, 264
Yutuwe, 172

Zeus Lykos, clan of, 311
Zucchi, Alberta, 138

ABOUT THE AUTHOR

Patrick Tierney spent eleven years researching and writing this book. He received a B.A. from UCLA in Latin American Studies in 1980. From 1983 to 1989 he joined archaeologists in discovering Inca sacrificial sites above 17,000 feet and documenting ongoing ritual killing in the Andes. His earlier book, *The Highest Altar: The Story of Human Sacrifice* (Viking, 1989), received favorable reviews in the *New York Times Book Review,* the *Los Angeles Times Book Review,* and other publications, was published in seven foreign editions, and is the basis of a *National Geographic* documentary that will be released in 2000. He speaks Spanish and Portuguese and is now a visiting scholar at the University of Pittsburgh's Latin American Studies Center.